Corrosion of Austenitic Stainless Steels
Mechanism, Mitigation and Monitoring

Charles Seale-Hayne Library

University of Plymouth

(01752) 588 588

LibraryandITenquiries@plymouth.ac.uk

Editors: H.S. Khatak ☐ Baldev Raj

Corrosion of Austenitic Stainless Steels
Mechanism, Mitigation and Monitoring

ASM
INTERNATIONAL
®
**The Materials
Information Society**

Narosa Publishing House
New Delhi   Chennai   Mumbai   Kolkata

EDITORS
H.S. Khatak
Head
Corrosion Science and Technology Division
Indira Gandhi Centre for Atomic Research
Kalpakkam-603102, India

Baldev Raj
Director
Materials, Chemical and Reprocessing Groups
Indira Gandhi Centre for Atomic Research
Kalpakkam-603102, India

Copyright © 2002 Narosa Publishing House
22 Daryaganj, Delhi Medical Association Road, New Delhi-110 002, India

Exclusive distribution in North America only by
ASM International, 9639 Kinsman Road, Materials Park,
Ohio, USA 44073-0002

ISBN 0-87170-752-7

Printed in India

Foreword

Materials science, like all other branches of the sciences, is full of many surprises and paradoxes. The corrosion behavior of austenitic stainless steels offers many examples for this. These materials, which have excellent resistance to general corrosion, are susceptible to many forms of localized corrosion like pitting, crevice attack and intergranular corrosion. Their vulnerability to stress corrosion cracking (SCC) is a classic instance of the paradox, that the stronger an alloy and greater its corrosion resistance, the more prone it is for failure by SCC.

My first encounter with the corrosion problems of austenitic steels was, in 1968, the intergranular SCC of welded structures in our first nuclear power plant at Tarapur (India); the cracks were detected towards the end of the construction, just before fuel loading and delayed the commissioning by nine months due to the extensive repairs that had to be carried out.

When I moved from Bhabha Atomic Research Center Mumbai to nucleate and initiate the Metallurgy Program at the newly created Reactor Research Center (since renamed the Indira Gandhi Center for Atomic Research, IGCAR) at Kalpakkam in 1974, the austenitic stainless steels along with the chromemoly ferritic steels became the major alloys for studies, characterization, and indigenous development for the Program. Austenitic stainless steels are so much a part and parcel of Fast Breeder Reactors (FBR), that the French called their reactor SUPERPHENIX "A Cathedral in Stainless Steel". As one of the major centers in the world engaged in the research, development, design and engineering of FBR, IGCAR has also established expertise and experience on all aspects of the metallurgy, properties and behavior of these steels. One area in which IGCAR has done remarkably original and path breaking research is the study of mechanisms, mitigation and monitoring of corrosion.

This book fulfils a need felt by students, teachers, researchers, fabricators and users of austenitic stainless steels in myriad industries for a ready source of information on all aspects of corrosion and its prevention as well as monitoring, in austenitic stainless steels. It is very gratifying to see that the book also documents in one place all the work done in IGCAR in this important area; the lists of references at the end of each chapter include a substantial number of papers and theses from the center. All the contributors are outstanding scientists whom I have known as colleagues and collaborators for varying durations from a few years to three decades; four of them including the two Editors were my doctoral students and one I guided for his M.Tech. thesis. That is why writing this foreword is not only a matter of immense joy but also of just pride for me.

PLACID RODRIGUEZ

Preface

Austenitic stainless steels are a family of unique engineering materials because of their applications in a variety of industries ranging from the paper and pulp industry, to industries such as chemical, petrochemical and nuclear industries. Their wide-ranging applications are facilitated by virtue of their adequate high temperature mechanical properties, good fabricability, and excellent corrosion resistance in a variety of service environments. The general corrosion resistance of austenitic stainless steels is attributed to the formation of a thin, adherent and self-healing 'passive film' developed on the surface in most environments. However, when this passive film is destroyed at some localized areas, dangerous manifestations of localized corrosion processes could result leading to unexpected and sudden failures. Corrosion-related failures form a substantial portion of the total failures of engineering components made of austenitic stainless steels leading to huge production losses and replacement cost in various industries.

Science and technology of fast breeder nuclear reactors is the mandate of the Indira Gandhi Centre for Atomic Research (IGCAR) at Kalpakkam, India. Austenitic stainless steels are the major materials of construction for the fast breeder reactors. In fast breeder reactors, these steels should be compatible with a host of environments ranging from liquid metal to withstanding hostile neutron bombardment and marine environment during fabrication, storage etc. With the mandate of better and better understanding of austenitic stainless steels for fast reactor applications, the laboratories for corrosion science and technology and nondestructive testing and evaluation were established in 1975. In the past two and half decades, these laboratories have done pioneering research in various aspects of physical, mechanical, corrosion and fabrication metallurgy. The wealth of information generated on the corrosion of austenitic stainless at the IGCAR is being presented in this book. This has been done with a view to share the knowledge and expertise available at IGCAR with the industries, researchers and academicians. The importance of this effort will be further appreciated considering the fact that such a comprehensive treatise on the topic of corrosion of the so-widely-used austenitic steels is not readily available. The book primarily focuses on the various corrosion aspects of AISI types 304 and 316 stainless steels and their modifications, since these varieties are used by most of the industries. Comprehensive literature review of the various forms of corrosion of austenitic stainless steels and the results of the research work carried out at our laboratories form the basis of this book. We hope that this treatise on corrosion of austenitic stainless steels would enable full exploitation of the potential of stainless steels by the industries and also cater to the needs of materials technologists students of metallurgy and scientists for many years to come.

The book starts off with a comprehensive review on the physical, mechanical and welding metallurgy of austenitic stainless steels along with discussions on their physical properties followed by chapter on uniform corrosion. The corrosion behaviour of austenitic stainless steels is sensitive to the various metallurgical parameters such as cold work, sensitization and high temperature phase transformation besides environmental factors. Role of these variables on localized corrosion such as pitting corrosion, crevice corrosion, intergranular corrosion and stress corrosion cracking in aqueous environments is detailed. Microorganisms influence corrosion behaviour by accelerating half-cell reaction and/or by causing changes in chemistry of environment at surface. These aspects are discussed in one of the chapters. Two chapters have been devoted to high temperature corrosion in liquid metals and in air/ oxygen. A highlight of this book is the chapter on surface modification for corrosion protection, where in different surface modification techniques such as application of coatings, thermo-chemical, kolsterisation, sol-gel, thermal spraying and vapour deposition processes, laser surface melting/alloying, and ion implantation, are discussed in adequate measure to improve the corrosion resistance of the austenitic stainless steel have been discussed. General guidelines for corrosion control which include criteria to be adopted at various stages such as in design, material selection, fabrication, storage, erection, operation are included in the book. A comprehensive review on the various methods for on-line detection and monitoring of corrosion using non-destructive testing techniques, such as eddy current, ultrasonic, radiography, acoustic emission, thermography etc., will be of use to the practicing engineers. Lastly, some important failures that have occurred in austenitic stainless steel components, the analysis of these failures, and steps taken to prevent such future failures completes this exceptional treatise on corrosion of austenitic stainless steels.

Each chapter is encoded with a broad spectrum of reference literature for serving the enhanced appetite of the reader. The complexity and diversity relating to corrosion aspects of austenitic stainless steels will continue to attract more skills in R&D and would demand radically different pathways for combating corrosion in the coming years. We indeed welcome and look forward for this to happen. It is hoped that this book would enrich the readers, enhance useful applications and contribute towards avoiding service failures. Authors of the book look forward to attracting young researchers to the fascinating and rewarding career to pursue research and development in this important field namely corrosion of austenitic stainless steels.

EDITORS

Acknowledgement

The editors/authors are extremely grateful and credit the following organizations for giving their permission to incorporate published information like technical content, photographs, sketches etc. in our book on "Corrosion of Austenitic Stainless Steels"

1. American Society for testing and materials, Pennsylvania, USA for Table 1.4 and Fig. 11.5.
2. American Ceramic Society, Ohio, USA, for the Fig. 14.16.
3. NACE International, Houston USA for Figs. 2.20, 2.21; Tables 2.6 and 2.9.
4. Elsevier Science, Amsterdam, Netherlands for Fig. 2.22.
5. Elsevier Science, Global Rights, Oxford, England for Figs. 1.4, 12.15, 14.6, 14.7, 14.4 and 14.15.
6. Engineering Materials Advisor Services Ltd., West Midlands, USA for Process Superposition Model, Process Competition and Process Interactive Model in Chapter 8.
7. Werkstoffe und Korrosion, Postfach, Germany for Figs. 2.14, 2.23 and Table 2.7.
8. John Wiley & Sons, Inc., New York for Figs. 2.12, 2.19.
9. Welding Journal, Miami, Florida for Fig. 1.27.
10. Allerton Press, Inc., New York, USA for Fig. 8.15.
11. Springer Verlag, Vienna, Austria for Fig. 5.3.
12. ASM International, Ohio, USA for Figs. 1.19, 1.20, 5.9 and 12.14.
13. Iron and Steel Institute of Japan, Tokyo for Fig. 11.6.
14. IOM Communications. London UK for Figs. 1.16, 1.22, 1.5 and 1.6.
15. Mc-Graw Hill Company, New York, USA for Fig. 2.9.
16. Kluwer Academic Publishers, Dordrecht, The Netherlands for Figs. 1.11, 1.18, 1.21 and 1.23.

Contributors

Baldev Raj
Director, Materials, Chemical and Reprocessing Groups
Indira Gandhi Centre for Atomic Research
Kalpakkam-603 102
E-mail: dmg@igcar.ernet.in

R.K. Dayal
Head, Aqueous Corrosion and Surface Studies Section
Corrosion Science and Technology Division
Indira Gandhi Centre for Atomic Research
Kalpakkam-603 102
E-mail: rkd@igcar.ernet.in

S.K. Dewangan
Scientific Officer
Division for Post Irradiation Examination and
Non-Destructive Testing
Indira Gandhi Centre for Atomic Research
Kalpakkam-603 102

Geogy George
Scientific Officer
Corrosion Science and Technology Division
Indira Gandhi Centre for Atomic Research
Kalpakkam-603 102
E-mail: geogy@igcar.ernet.in

Hasan Shaikh
Scientific Officer
Corrosion Science and Technology Division
Indira Gandhi Centre for Atomic Research
Kalpakkam-603 102
E-mail: hasan@igcar.ernet.in

P. Kalyanasundaram
Head, Division for Post Irradiation Examination and
Non-Destructive Testing
Indira Gandhi Centre for Atomic Research
Kalpakkam-603 102
E-mail: pks@igcar.ernet.in

U. Kamachi Mudali
Scientific Officer
Corrosion Science and Technology Division
Indira Gandhi Centre for Atomic Research
Kalpakkam-603 102
E-mail: kamachi@igcar.ernet.in

K.V. Kasi Viswanathan
Head, Post Irradiation Examination and Remote
Handling Section
Indira Gandhi Centre for Atomic Research
Kalpakkam-603 102
E-mail: kasi@igcar.ernet.in

H.S. Khatak
Head, Corrosion Science and Technology Division
Indira Gandhi Centre for Atomic Research
Kalpakkam-603 102
E-mail: khatak@igcar.ernet.in

P. Muraleedharan
Scientific Officer
Corrosion Science and Technology Division
Indira Gandhi Centre for Atomic Research
Kalpakkam-603 102
E-mail: pmurali@igcar.ernet.in

N.G. Muralidharan
Scientific Officer
Division for Post Irradiation Examination and Non
Destructive Testing
Indira Gandhi Centre for Atomic Research
Kalpakkam-603 102
E-mail: ngm@igcar.ernet.in

S. Ningshen
Scientific Officer
Corrosion Science and Technology Division
Indira Gandhi Centre for Atomic Research
Kalpakkam-603 102
E-mail: ning@igcar.ernet.in

N. Parvathavarthini
Scientific Officer
Corrosion Science and Technology Division
Indira Gandhi Centre for Atomic Research
Kalpakkam-603 102
E-mail: npv@igcar.ernet.in

M.G. Pujar
Scientific Officer
Corrosion Science and Technology Division
Indira Gandhi Centre for Atomic Research
Kalpakkam-603 102
E-mail: pujar@igcar.ernet.in

N. Raghu
Scientific Officer
Division for Post Irradiation Examination and Non-
Destructive Testing
Indira Gandhi Centre for Atomic Research
Kalpakkam-603 102

S. Rajendran Pillai
Scientific Officer
Corrosion Science and Technology Division
Indira Gandhi Centre for Atomic Research
Kalpakkam-603 102
E-mail: srp@igcar.ernet.in

V.R. Raju
Scientific Officer
Corrosion Science and Technology Division
Indira Gandhi Centre for Atomic Research
Kalpakkam-603 102
E-mail: raju@igcar.ernet.in

Rani P. George
Scientific Officer
Corrosion Science and Technology Division
Indira Gandhi Centre for Atomic Research
Kalpakkam-603 102
e-mail: rani@igcar.ernet.in

P. Shankar
Scientific Officer
Physical Metallurgy Section
Materials Characterisation Group
Indira Gandhi Centre for Atomic Research
Kalpakkam-603 102
E-mail: pshankar@igcar.ernet.in

Contents

1. Introduction to Austenitic Stainless Steels

Geogy George[1] and Hasan Shaikh[1]

Abstract The family of austenitic stainless steels has a wide variety of grades precisely tailored for specific applications such as household and community equipment, transport, food industry, industrial equipment, chemical and power engineering, cryogenics, and building industry. The optimum choice of the grades would depend on service needs and this would require a clear understanding of the metallurgical parameters, which control the microstructure and thus the mechanical properties, formability and corrosion resistance. This chapter, in brief, deals with the physical metallurgy, welding metallurgy, and physical and mechanical properties of austenitic stainless steels. In the physical metallurgy of stainless steels the tendency of alloying elements to form different phases, the transformation of austenite to martensite during cooling or straining, hardening processes and formation of intermetallic phases, have been discussed. The influence of chemical composition and temperature on the various physical properties of austenitic stainless steel such as coefficient of expansion, thermal conductivity and magnetic permeability is highlighted. Variation in mechanical properties, such as tensile, fatigue and creep strengths of austenitic stainless steels with temperature, composition and microstructure has been discussed. The mechanisms to strengthen the austenitic stainless steels by appropriate thermo-mechanical treatments, grain refinement etc. have also been addressed. Austenitic stainless steels lend themselves remarkably to deep drawing and cold rolling, where their work-hardening characteristics enable high strength levels to be attained. Weldability is excellent, and welds, which do not transform to martensite during air-cooling, have mechanical properties similar to base metal.
Key Words Austenitic stainless steels, high nitrogen steels, physical metallurgy, physical properties, mechanical properties, welding, martensite.

HISTORICAL BACKGROUND TO STAINLESS STEELS

As a class of materials, stainless steels stand apart and are considered the backbone of modern industry since they find wide applications in chemical, petrochemical, off-shore, power generation and allied industries. In 1889, Riley of Glasgow discovered that additions of nickel significantly enhanced the tensile strength of mild steel, and in 1905, Portevin observed that steels containing more than 9% chromium were resistant to acid attack. The transition from the laboratory to the first attempts to confirm the practical applications of stainless steels took place principally from 1910 to 1915. To cite the pioneers of this work would go beyond the scope of this introduction; nevertheless, a few

[1]Scientific Officers, Corrosion Science and Technology Division, Indira Gandhi Centre for Atomic Research, Kalpakkam-603 102, India.

important names must be mentioned: the Englishman Brearley for martensitic steels, the Americans Dansitzen and Becket for ferritic steels, the Germans Maurer and Strauss for austenitic steels [1]. The term "stainless" (inoxydable in French or rostfrei in German) is now popularly used for iron alloys containing greater than 12 wt. % Cr. In a relatively short span of time since the discovery, the applications of stainless steel have grown rapidly with its image changing from that of an expensive, high-technology wonder alloy to that of a cost-effective, everyday material of construction.

The design of stainless steel alloys has been motivated primarily by chemical, mechanical and thermal stability considerations. The base for the various stainless steels is the binary Fe-Cr system [2] (Fig.1), the properties of which are modified by the addition of several major alloying elements such as Ni, Mo, Mn etc. as well as minor ones such as C and N. Fe-Cr-Ni alloys are the most predominantly used austenitic stainless steels. Important phase relationships in Fe-Cr-Ni stainless steels can be considered to stem from the properties of the binary Fe-Cr and Fe-Ni phase diagrams. A convenient way of understanding the phase relationship in the Fe-Cr-Ni ternary system is by the use of cross-sections through the ternary diagram, such that the proportion of one element is constant. Fig. 2 shows a section of the Fe-Cr-Ni diagram at a constant Fe content of 70% [3]. It is clear from the diagram that austenite is the stable phase in the Ni-rich side of the diagram while delta-ferrite is the equilibrium phase in the Cr-rich side.

The important factors, which must be considered in the design of the various types of stainless steel, are:

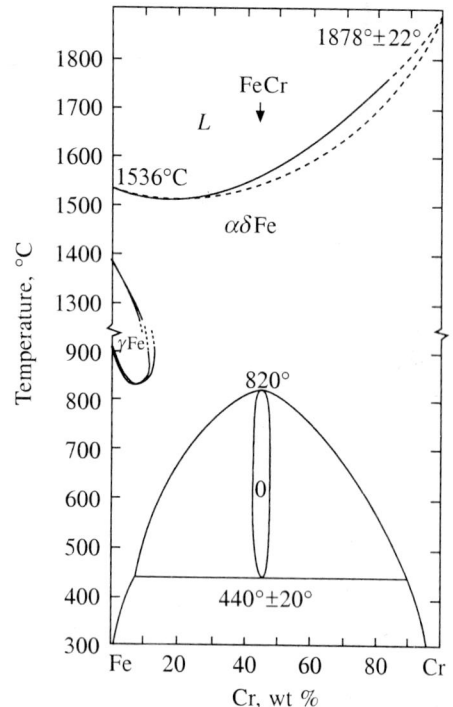

Fig. 1. Fe-Cr binary phase diagram [2].

(i) Corrosion and oxidation resistance in the operating environment
(ii) Mechanical and physical properties
(iii) Fabrication characteristics from the point of view of both hot and cold working
(iv) Welding; many of the stainless steels are required to be readily weldable, and welding must not impair the corrosion resistance, creep resistance or general mechanical properties.

There are many different stainless steels, and the main types, are grouped according to their metallurgical structure as follows [4].

Martensitic Stainless Steels

They are Fe-Cr-C alloys with or without addition of other alloying elements. Chromium content is 12-18 wt. % and carbon is 0.1-1.2 wt. % [5]. Several other additions are also made such as Mo, V, Nb, Ti and Cu to get certain desirable properties. These alloys are austenitic up to 1050 °C, but transform to martensite on cooling. The additions of alloying elements lower the martensite start (M_s) and finish

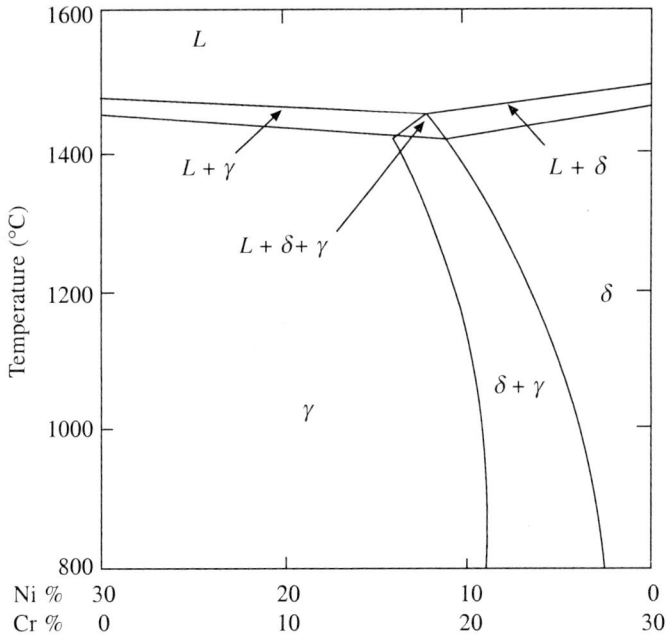

Fig. 2. **Vertical section of Fe-Cr-Ni phase diagram showing the variation of solidification mode with composition for a constant Fe content of 70% [3].**

(M_f) temperatures. Therefore, controlled addition of alloying elements is necessary to maintain the M_s temperature at a reasonably high value above room temperature.

These stainless steels are characterised by very high strength and low toughness. Tempering is done to increase the toughness. Typical applications include turbine blades, springs, aircraft fittings, surgical instruments, knives, cutlery, razor blades and other wear-resisting parts. Typical examples of martensitic stainless steels are AISI Types 403, 410, 420 and 431 and their compositions are shown in Table 1.

Table 1. **Chemical composition of martensitic stainless steel grades [7]**

Grade AISI	C Max	Si Max	Mn Max	P Max	S Max	Ni	Cr
403	0.15	0.50	1.0	0.040	0.030	–	11.5–13.0
410	0.15	1.0	1.0	0.040	0.030	–	11.5–13.5
420	0.15 min	1.0	1.0	0.040	0.030	–	12.0–14.0
431	0.20	1.0	1.0	0.040	0.030	1.25–2.50	15.0–17.0

The fabricability of martensitic stainless steels is poor because of a hard microstructure. Therefore, a special class of these steels has been developed that can be termed as controlled transformation steels. The carbon content is limited to a maximum value of 0.1 wt. % and chromium is in the range of 16-19 wt. %. Substantial alloying additions (Ni, Co, Mn, Mo and Cu) are required to depress the M_s temperature to well below 0°C. The alloys remain austenitic at room temperature and are amenable to various forming operations. The martensite is formed on freezing the alloy below M_s temperature. These alloys are heat treated to achieve precipitation strengthening. Typical examples of this class of stainless steels are 17-7PH, PH 15-7Mo, PH 14-8Mo and their compositions are listed in Table 2.

Table 2. Typical chemical composition of some precipitation-hardened stainless steels [7]

Grade	C	Mn	Si	Cr	Ni	Mo	Al	N
17-7PH	0.07	0.50	0.30	17.0	7.1	–	1.2	0.04
PH 15-7Mo	0.07	0.50	0.30	15.2	7.1	2.2	1.2	0.04
PH 14-8Mo	0.04	0.02	0.02	15.1	8.2	2.2	1.2	0.005
AM-350	0.10	0.75	0.35	16.5	4.25	2.75	–	0.10
AM-355	0.13	0.85	0.35	15.5	4.25	2.75	–	0.12

Precipitation Hardenable Stainless Steels

They are chromium-nickel grades that can be hardened by an aging treatment at a moderately elevated temperature [6]. These grades may have austenitic, semi austenitic, or martensitic crystal structures. Semi austenitic structures are transformed from a readily formable austenite to martensite by high temperature austenite-conditioning treatment. Some grades use cold work to facilitate transformation. The strengthening effect is achieved by adding such elements as copper and aluminium, which form intermetallic precipitates during aging. In this solution—annealed condition, these grades have properties similar to those of the austenite grades and therefore readily formable. The precipitation-hardened grades must not be subjected to further exposure to elevated temperature by welding or during service, because overaging of the precipitates can result in loss of strengthening.

Ferritic Stainless Steels

The ferritic stainless steels are Fe-Cr alloys with 15-30 wt. % Cr, low C, no Ni and often Mo, Al, Nb or Ti. The formability and weldability of these steels are poor but they possess moderate to good corrosion resistance. However, these alloys are prone to high temperature embrittlement. Modern melting and refining techniques like Vacuum-Oxygen-Decarburisation and Argon-Oxygen-Decarburisation have achieved considerable reduction in C and N contents in these alloys. The steels with low interstitial content have improved formability, weldability and toughness. Some typical steels in this category are AISI Types 405, 409, 410S, 430, 434 and 446 whose compositions are shown in Table 3.

Table 3. Chemical composition of important ferritic stainless steel grades [7]

Grade AISI	C max	Si max	Mn max	P max	S max	Ni	Cr	Others
405	0.08	1.0	1.0	0.040	0.030	–	11.5–14.5	Al 0.10–0.30
409	0.08	1.0	1.0	0.045	0.045	–	10.5–11.75	Ti 6×Cmin but 0.75 max
410S	0.08	1.0	1.0	0.040	0.030	0.60 max	11.5–13.5	–
430	0.12	1.0	1.0	0.040	0.030	–	16.0–18.0	–
434	0.12	1.0	1.0	0.040	0.030	–	16.0–18.0	Mo 0.75–1.25
446	0.20	1.0	1.5	0.040	0.030	–	23.0–27.0	N 0.25 max

Austenitic Stainless Steels

These stainless steels contain 18–25 wt. % Cr and 8–20 wt. % Ni and low C. These steels may also have additions of Mo, Nb or Ti and are predominantly austenitic at all temperatures, although depending on composition and thermomechanical history some delta-ferrite may be present. The austenitic alloys constitute the largest group of stainless steels in use, making up 65 to 70 % of the total. They occupy their dominant position not only because of their excellent corrosion resistance, but also because of an extensive inventory of ancillary properties, which include strength at elevated temperatures, stability at cryogenic temperatures and ease of fabricability including weldability. Some representative Fe-Cr-Ni stainless steels, arranged in order of Ni and Cr concentrations, are shown in (Fig. 3). Table 4 lists the chemical compositions of important austenitic stainless steels and the specifications used for austenitic stainless steel by different countries.

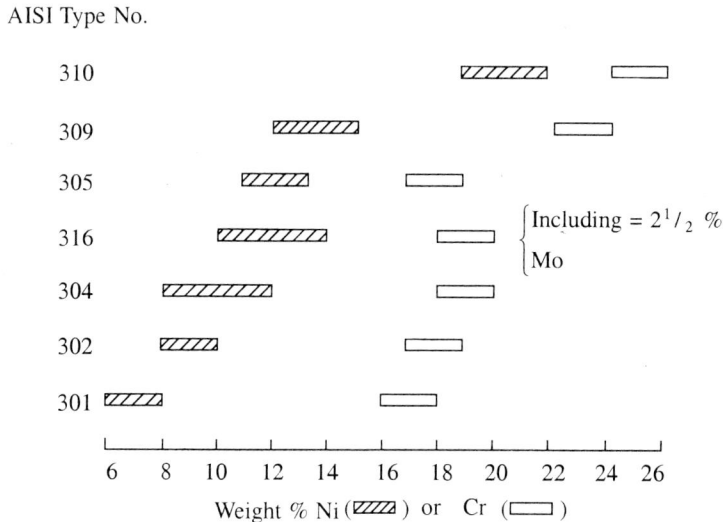

Fig. 3.   **Some representative Fe-Cr-Ni stainless steels arranged in order of Ni and Cr contents [7].**

High Nitrogen Stainless Steels High-nitrogen stainless steels are becoming an increasingly important new class of engineering materials with their better property combinations such as strength, toughness, creep resistance, non-ferromagnetic behaviour, corrosion resistance and stress corrosion cracking resistance [8]. These steels are considered 'high nitrogen' if they contain more than 0.08 wt % N with a ferritic matrix or 0.4 wt % N with an austenitic matrix [8]. The solubility of nitrogen in a Fe-Cr-Ni alloy is much lower than Fe-Cr-Mn alloys with comparable chromium content as illustrated in Fig. 4 [9]. The high nitrogen austenitic stainless steel having the composition 18 % Cr, 18 % Mn, 0.5–0.6 % N with very low carbon has the highest product of strength and toughness [$K_{IC} \cdot \sigma_{0.2} = 3 \times 10^5$ MN^2. m$^{-7/2}$] (Fig. 5) [8]. Yield strengths of 2400 MPa can be achieved

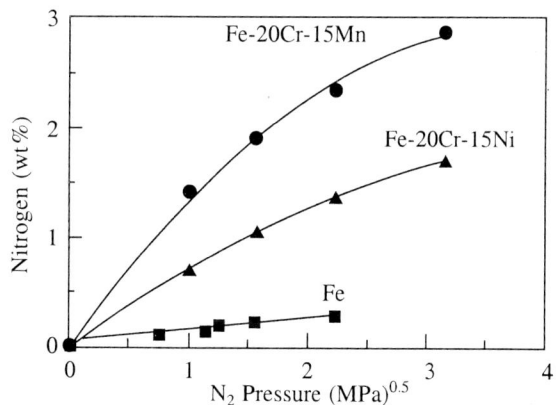

Fig. 4.   **Nitrogen solubility in liquid Fe-based alloys at 1873 K as a function of nitrogen-gas pressure [9].**

Table 4. Important austenitic stainless steel specifications and chemical compositions [78]

JIS	AISI	BS	DIN	N F	TOCT	C Max	Si Max	Mn Max	S Max	P Max	Ni	Cr	Mo Max	N Max	Other
										Chemical Composition (%)					
SUS201						0.15	1.0	5.5–7.5	0.06	0.03	3.5–5.5	16.0–18.0	–	0.25	–
	201					0.15	1.0	6.5–7.5	0.06	0.03	3.5–5.5	16.0–18.0	–	0.25	–
SUS202	202					0.15	1.0	7.5–10.0	0.06	0.03	4.0–6.0	17.0–19.0	–	0.25	–
						0.15	1.0	7.5–10.0	0.06	0.03	4.0–6.0	17.0–19.0	–	0.25	–
		284S16				0.07	1.0	7.0–10.0	0.06	0.03	4.0–6.5	16.5–18.5	–	0.15–0.25	–
					12X17 T9AH4	0.12	0.80	8.0–10.5	0.035	0.035	3.5–4.5	16.0–18.0	–	0.15–0.25	–
SUS301	301					0.15	1.0	2.0	0.045	0.03	6.0–8.0	16.0–18.0	–	–	–
SUS 301J1		301S21				0.08–0.12	1.0	2.0	0.045	0.035	7.0–9.0	16.0–18.0	–	–	–
			X12Cr Ni177			0.15	1.0	2.0	0.045	0.03	6.0–8.0	16.0–18.0	–	–	–
				Z12CN 17.07		0.12	1.5	2.0	–	–	6.0–8.0	16.0–18.0	0.8	–	–
						0.08–0.15	1.0	2.0	0.04	0.03	6.0–9.0	16.0–18.0	–	–	–
SUS302	302					0.15	1.0	2.0	0.045	0.03	8.0–10.0	17.0–19.0	–	–	–
SUS 302B	302B					0.15	2.0–3.0	2.0	0.045	0.03	8.0–10.0	17.0–19.0	–	–	–
						0.15	1.0	2.0	0.045	0.03	8.0–10.0	17.0–19.0	–	–	–
						0.15	2.0–3.0	2.0	0.045	0.03	8.0–10.0	17.0–19.0	–	–	–
		302S31				0.12	1.0	2.0	0.045	0.03	8.0–10.0	17.0–19.0	–	–	–
				Z10CN 18.09		0.12	0.75	2.0	0.04	0.03	7.5–9.5	17.0–19.0	–	–	–
SUS303	303					0.15	1.0	2.0	0.2	0.15	8.0–10.0	17.0–19.0	0.6	–	–
SUS 303Se	303Se					0.15	1.0	2.0	0.2	0.06	8.0–10.0	17.0–19.0	0.6	–	Se ≤0.15
		303S31				0.15	1.0	2.0	0.2	0.15	8.0–10.0	17.0–19.0	0.6	–	–
			X10Cr NiS189			0.15	1.0	2.0	0.2	0.06	8.0–10.0	17.0–19.0	0.6	–	Se ≤0.15
						0.12	1.0	2.0	0.06	0.15–0.35	8.0–10.0	17.0–19.0	1.0	–	–
						0.12	1.0	2.0	0.06	0.15–0.35	8.0–10.0	17.0–19.0	–	–	–
				Z10CNF 18.09		0.12	1.0	2.0	0.05	0.06	8.0–11.0	17.0–19.0	–	–	–
					12X18 H10E	0.12	0.80	2.0	–	–	9.0–11.0	17.0–19.0	–	–	–
SUS304	304					0.08	1.0	2.0	0.045	0.03	8.0–10.5	18.0–20.0	–	–	–
SUS304N1						0.08	1.0	2.5	0.045	0.03	7.0–10.5	18.0–20.0	–	0.10–0.25	–
SUS304N2						0.08	1.0	2.5	0.045	0.03	7.5–10.5	18.0–20.0	–	0.15–0.30	–
304LN						0.03	1.0	2.0	0.045	0.03	8.5–11.5	17.0–19.0	–	–	–
						0.08	1.0	2.0	0.045	0.03	8.0–10.5	18.0–20.0	–	–	–
	304L					0.03	1.0	2.0	0.045	0.03	8.0–12.0	18.0–20.0	–	0.1–0.16	–

(Contd)

JIS	AISI	BS	DIN	N F	TOCT	Chemical Composition (%)									
						C Max	Si Max	Mn Max	S Max	P Max	Ni	Cr	Mo Max	N Max	Other
	304N					0.08	1.0	2.0	0.045	0.03	8.0–10.5	18.0–20.0	–	0.1–0.16	–
	XM21 ASTM					0.08	1.0	2.0	0.045	0.03	7.0–10.5	18.0–20.0	–	0.1–0.3	–
	304LN ASTM					0.03	1.0	2.0	0.045	0.03	8.0–12.0	18.0–20.0	–	0.1–0.16	–
		304S11				0.03	1.0	2.0	0.045	0.03	9.0–12.0	17.0–19.0	–	–	–
		304S15				0.06	1.0	2.0	0.045	0.03	8.0–11.0	17.5–19.0	–	–	–
		304S31				0.07	1.0	2.0	0.045	0.03	8.0–11.0	17.0–19.0	–	–	–
			X5Cr Ni1810			0.03	1.0	2.0	0.045	0.03	8.0–12.0	17.0–19.0	–	–	–
			X2Cr Ni1911			1.0	2.0	0.040	0.03	0.03	9.0–12.0	17.0–19.0	–	–	–
				Z6CN 18.09		0.07	1.0	2.0	0.040	0.03	8.0–11.0	17.0–19.0	–	–	–
				Z2CN 18.10		0.03	1.0	2.0	0.040	0.03	9.0–11.0	17.0–19.0	–	–	–
				Z5CN 18.09AZ		0.06	1.0	2.0	0.040	0.03	8.0–11.0	17.0–20.0	–	0.12–0.20	–
				Z2CN 18.10AZ		0.05	1.0	2.0	0.040	0.03	9.0–12.0	17.0–19.0	–	0.12–0.20	–
					08X18 H10	0.04	0.8	2.0	–	–	9.0–11.0	17.0–19.0	–	–	–
					03X18 H11	0.04	0.8	2.0	–	–	10.0–12.0	17.0–19.0	–	–	–
SUS305	305					0.12	1.0	2.0	0.045	0.03	10.5–13.0	17.0–19.0	–	–	–
SUS305JI						0.08	1.0	2.0	0.045	0.03	11.0–13.5	16.5–19.0	–	–	–
						0.12	1.0	2.0	0.045	0.03	10.5–13.0	17.0–19.0	–	–	–
						0.10	1.0	2.0	0.045	0.03	11.0–13.0	17.0–19.0	–	–	–
		305S19	XcrNi 1911			0.07	–	–	–	–	10.5–12.0	17.0–19.0	–	–	–
				Z8CN 19-12		0.08	1.0	2.0	0.040	0.03	11.0–13.0	17.0–19.0	–	–	–
SUS309S	309S					0.08	1.0	2.0	0.045	0.03	12.0–15.0	22.0–24.0	–	–	–
				Z10CN 24-13		0.08	1.0	2.0	0.045	0.03	12.0–15.0	22.0–24.0	–	–	–
						0.10	1.0	2.0	0.040	0.03	12.0–14.0	22.0–24.0	–	–	–
SUS310S	310S					0.08	1.5	2.0	0.045	0.03	19.0–22.0	24.0–26.0	–	–	–
				Z12CN 25-20		0.08	1.2	2.0	0.045	0.03	19.0–22.0	24.0–26.0	–	–	–
						0.08	1.2	2.0	0.040	0.03	19.0–21.0	24.0–26.0	–	–	–
					10X23 H18	0.10	1.0	2.0	0.035	0.02	17.0–20.0	22.0–25.0	–	–	–
					20X23 H18	0.20	1.0	2.0	0.035	0.02	17.0–20.0–	22.0–25.0	–	–	–

(Contd)

Table 4. (Contd)

JIS	AISI	BS	DIN	N F	TOCT	Chemical Composition (%)									
						C Max	Si Max	Mn Max	S Max	P Max	Ni	Cr	Mo Max	N Max	Other
SUS316						0.08	1.0	2.0	0.045	0.03	10.0–14.0	16.0–18.0	2.0–3.0	–	–
SUS316J1						0.08	1.0	2.0	0.045	0.03	10.0–14.0	17.0–19.0	1.2	2.75	Cu 1.0–2.5
SUS316L						0.03	1.0	2.0	0.045	0.03	12.0–15.0	16.0–18.0	2.0–3.0	–	–
SUS316J1L						0.03	1.0	2.0	0.045	0.03	12.0–13.0	17.0–19.0	1.2–2.75	–	Cu 1.0–2.5
SUS316N						0.08	1.0	2.0	0.045	0.03	10.0–14.0	16.0–18.0	2.0–3.0	0.12–0.22	–
SUS316LN						0.03	1.0	2.0	0.045	0.03	10.0–14.0	16.0–18.0	2.0–3.0	0.12–0.22	–
	316					0.08	1.0	2.0	0.045	0.03	10.0–14.0	16.0–18.0	2.0–3.0	–	–
	316L					0.03	1.0	2.0	0.045	0.03	10.0–14.0	16.0–18.0	1.2–2.75	–	–
	316N					0.08	1.0	2.0	0.045	0.03	10.0–14.0	16.0–18.0	2.0–3.0	0.16–0.26	–
	316LN					0.03	1.0	2.0	0.045	0.03	10.5–14.5	16.5–18.5	2.0–3.0	0.12–0.22	–
		316S33				0.07	1.0	2.0	0.045	0.03	11.0–14.0	16.5–18.5	2.5–3.0	–	–
		316S31				0.07	1.0	2.0	0.045	0.03	10.5–13.5	16.5–18.5	2.0–2.5	–	–
		316S11				0.03	1.0	2.0	0.045	0.03	11.0–14.0	16.5–18.5	2.0–2.5	–	–
		316S13				0.03	1.0	2.0	0.045	–	11.5–14.0	16.5–18.5	2.5–3.0	–	–
						0.07	–	–	–	–	10.5–13.5	16.5–18.5	2.0–2.5	–	–
			XcrNiMo 17122			0.07	1.0	2.0	–	–	10.5–13.5	16.5–18.5	2.0–2.5	–	–
			X5CrNiMo 1810			0.03	–	–	–	–	11.0–14.0	16.5–18.5	2.0–2.5	–	–
			X2CrNiMo 17132	Z6CND 17.11		0.07	1.0	2.0	0.040	0.03	10.0–12.5	16.0–18.0	2.0–2.5	–	–
				Z6CND 17.12		0.03	1.0	2.0	0.040	0.03	10.5–13.0	16.0–18.0	2.0–2.5	0.10–0.20	–
					03X17 H14M2	0.03	0.8	1.0–2.0	0.035	0.02	13.0–15.0	16.0–18.0	2.0–2.3	–	–
SUS317						0.08	1.0	2.0	0.045	0.03	11.0–15.0	18.0–20.0	3.0–4.0	–	–
SUS317L						0.03	1.0	2.0	0.045	0.03	11.0–15.0	18.0–20.0	3.0–4.0	–	–
SUS317J1						0.04	1.0	2.5	0.045	0.03	15.0–17.0	16.0–19.0	3.0–6.0	–	–
	317	317S16				0.08	1.0	2.0	0.045	0.03	11.5–15.0	18.0–20.0	3.0–4.0	–	–
	317L	317S12				0.03	1.0	2.0	0.045	0.03	11.0–15.0	18.0–20.0	3.0–4.0	–	–
						0.06	1.0	2.0	0.045	0.03	12.0–15.0	17.5–19.0	3.0–4.0	–	–
						0.03	1.0	2.0	0.045	0.03	14.0–17.0	17.5–19.0	3.0–4.0	–	–
SUS321	321	321S31				0.08	1.0	2.0	0.045	0.03	9.0–13.0	17.0–19.0	–	–	Ti ≤ 5 C
						0.08	1.0	2.0	0.045	0.03	9.0–12.0	17.0–19.0	–	–	Ti ≤ 5 C
						0.08	1.0	2.0	0.045	0.03	9.0–12.0	17.0–19.0	–	–	Ti = 5C–
0.8			X6CrNiTi 1810			0.08	1.0	2.0	0.045	0.03	9.0–11.0	17.0–19.0	–	–	Ti=5C
0.8															–
0.6				Z6CNT 18.10		0.08	1.0	2.0	0.045	0.03	9.0–11.0	17.0–19.0	–	–	Ti = 5 C–

(Contd)

JIS	AISI	BS	DIN	N F	TOCT	Chemical Composition (%)									
						C Max	Si Max	Mn Max	S Max	P Max	Ni	Cr	Mo Max	N Max	Other
0.6					08X18 H10T	0.08	0.8	2.0	0.045	0.02	9.0–11.0	17.0–19.0	–	–	Ti = 5C –
SUS329J1						0.08	1.0	1.5	0.040	0.03	3.0–6.0	23.0–28.0	1.0–3.0	–	–
SUS329J2	329					0.03	1.0	1.5	0.040	0.03	4.5–7.5	22.0–26.0	2.5–4.0	0.08–0.2	–
						0.10	1.0	2.0	0.040	0.02	3.0–6.0	25.0–30.0	–	–	–
SUS347		347S31				0.08	1.0	2.0	0.045	0.03	9.0–13.0	17.0–19.0	–	–	–
	347					0.08	1.0	2.0	0.045	0.03	9.0–13.0	17.0–19.0	–	–	–
			X6CrNiNb 1810			0.08	1.0	2.0	0.045	0.03	9.0–12.0	17.0–19.0	–	–	–
				Z6CNNb 18.10		0.08	1.0	2.0	0.040	0.03	9.0–11.0	17.0–19.0	–	–	–
					08X18 H125	0.08	1.0	2.0	0.040	0.03	9.0–11.0	17.0–19.0	–	–	–
						0.08	0.8	2.0	0.02	0.035	11.0–13.0	17.0–19.0	–	–	–
SUS384						0.08	1.0	2.0	0.045	0.03	15.0–17.0	17.0–19.0	–	–	–
	384			Z6NC 18.16		0.08	1.0	2.0	0.045	0.03	15.0–17.0	17.0–19.0	–	–	–
						0.08	1.0	2.0	0.040	0.03	15.0–17.0	17.0–19.0	–	–	–

through cold work due to the high work hardening coefficient in these steels with added benefit of absence of deformation induced martensite (Fig. 6) [10]. Some of the Ni-free high nitrogen stainless steels are given in Table 5 [11].

Duplex Stainless Steels

These steels contain austenite and ferrite in equal proportions and are characterised by superior toughness, as compared to fully ferritic stainless steels, and excellent corrosion resistance. The balance between ferrite and austenite is achieved by adjusting the amounts of Cr (18-26-wt. %), Ni (5–6-wt. %), Mo (1.5–4 wt. %) and nitrogen. Table 6 shows typical compositions of some duplex stainless steels.

This chapter deals with physical metallurgical aspects and the mechanical and physical properties of austenitic stainless steels. The corrosion properties of austenitic stainless steel are discussed in the remaining chapters in the book.

2.0   AUSTENITIC STAINLESS STEELS

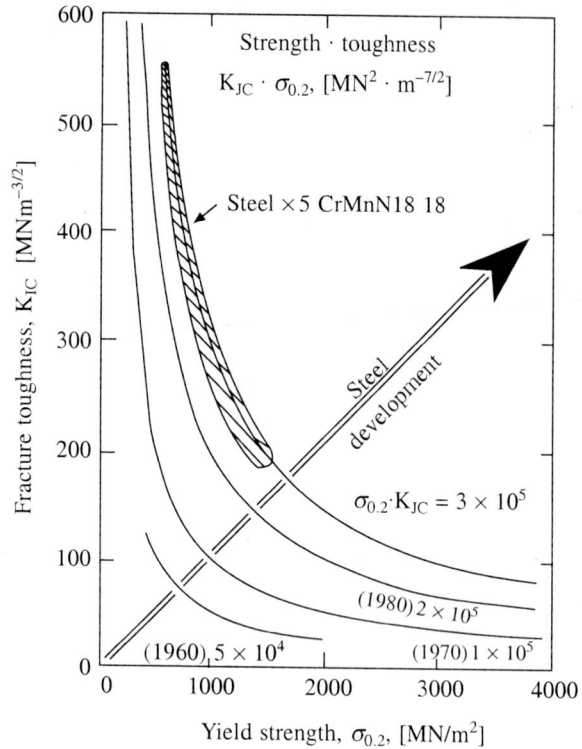

Fig. 5.   Chronological development of commercial steels with progressively better values of the product of strength and toughness [8].

The wrought austenitic steels are either single (γ) or duplex ($\gamma + \alpha$ or δ) phase structures, where γ refers to the face-centred-cubic (fcc) austenite and α or δ refers to the body-centred-cubic (bcc) ferrite respectively, at the usual solution treatment temperatures of above 1000 °C. This is evident from (Fig. 7) which shows part of the isothermal section of the iron-chromium-nickel equilibrium diagram at 1100 °C [12]. Type 310, 20Cr- 25Ni -Nb and 12R72HV steels fall within the single-phase (γ) field whilst the other steels listed in the figure lie just within the two-phase ($\gamma + \alpha$) field. The effect of various austenitising and ferritising elements on austenite and ferrite can be expressed in terms of Ni equivalent and Cr equivalent respectively [13]. The following nickel and chromium equivalents can be used to locate typical compositions of the commercial steels in the isothermal section of the Fe-Cr-Ni diagram [14].

$$\text{Ni equivalent (wt. \%)} = \% \text{ Ni} + \% \text{ Co} + 0.5\% \text{ Mn} + 30\% \text{ C} + 0.3\% \text{ Cu} + 25\% \text{ N} \qquad (1)$$

$$\text{Cr equivalent (wt. \%)} = \% \text{ Cr} + 2\% \text{ Si} + 1.5\% \text{ Mo} + 5\% \text{ V} + 5.5\% \text{ Al} + 1.75\% \text{ Nb}$$

$$+ 1.5\% \text{ Ti} + 0.75\% \text{ W} \qquad (2)$$

Apart from removing the γ-stabilizing elements, nitrogen and carbon, from solid solution, alloying elements like titanium and niobium act as α-and σ- stabilizing elements in their own right. Consequently, their α- and σ-forming effects can be significantly enhanced [15, 16].

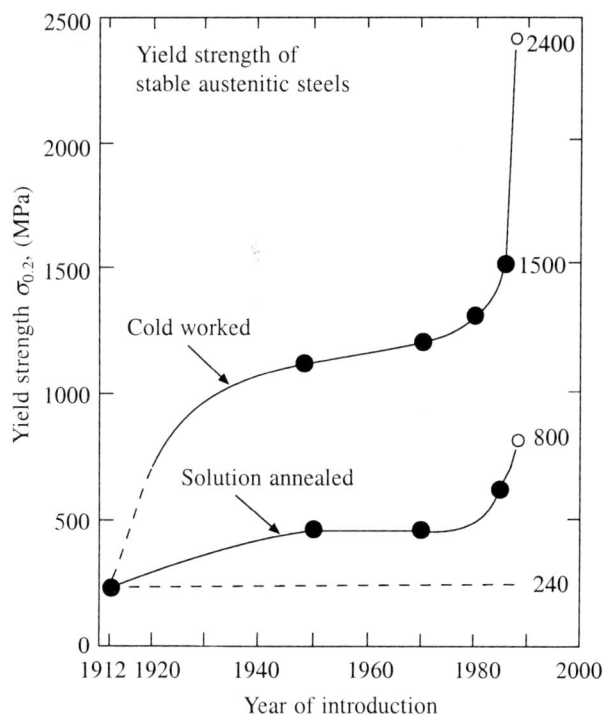

Fig. 6. **Effect of cold work (without formation of deformation induced martensite) on yield strength in stable austenitic stainless steels [8].**

Table 5. **Chemical compositions of some Ni-free high nitrogen austenitic stainless steels [11]**

Steel	*Composition (wt %)*					
	Cr	*Mo*	*Mn*	*N*	*C*	*Si*
NML-B2	20.90	–	13.80	0.69	0.12	0.70
NML-B4	22.07	–	16.63	1.01	0.04	0.40
NML-B5	17.90	–	16.93	0.69	0.05	0.28
VSG Steel	18.00	2.00	18.00	0.90	0.10	–
Swiss A	17.10	3.20	11.40	0.92	0.008	1.20
Swiss D	16.40	4.20	11.80	0.98	0.009	1.60
Swiss G	11.80	7.80	11.10	0.80	0.016	1.40
Retaining Ring Steel	18.00	–	18.00	0.5-0.6	Very low	–

Constitution of Austenitic Steels

The constitution of the austenitic steels at ambient temperature, following rapid cooling from the solution treatment temperature, can be predicted using the Schaefler diagram (Fig. 8) which shows the phase fields in terms of the nickel and chromium equivalents [13]. The high temperature phases in majority of the austenitic steels can be made stable at ambient temperature either by adjusting the chemical composition or by rapid cooling. However, partial transformation of the $\gamma \to \alpha'$ martensite

Table 6.   Typical chemical composition of some duplex stainless steels [7]

UNS No.	Cr	Ni	Mo	N	Others
S32304	23	4	0.2	0.1	–
S31803	22	5	2.7	0.14	–
S32760	25	7	3.6	0.24	0.7Cu, 0.7W
S32550	25	6.5	3.8	0.26	1.5Cu
S32750	25	7	3.8	0.27	–

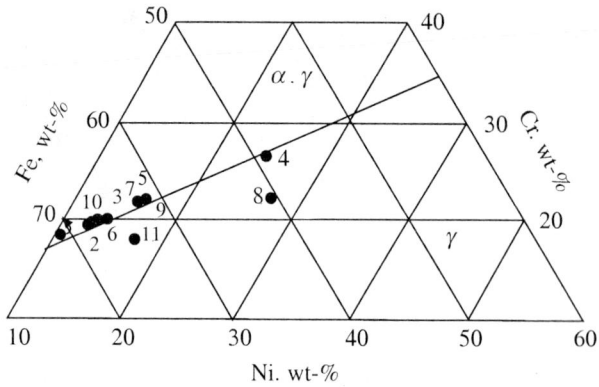

1	AISI 301	5	AISI 316	9	M316
2	AISI 302	6	AISI 321	10	FV548
3	AISI 304	7	AISI 347	11	12R72HV
4	AISI 310	8	20Cr-25 Ni-Nb		

Fig. 7.   Part of isothermal section of Fe-Cr- Ni equilibrium diagram at 1100 °C [12].

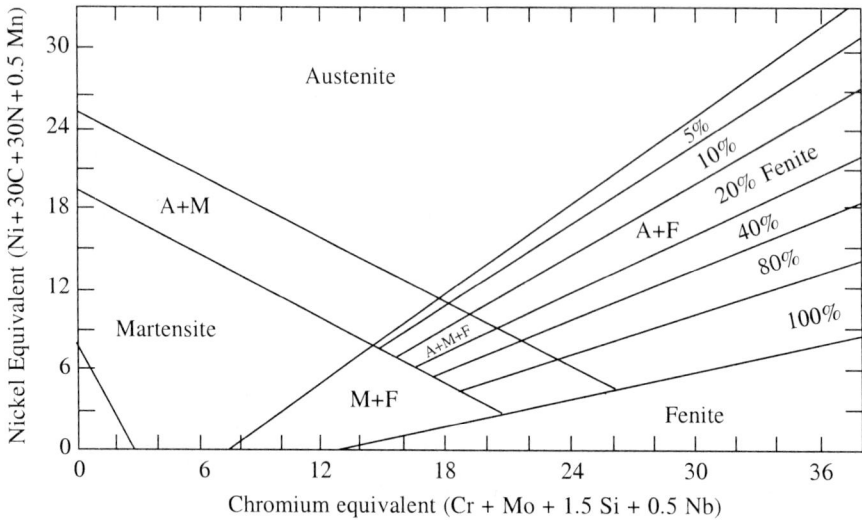

Fig. 8.   Schaeffler diagram [13].

phase occurs in AISI type 301 stainless steel during cooling due to the relatively high M_s (α') temperature of this steel [17].

The addition of nickel to 18% Cr steels enlarges the gamma loop considerably [12, 19, 20, 21]. Increasing nickel has two main effects on the constitution and microstructure, which are as follows:

(i) It increases the amount of austenite present at the solution-treatment temperature. However, at low nickel contents this austenite may transform wholly or partially to martensite on rapid cooling to room temperature [4].

(ii) It decreases the M_s temperature such that, with about 8% Ni, the M_s temperature is just below room temperature and stable austenite is retained after cooling from the solution-treatment temperature to room temperature [21, 22, 23].

An 18Cr-8Ni carbon-free steel is a borderline with respect to a fully austenitic structure and may contain a little delta ferrite [24]. About 12% Ni is required to produce a fully austenitic structure at solution-treatment temperatures of about 1050 °C. An 18 Cr-8 Ni-0.1C alloy is fully austenitic above about 900 °C because carbon is a powerful austenite-forming element. However, the M_s temperature is only just below room temperature, leading to partial transformation of austenite to martensite either during a refrigeration treatment or during cold working [4]. The interaction between chromium and nickel in promoting the formation of stable austenite in 0.1 % C steels, after cooling from 1050 to 1100 °C, is therefore of utmost importance (Fig. 9) [24]. Some of the main effects are:

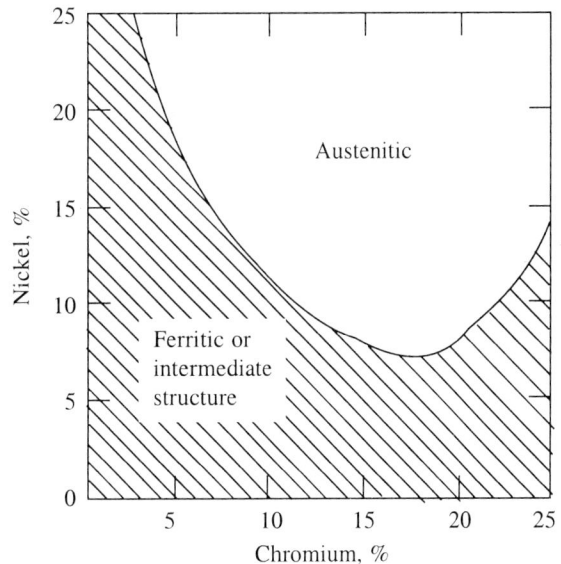

Fig. 9. **Effect of nickel and chromium equilibrium on constitution of 0.1 % C.**

(i) At low chromium contents, chromium acts as an austenite stabilizer by expanding the gamma phase up to the minimum in the gamma loop.

(ii) At 18% Cr, a minimum nickel content is required to promote a fully austenitic structure, which is stable at room temperature.

(iii) With more than 18% Cr, the ferrite–forming tendency of chromium predominates and increasing nickel is required to eliminate delta ferrite, although the austenite becomes increasingly stable with respect to martensite formation.

The ferrite-stabilizing character of molybdenum is illustrated in Fig. 10 which shows the room temperature structure of an 18% Cr-8% Ni-2% Mo stainless steel to be dual phased, consisting of both austenite (A) and ferrite (F). In order to maintain a fully austenite structure the nickel content of an 18% Cr- 2% Mo steel must be greater than 10%. Molybdenum also promotes the formation of intermetallic phases, particularly sigma, which causes room temperature embrittlement [25]. It extends the range of stability of this phase and shifts the A/A+F boundary to lower chromium contents. Like carbon, nitrogen is also a strong γ stabilizer, whose influence is shown in Fig. 11.

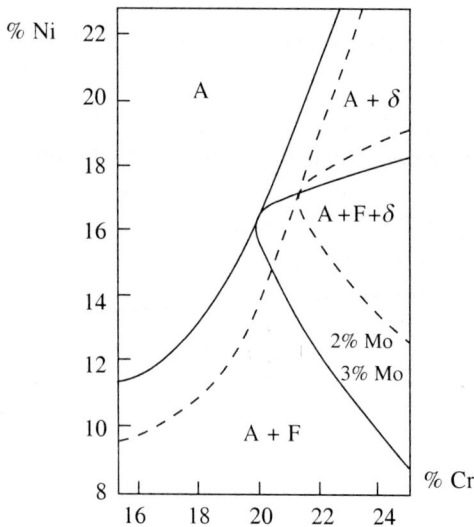

Fig. 10. Effect of molybdenum on the structure of Fe-Cr-Ni aloys air cooled from 1100 to 1150 °C [25].

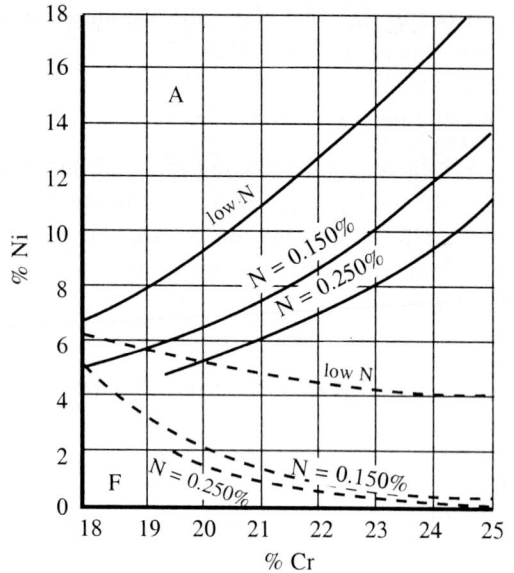

Fig. 11. Influence of nitrogen on the austenite (A) and ferrite (F) phase boundaries, and on the structure of Fe-Cr-Ni stainless steels [25].

The effect of other alloying elements is also important in that, depending upon whether they are austenite or ferrite-forming elements, they will decrease or increase the tendency for delta-ferrite formation at the solution-treatment temperature. Many workers [14] have considered the effect of alloying elements on the phase stabilities of austenitic stainless steels by using chromium and nickel equivalent compositions and superimposing them on the Schaeffler diagram [13].

Transformation of Austenite to Martensite

Austenite in the lower range of highly alloyed stainless steels may be transformed to martensite. This can occur either in the solution-treated condition when the M_s temperature is above room temperature or it may occur during refrigeration in more stable alloys in which the M_s temperature is below room temperature. Martensite may also be formed by deformation, above room temperature in the case of unstable steels and below room temperature in the case of stable steels, depending on M_d temperature.

Apart form cobalt, almost all alloying elements depress the M_s temperature [21, 22, 23]. Recently, linear equations relating the M_s temperature to the composition have been developed for austenitic stainless steels [26]. This type of relationship is important, particularly if used to establish the M_d temperature, in assessing the cold formability of austenitic stainless steels [27]. Equations have also been established for austenitic steels relating the M_{d30} temperature, at which 50% of martensite is produced under the action of a true strain of 0.30, to the composition of the steel [28]. M_d is always higher than M_s.

$$M_s \,(°C) = 1302 - 42(\%Cr) - 61(\%Ni) - 33(\%Mn) - 28(\%Si) - 1667(\%C + \%N) \qquad (3)$$

$$M_d \,(°C) = 413 - 462 \,(\%C + \%N\,) - 0.2 \,(\%Si) - 8.1 \,(\%Mn) - 13.7 \,(\%Cr)$$

$$- 0.5 \,(\%Ni) - 18.5 \,(\%Mo) \qquad (4)$$

Carbide and Nitride Precipitation

Unstabilized grades Austenitic stainless steels can contain up to 0.15% of carbon. The solubility limit for carbon in a type 18-8 alloy is indicated in Fig. 12 by the line separating the single-phase γ region from the γ + carbide field. The carbides correspond to the type $M_{23}C_6$, where M is principally Cr, but it can be partially replaced by Fe, Mo and Ni, whence the general designation $(Cr, Fe, Mo, Ni)_{23}C_6$.

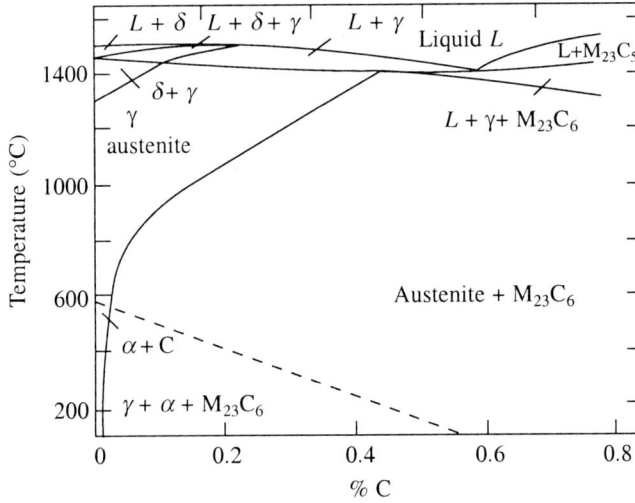

Fig. 12. **Solubility of carbon with respect to $M_{23}C_6$ carbides (M = Cr, Fe, Mo, Ni) in an 18% Cr-8% Ni stainless steel.**

After austenitizing at around 1100 °C, the carbon is retained in solution only by rapid cooling. If the annealed alloy is held in the temperature range between 450 and 850 °C, either during service or

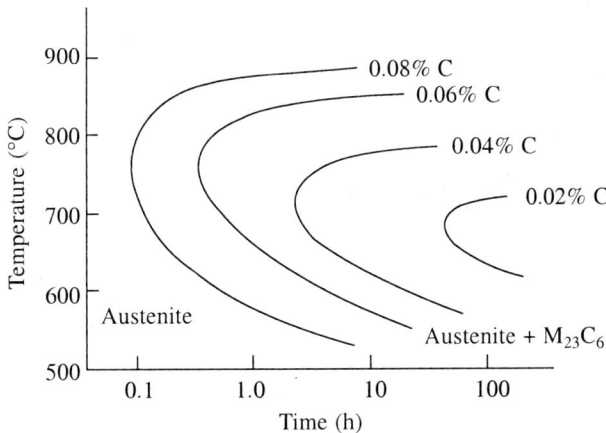

Fig. 13. **Influence of C content on the kinetics of $M_{23}C_6$ precipitation a type 18-10 austenitic stainless steel [25].**

cooled slowly after a welding operation, the excess carbon precipitates at grain boundaries in the form of chromium-rich $M_{23}C_6$ carbides (Fig.13). The $M_{23}C_6$ carbides precipitate initially at grain and incoherent twin boundaries, before forming with the austenite grains.

Molybdenum decreases the solubility of carbon in austenite and accelerates the $M_{23}C_6$ precipitation [5]. An increase in nickel content has a similar effect [30], while nitrogen retards the precipitation and coalescence of $M_{23}C_6$ (Fig.14) [29].

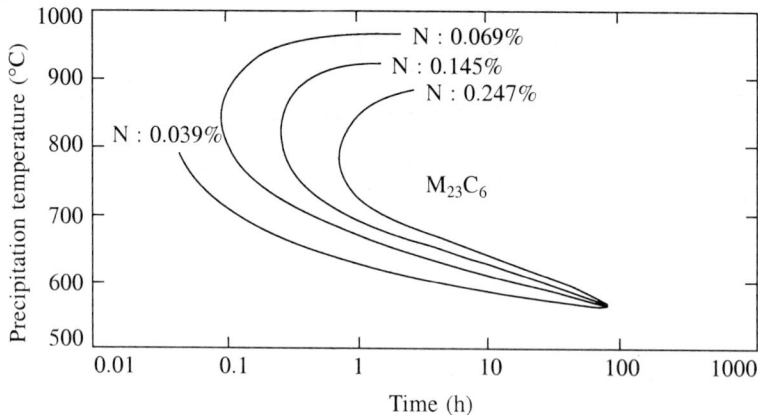

Fig. 14.   Influence of nitrogen on the kinetics of $M_{23}C_6$ precipitation in a 17 Cr-13 Ni-5 Mo-0.05 C steel [29].

Stabilized grades The addition of titanium or niobium retards the precipitation of chromium-rich $M_{23}C_6$ carbides, thus increasing the resistance to intergranular corrosion [5]. The austenite is depleted in carbon due to the selective formation of Ti(C, N) and Nb(C, N) carbonitrides and $Ti_4C_2S_2$ carbosulfides [5].

The following relations give the solubility of titanium and niobium carbides in 18Cr-12Ni steel [5]:

$$\log [Ti] [C] = 2.97 - 6780/T$$

$$\log [Nb][C] = 4.55 - 9350/T$$

Corresponding to the general equation:

$$\log [M][X] = A - H/RT$$

where A is a constant, H is the heat of dissolution, R is the perfect gas constant and T is the absolute temperature [25].

The M(C, N) particles precipitate essentially within the grains. However, intergranular precipitation occurs under certain conditions, particularly at high austenitising temperatures. This phenomenon is observed in weld zones of 18 Cr-10 Ni-titanium stabilised alloys.

Intermetallic phases Alloys containing transition elements A, such as Fe, Ni, Mn, Co, etc., together with transition elements B, of the type Cr, Ti, V, etc., can form intermetallic phases with formula ranging from A_4B to AB_4. On high temperature exposure, austenitic stainless steels are known to result in precipitation of a host of secondary phases [31]. Some of these phases commonly occur and are well understood with respect to their impact on mechanical and corrosion properties. This section discusses some of these phases such as sigma, chi, carbides and R-phase, Lave's, G-phase, mu phase, Z-phase etc., also precipitate in austenitic stainless steels [32, 33].

Sigma phase Sigma phase has a body centered tetragonal structure. The values of the Cr and Ni equivalents can be used to evaluate the possibility of sigma phase formation in a Fe-Cr-Ni alloy at high temperature (Fig. 15). The propensity of sigma phase precipitation in austenitic stainless steels depends on the chemical composition of the residual austenite after precipitation of carbides and nitrides, which always form first [31].

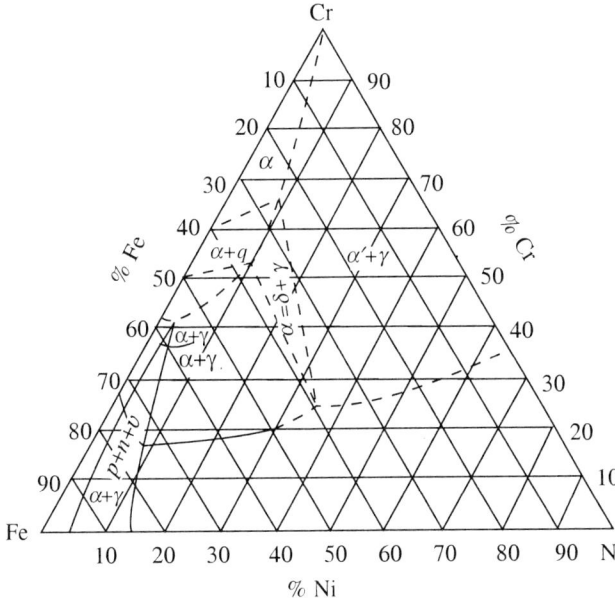

Fig. 15.   A section of the Fe-Cr-Ni ternary equilibrium diagram at 650°C [25].

The tendency to sigma phase formation of an austenitic stainless steel can be known from the formula proposed by Hull [34]. The formula for Equivalent Chromium Content (ECC) is,

$$\text{ECC} = \% \ \text{Cr} + 0.31 \ \% \ \text{Mn} + 1.76 \ \% \ \text{Mo} + 0.97 \ \% \ \text{W} + 2.02 \ \% \ \text{V} + 1.58 \ \% \ \text{Si} + 2.44 \ \% \ \text{Ti}$$

$$+ \ 1.7 \ \% \ \text{Nb} + 1.22 \ \% \ \text{Ta} - 0.266 \ \% \ \text{Ni} - 0.177 \ \% \ \text{Co}.$$

If the equivalent Cr content (ECC) is greater than 17–18 wt %, the steel is susceptible to sigma formation. This equation was modified by Gill et al. to account for the strong influence of carbon [35]. As per these authors, normalised equivalent chromium content, NECC = ECC/% C. This suggests that sigma phase precipitation becomes easier as the carbon content of the matrix reduces [36]. Fig. 16 shows the influence of various alloying elements on the kinetics of sigma phase precipitation. Chromium, molybdenum, titanium and niobium all promote sigma formation, while the precipitation rate is also accelerated by the addition of 2 to 3% of silicon [37]. Incorporation of nitrogen in the weld deposit avoids/delays nucleation of sigma and chi phases [38]. The presence of delta-ferrite and low interstitial content affect the growth kinetics of sigma and other intermetallic phases but not the total content of these phases [39].

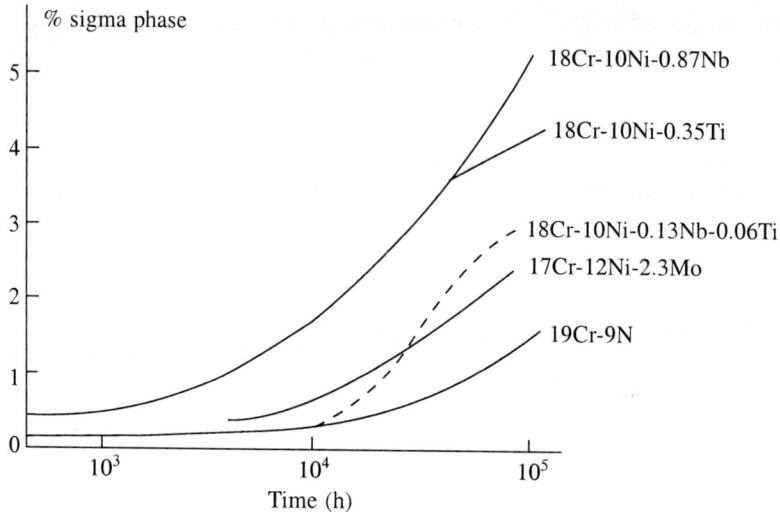

Fig. 16.   Rate of sigma phase precipitation at 700 °C [37].

Cold work decreases the incubation period for sigma phase formation [25]. On the contrary, an increase in grain size, due to annealing at a very high temperature, retards sigma phase precipitation [37]. The presence of delta ferrite, particularly in welds, can reduce incubation period for sigma formation in an austenitic stainless steel [5]. Stress accelerates sigma phase precipitation and extends its range to lower temperature [40].

The precipitation of sigma phase is controlled both by the rate of diffusion of chromium and other sigma-forming elements and by the mode of nucleation [40]. The chemical composition of sigma phase, determined for different types of austenitic steels (17 Cr-12 Ni-2.5 Mo-Ti, 25 Cr-20 Ni-0.03 & 0.13 C- 0.6 & 2 Si) exposed for times between 10 and 5000 hours at temperatures from 650 to 900 °C, was found to vary with time and temperature [41]. The compositions of delta-ferrite and sigma phase are close to each other [42]. Hence, delta-ferrite, in an austenitic stainless steel weld metal, easily transforms to sigma phase by a crystallographic re-orientation [42]. Heat input during welding has a significant say in the precipitation kinetics of sigma and other intermetallic phases. Higher heat input to the weld metal retards the decomposition kinetics of delta-ferrite and thus the precipitation kinetics of sigma phase [43]. Kokawa et al. reported faster precipitation kinetics of sigma phase in vermicular ferrite than in lacy ferrite [44]. Sigma phase is known to affect the tensile and creep ductilities of the stainless steel [45, 46].

Chi phase The importance of chi phase in austenitic stainless steel has been lucidly brought out by Weiss et al. [47]. Chi phase has a body centred cubic structure and is a stable intermetallic compound containing Fe, Cr and Mo [47]. Chi is a carbon-dissolving compound of the type $M_{18}C$ [48]. The composition of chi can vary appreciably with a high tolerance for metal atom interchange [48]. Upon addition of carbon, the metal atom proportion within the chi phase is shifted towards Mo at the cost of Fe and Cr i.e. towards the strongest carbide former [49]. The precipitation diagrams for Mo-containing and Ti and Nb stabilised stainless steels are shown in Figs. 17 and 18, respectively [25].

Weigand and Doruk reported that chi and lave's phases form simultaneously with carbides [50]. Presence of delta-ferrite in the steel favoured the precipitation of sigma and chi phases [50]. Solomon and Devine showed that chi phase precipitates at lower temperatures of aging as compared to sigma phase [51].

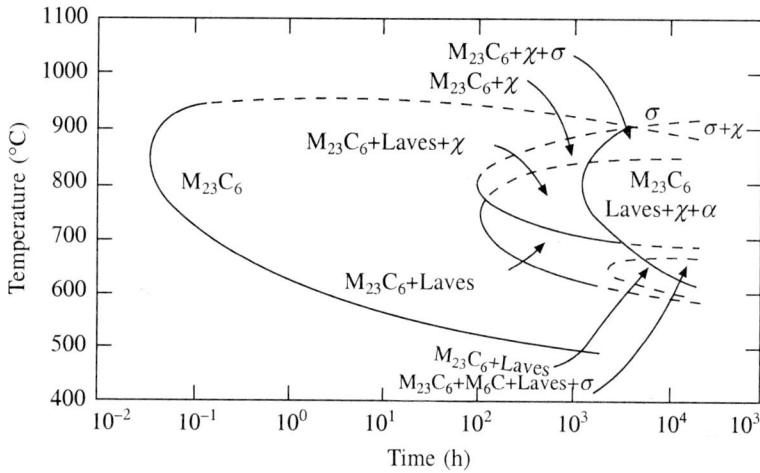

Fig 17.   **TTT diagram for precipitation in an 18 Cr-12 Ni-2 Mo austenitic stainless steel [5].**

Weiss and Stickler showed that at liquid nitrogen temperatures, presence of $M_{23}C_6$ led to a sharp decrease in impact strength while the presence of chi phase did not lead to a further significant drop [47]. Shankar et al. showed that copious precipitation of chi phase beyond 100 hours of aging, at 850 °C, in nitrogen containing AISI type 316L stainless steel led to a sharp decrease in the tensile ductility [52]. However, good resistance to brittle microcracking in presence of chi phase and its interfaces has been observed during creep crack growth [48, 53].

R-phase It is a Fe-Cr-Mo intermetallic phase with a hexagonal structure having unusually large lattice spacings [54]. Dyson and Keown considered the atomic movements in ferrite to accommodate the R-phase structure [55]. They showed that only small atomic movements and lattice strains were required for R-phase to form from ferrite. Tavassoli et al. reported its presence on aging Mo-bearing alloys [54]. R-phase is reported to precipitate inside the delta-ferrite of the weld metal, with lath morphology [53, 54]. Formation of R-phase has been reported during stress relief of austenitic weldments containing higher amounts of ferritisers in the weld metal [53].

Carbides $M_{23}C_6$ carbides are face centered cubic structured precipitate [48]. Carbide precipitation usually precedes the formation of intermetallic phases [47, 56]. As the formation of intermetallic phases increases, the carbides redissolve due to thermodynamic considerations, to replenish the matrix in Cr, Mo, C & N. During this period, the precipitation rate of intermetallic phases decreases after which an increase is again observed as shown in Fig. 19 [56, 57]. The precipitation kinetics of $M_{23}C_6$ carbide phase in an AISI type 316L weld metal [57] is shown in Fig. 20. Precipitation of these carbides at grain boundaries is known to impair impact property more than any other mechanical property [47]. Their precipitation is extremely deleterious to the localised corrosion behaviour of austenitic stainless steels, as discussed extensively in this book.

Solid solution hardening The interstitial alloying elements N, C and B produce considerable strengthening

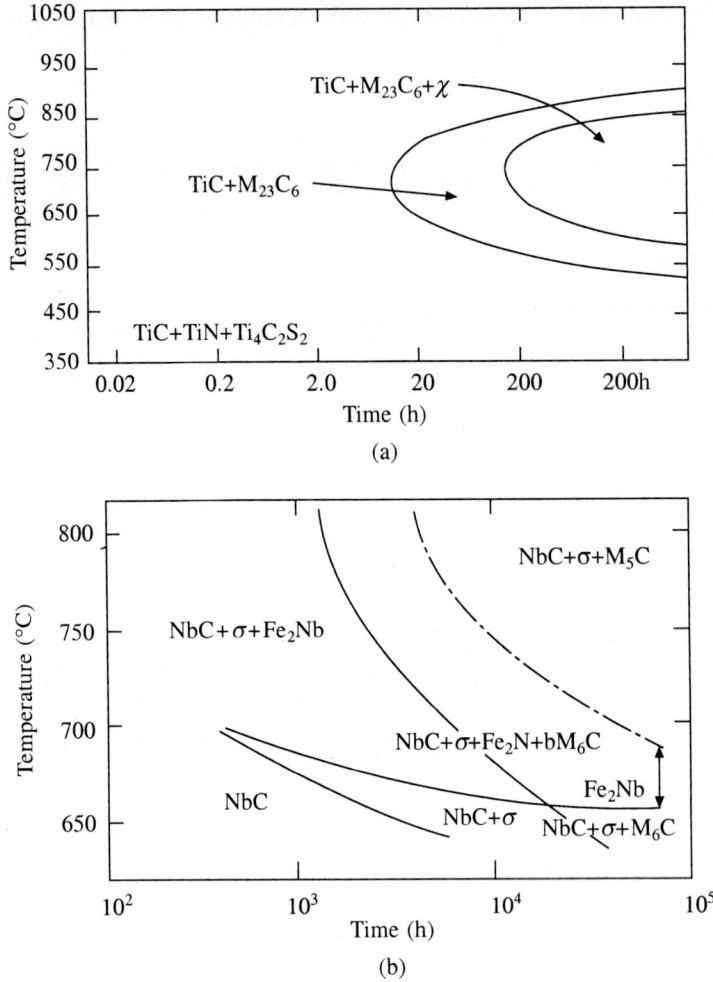

Fig. 18.   TTT diagram for precipitation in (a) an 18 Cr-10 Ni-Ti austenitic stainless steel and (b) an 18 Cr-10 Ni-0.9 Nb austenitic stainless steel

(Fig. 21) [25]. Increase in yield strength caused by substitutional solid solution elements, particularly by austenite stabilizers, is moderate. Hardening is due to the inhibition of dislocation movement by the lattice distortion associated with solute atom [5]. The most effective method of increasing the yield strength of austenitic stainless steels is by introduction of nitrogen, an addition of 0.1% leading to a gain of about 50 MPa [5].

Hardening by grain refinement For austenitic stainless steel, strengthening can be obtained by grain refinement, according to the Hall-Petch relation:

$$\sigma = \sigma_0 + Kd^{-1/2}$$

where d is the mean grain diameter, σ the yield stress, and σ_0 and K are temperature dependent constants for the material considered. The hardening due to grain refinement is due to the difficulty

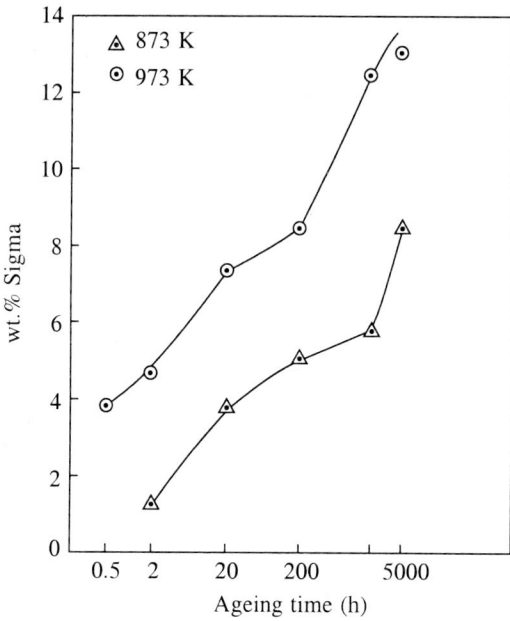

Fig. 19. **Growth kinetics of sigma at 873 K and 973 K [57].**

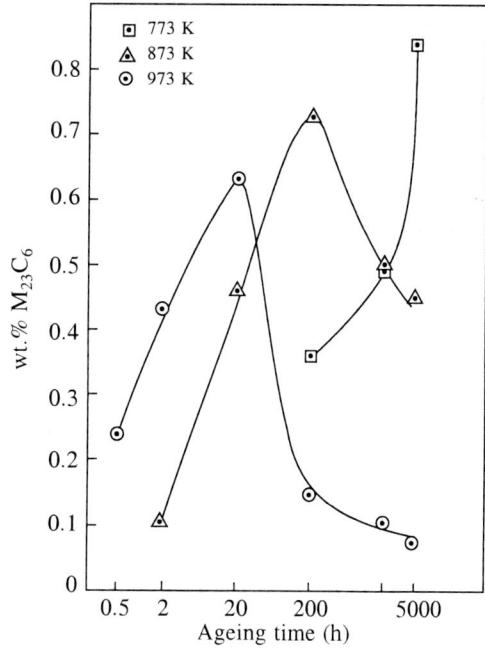

Fig. 20. **Amount of $M_{23}C_6$ formed during aging [57].**

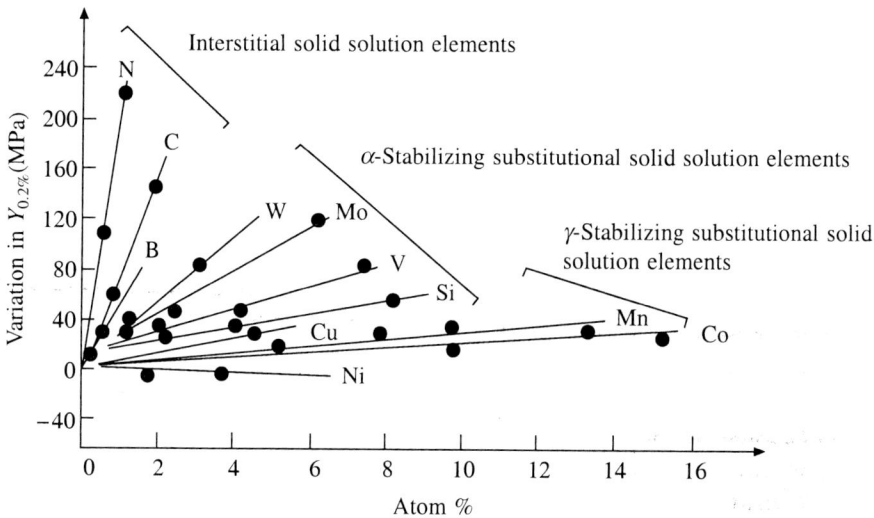

Fig. 21. **Effect of solid solution strengthening in the austenite [25].**

in propagating plastic strain from one grain to another, and therefore depends on the available slip systems [29]. The contribution of grain boundaries to yield strength is enhanced by nitrogen as it increases the K coefficient (the dislocation locking parameter) in the Hall-Petch equation [10].

Strain hardening The yield strength and tensile strengths are increased by cold work while the ductility is lowered. The greater the amount of plastic strain, the higher is the stress required to deform the material further. This phenomenon is known as strain (or work) hardening. The cause for strain hardening is the increased difficulty of dislocation movement, as their density increases with deformation, due to their interaction with each other or with vacancies and other crystal defects.

Certain elements increase the already high work-hardening rate of the austenitic steels. Low nickel grades are less stable and will tend to gradually transform to martensite during cold working, leading to pronounced hardening (Fig. 22) [30].

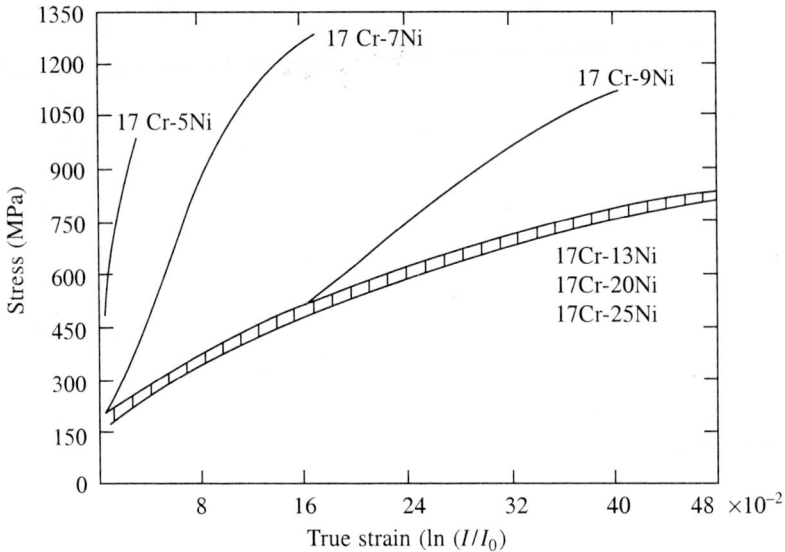

Fig. 22.   Effect of nickel content on the true stress-strain curves for 17% Cr austenitic stainless steel [30].

Low carbon austenites(0.02%) work harden faster than those with larger amounts of this element (> 0.06%). Copper reduces strain hardening, whereas nitrogen and silicon increase it [5]. In unstable steels, apart from alloy chemistry, which determines the M_d temperature, the quantity of martensite formed depends on the amount of strain and the deformation temperature. An increase in strain rate also leads to more rapid hardening (Fig. 23) [25].

Precipitation hardening Intragranular precipitation of particles based on elements such as C, N, B, V, Nb, or Ti is an important strengthening mechanism in austenitic stainless steels. Fine precipitates uniformly distributed in the matrix act as efficient obstacles to dislocation movement [29]. After solution treatment, the precipitation of TiC and TiN occurs within the grains, and can be used to increase the creep strength. Austenitic grades with large boron content on cold working can develop a uniform dispersion of fine and stable $M_{23}(C, B)_6$ particle [25].

Welding Metallurgy
Austenitic stainless steels can generally be readily welded, since no hard structures are formed in the heat-affected zone. However, a number of detrimental effects can occur. They are :

(i) A fully austenitic weld metal can produce hot cracking because of the stresses set up during

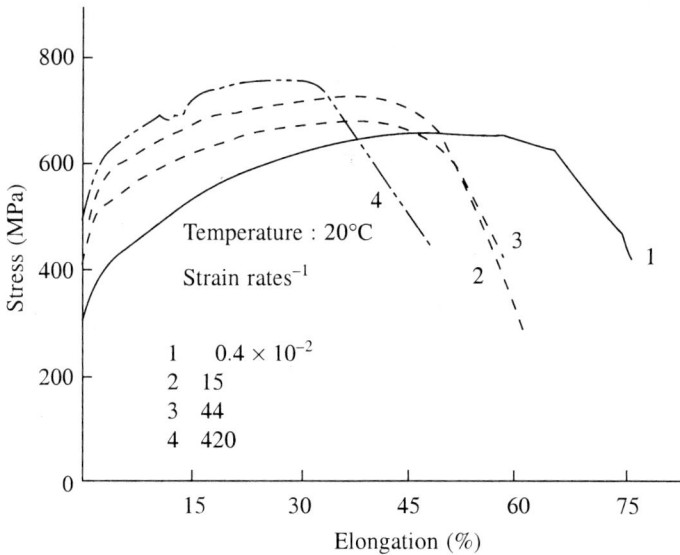

Fig. 23. Stress-strain curves for an 18 Cr-12 Ni-2 Mo austenitic stainless steel deformed at different strain rates [25]

contraction that accompanies solidification of the weld. This can be overcome by ensuring that the weld metal contains a little delta ferrite [4].

(ii) Various forms of liquation cracking can occur in both the weld metal and heat-affected zone close to the weld metal, if low melting point phases, e.g. borides, are present. The problems in fully austenitic welds can be minimized by reducing the joint restraint and heat input during welding, by decreasing the concentrations of the detrimental elements and trace impurities in the steels, and by using consumable electrodes with balanced compositions such that 5–10% δ-ferrite is produced in the weld deposit [58].

(iii) In stabilized steels, especially those containing niobium, the high temperatures in the heat-affected zone of a weld can dissolve some NbC. Subsequent strain induced precipitation of the NbC can occur during a post-weld stress-relieving treatment, and this can lead to a form of low-ductility creep-rupture cracking [59, 60]. This can be overcome by stress relieving at higher temperatures at which the NbC overages. Alternately, a full solution treatment may be used.

(iv) During welding, parts of the heat-affected zone are heated in the range in which $Cr_{23}C_6$ precipitates at the austenite grain boundaries. This locally lowers the chromium content, so that preferential corrosive attack occurs in the chromium-depleted zone. This is known as weld decay. Remedial measures could be full solution treatment at 1050°C to dissolve any grain-boundary carbides precipitated in the heat-affected zone or annealing at about 900°C to allow chromium to diffuse from bulk into the impoverished zone ('healing' treatment) [4]. Austenitic stainless steels stabilized with Ti or Nb are susceptible to knife line attack [4]. Because stabilization is entirely effective, there has been a trend to produce austenitic stainless steels with very low carbon contents of 0.03% maximum [61]. In the absence of efficient stabilization, low heat input during welding may be used to minimize the time for which the heat-affected zone is in the sensitization temperature range, and to decrease the width of the sensitized region [4].

The possibility of formation of delta-ferrite during cooling of weld metal and the primary solidification mode can be known from the 70% iron isopleth (Fig. 2). The amount of delta-ferrite that is retained in the weld metal can be known from the constitution diagrams, if the chemical composition is known. Schaefler diagram is the most popular of these constitution diagrams. Fig. 24 locates important stainless steels on this diagram with respect to its propensity to form delta-ferrite [62]. However, Schaefler diagram does not account for the influence of Nitrogen, a very potent austenitiser. Effect of nitrogen has been accounted for in the diagram proposed by W.T.Delong (Fig. 25) [63]. The latest constitution diagram that has been produced is by the Welding Research Council in 1992 (Fig. 26) [64].

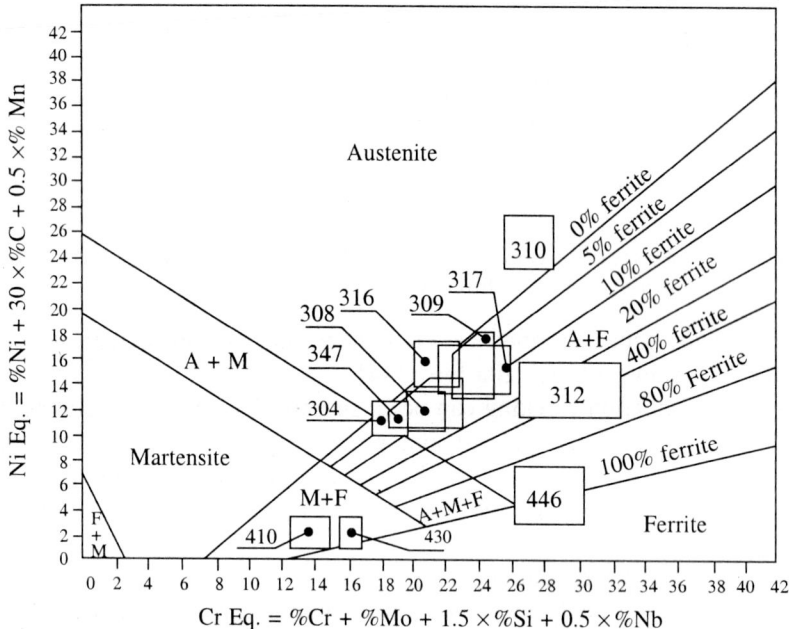

Fig. 24.   Schaeffler diagram—Position of some common grade [5]

The amount of delta-ferrite to be retained in the weld metal is governed by the needs of service. Generally, delta-ferrite embrittles a weld metal [65] and deteriorates corrosion property [66], details of which are discussed elsewhere in the book. However, Shaikh et al. recently reported that embrittlement of austenitic stainless steel weld metal was not due to presence of delta-ferrite but because of cold work present in the weld metal [67].

Physical Properties
The physical properties of austenitic stainless steels that are considered are in regards to the functional properties of stainless steels and related alloys [68]. The major physical properties considered are melting range, density, coefficient of expansion, modulus of elasticity, electrical resistivity, thermal conductivity, specific heat and magnetic permeability.

1. *Melting range:* Table 7 shows the solidus and the liquidus temperatures for various grades of austenitic stainless steels. These values were determined by differential thermal analysis (DTA).

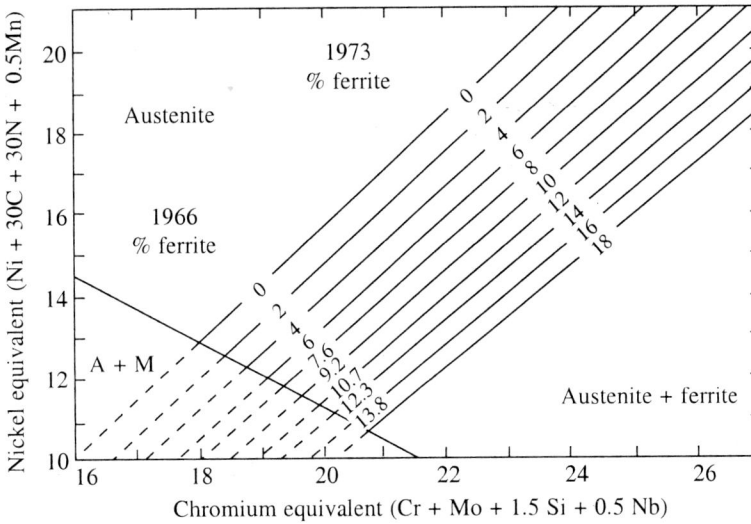

Fig. 25. W.T. DeLong diagrams for welds in stainless steels. If the nitrogen content of the metal has not been determined by analysis, the following values will be taken as a function of different welding processes: covered electrode welding, under shielding gas nonconsumable electrode (G.T.A.W.), plasma, under solid flux: N = 0.06%; under shielding gas with consumable electrode (G.M.A.W.): N = 0.08%; with self-shielded flux-cored wire: N = 0.12% [63].

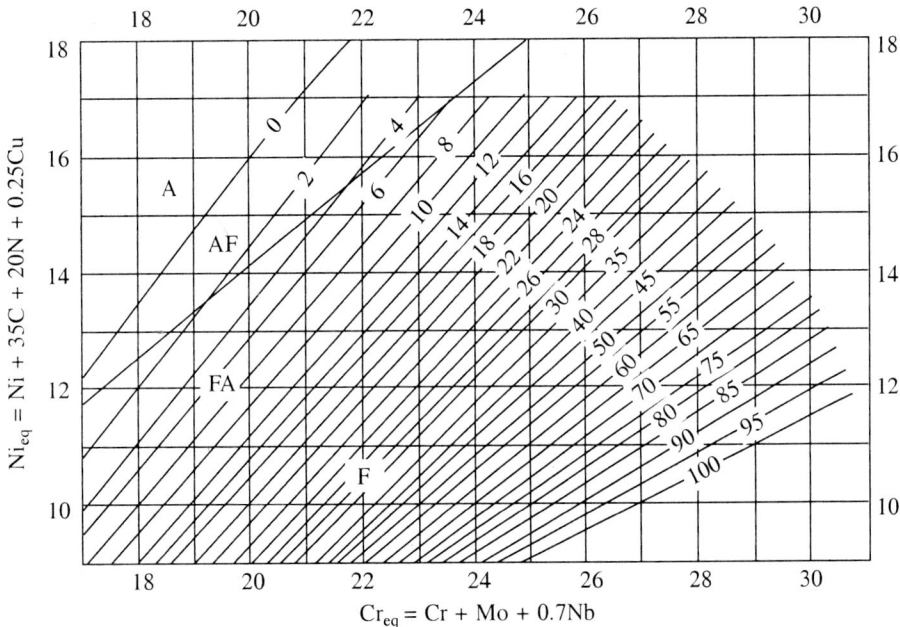

Fig. 26. WRC-1992 diagram [64].

The magnitude of the interval between the solidus and liquidus temperatures varies considerably for the different grades. This interval is greater for the highly alloyed grades and elements like niobium and molybdenum tend to increase it [5].

Table 7.　Melting range for a number of austenitic stainless steels [5]

AISI	Solidus (°C)	Liquidus (°C)	Solidus-Liquidus (°C)
202	1398	1454	56
302	1400	1447	47
304	1405	1448	43
304L	1394	1440	46
305	1400	1435	35
310	1350	1395	45
314	1322	1388	66
316	1392	1444	52
316L	1405	1445	40
316 Ti	1378	1432	54
316 Nb	1370	1431	61
321	1398	1448	50
347	1394	1446	52

2. *Density:* The density of a number of alloys at room temperature is given in Table 8. The density varies little in the 18-10 Cr-Ni and 13–17% Cr range of stainless steels. For these steels, it is of the order of 7.7 to 7.9 g/cm^3. [5].

3. *Coefficient of expansion:* Table 9 gives the mean coefficients of linear expansion for a series of grades at high temperatures and Table 10 at low temperature [69]. The coefficient of expansion of the 18–10 type austenitic steels is significantly high, of the order $17 \times 10^{-6}/°C$, which increases the problems related to changes in dimension during the heating and cooling of products. The coefficient of thermal expansion increases with the rise in temperature. For weld deposits of stainless alloys, Elmer, Olson and Matlock have studied the influence of composition and structure on the coefficient of expansion (Fig. 27) [70].

Table 8.　Density of the principal stainless steels (various sources) [5]

Designation AISI	Density (g. cm^{-3})
201	7.7
202	7.7
301	7.7
302	7.9
303	7.9
304	7.9
304L	7.9
305	7.9
308	7.9
310	7.9
316	7.9
321	7.9
347	7.8

4. *Elastic modulus:* Table 11 gives the values of both Young's modulus and the shear or torsional elastic modulus for a series of grades. Nunes and Martin [71] have shown that strain hardening causes an increase in the elastic modulus for unstable austenite grades. On the other hand, for stable steels, the modulus decreases to a strain hardening level of 80 to 85% and then increases if strain hardening is continued [5]. Rise in temperature will result in the decrease in elastic modulus of austenitic stainless steels as shown in Table 12.

5. *Electrical resistivity:* Table 13 [69] gives the resistivity values of austenitic grades at room temperature. Table 14 [69] shows its variations at high and low temperature. Resistivity changes little in the standard range of 13 to 18 % Cr grades. Resistivity increases with temperature for austenitic stainless steels as shown in Table 14.

Table 9. Mean coefficient of expansion in $10^{-6} \cdot °C^{-1}$ of some stainless steels—variations as a function of temperature (various sources) [5]

Grade	Mean coefficient of expansion $(10^{-6} \cdot °C^{-1})$				
AISI	20–200 °C	20–400 °C	20–600 °C	20–800 °C	20–1000 °C
304	17	18	19	19.5	20.0
316	16.5	17.5	18.5	19.0	19.5
314	15	16	17	18	19

Table 10. Mean coefficient of expansion in $10^{-6} \cdot °C^{-1}$ of some stainless steels at low temperatures [5]

Grade	Temperature (°C)			
AISI	−184 to 21 °C	−129 to 21 °C	−73 to 21 °C	−18 to 21 °C
301	13.7	14.1	14.8	15.7
304	13.3	13.9	14.8	15.7
316	12.8	13.3	14.1	14.8
347	13.5	14.6	15.3	15.7
310	12.6	13.5	14.1	14.4

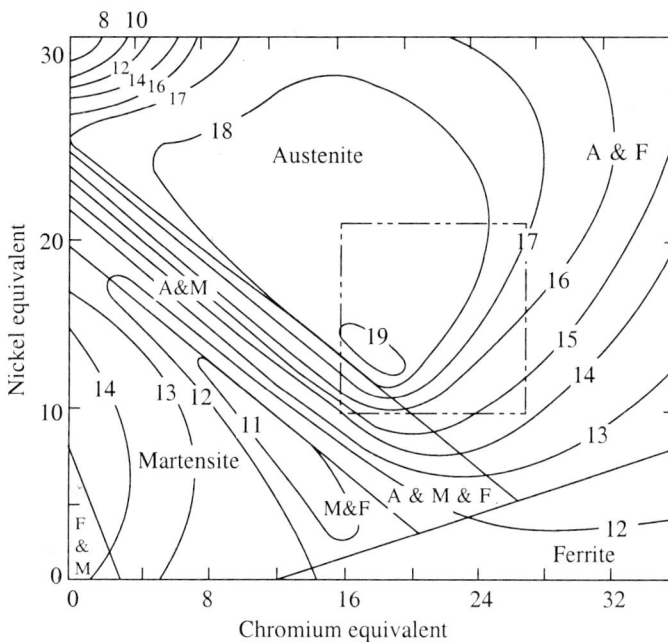

Fig. 27. Mean coefficient of expansion of alloys deposited by welding (in$10^{-6}/°C$ between 0 and 400 °C) superimposed on a Schaeffler diagram [70].

6. *Thermal conductivity and Specific heat:* Table 13 and 15 [69] give the specific heat values for a range of alloys at room and high temperatures. The thermal conductivity and specific heat increase with temperature [5].

Table 11. Modulus of elasticity of some stainless steels [5]

Grade AISI	Young's modulus (KN/mm^2)	Shear modulus (KN/mm^2)
302	193	79
304	193	79
310	193	73
316	196	78

Table 12.   Changes in elastic modulus of some stainless steels as a function of temperature; static values [5]

Grade AISI	Temperature (°C)						
	−196	20	100	200	400	600	800
302	200	193	191	183.5	168.5	153.5	139
304	208	193	191	183	168	148	128
310	–	193	192	184	173	155	134
316	–	193	192	185	168.5	151	132
321	–	193	192	182	166	151	132
347	208	193	184	168	152	152	134

Young's modulus E (KN/mm^2)

Grade AISI	Temperature (°C)					
	20	100	200	400	600	800
304	79	75	72	64	54	50
310	73	72	70	66	59	50
316	78	76	73	65	59	52
321	76	74	72	64	58	52

Shear modulus G (KN/mm^2)

7. *Magnetic permeability:* Table 16 gives the permeability values for different austenitic steels. They are non-magnetic when their structure is fully austenitic and have Curie points much lower than room temperature. When austenite transforms to martensite by strain hardening, the metal becomes ferro-magnetic, as seen in AISI 301. Similarly, AISI 308 type ingot can contain up to 15% of ferrite content and will thus be ferro-magnetic. The transformation of ferrite to σ- phase by heating in the 873–1173 K temperature range also causes the material's magnetism to disappear. The 18-10 type austenitic stainless steels retain their non-magnetism at much lower temperature. But the less stable austenitic grades like the 18–8 type may transform partially to martensite at low temperature and thus become ferro-magnetic. Nickel-rich austenitic stainless alloys remain austenitic at low temperature and have Curie points, the level of which depends on their composition [5].

Table 13. Resistivity, thermal conductivity and specific heat of some stainless steels at room temperature [5]

Grade AISI	Electrical resistivity ($\mu\Omega.cm$)	Thermal conductivity ($W.m^{-1}K^{-1}$)	Specific heat ($J.kg^{-1}K^{-1}$)
201 202	69	14.6	500
301 302 303 304 305	72	14.6	500
310 314	90	14.6	500
316 316Ti 316Cb	74	14.6	500
321 347	72	14.6	500

Table 14. Variation in resistivity ($\mu\Omega \cdot cm$) of some stainless steels as a function of temperature [5]

Grade AISI	Temperature (°C)							
	−196	−78	20	200	400	600	800	1000
301	–	–	72	83	94	105	114	–
302	–	–	72	84	96	106	115	119
304	55	65	72	85	98	111	120	–
310	–	–	90	100	110	120	125	130
316	60	68	74	85	98	108	–	–
321	–	–	72	90	103	115	123	–
347	52	60	72	88	97	110	119	–

Table 15. Influence of temperature on the specific heat of some stainless steels ($J\ kg^{-1}K^{-1}$) [5]

Grade AISI	Temperature (°C)							
	−196	−78	20	200	400	600	800	1000
301	285	394	456	527	571	595	628	695
304	136	408	444	–	–	–	–	–
316	284	393	452	523	561	582	628	722
347	285	393	452	520	561	595	636	741
314	–	–	502	544	586	627	710	795

Mechanical Properties

Tensile strength and Toughness characteristics Pickering [4] and Irvine et al. [72] derived empirical relationships between compositional and microstructural parameters, and tensile properties of austenitic stainless steels, as shown by the following equations:

Table 16. Magnetic permeability of various austenitic stainless steels [5]

Grades	Permeability (H 100/1000 Oe)
Z8CN 18.12	
Annealed	1.001 to 1.005
90% cold work	1.1
Z 12CN 17.07	
Annealed	1 to 1.1
Heavily cold worked	10 to 20
Z 6CND 17.11	1.1
Z 6NCTD V 25.15	1.005
NC 15 Fe	1.005
Annealed or cold worked	
KC20N 16 FeD	<1.05

$$0.2\% \text{ Proof stress (MN m}^{-2}) = 15.4 \{4.4 + 23(C) + 1.3(Si) + 0.24(Cr) + 0.94(Mo) + 1.2(V)$$

$$+ 0.29(W) + 2.6(Nb) + 1.7(Ti) + 0.82(Al) + 32(N)$$

$$+ 0.16(\delta\text{-ferrite}) + 0.46d^{-1/2}\} \tag{5}$$

$$\text{Tensile strength (MNm}^{-2}) = 15.4 \{29 + 35(C) + 55(N) + 2.4(Si) + 0.11(Ni) + 1.2(Mo)$$

$$+ 5.0(Nb) + 3.0 (Ti) + 1.2(Al) + 0.14(\delta\text{-ferrite}) + 0.82\ t^{-1/2}\} \tag{6}$$

where δ-ferrite is the percentage of δ-ferrite, d the mean linear intercept of the grain diameter (mm), t the twin spacing (mm) and the brackets indicate the alloying addition in weight percent.

The relationships in equation (5) and (6) indicate that high proof and tensile strengths are correlated with the high carbon specification of stainless steel. Type 316 steel and stabilised steels (types 321 & 347) containing high titanium and niobium have high tensile strengths. Chromium has a positive effect on property [5]. Molybdenum and silicon increase the strength either by solid solution hardening or by their effect on the stacking fault energy [5]. The twin spacing does not affect the proof stress because the stacking fault energy, which controls the work-hardening rate, has little or no effect at the low strains at which the proof stress is measured [4]. The twin spacing is much more important than the grain size in controlling the tensile strength because the effect of stacking fault energy on the work-hardening rate, and hence on the tensile strength, is quite significant [5]. However, in high stacking fault energy austenites, in which there are relatively few twins, the tensile strength will depend on the grain size, following a Hall-Petch type of relationship. In this case, increasing the grain size decreases the proof stress value. δ-ferrite increases the proof stress and tensile strength values by a dispersion-strengthening effect [5]. Table 17 gives the mechanical properties of the main austenitic stainless steels at room temperature in the annealed state.

The austenitic stainless steels retain a high ductility and good impact strength at low temperature, which makes them particularly useful for cryogenic applications [77]. The tensile strength greatly increases at low temperatures. At high temperatures, the yield strength and the U.T.S. decrease for austenitic steels as shown in Table 18.

Fatigue The fatigue limit for a material is the maximum alternating stress that may be applied

Table 17.   Mechanical properties at room temperature of austenitic stainless steels [5]

Grade AISI	Mechanical properties			Solution treatment (water quench) (°C)
	0.2% Y. S. (MPa)Min.	UTS (MPa)	El %	
302	215	490–690	45	1050
304	195	490–690	45	1050
304L	185	470–670	45	1050
321	205	500–700	40	1075
347	205	500–700	40	1075
316	205	500–700	45	1075
316L	195	480–680	45	1075
316 Ti	215	510–710	40	1075
316 Cb	215	510–710	40	1075
304N	250	550	45	1025
316N	280	600	45	1050
309	240	540	30	1120
314	240	540	30	1120

Table 18.   Changes in tensile properties of four austenitic stainless steels as a function of temperature [5]

Grade AISI	Properties	Temperature °C							
		20	100	200	300	400	500	600	700
304	0.2%Y.S. (MPa)	247	243	169	148	136	133	125	109
	UTS (MPa)	599	496	456	449	443	416	367	268
	El (%)	62.6	56.1	46.4	41.6	43.1	41.7	41.1	47.7
316	0.2%Y.S. (MPa)	254	200	172	161	157	144	141	125
	UTS (MPa)	588	493	483	479	472	457	421	327
	El (%)	60.1	52.1	46.0	41.9	41.9	41.7	42.6	49.6
321	0.2%Y.S. (MPa)	234	206	194	163	161	152	145	138
	UTS (MPa)	588	506	452	435	436	391	376	269
	El (%)	53.9	47.5	42.0	42.0	36.4	34.8	36.0	48.4
347	0.2%Y.S. (MPa)	250	213	195	179	168	157	155	144
	UTS (MPa)	609	540	475	451	448	422	387	292
	El (%)	49.2	46.8	40.7	36.8	35.3	34.1	35.3	49.4

indefinitely without causing fracture. Fig. 28 [73] schematically shows the principal parameters and phenomena taken into consideration as regards to fatigue. The fatigue limit for austenitic steels is 0.4 times the tensile strength. Thus, this is of the order of the yield strength. In case of low cycle plastic fatigue, austenitic stainless steels undergo considerable strain hardening in cyclic imposed deformation. The fatigue limit also increases with strain hardening, in proportion to the tensile strength. This proportionality is maintained up to strengths of the order of 1100 MPa [74].

For nuclear engineering applications, complete absence of crack initiation is required for stainless steels for which the calculation codes provide curves relating the amplitude of the alternating strain

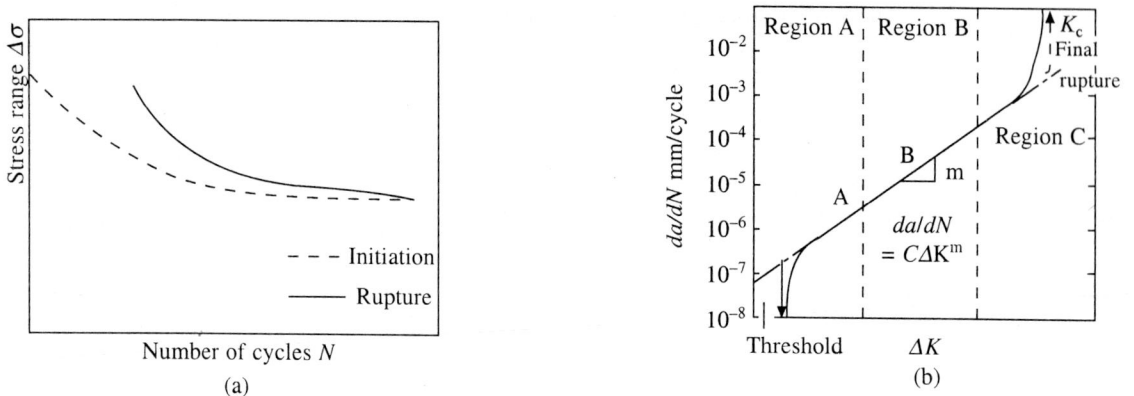

Fig. 28.   (a) "Stress-number of cycles" curves giving the number of cycles to initiation and fracture. (b) The crack propagation rate da/dN as a function of stress intensity factor ΔK. Region A: low crack rate (threshold), Region B: intermediate range (Paris relation), Region C: high crack rate (K_c).

to the number of admissible cycles as shown in Fig. 29 [5]. Fatigue strength increases with decreasing temperatures.

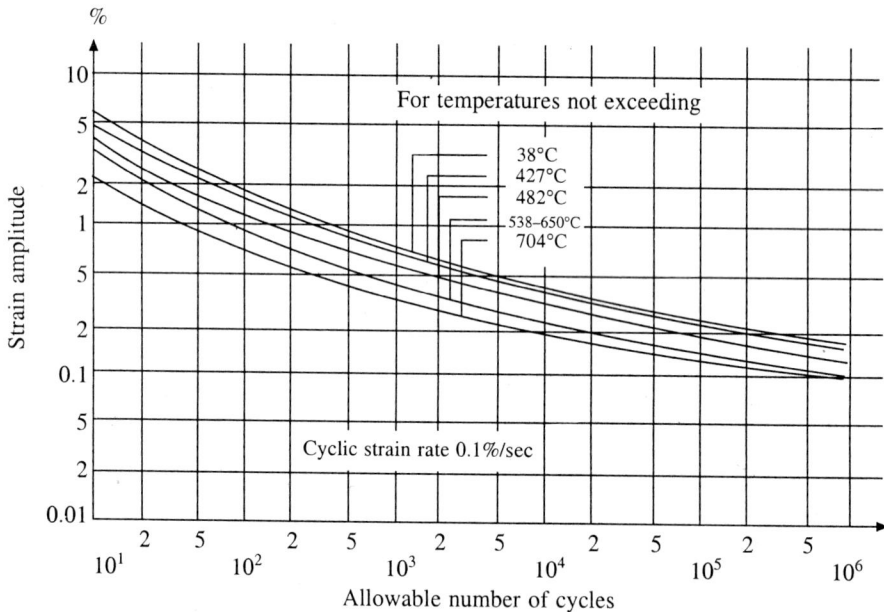

Fig. 29.   Fatigue limit curves for imposed strain testing for the AISI 304 and AISI 316 austenitic steels (transcribed from the Boiler and Pressure Vessel Code-ASME-Code Case 1592).

Creep The rate-controlling creep deformation processes in austenitic steels, at a given temperature, are dependent on the applied stress. The deformation at high stresses occurs primarily by a dislocation climb or glide process whereas other mechanisms such as grain boundary sliding, solute drag and vacancy diffusional processes determine the creep rate at lower stress [17]. The dependence of the creep deformation rates on temperature and applied stress and the activation energies for the different processes have been tabulated elsewhere [75].

The austenitic stainless steels, by virtue of their microstructure, are the ones that best resist high temperature creep. The alloying elements have a strong influence on the creep resistance of austenitic steels. Titanium strongly increases the creep resistance of austenitic stainless steels. The titanium content giving the best creep resistance lies between 0.25 and 0.5%, which corresponds to a Ti/C ratio higher than the stoichiometry of the carbide TiC. The titanium action depends on the creep temperature. At low temperatures, as per the mechanisms suggested by Williams and Harris [25, 76] the deformation occurs within the grains and the small carbide precipitates nucleated on the dislocation will prevent further deformation or else the titanium atom in the solution will restrict the motion of the dislocations to which they are strongly bound [5]. At high temperature, the intergranular carbides ensure the resistance of grain boundaries. Swindeman and Binkman [77] have shown that increasing the niobium content from 20 to 100 ppm in a AISI type 304 alloy distinctly increases creep life while decreasing ductility (Fig. 30) [77].

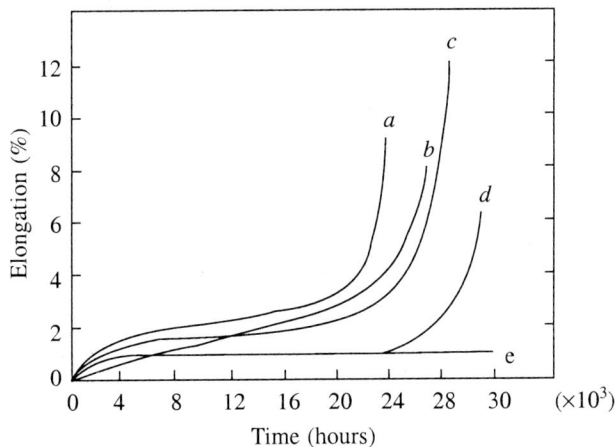

Fig. 30. **Comparison of creep behavior at 866 K and under 117 MPa stress as a function of Nb content of a AISI type 304 steel.** a = **20 ppm,** b = **30 ppm,** c = **50 ppm,** d = **80 ppm,** e = **100 ppm. After Swindeman and Brinkman [77].**

Vanadium, like niobium and titanium, increase creep life but to the prejudice of the ductility [5]. Molybdenum improves the creep properties of stainless steels, as it is a substitutional element as well as a carbide former [5]. Nitrogen increases creep life but reduces the secondary creep rate and fracture ductiliity. Boron has a beneficial effect on the creep strength of stainless steels containing Molybdenum [5].

Low strain hardening enhances resistance to creep particularly at low temperature. For each temperature, there is an optimum amount of strain; the strain value decreases as the temperature is raised. The influence of grain size (G) depends on the temperature [5]. Large grain size is preferable at high temperature for better creep strength (G > 3 for AISI 316 steel).

CONCLUSION

Austenitic stainless steels are most commonly used in major activity sectors such as house hold and community equipments, transport, food industry, industrial equipments, chemical and power engineering, cryogenics, and building industry.

The family of austenitic stainless steels has a wide variety of grades precisely tailored for a specific application. The optimum choice of the grades would depend on service needs and this would require a clear understanding of the metallurgical parameters, which control the microstructure and thus the mechanical properties, formability and corrosion resistance.

This chapter, in brief, has dealt with the physical metallurgy, welding metallurgy, physical and mechanical properties of austenitic stainless steels. In the physical metallurgy of stainless steels the effect of alloying elements in forming different phases, the transformation of austenite to martensite during cooling or straining, hardening processes and formation of intermetallic phases, have been discussed. The influence of chemical composition and temperature on the various physical properties of austenitic stainless steel has been emphasised. Austenitic steels are distinguished by a higher coefficient of expansion, a lower thermal conductivity and, when the structure is entirely austenitic, a non-magnetism that is retained at low temperatures.

The mechanical properties, like tensile strength, fatigue and creep strengths of austenitic stainless steels vary with temperature, composition and microstructure. Austenitic stainless steels have particularly low yield strengths, and several processes are used to improve them: appropriate thermomechanical treatments, hardening with nitrogen, precipitation hardening. Their creep strength is excellent up to 973 K, and can be further improved by alloying with N, Mo, Nb, Ti, W, V or B. In austenitic steels, at temperatures as low as –73 K, toughness is relatively unaffected, whereas their strength increases. They lend themselves remarkably to deep drawing and cold rolling, where their work-hardening characteristics enable high strength levels to be attained. Weldability is excellent, and welds, which do not transform to martensite during air-cooling, have mechanical properties similar to base metal.

REFERENCES

1. Maurer E. and Strauss B., *Kruppsche Monatsch*, 1920, p. 120.
2. Rivlin, V.G. and Raynor, G.V., *International Metals Reviews*, 1, 1980, p. 21–38.
3. Kujanpaa, V., Suutala, N., Takalo, T. and Moisio, T., *Welding Research International*, 9, 1979, p. 55.
4. Pickering, F.B., *International Metals Reviews*, 1976, p. 227.
5. *Stainless Steels*, Scientific Editors: Lacombe, P., Baroux, B. and Beranger, G., Les Editions de Physique Les Ulis 1993.
6. *Metals Handbook*, 9th Edition, Vol. 13, Corrosion, p. 550.
7. Gill, T.P.S., : Proceedings of the Course on *Welding of Stainless Steels and Corrosion Resistant Alloys*, Indian Institute of Welding, Madras, 1997, p. 1.
8. Speidel, M.O., Proceedings of the 1st International Conference on *High Nitrogen Steels "HNS 88"*, Lille, May 1988, edited by Foct, J. and Hendry, A., The Institute of Metals, London, 1989, p. 92.
9. Simmons, J.W., *Materials Science and Engineering*, A207, 1996, p. 159.
10. Uggowitzer, P.J. and Hazenmoser, M., Proceedings of the 1st *International Conference on High Nitrogen Steels "HNS 88"*, Lille, May 1988, edited by Foct, J. and Hendry, A., The Institute of Metals, London, 1989, p. 174.
11. Santhi Srinivas, N.V. and Kutumbarao, V.V., *The Banaras Metallurgist*, 14 & 15, 1997, p. 148–161.
12. Cook, A.J. and Brown, B.R. : *Journal of Iron & Steel Institute*, 345, 1952, p. 171.
13. Schneider, H. : *Foundry Trade F*, 108, 1962, p. 562.
14. Briggs, J.Z. & Parker, T.D., ' *The super- 12% Cr steels*', Climax Molybdenum Co., 1965.
15. D.T. Llewellyn & V.J. McNeeley : *Sheet Metal Ind.*, 49, 1972, p. 17.
16. Wilson, F.G. and Pickering, F.B. : *Journal of Iron and Steel Institute*, 204, 1966, p. 628.

17. Harries, D.R., Proceedings of the International Conference on *Mechanical Behaviour and Nuclear Applications of Stainless Steel at Elevated Temperatures*, Varese, Italy, The Metals Society, London, 1981, p. 1.
18. Zappfe, C., *Stainless Steels*, American Society for Metals, 1949.
19. Pryce, L. and Andrews, K.W., *Journal of Iron and Steel Institute*, 195, 1960, p. 415.
20. Hattersley, B. and Hume-Rothery, W., *Journal of Iron and Steel Institute*, 207, 1966, p. 683.
21. Monkman, F.C. et al: *Metal Progress*, 71, 1957, p. 94.
22. Eichelman, G.H. and Hull, F.C., *Transactions of the American Society for Metals*, 45, 1953, p. 77.
23. Angel, T., *Transactions of the American Society for Metals*, 173, 1954, p.165.
24. Keating, F.H. *'Chromium-Nickel Austenitic Steels'*; Butterworths, London, 1956.
25. Marshall P., *Austenitic Stainless Steels–Microstructure and Mechanical Properties*, Elsevier, 1984.
26. Harries, D.H., *Personal Communications*, UKAEA, 1974.
27. Gladman, T., et al.: *Sheet Metal Ind.*, 51, (5), 1974, p. 219.
28. Gladman, T., et al.: *Unpublished work*, 1972.
29. Degalaix S., Foct J., *Mem. Sci. Rev. Met.*, 1987, p. 645–653.
30. Pickering F.B., Proceedings of the Conference on *Stainless Steels 84*, Gothenburg, 1984, p. 2.
31. Barick J., *Material Science and Technology*, 4, 1988, p. 5.
32. Robinson P.W. and Jack D.H., Proceedings of the *Conference on New Developments in Stainless Steel Technology*, Ed: Lula, R.A., American Society for Metals, 1985, p. 71.
33. Powell D.J., Pilkington R., Miller D.A., Proceedings of the *Conference on Stainless Steels 84*, Gothenburg, 1984, p. 382.
34. Hull, F.C., *Welding Journal*, 52, 1973, p. 104–s.
35. Gill, T.P.S., M. Vijayalakshmi, J.B. Gnanamoorthy and K.A. Padmanabhan, *Welding Journal*, 65, 1986, p. 122-s.
36. Slattery, G.F., Keown, S.R. & Lambert, M.E., *Metals Technology*, 10, 1983, p. 373.
37. Minami Y., Kimura M., Ihara Y., *Material Science & Technology*, 2, 1986, p. 795–806.
38. Leitnaker, J.M., *Welding Journal*, 61, 1982, p. 9-s.
39. Gill, T.P.S., *Ph.D Thesis*, Indian Institute of Technology, Madras, 1984.
40. Barick J., *Metal Science*, 14A, 1983, p. 635–641.
41. Barcik J., Brzycka B., *Metal Science* 17, 1983, p. 256–260.
42. Barcik J., *Journal Applied Crystallography*, 16, 1983, p. 590.
43. Verma, D.D.N., *M.S. Thesis*, Indian Institute of Technology, Madras, 1983.
44. Kokawa. H., Kuwana, T. and Yamamoto, A., *Welding Journal*, 68, 1989, p. 92-s.
45. Shaikh, H., Pujar, M.G., Sivaibharasi, N., Sivaprasad, P.V. and Khatak, H.S., *Materials Science and Technology*, 10, 1994, p. 1096.
46. Mathew, M.D., Sasikala, G., Gill, T.P.S., Mannan, S.L. and Rodriguez, P., *Materials Science and Technology*, 10, 1994, p. 1104 .
47. Weiss, P. and Stickler, R., *Metallurgical Transactions A*, 31A, 1972, p. 851.
48. Lai, J.K. and Haigh, J.R., *Welding Journal*, 58, 1979, p. 1-s.
49. Goldschmidt, H.J., *Interstitial Alloys*, Butterworths, London, 1967.
50. Weigand, H. and Doruk, M., *Arch. Eissenhuttenw*, 8, 1962, p. 559.
51. Solomon, H.D. and Devine, T.M., *ASTM STP*, 672, 1979, p. 430.
52. Shankar, P., Shaikh, H., Sivakumar, S., Venugopal, S., Sundararaman, D. and Khatak, H.S., *Journal of Nuclear Materials*, 264, 1999, p. 29.
53. Thomas R.G. and Keown, S.R., Proceedings of the *International Conference on Mechanical Behaviour and Nuclear Applications of Stainless Steels at Elevated Temperatures*, Varese, Italy, Metals Society, London.
54. Tavasolli, A.A., Bisson, A and Soulat, P., *Metals Science*, 18, 1984, p. 345.
55. Dyson, D.J. and Keown, S.R., *Acta Metallurgica*, 17, 1969, p. 1095.
56. Shaikh, H., Khatak, H.S., Seshadri, S.K., Gnanamoorthy, J.B. and Rodriguez, P., *Metallurgical and Materials Transactions A*, 26A, 1995, p. 1859.
57. Gill, T.P.S., Vijayalakshmi, M., Rodriguez, P. and Padmanabhan, K.A., *Metallurgical Transactions A*, 20A, 1989, p. 1115.

58. Gooch, T.G., and Honeycombe, J., *British Welding Journal*, 2,1970, p. 375.
59. K.J. Irvine et al.: *Journal of Iron & Steel Institute*, 196, 1960, p. 166.
60. N.E. Moore and J.A. Griffiths: *Journal of Iron and Steel Institute*, 197, 1961, p. 29.
61. D.G. Berwick: *Metallurgia*, May 1966, p. 218.
62. Schaeffler A.L., *Metal Progress*, 56, 1949, p. 680.
63. Delong W.T., *Metal Progress,* 77, 1960, p. 98.
64. Kotecki, D.J. and Siewerd, T.A., *Welding Journal*, 71, 1992, p. 171-s.
65. Ward, A.L., *Nuclear Technology*, 24, 1974, p. 201.
66. Gill, T.P.S., Gnanamoorthy, J.B. and Padamnabhan, K.A., *Corrosion*, 43, 1987, p. 203.
67. Shaikh, H., Vinoy, T.V. and Khatak, H.S., *Materials Science and Technology*, 14, 1998, p. 129
68. Hochmann J., *Aciers in oxy dables et Refractaries* (Techniques de l'Ingenieur) 1974–1981.
69. *Documentation International Nickel*, B 24-25-26-31.
70. Elmer J.W., Olson D.L., Matlock D.K., *Welding Journal*, 61, 1982, p. 293-s.
71. Nunes, J., Martin, A., *Journal of Material Science*, 1975, p. 641.
72. Irvine K.J. et. al., *Journal of Iron & Steel Institute*, 207, 1969, p. 1017.
73. Bathia C., Bailon J.P., *La Fatigue des materiaux et des structures,* Maloine Publisher.
74. Colombier L., Hochmann J., *Aciers inoxydables et Refractaires,* Dunod, 1965.
75. J.M. Silcock and G. Willoughby : Proceedings of the *Conference on Creep strength in Steel and High Temperature Alloys*, The Metals Society, London, 1974, p. 1.
76. Williams, T.M., Harries, D.R., *Meeting on Creep Strength in Steels and High Temperature Alloys,* Sheffield 1972.
77. Swindeman R.W., Brinkman C.R., *Pressure Vessels and Piping: Design Technology,* ASME 1982.
78. Albert. S. Melilli Editor, *Handbook of Comparative World Steel Standards.*

2. Uniform Corrosion of Austenitic Stainless Steels

S. Ningshen[1] and U. Kamachi Mudali[1]

Abstract Uniform corrosion aspect of austenitic stainless steels is reviewed. The choice of austenitic stainless steels is steadily increasing with the introduction of newer alloys and by the modification of traditional ones to improve one or more of their properties. This paper encompasses the spectrum of uniform corrosion problems of austenitic stainless steels in various corrosive environments. The application of existing and advanced austenitic alloys for various chemical media and the various testing techniques employed for assessing the uniform corrosion are highlighted.

Key Words General/unifom corrosion, austenitic steels, electrochemical technique, mineral acid, alkalies, seawater, methanol.

INTRODUCTION

The simplest form of corrosion is "uniform" or "general" corrosion. "Uniform" or "general" corrosion is the most common form of corrosion and it represents the greatest destruction on metal on a tonnage basis. The term "uniform" or "general" corrosion is used to describe the corrosion damage that proceeds in a relatively uniform manner over the entire surface of an alloy. It is an even rate of metal loss over the exposed surface. It is characterized by a chemical or electrochemical reaction or metal loss due to chemical attack or dissolution that proceeds uniformly over the entire exposed surface or over a large area. During such process, the material becomes thinner as it corrodes until its thickness is reduced to the point at which failure occurs. Since this type of corrosion is intense, corrosion rates for materials are often expressed in terms of metal thickness loss per unit of time. One common expression is "mils per year" or sometime, millimeters per year. Because of its predictability, low rates of corrosion are often tolerated. Catastrophic failures due to uniform corrosion are rare if planned inspection and monitoring is implemented. For most chemical process industries, corrosion rates of less than 3 mils per year (mpy) are considered acceptable, while rates between 2 and 20 mpy are routinely considered acceptable for many engineering applications. In severe environments, materials exhibiting high corrosion rate between 20 and 50 mpy might be economically justifiable. Materials, which exhibit rate of uniform corrosion beyond this, are usually unacceptable [1]. A schematic illustration of uniform corrosion is depicted (Fig. 1). The corrosion rates of most commonly used austenitic stainless steels in a wide variety of chemical environments can be found in the literatures [2–5].

[1]Corrosion Science and Technology Divison, Metallurgy and Materials Group, Indira Gandhi Centre for Atomic Research, Kalpakkam-603 102, India.

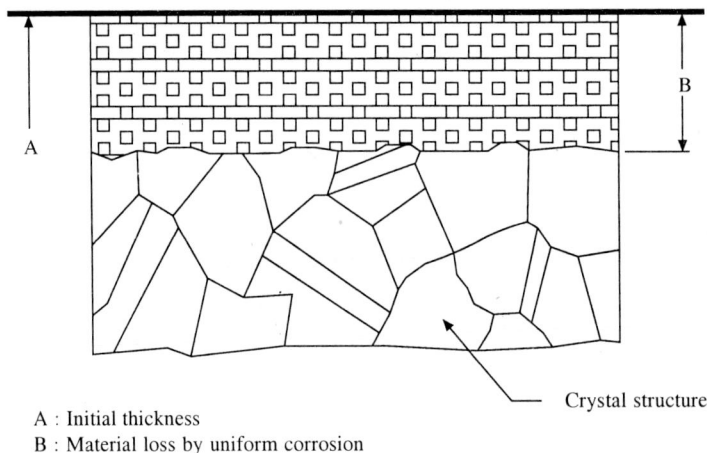

A : Initial thickness
B : Material loss by uniform corrosion

Fig. 1. Schematic illustration of uniform corrosion.

Mixed potential theory [6] provide a useful criteria, in terms of readily measurable electrochemical parameters, for determining whether a given austenitic stainless steel in a given environment will exhibit uniform corrosion or not. The mixed potential, which is commonly referred to as the "corrosion potential," denoted by the symbol E_{corr}, is the potential at which the total rate of all the anodic reactions equal to the total rates of all the cathodic reactions. The current density at E_{corr} is called the "corrosion current density" i_{corr}, and is a measure of corrosion. The schematic figure showing the extrapolation of the measurement of i_{corr} is shown in Figure 2. Some examples are illustrated in Figure 3 in terms of the relative position of the corrosion potential, E_{corr} with respect to the anodic polarization curve, ABCDE [7]. E_{corr} represents the intersection of the anodic and cathodic polarization curve and the current density, i represents the corrosion rate. The most desirable situation is represented by E_{corr-4}/i_4, which is typical of stainless steels in water. While some uniform corrosion is occurring at a rate proportional to i_4, the corrosion rate is very low and the stainless is considered to be passive in this condition. However, low uniform corrosion rates can be obtained under conditions where the stainless steel is not passive. For example, the corrosion rate obtained under conditions defined by E_{corr-1}/i_1 is low and may be acceptable for certain industrial operations, such as containment of relatively dilute sulfuric acid in stainless steels equipment. More concentrated sulfuric acid will give higher corrosion rates typified by E_{corr-2}/i_2. Uniform corrosion can also occur in the transpassive region, E_{corr-5}/i_5, as such for stainless steel corroding in highly concentrated nitric acid environments. In this regime, severe intergranular corrosion may also occur.

It should be noted that in some environments, such as strong solutions of hydrochloric and hydrofluoric acid, the active-passive behaviour characterized by curve ABCDE will not develop, and the current density simply increases with increasing potential, as shown in the curve, ABF in Figure 3. High corrosion rates typified by E_{corr-3}/i_3, can result under these conditions, particularly in the presence of oxidizers (e.g., Fe^{+3} or Cu^{+2}) that raises E_{corr} in the noble direction. Anodic protection is ruled out under such conditions, since raising the potential in the noble direction increases the current density and hence the corrosion rate.

Most cases of uniform corrosion of stainless steels can be rationalized in terms of the simple criteria described in Figure 3, and can be predicted by weight loss technique or by electrochemical techniques. In essence the deterioration of austenitic stainless steels by uniform corrosion is a predictable

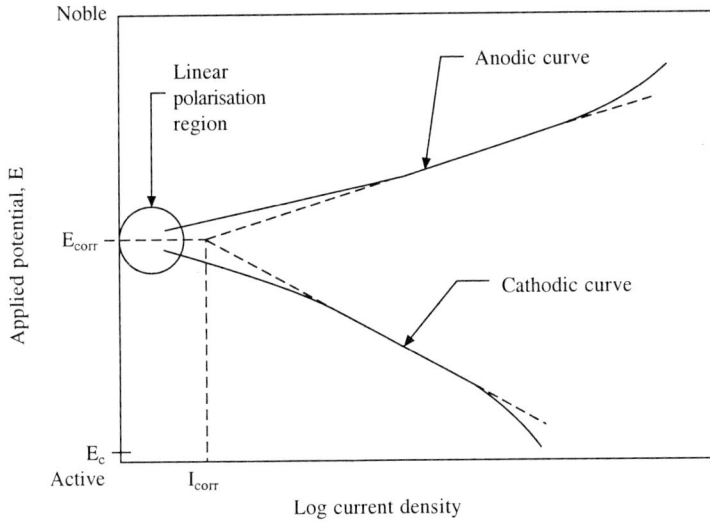

Fig. 2. Schematic figure showing the extrapolation of the measurement of I_{corr}.

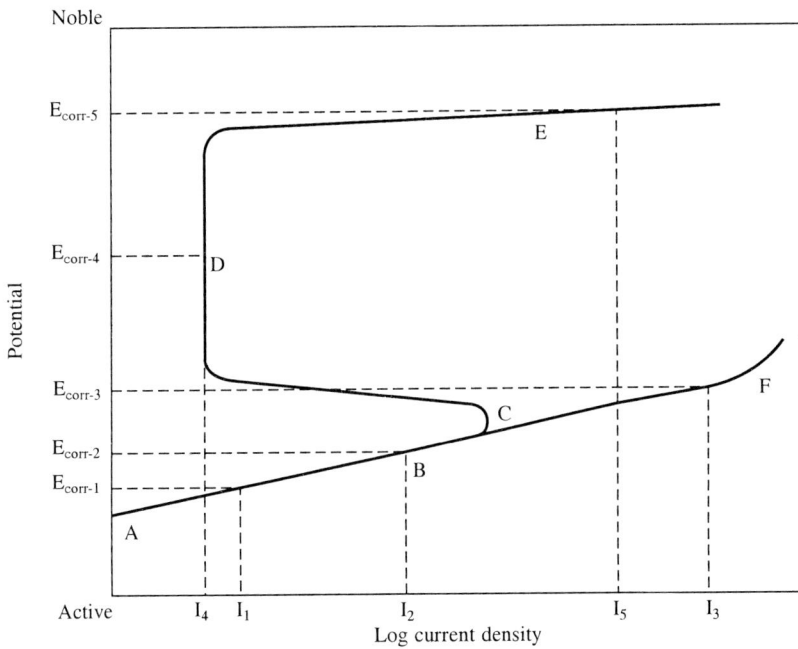

Fig. 3. Polarization curves showing some examples of uniform corrosion [7].

phenomenon in many cases. Their selection for resistance to uniform corrosion derive primarily from economic considerations, such as the survival of the low cost material for the full design life of given equipment.

TEST METHODS FOR MEASUREMENT AND EXPRESSION OF GENERAL CORROSION

The rate of uniform corrosion can be measured from the corrosion rate. Methods available for measurement of corrosion rate are: (1) the electrochemical techniques, (2) weight loss tests method and (3) electrical resistance measurement method.

ELECTROCHEMICAL TECHNIQUES

Measurement of Corrosion Rate by Tafel Extrapolation of Corrosion Current Density

This technique uses data obtained from cathodic or anodic polarization measurements. Cathodic polarization data are preferred, since these are easier to measure experimentally. A schematic diagram for conducting such polarization measurement is shown in Figure 2. In actual practice, an applied polarization curve becomes linear on a semi logarithmic plot at approximately 50 mV more active than the corrosion potential. This region of linearity is referred to as Tafel region. The corrosion current density, i_{corr}, cannot be measured directly, since the current involved is the one that flows between numerous microscopic anodic and cathodic sites on the surface of the corroding metal. One way to measure i_{corr} is by the extrapolating on of certain linear segments of measured potential-current density curves. Extrapolating the linear segments of either the measured anodic or cathodic curve and intersecting these extrapolated lines with a line drawn from E_{corr} parallel to current density axis yield, i_{corr}, as shown in the Figure 2. Several practical procedures are available in literature for measuring i_{corr} by extrapolation technique, including both potentiostatic and galvanostatic [8–10]. The corrosion current density, i_{corr}, can be converted to corrosion rate by the relationship

$$R_{mm/y} = 0.0033 i_{corr} \frac{e}{\rho} \tag{1}$$

where $R_{mm/y}$ = corrosion rate (millimeter/year); i_{corr} = corrosion current density ($\mu A/cm^2$); e = equivalent weight (chemical) of the metal; and ρ = density of the metal (g/cm^3).

This equation describes the corrosion rate of a pure metal that has a certain density and equivalent weight. Since stainless steels comprise a number of major alloying elements of different densities and equivalent weights, a computation must be made of the partial distributions of the various alloying elements. Such computations have been made for a number of stainless steels and higher alloys, and are listed in Table 1. The conversion factor K, for each alloy listed in Table 1, multiplied by the measured i_{corr}, yields the corrosion rate; that is,

$$R_{mm/y} = K i_{corr} \tag{2}$$

The extrapolation technique for measuring i_{corr} depends on the ability to identify the linear (Tafel) region. Corrodents in which more than one reduction reactions are operative or in which the rate of arrival of the reducible species (e.g., hydrogen ions) at the cathode surface determine the reduction rate (known as concentration polarization) exhibit less distinct linear regions, making extrapolation less certain. These disadvantages can largely be overcome by the linear (resistance) polarization technique. With this technique it is possible to measure extremely low corrosion rates, and it can also be used for continuously monitoring the corrosion rate of a system.

Table 1 **Conversion factors of some austenitic alloys used for corrosion rate calculation [7]**

Alloy	Conversion Factor, K
Type 304	0.01346
Type 316	0.01397
Alloy 800	0.01346
Alloy 600	0.01219
Alloy 625	0.01473

Measurement of Corrosion Rate by Linear (Resistance) Polarization

The value of i_{corr} can also be measured by another technique, generally known as "linear polarization." This technique is based on the fact that at a potential very close to $E_{corr} \pm 10$ mV, the slope of the potential/applied current curve is approximately linear as shown in Figure 4. This slope, $\Delta E/\Delta i$, has the units of resistance given in ohms (volts/amperes or millivolts/milliamperes); the disadvantages of the Tafel extrapolation method can largely overcome by using linear-polarization analysis. Accepting the constancy of the slope $\Delta E/\Delta i$, it has been shown [11] that i_{corr} is related to the inverse of this slope by the equation (Stern Geary Equation):

$$i_{corr} = \left[\frac{\beta a \beta c}{2.3(\beta a + \beta c)} \right] \frac{\Delta i}{\Delta E} \tag{3}$$

where βa and βc are the anodic and cathodic Tafel slopes, respectively. It is generally accepted that the quantity $\beta a \beta c / 2.3(\beta a + \beta c)$ is a constant (K). Hence, i_{corr} is given by the relationship [1].

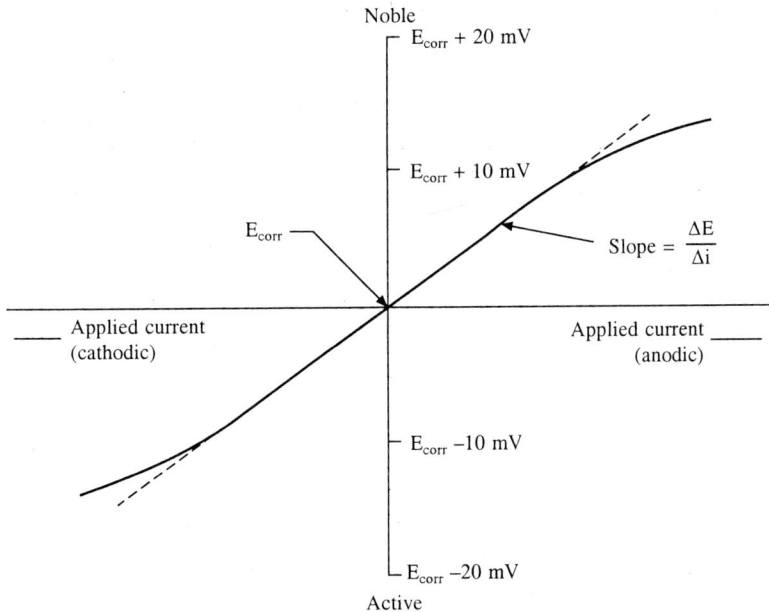

Fig. 4. Measurement of corrosion rate by linear polarization.

$$i_{\text{corr}} = K\left(\frac{\Delta i}{\Delta E}\right) \tag{4}$$

An appraisal employing mathematical and graphic methods has led to the conclusion that the assumptions of linearity are sufficient for the technique to be valid in many practical corrosion systems [12]. A schematic of the linear polarization corrosion rate measurement system is shown in Figure 5. Accordingly, it is accepted that linear polarization method is a useful, electrochemical technique for measuring corrosion rates.

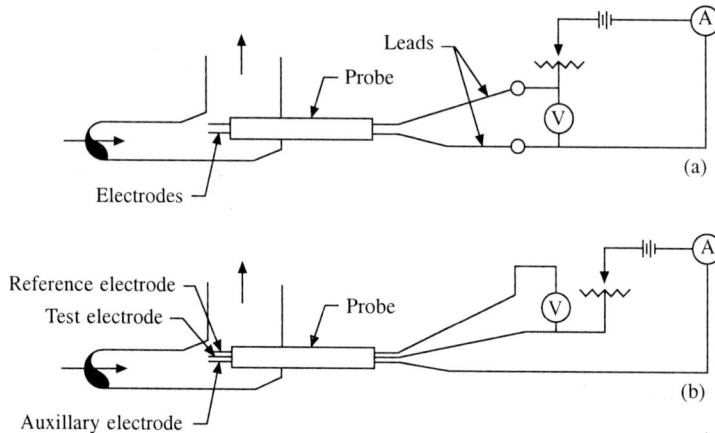

Fig. 5.   **Linear polarization corrosion rate measurement systems: (a) Two-electrode system (b) Three-electrode system.**

Measurement of Corrosion Rates by Electrochemical Impedance Techniques

Electrochemical impedance technique offers the advantage that corrosion rates can be measured for even electrolyte of low conductivity and the conductivity does not affect the accuracy of the measurement. The value of i_{corr} can also be measured by electrochemical impedance techniques [13]. Impedance, measured in ohms, is defined as the total opposition offered by an electric circuit to the flow of an alternating current of a single frequency; it is a combination of resistance and reactance. Expressions for reactance employ the complex number $j = \sqrt{-1}$, and it can be shown that impedance, Z, can be expressed in complex number notations as:

$$Z = Z' + jZ'' \tag{5}$$

Where Z' and Z'' represent its real and imaginary components. The entities Z' and $-jZ''$ represent the axes –of the Nyquist diagram, which describes points representing the magnitude and direction of the impedance vector at a particular angular frequency, ω. For resistor R_p and capacitor C_p connected in parallel, the complex response represented by the Nyquist diagram is a semicircle, as shown in Figure 6.

A corroding electrode is considered to behave in a manner similar to that of a Randles type of equivalent circuit [14], shown in Figure 6(a). R_Ω represents resistance of the solution and of the corrosion product film. The parallel combination of the resistor R_t and capacitor C_{dl} represents the corroding interface. C_{dl} is the electrochemical double-layer capacitance, and R_t is the charge transfer

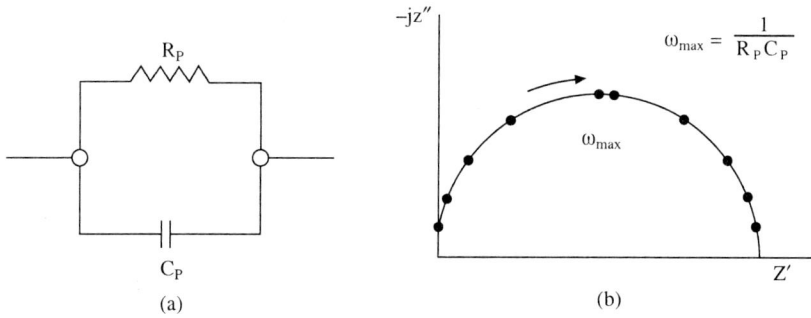

Fig. 6. Parallel resistor capacitor combination (a) and its complex response (b).

resistance. R_t is a measure of the electron transfer across the surface and determines the rate of the corrosion reaction. Provided that corrosion is under activation potential (and not under diffusion control), the quantity R_t is equivalent to the slope of $\Delta E/\Delta i$, in the Stern-Geary equation (3). Substituting R_t we get the expression:

$$i_{corr} = \frac{\beta a \beta c}{2.303(\beta a + \beta c)} \cdot \frac{1}{R_t}$$ (6)

$$i_{corr} = \frac{constant}{R_t}$$ (7)

The quantity R_t can be obtained from Nyquist plot shown in Figure 6(b). Complication arises when corrosion is under diffusion control. To account for diffusion effects, it is necessary to include an additional circuit element, W, in series with R_t, as shown in Figure 7c. This element is known as the "Warburg" impedance, describes the impedance of the diffusion-related processes. Under these conditions, the value of R_t is obtained by extrapolation of the semicircle to the real axis (Figure 7d).

Corrosion Rate Measurement by Electrochemical Noise (ECN)
Free-corrosion potential is not an unvarying parameter, but rather a noisy signal superimposed on a steady potential. This electrochemical noise, is fluctuating in potential and current, which spontaneously occur when metal, is corroding, caused by random processes in the corrosion reaction. Thus there is a correlation between corrosive attack and low frequency fluctuation of electrode potential. The analysis of this noise yields a parameter that is a function of corrosion rate; and the frequency, slope, and amplitude of the noise enable a distinction to be made between uniform corrosion and localized corrosion such as pitting and crevice attack etc [15]. Figure 8 is a block diagram of ECN monitoring system.

The fundamental processes that generate electrochemical noise are concerned with fluctuations in the rate of the electrochemical reaction. This essentially is a process, which creates current noise. Thus, current noise bears a fairly direct relationship to the underlying physical processes, and in general as rates of reaction increase, the current noise increases. In contrast, the potential noise is a secondary manifestation of the current noise. R_p and i_{corr} can be estimated by dividing the potential noise by current noise. If solution resistance effects are ignored, R_p at low frequencies is given by [16].

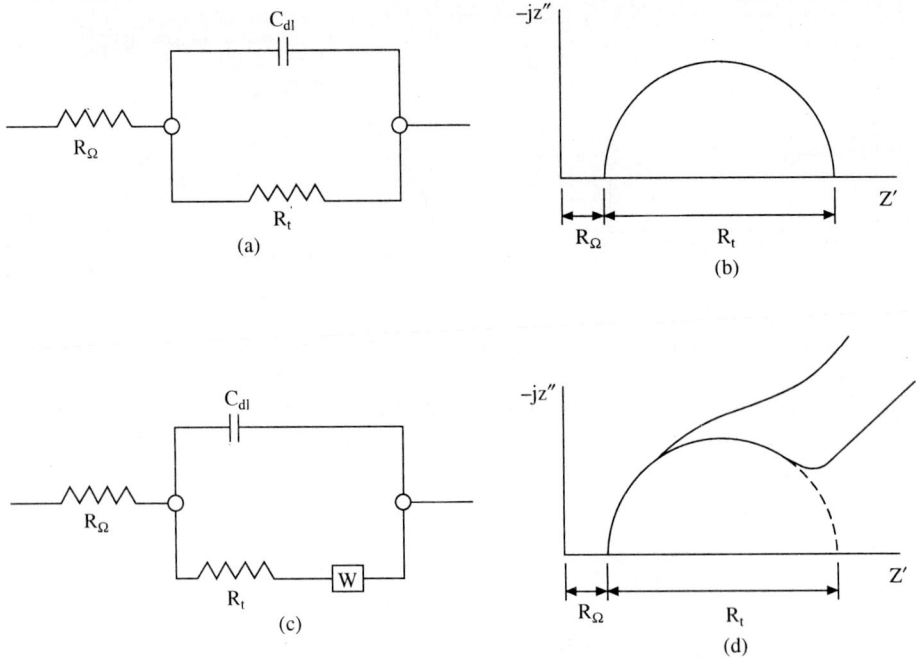

Fig. 7. Randles-type equivalent circuit without (a) and with (c) the Warburg impedance, W, and the corresponding complex plane response (b) and (d), respectively.

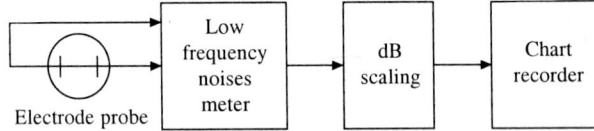

Fig. 8. Analogue instrumentation for on-line ECN monitoring.

$$\sqrt{\overline{E_n^2}} = \sqrt{\overline{I_n^2}} \times R_p \quad \text{or} \quad R_p = \frac{\sqrt{\overline{E_n^2}}}{\overline{I_n^2}} \qquad (8)$$

where $\overline{I_n^2}$ is the mean squared noise current and $\overline{E_n^2}$ = mean squared noise potential.

The resultant parameter is known as the electrochemical noise resistance, and several studies have shown that it gives a good indication of the corrosion rate. Also, we can calculate i_{corr} from the noise resistance by the equation [18]

$$i_{corr} = \frac{const}{R_p} = \frac{\sqrt{\overline{I_n^2}}}{\sqrt{\overline{E_n^2}}} \qquad (9)$$

These estimate i_{corr} are derived from a statistical analysis of the expected result of adding many independent events.

Corrosion Rate by Weight Loss Measurement

The rate of uniform corrosion can be measured using corrosion coupon testing by weight loss measurement. Coupon corrosion testing is predominantly designed to investigate uniform corrosion. ASTM Designation: G31 gives a definite guideline for carrying out such an experiment. This practice describes accepted procedures, which includes specimen preparation, apparatus, test conditions, method of cleaning specimens, evaluation of results, calculation and reporting of corrosion rates. A good corrosion rate expression should involve (i) familiar units, (ii) easy calculation with minimum opportunity for error, (iii) ready conversion to life in years, (iv) penetration, and (v) whole numbers without cumbersome decimals.

Corrosion rates have been expressed in a variety of ways in the literature; such as percent weight loss, milligram per square centimeter per day, and grams per square inch per hour. These do not express corrosion resistance in terms of penetration. The expression mils per year is the most desirable way of expressing corrosion rate. This expression is readily calculated from weight loss of the metal or the alloy specimen during the corrosion test [1]. The conversion from other units to obtain mils per year is given in Table 2. As per ASTM G31 calculating corrosion rates requires several pieces of information and several assumptions: (i), the use of corrosion rates implies that all mass loss has been due to uniform corrosion and not due to localized corrosion. (ii), the use of corrosion rates also implies that the material has not been internally attacked as by dezincification or intergranular corrosion and (iii), internal attack can be expressed as corrosion rates if desired. However, in such a case the calculation must not be based on weight loss (except in qualification tests such as practice A 262), which is usually small but on microsections, which show depth of attack.

Table 2 Conversion from other corrosion rate units to obtain mils per year [7]

Unit to Be Converted	Multiplier
Inches per year	1000
Inches per month	12.1000
Millimeters per year	39.4
Micrometers per year	0.039
Milligrams per square decimeters per day (mdd)	1.44/density
Grams per square meter per day	14.4/density

Assuming that localized or internal corrosion is not present, the average corrosion rate can be calculated by the following equations:

$$\text{Corrosion rate} = (K \times W)/(A \times T \times D) \qquad (10)$$

where K is a constant; T the time of exposure in hours to the nearest 0.01 h; A the area in cm^2 to nearest 0.01 cm^2; W the mass loss in g to nearest 1 mg (corrected for any loss during cleaning) and D the density in g/cm^3.

Many different units are used to express corrosion rates. Using the units for T, A, W and D from Table 3, the corrosion rate can be calculated in variety of units with appropriate value of K given in Table 3.

Table 3 Corrosion rate units with appropriate value of K

Corrosion Rate Units Desired	Constant (K) in Corrosion Rate Equation
Mils per year (mpy)	3.45×10^6
Inches per year (ipy)	3.45×10^3
Inches per month (ipm)	2.87×10^2
Millimetres per year (mm/y)	8.76×10^4
Micrometres per year (μm/y)	8.76×10^7
Picometres per second (pm/s)	2.78×10^6
Grams per square per hour (g/m^2.h)	$1.00 \times 10^4 \times D^A$
Milligrams per square decimeter per day (mdd)	$2.40 \times 10^6 \times D^A$
Microgams per square metre per second (μg/m^2.s)	$2.78 \times 10^6 \times D^A$

[A]Density is not needed to calculate the corrosion rate in these units. The density in the constant K cancels out the density in the corrosion rate equation.

A rapid and ready conversion for several corrosion rates can be made by means of a monograph (Fig. 9). Mathematical computations are not necessary, and the accuracy is good. The monograph is particularly helpful when data in milligrams per square decimeter per day are encountered.

This permits conversion of mils per year, inches per year, inches per month, and milligrams per square decimeter per day (mdd) from one to another. The first three names are directly converted on the scale A.

These are then converted to mdd by means of C scale and the B scale for density. The mdd does not consider or include the density or type of material involved. Density is given as grams per cubic centimeter.

Corrosion Rate Measurement by Electrical Resistance Method

Weight-loss measurements indicate the average corrosion rate over a period of time. Electrical resistance measurement is comparatively a better technique than the weight loss method. In this method, a coupon of material, identical to the alloy whose corrosion rate to be measured is exposed to the corrodent and periodically withdrawn to measure its loss of weight, which directly relates to corrosion rate [17]. Its operation is based on the increase in electrical resistance of an exposed corrosion coupon material. A metallic conductor sensing probe, generally a thick wire, strip, or tube of the same material as the equipment under test, is exposed to the process stream. The electrical resistance of this probe is compared with that of an identical reference probe that is shielded from the corrodent. As the exposed probe corrodes, its electrical resistance increase, and this change is related to the extent of corrosion. This method is fast and sensitive. It is quite similar to the weight-loss coupon method, but it enjoys the great advantage of permitting continuous monitoring without removing the coupon. It is also superior to weight-loss method because errors caused by removal of the corrosion products is eliminated and continuous monitoring can indicate the effect of process variable on corrosion rate. Electrical resistance probes can serve as an accurate measure of the corrosion rate only when the corrosion is uniform.

Corrosion rate

mpy	ipy	ipm	mdd

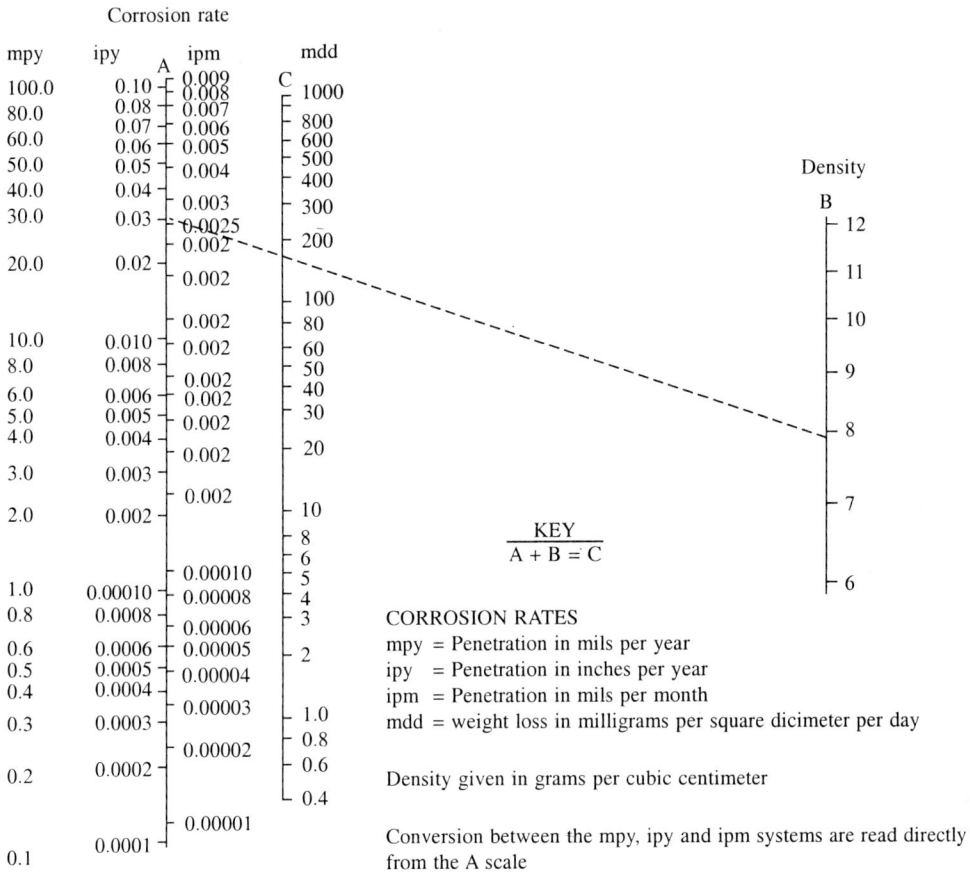

Fig. 9. Nomograph for mpy, ipy, ipm, and mdd [1].

CORROSION RATES
mpy = Penetration in mils per year
ipy = Penetration in inches per year
ipm = Penetration in mils per month
mdd = weight loss in milligrams per square dicimeter per day

Density given in grams per cubic centimeter

Conversion between the mpy, ipy and ipm systems are read directly from the A scale

KEY
A + B = C

UNIFORM CORROSION BEHAVIOR OF AUSTENITIC STAINLESS STEELS IN MINERAL ACID

Most of the severe uniform corrosion problems are encountered in mineral acids or their derivatives. This paper describes the effects of sulfuric, hydrogen chloride gas, hydrochloric, phosphoric and nitric acid on the uniform corrosion of the austenitic stainless steels.

Sulfuric Acid

Of all inorganic acids, sulfuric acid (H_2SO_4) is used in largest volume and is generally considered to be one of the most important chemicals in the industries. Many metallic materials and alloys are corroded by sulfuric acid because of its low pH. In the middle range of concentrated sulfuric acid has the highest concentration of H^+ ions, resulting in a strong corrosivity (0.5% $H_2SO_4 \rightarrow$ pH $\approx$ 2.1, 5% $H_2SO_4 \rightarrow$ pH $\approx$ 1.2, 50% $H_2SO_4 \rightarrow$ pH $\approx$ 0.3). Depending on the concentration and temperature sulfuric acid can be either a reducing acid or an oxidizing acid. Traces of impurities, e.g. atmospheric oxygen, Fe^{3+} salts, SO_3 etc., can completely change the character of sulfuric acid, turning a reducing solution into oxidizing.

The austenitic Cr-Ni steels achieve their corrosion resistance by the formation of a passive layer on their surface. This layer can also develop under oxidizing conditions in sulfuric acid, and consists of iron oxide and chromium oxide with incorporated sulfates, which improve its stability. At higher acid flow rates and under reducing conditions, the protective layer is destroyed or its formation inhibited. Sometimes, quite a considerable increase in corrosion is associated with this situation. The corrosion rates of a number of austenitic stainless steels in different concentrations of the under aerated sulfuric acid at 293 K are shown in Figure 10, and the temperature effect on corrosion rate is given in Figure 11 [18]. These graphs show that 18Cr-8Ni austenitic steels has the lowest resistance in sulfuric acid, and only a limited increase in the corrosion resistance is possible by alloying.

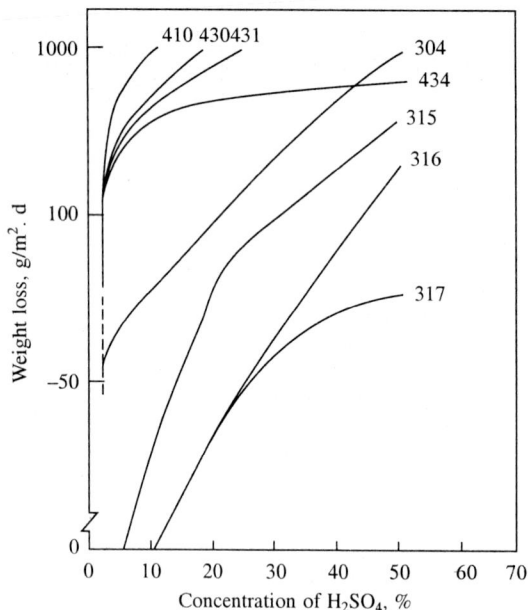

Fig. 10. Corrosion rates of various austenitic stainless steels in different concentrations of sulfuric acid at 293 K [18].

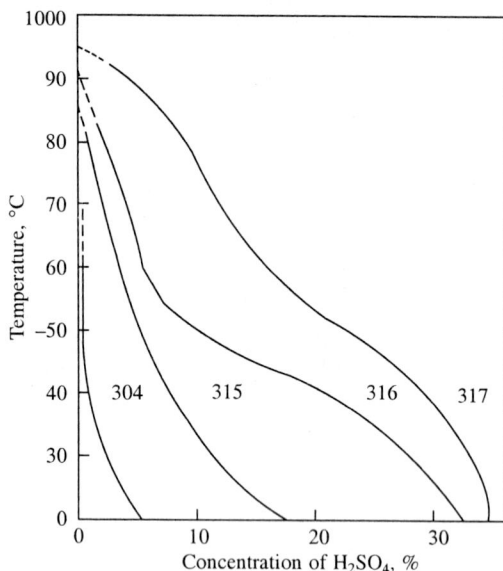

Fig. 11. Concentrations and temperatures of sulfuric acid solutions required to give a corrosion rate of 25 $g/m^2/d$ for various stainless steels [18].

A noticeable improvement can be achieved by higher Ni-contents (Fig. 12). It is evident from Figures 10 and 11 that the austenitic grades, particularly the molybdenum grades, have a significant resistance at low concentrations and at higher temperatures. Three strategies have been employed to tackle the corosion problems that arises in sulfuric acid at higher concentration and temperatures: (i), the use of anodic protection; (ii), additions of oxidizing agents to the sulfuric acid solutions; and (iii), the use of more corrosion resistant alloys.

Anodic protection has been used to protect AISI type 316 stainless steel heat exchangers handling sulfuric acid at concentrations in the range of 10–98 % and temperature up to 373 K [19]. This reduces the corrosion rate of AISI 316 stainless steel from 200 mpy in the unprotected state to 1 mpy in the protected state. Anodic protection, imparted through galvanic platinum cathodes, has been used for stainless steels tanks containing sulfuric acid [20]. Additions of oxidizing agents such as nitric acid or cupric ions, to the sulfuric acid also extend the range of usefulness. In terms of mixed potential theory, the addition of these oxidizers changes the range of corrosion potential [21].

Regarding the effect of oxygen in sulfuric acid, it is generally accepted that austenitic stainless steels in oxygenated solutions of dilute sulfuric acid at ambient temperatures exhibit a passive behaviour. However, there are instances where the introduction of small amount of oxygen into the deaerated acid will increase the corrosion rate due to oxidizing nature of the solution. Industrial practices favor ferric or cupric sulfate, or nitric acid additions wherever possible. However, the effect of the ferric ion is not significant at acid concentrations of 96% or more [22]. In austenitic stainless steels, increasing the chromium content of the alloy increases the corrosion rate in sulfuric acid [23]. However increasing the nickel, molybdenum, and copper content of the austenitic alloys results in major increase in corrosion resistance in sulfuric acid environment. Copper additions to austenitic alloys also significantly improved its corrosion resistance in sulfuric acid. An example is shown in Figure 13 for 20 Cb-3 stainless steel, which contains 3–4% Cu. The exact mechanism by which alloyed copper improves the corrosion resistance in stainless steels in sulfuric acid remains to be established, but it is believed to involve the formation of copper layer on the surface [24, 25].

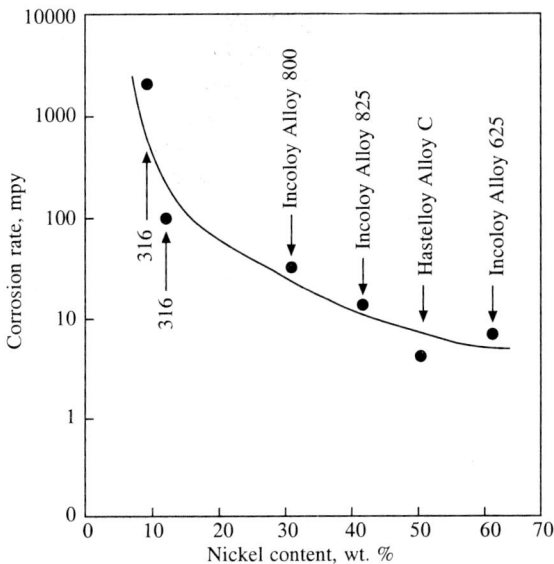

Fig. 12. Effect of nickel content on the corrosion rate of austenitic stainless steels and higher alloy in a 15% H_2SO_4 at 383 K [7].

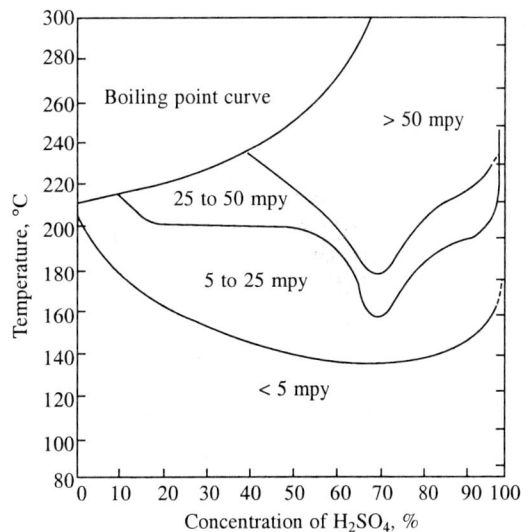

Fig. 13. Iso-corrosion chart for 20Cb-3 stainless steel in reagent-grade sulfuric acid [24, 25].

Copper-containing stainless steels and higher alloys such as AISI types 904L (1.5% Cu), 20 Cb-3 (3.5% Cu), alloy 825 (2% Cu and CN-7M(3.5% Cu) steels are often used in chemical industry for sulfuric acid service. It should also be noted that pure sulfuric acid in high concentrations (e.g., oleum) is not particularly corrosive to austenitic stainless steels [2].

Figure 14 shows the corrosion rates of AISI type 316 Ti and 321 stainless steels in sulfuric acid in the concentration range 67 to 98% at temperatures above 353 K. The corrosion rates decrease as the concentration increases and increase as the temperature rises. In practice, this material cannot be used in hot sulfuric acid under long-term exposure. Figure 15 shows the dependence of the corrosion rates on the acid concentration at 373 K. At an acid concentration of 100% and 373 K, the material is practically resistant, with corrosion rate at 0.03 mmy^{-1} [26, 27]. In highly concentrated hot sulfuric acid, the corrosion rates falls as the metal sulfate content of the acid increase. Figure 15 also shows the influence of the metal sulfate content at 373 K. This fact is of interest only in a static medium, because in agitated or flowing acid, fresh acid is constantly supplied.

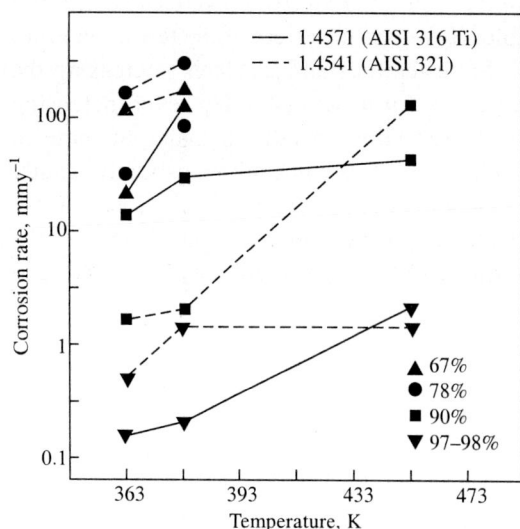

Fig. 14. Influence of the temperature and concentration on the corrosion rate of AISI 316 Ti and AISI 321. Sulfuric acid concentration [26]. ▲ 67%, ● 78%, ■ 90%, ▼ 97–98% [26, 27].

The decreasing corrosion rate of austenitic Cr-Ni steels with increasing concentration of metal sulfates and impurities resultant there of, is due to the diffusion polarization of both cathodic and anodic reactions [26, 27]. Agitating the corrosion solution can increase the corrosion resistance of these Cr-Ni steels, and the passive state, if present, can be stabilized or its range extended. AISI type 316 austenitic stainless steel can be used for handling sulfuric acid below 7% concentration at room temperature, and AISI type 317 can be used for handling up to 333 K. AISI type 304, 310, 316 and 317 stainless steels can be used for handling acid above 90% concentration at room temperature [28]. For handling concentration above 90% H_2SO_4, mild steel is equally good at much lower cost. The corrosion rates of stainless steels for sulfuric acid of 92% or higher concentration are almost nil, compared with 0.15 to 1.0 mm y^{-1} (6 to 39 mils/yr) for steel. Therefore, austenitic stainless steels are used for piping, tanks etc., only where product purity necessitates it. Molybdenum stretches the passive region, making AISI grades 316 and 317 stainless steels as acceptable material of construction for handling H_2SO_4 above 90 % concentration at ambient temperature. The upper temperature limit of stable passivity for grades AISI type 304 and 316 stainless steels in 93% acid is approximately 313 K. For 98.5% H_2SO_4, the upper limit is 343 K. As concentration goes up beyond 99%, susceptibility to corrosion rate decrease rapidly, allowing the use of stainless steels above 373 K. Figure 16 shows that, an alloying addition of 0.2% Pd reduces the corrosion rates of austenitic stainless steels in 20% H_2SO_4 at 293 K to below 1 mm y^{-1}. Deaeration reduces the corrosion resistance of austenitic stainless steel, particularly for dilute acids [29].

Hydrogen Chloride Gas and Hydrochloric Acid

Austenitic stainless steels are suitable for dry hydrogen chloride gases at room temperature. However, the presence of moisture changes the behavior to that of corrosion by hydrochloric acid. Hydrochloric acid is soluble in water in any proportion with the development of heat. The acid is then completely dissociated into ions, which is the basis for the high aggressivity toward metals and alloys, which have a negative standard potential with the evolution of hydrogen. The austenitic stainless steels with about 18% Cr and 8 to 10% Ni, with or without Ti or Nb, for example AISI type 302, 304, 321 and 347 stainless steels are generally attacked in hydrogen chloride gas. Table 4 shows the uniform corrosion behavior of these steels towards dry hydrogen chloride gas. Chromium-nickel steels of type 18–8 are used as material for valve casings and spindles for hydrogen chloride gas at room temperature [30].

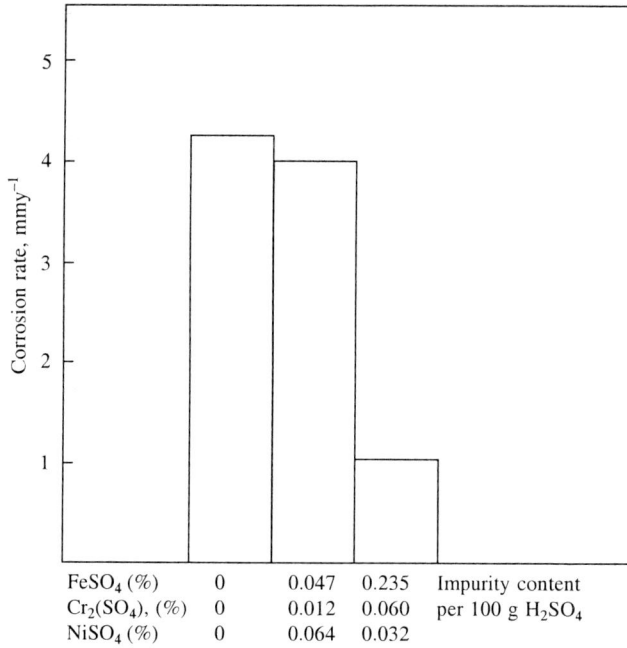

FeSO$_4$ (%)	0	0.047	0.235	Impurity content
Cr$_2$(SO$_4$), (%)	0	0.012	0.060	per 100 g H$_2$SO$_4$
NiSO$_4$ (%)	0	0.064	0.032	

Fig. 15. Influence of the metal sulfate content in 90 % sulfuric acid at 373 K on the corrosion rate of AISI 321 [26, 27].

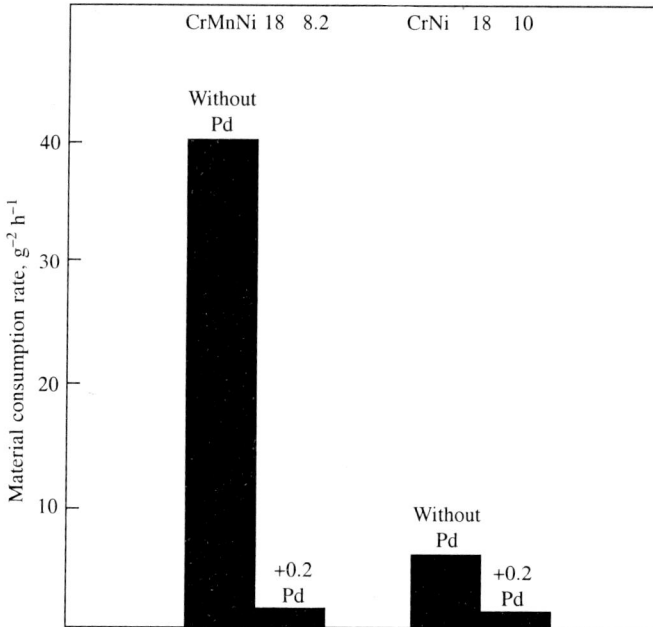

Fig. 16. Corrosion behavior of CrNi-steel 18-10 and CrMnNi-steel 18-8-2 with or without a Pd-content in 20% sulfuric acid at 293 K [29].

Air containing hydrogen chloride in chemical works attack steel with the AISI type 304 L at the corrosion rate of 0.142 mm y^{-1} at room temperature and a test duration of 184 days [31]. Chemical brightening of austenitic Cr-Ni steel is possible with hydrogen chloride gas [30]. The effect of nickel content on the corrosion rate of austenitic stainless steels and higher alloys in 15% HCl solution at 339 K (Figure 17) shows that with the increase in nickel content the corrosion resistance of these alloys decreases [31].

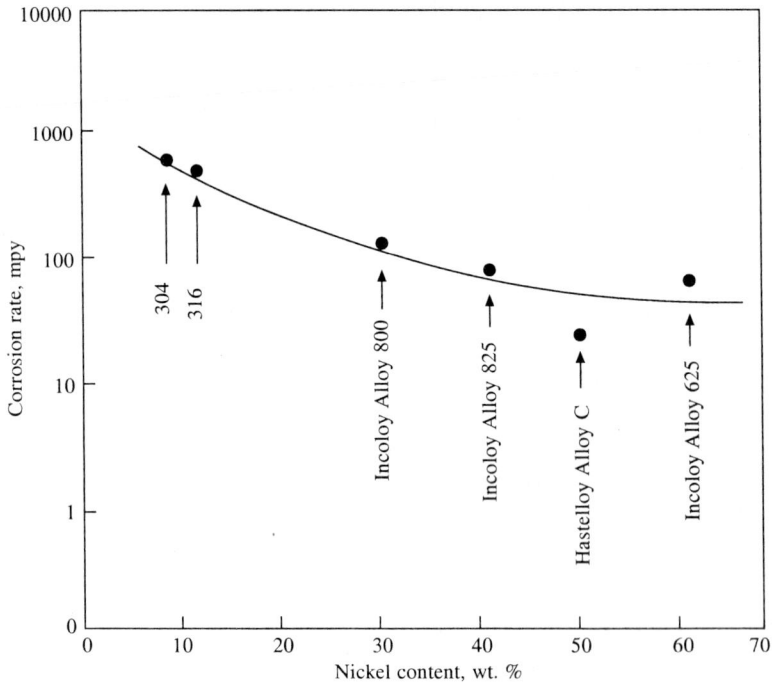

Fig. 17.   Effect of nickel content on the corrosion rate of austenitic stainless steels and higher alloys in a 15% HCl at 339 K [33].

Investigations on AISI type 304 stainless steel in a test environment of 60% by volume H$_2$O and 40% by volume of HCl, between 368 K and 453 K, showed corrosion rates that is listed in Table 5 [32]. The high corrosion rates at lower temperatures result from dew point effects [32]. A lot of experiments have ben carried out on the corrosion behaviour of austenitic Cr-Ni steels in hydrogen flue gas environment [33]. The studies include the examination of the influence of HCl on the rate of the corrosion and corrosion enhanced cracking susceptibility of such steels, employed in the construction of pulverized coal-fired boilers. A study on the corrosion behavior of AISI type 310 stainless steel in combustion gases with (100 ppm) and without HCl was performed at 463 K. The addition of HCl led to a two-fold increase in weight gain because HCl promotes corrosion, particularly by destroying the passive layer [36]. The effect of water vapor is not appreciable at 473 to 573 K, provided the temperature is at least above dew point [37, 38]. AISI type 316 stainless steel is therefore used for cracked gases containing hydrogen chloride. Hydrogen chloride produced by vaporizing of 31% hydrochloric acid, attacked austenitic chromium-nickel steels [39].

Table 4   Corrosive action of hydrogen chloride gas on austenitic chromium-nickel steel

Type of Steel	Temperature, K	Corrosion Rate, mm y^{-1}	Remarks	Literature
AISI 304	room temp.	0.08	aerated	[31]
AISI 304	478	1.25	greatly aerated	[31]
AISI 304	773	0.28	21 *d* test duration	[31]
AISI 304	863	0.51 to 0.63	in hydrogen atmosphere	[31]
CrNi-steel 18–10 and 18-10 Ti	313	0.55 to 1.10	–	[36]
CrNi-steel 18-8	643	0.05	–	–

Hydrochloric acid of any concentration has a reducing and activating effect on austenitic Cr-Ni steels. The passive film is destroyed and the steels are dissolved with the evolution of hydrogen [40]. Oxidizing agents intensify the attack [41]. The steels are even unsuitable for use in cold condition. Very dilute hydrochloric acid leads to pitting corrosion. Some corrosion rates of Cr-Ni austenitic stainless steels are shown in Table 6.

Table 5   Corrosion rates of AISI 304 L in 60% H$_2$O + 40% HCl at various temperatures [34]

Temperature, K	Corrosion rate, mm y^{-1}	Temperature, K	Corrosion rate, mm y^{-1}
773	0.483	448	0.483
673	0.015	443	0.010
573	0.010	433	0.027
483	–	413	–
473	0.007	393	–
463	0.005	383	0.485
453	0.004	368	12.100

Table 6 Corrosion rates of austenitic chromium-nickel steels in hydrochloric acid [40]

Type of Steel	Acid concentration, %	Test Temperature, K	Corrosion rate, mm y^{-1}
Steel 18-8	0.5	293	< 1.1
AISI 302	3.6	297	1.65
Steel 18-9 Ti	3.6	293	0.22
AISI 302	10.3	291	2.13
AISI 304	15	room temp.	3.9–4.5
AISI 302	25	297	31.1
AISI 303	0.36	343	1.4
Steel 18 Cr 8–9 Ni	0.5	boiling	> 11

As shown in Table 5, steels of this type have just sufficient resistance at room temperature to hydrochloric acid not exceeding a concentration of 50 % with regard to the uniform attack [42]. Weld seams are attacked by boiling 20.5% acid about 3 times more severely than the base material [43]. To reduce the attack by hydrochloric acid on austenitic Cr-Ni steels, inhibitors, for example dibenzyl sulfoxide, phenylthiourea or quinoline [44], are used. However, these are only moderately effective. 0.2% Rodine®213 or Cronox® reduce corrosion rate of steel in 10% hydrochloric acid at 344 K by 65 to 70% [45]; however, as a general rule, the conventional inhibitors for the Fe/HCl system are of only moderate efficiency for the austenitic Cr-Ni steels [46].

Although it is true that the attack by hydrochloric acid can be slightly diminished by alloying with 1 to 3% molybdenum [47], these steels have only moderate resistance even to dilute and cold hydrochloric acid and in cold condition. They are unsuitable at higher concentrations and temperatures [48]. Although the corrosion is uniform, in some cases pitting corrosion and stress corrosion cracking have to be expected. The Cr-Ni-Mo steels of AISI type 316 and 317 stainless steels are therefore not suitable for hydrochloric acid service [49]. The Cr-Ni steels which contain molybdenum and copper, are somewhat resistant to hydrochloric acid as long as no oxidizing substances are present in the acid, for example, ferric chloride, copper chloride or air [50]. In the cold condition 0.5% acid hardly attacks [51].

At 353 K, the corrosion rate remains under 0.04 mm y^{-1} [52]; only at the boiling point does it reach 0.84 mm y^{-1}. The 5 to 20% acid causes a loss of 0.126 to 0.176 mm y^{-1} at 293 K [36]. An alloying addition of 0.3% tin considerably reduced the attack of 2% hydrochloric acid on 18Cr-20 Ni-2Mo steel at 373 K [53].

Nitric Acid

Nitric acid is one of the strongest mineral acids and is also a liquid with vigorous oxidizing properties especially in the anhydrous form. The choice of metals and alloys for nitric acid service is quite limited. For most plant application, the choice is usually between only two general classes of materials - stainless steels and high silicon irons. The choice is further limited because minimum chromium content is required in the stainless steels materials. In nitric acid, austenitic stainless steels containing 18% chromium generally exhibit passive behavior over wide ranges of concentration and temperature. The corrosion rates of AISI type 410, 430 and 304 stainless steels in boiling nitric acid solutions are shown in Figure 18. Problems associated with intergranular corrosion of sensitized materials have led to the wide spread use of AISI type 304L stainless steel for all types of equipment handling nitric acid. The corrosion rates of AISI 304L stainless steel are given in Figure 19.

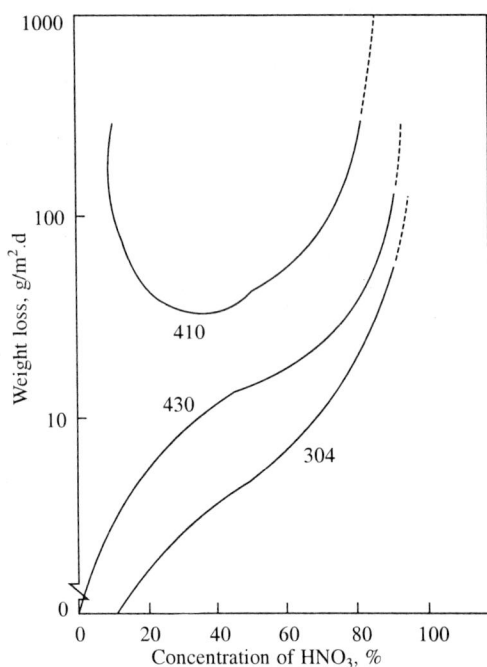

Fig. 18. Corrosion rate of various austenitic stainless steels in boiling nitric acid solution [18].

The most important ingredients for resistance to nitric acid is chromium. As the chromium content increases, the corrosion rate decreases. The minimum amount of chromium generally accepted is 18%. This makes the austenitic stainless steels very well suited for practically all concentrations and temperatures. The addition of molybdenum to stainless steels, as in AISI type 316, does not improve corrosion resistance to nitric acid. For lower concentration (<60%), the corrosion resistance of austenitic stainless steels is determined by chromium content. For higher concentrations, with Cr^{6+} ions, the corrosion behavior is determined by other elements. At this concentration, the acid shows powerful oxidizing effects. Since the electrochemical potential is near the transpassive region, which increases the corrosion rate.

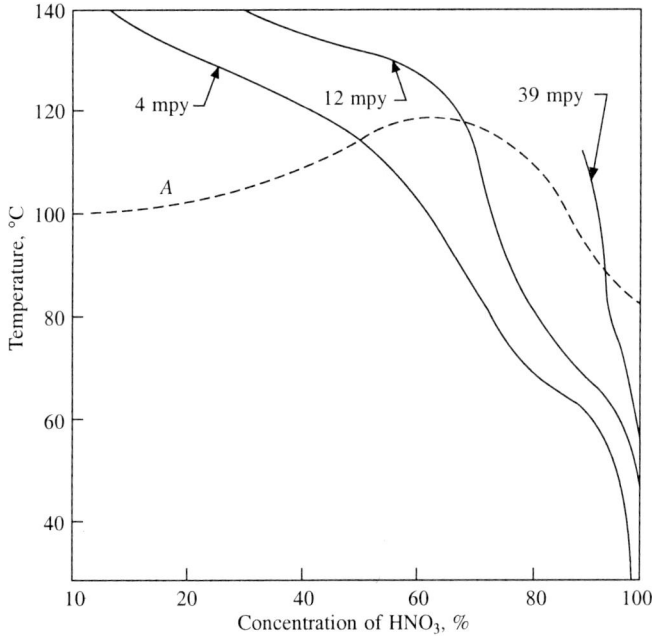

Fig. 19. Corrosion behavior of type AISI 304L stainless steel in nitric acid solutions of various concentrations and temperatures. Curve A is the boiling point curve [7].

The austenitic steel AISI type 304L is generally used in nitric acid plants. By reducing the contents of Si, P and S, the corrosion behavior of the steel AISI type 304L (HNO_3 grade) at higher concentrations of nitric acid can be improved. The use of austenitic 18Cr-8Ni steels is recommended in nitric acid plant for oxidation of ammonia and further processing to 60% acid [54].

Fe-0.01C-18Cr-15Ni-4Si stainless steel was developed to keep intergranular and uniform corrosion low when using steels for industries processing nitric acid. However, since austenitic steels with high silicon and low carbon content tend to crack in the weld region, the filler material must be alloyed to retain 4 to 10% δ-ferrite in the weld metal [55, 56].

Stainless steels, such as AISI type 304 L, are preferentially attacked by hot nitric acid in the crevices which form, for example, during welding or lining, or between the steel and the plastic sealing discs. While the AISI type 304 L stainless steel corrodes at the rate of 0.05 and 0.07 mm y^{-1} after 100 and 500 h respectively in 15% nitric acid at 367 K, rates of 0.27 and 0.54 mm y^{-1} respectively occur after 500 h in the crevice at the weld and between the steel and the sealing disc. The relationship between cold working, precipitation behavior and corrosion resistance in boiling 5 mol L^{-1} nitric acid was investigated on specimens taken from a pipe made of steel X 2 CrNiSi 18–15 (Fe-0.009C-4.24Si-1.33Mn-0.008P-0.001S-17.32Cr-14Ni-0.01 Mo-0.13 Nb-0.03Ti-0.012N). Cold worked (4 to 20%)

steels subsequently heat-treated at 973 K showed higher materials consumption rate. This is attributed to precipitates ($M_{23}C_6$, Cr-rich carbide and $Cr_5Ni_3FeSi_2$) produced at the grain boundaries by deformation and subsequently annealing, the amount of which increases with the degree of deformation. The specimens were annealed beforehand at 1393K/10min/water to establish a recrystallized, precipitation-free starting structure. The material consumption rates per unit area of X2 CrNiSi 18 15 steel are shown as a function of the treatment state and test duration in Table 7. As a result of galvanic contact of the unworked and deformed specimens with the same treatment state, the material consumption rate of the deformed specimens was 2 to 3 times higher than that of the unworked specimen. Deformations of the steel and stress-induced precipitates impair the corrosion resistance in nitric acid [57].

Table 7 Material consumption rates of specimens of steel X2 CrNiSi 18 15 with different treatment condition in 5 mol L^{-1} boiling HNO_3[57]

Treatment condition	Surface, cm^2	Test Duration, h	Material Consumption Rate, $g\ m^{-2}\ h^{-1}$
Non-deformed Non-sensitized	16.1 (16.1)*	48	0.08 (0.10)*
Deformed (20%) Non-sensitized	12.1 (12.3)*	72	0.12 (0.28)*
Non-deformed Sensitized	16.2 (16.4)*	48	0.08(0.09)*
Deformed (20%) Sensitized	11.6 (11.6)*	72	0.08(0.17)*

* Values of specimens in galvanic contact.

Steel with good strength, toughness and weldability as well as corrosion resistance, usually contain 4–6% Mn, 6–9.5% N, 20–21.5 % Cr, 0.25–0.35%N and possibly up to 2.5% Mo and 1% Nb. A steel of this type (completely austenitic and non-magnetic) corrodes at a rate of 0.15 mm y^{-1} (tempered) and 0.17 mm y^{-1} (sensitized) in boiling nitric acid (Huey test) [58].

The resistance of the austenitic 17Cr–14Ni–4Si steel to boiling 65% nitric acid can be further improved by the addition of niobium, zirconium or tantalum. However, nitrogen and titanium additions cause deterioration. With its good mechanical properties and weldability, the steel Fe ≤ 0.05C-17Cr-14Ni-4Si-0.8 Nb exhibits an excellent resistance to concentrated nitric acid [59].

The corrosion rate of 17Cr-14Ni-4Si steel in boiling 65% nitric acid only becomes independent of the heat treatment temperature between 870 to 1220 K if about 1% niobium, zirconium or tantalum is alloyed with the steel [60]. The material consumption rate is between 0.4 and 0.6 g $m^{-2}\ h^{-1}$ for this alloy.

According to Huey test, AISI type 304 can be used for waste tanks in the nuclear industry if the steel is heated at 813 K for 1 to 10 h (no longer), since the corrosion rate in boiling 65% nitric acid does not exceed 0.24 mm y^{-1}. The AISI type 304 L having a corrosion rate of 0.15 mm y^{-1} can be used if it is tempered at 513 or 703 K for 100 h, the heat treatment temperature has a particular influence on the corrosion rate of the steel AISI type 304 L. [61]. Similarly (Fig. 20), cold working has a varying influence on the corrosion in 65% boiling nitric acid of the following non-sensitized steels: solution annealed at 1350 K for 2 h and rapidly quenched for AISI type 304 and 304 L as well as on AISI type 316 (Figure 20) [62]. Corrosion rate of AISI 304 in 65% boiling nitric acid reduced from 1.1 mm y^{-1} to on increasing the cooling rate from 1 K s^{-1} to 55 K s^{-1} [63].

Corrosion of AISI type 304 stainless steel in boiling 65% nitric acid (Huey test) increased considerably at 923 K by heat treatment for increasing duration, but decreased at 1023 K (Table 8) [64]. AISI type 304 and 304 L of comparable composition gave corrosion rates which approximately coincide after 240 h in the Huey test 0.46 mm y^{-1} for AISI type 304 (in the delivery state) and 0.31 mm y^{-1} for AISI type 304 L (20 mim at 950 K) and 0.61 mm y^{-1} (1 h at 950 K) [65]. AISI 304 (as sheet or pipe), heat treated for 20 min or 1 h at 950 K, corroded in boiling 65% nitric acid at a rate between 0.20 and 0.22 or 0.26 mm y^{-1}.

Even small amounts of boron reduce the corrosion rate of AISI type 304 in boiling 65% nitric acid (Figure 21). An increase in corrosion rate caused by sensitizing heat treatment between 922 and 1033 K is largely cancelled out by the presence of boron [66]. Boron content of 4 ppm had no influence on the corrosion behavior of a solution-annealed steel. The corrosion rate of this steel after 240 h was 0.065 mm y^{-1} with and without boron [66].

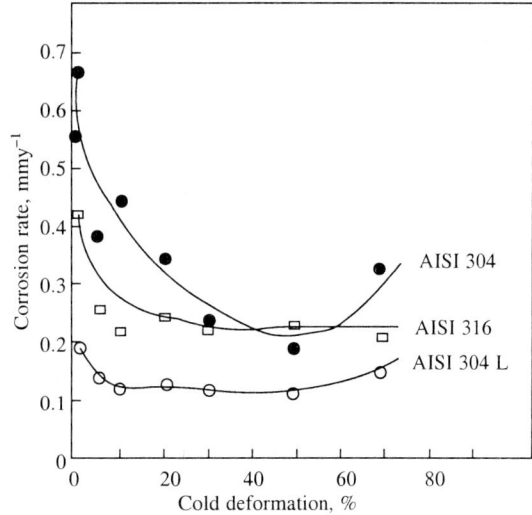

Fig. 20. **Corrosion rate of non-sensitized stainless steel in boiling 65% HNO_3 as a function of the degree of cold working [62].**

Both sulfur (0.03 %) and phosphorous (0.06%) have no influence on the corrosion behavior of AISI type 304 stainless steel in boiling 65% nitric acid (Huey test). Sulfur and phosphorous addition do not significantly affect the chromium depletion process and these elements do not stimulate the corrosion process. The material consumption rates were about 0.27 and 0.10 g m^{-1} h^{-1}, independent of the heat treatment of 100 to 1000 h at 923 and 973 K [67].

Table 8 **Corrosion rate of AISI 304 in boiling 65% HNO_3 for various heat treatment conditions [64]**

Heat Treatment		Corrosion Rate,
Temperature, K	Duration, h	mm y^{-1}
–	–	0.26
923	1	0.88
	2	4.2
	5	8.8
	50	21.4
1023	0.5	1.5
	1	1.3
	2	1.1
	5	1.1

The comparative corrosion rate of Ti and some metal alloys with the commercial and nuclear grade AISI type 304 L stainless steel in boiling nitric acid medium at different testing time durations is shown in Figure 22. The specimens of commercial AISI type 304 L stainless steel showed a higher corrosion rate to those of nuclear grade AISI type 304 L stainless steel. The high corrosion rate of commercial AISI type 304L stainless steel is characterized by changes in the surface appearance, with the formation of a brown oxide layer and a rough surface owing to the spalling of grain. The reduction of carbon, impurities and boron content improve the corrosion resistance [68].

The corrosion rates as a function of Si contents are shown in Figure 24. The austenitic 18Cr-8Ni steel with normal Si content of 0.5 to 1% and a carbon content slightly above 0.15% is resistant in up to 80% nitric acid concentration at 298 K (Figure 23). With more than 4% Si and a carbon content between 0.12 and 0.13%, this steel are practically resistant in the entire range of nitric acid [69].

Contact between the steel AISI type 304 stainless steel and platinum, gold or graphite in 65% nitric acid does not noticeably intensify the corrosion rate either at 293 K (from 0.008 with and without Pt contact to 0.007 with graphite and 0.002 mm y^{-1} with Au contact) or 373 K, but merely leads to a shift in the potential to more positive values, although these still remain within the passive range [70].

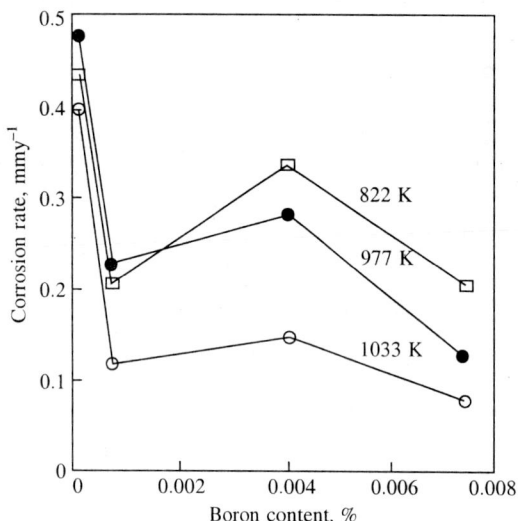

Fig. 21. **Influence of the boron content in AISI 304 on the corrosion rate in boiling 65% HNO$_3$ as a sensitization temperature [66].**

Phosphoric Acid

Phosphoric acid (H$_3$PO$_4$) is widely used in the manufacture of phosphate chemicals (fertilizer industry, production of soft drinks, corrosion inhibitors, coatings etc). Pure phosphoric acid is a reducing acid and is not very aggressive. Corrosion by phosphoric acid depends on the methods of manufacture and the impurities present. Commercial phosphoric acid contains fluorides, chloride, sulfates, and heavy metal ions as impurities, which significantly increase its corrosivity and makes its corrosion characteristic unpredictable. Chloride contamination significantly increases corrosion of austenitic stainless steel and demands the use of nickel-based alloys. Other variables include the concentration, temperature, aeration, etc.

Austenitic stainless steels in phosphoric acid solutions, generally, exhibit active passive behavior. The presence of oxidizers such as Fe^{+3} and Cu^{+2} often give rise to passive behavior and the corrosion is determined by the metallurgical and environmental conditions [71].

The corrosion rates of the various austenitic stainless steels in pure boiling solutions of phosphoric acid are shown in Figure 25, with AISI types 316 and 317 stainless steels exhibiting the highest corrosion resistance of the stainless steels tested [5]. Figure 26 shows that alloying with nickel and molybdenum gives rise to significant improvements in corrosion resistance. Since, in industry, phosphoric acid is often encountered in an impure state containing numerous contaminants that can affect the corrosion rate, testing in process or simulated environments is desirable [33].

Two impurities that increase the corrosivity of phosphoric acid are chlorides and fluorides. Magnesium reduces the corrosivity of phosphoric acid by forming complexes with fluoride and can be used as inhibitor [71].

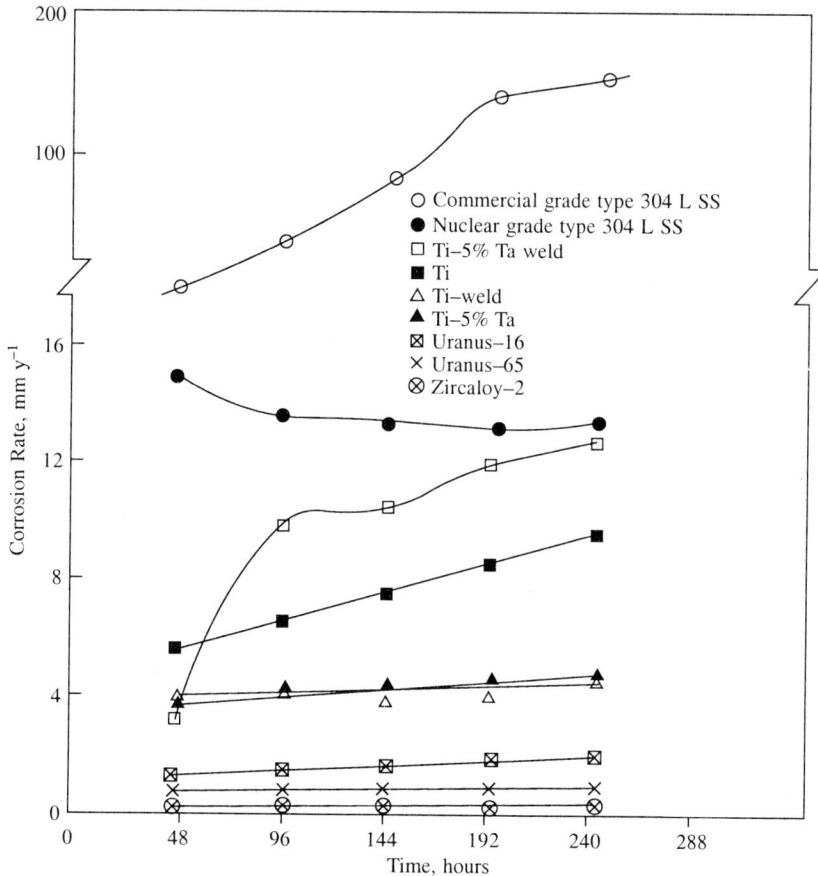

Fig. 22. Comparative corrosion rate of AISI type 304 stainless steel and other metal alloys in boiling nitric acid [68].

The most effective alloying constituents for improving the corrosion resistance in phosphoric acid are chromium and molybdenum. Both alloying elements are responsible for a good resistance of the material to pitting corrosion as well. These facts indicates that the resistance of molybdenum-free Cr-Ni austenitic steels in phosphoric acid will sometimes be below that of the high-alloy ferritic or of duplex steels. AISI type 301 and 302 stainless steels thus cannot be used either in pure phosphoric acid with 45–55% P_2O_5 or in one containing 20–30% P_2O_5, and cannot be used at all in crude acids of this type. AISI type 304 stainless steel has somewhat better resistance. It also cannot be used in the crude acid mentioned, but can be used in pure acid with 20-30% P_2O_5 (27.6-41.4 % H_3PO_4) at room temperature [72]. AISI type 304 stainless steel can be used in pure phosphoric acid and only in such in the entire concentration range up to about 353 K, and in pure boiling phosphoric acid of up to 50% strength (36.23% P_2O_5) [73]. Under these conditions, corrosion rates of up to 0.1 mm y^{-1} or slightly above can be expected.

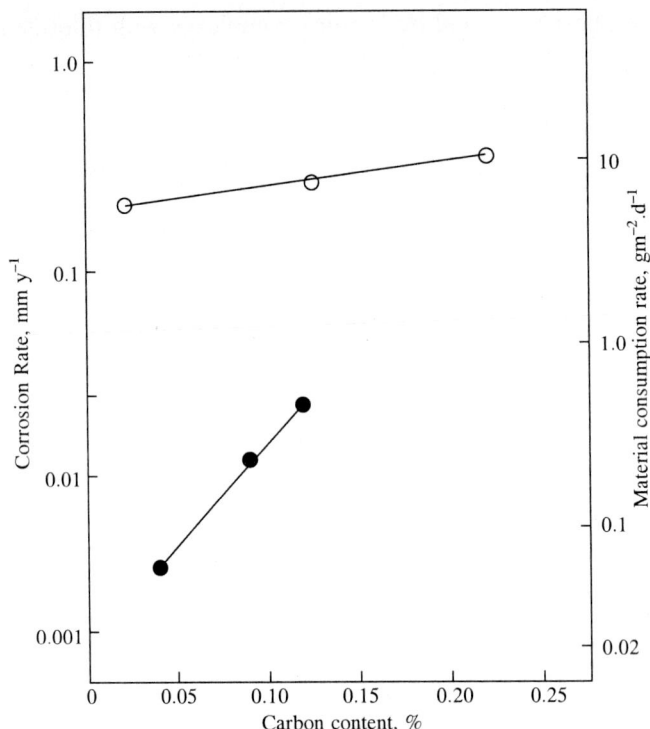

Fig. 23.   Influence of the carbon content on the corrosion rates of austenitic 18Cr-8Ni with Si content of about 1% and >6% HNO$_3$ at 298 K, test duration 720 h (1) 1%, (2) > 6% [69].

In industrial phosphoric acid, the austenitic Cr–Ni and Cr–Ni–Mo steels tend to assume unstable state. Passivation and activation proceed slowly in them, hence, it is difficult to determine the corrosion rate by measuring the weight losses. The corrosion rates for AISI type 304 stainless steel in aerated and non-aerated phosphoric acid in the range from 10 to 80% and up to 339 K are reported as being not more than 0.05 mm y^{-1} [74]. At higher concentrations, the corrosion increases rapidly in pure acid. In 85% acid at 339 K, the corrosion rate is 0.03 mm y^{-1} [75]. Electrochemical studies in an aqueous solution of 40% H$_3$PO$_4$ and 4% H$_2$SO$_4$ with addition of 3000 ppm Cl$^-$ + 24 gL^{-1} SiC 18Cr-10Ni steel with 1.5 and 4% Mo, showed a significant decrease in the corrosion rate; a jump from 0 to 1.5% Mo in particular being very effective. The corrosion current dropped by a factor of 100 by addition of 1.5% Mo, and by a factor of 60 by addition between 1.5% Mo and 4% Mo [76].

The effect of halogens on the corrosion behavior of AISI type 304 and 316 stainless steels in H$_3$PO$_4$ is summarized in Table 9. In principle, the attack on the Mo-free austenitic steels is greater than on the Mo-containing steels, although both these can be referred to as having good resistance (corrosion rates not more than 0.1 mm y^{-1}). In 30 and 50% H$_3$PO$_4$, the corrosion rate decreases in the sequence F$^-$ > Cl$^-$ > Br$^-$, while in 70 and 85% H$_3$PO$_4$ the sequence is of in the order of Cl$^-$ > F$^-$ > Br$^-$ [77].

The corrosion rate for AISI type 304 and 304 L stainless steels in 1% H$_3$PO$_4$ + 4% NaCl was only 0.15 mm y^{-1}. In the solution containing 0.2% H$_3$PO$_4$, uniform corrosion occurred after 10 days, although corrosion rates of 0.14 mm y^{-1} for steel AISI type 304 and 0.03 mm y^{-1} for AISI type 304 L in principle demonstrate that these steels are resistant to uniform corrosion.

An overview of the corrosion behavior of Cr-Ni-Mo steels with increasing Mo content is given in Figure 27 for a steel with 2% Mo and in Figure 28, for a steel with 5% Mo. Isocorrosion diagrams for these alloys were developed in chemically pure phosphoric acid without aeration. The resistance of these steels to phosphoric acid increases as the chromium content increases. The increase is significant with increasing molybdenum content [78].

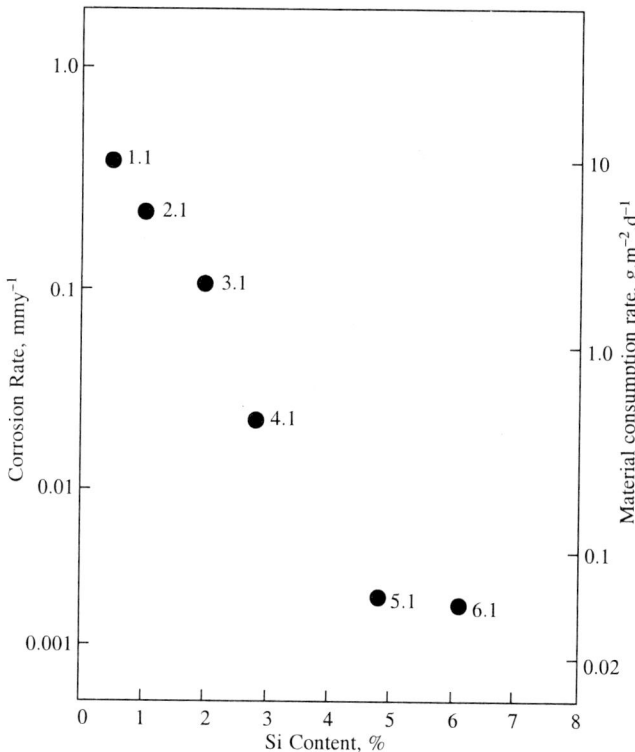

Fig. 24. Influence of the Si content on the corrosion rates of austenitic 18Cr-8Ni steel with about 0.03% HNO$_3$ at 298 K, test duration 720 h [69].

According to these diagrams, AISI type 316 stainless steel can be used in all concentrations of pure phosphoric acid and only in this acid up to 353 K, and in only up to about 50% boiling pure phosphoric acid if the corrosion rate of 0.1 mm y^{-1} may not be exceeded [79, 80]. The use of pure acid is also the reason why the difference in the corrosion resistance of the two steels is not so marked (0.1 mm y^{-1} curve). The differences becomes more visible if phosphoric acid contains impurities such as Cl$^-$ and F$^-$ ions and sulfuric acid which stimulate corrosiveness, while Fe$_2$O$_3$, Al$_2$O$_3$, SiO$_2$, CaO and MgO inhibit corrosion [81]. Studies on AISI type 316 L stainless steel, which is often used in chemical equipment construction, in respect of modification of its corrosion resistance by cold working shows that uniform corrosion in boiling 50% phosphoric acid is not adversely influenced [82]. Figure 29 shows the corrosion rates for AISI type 316 stainless steel as a function of its HF and H$_2$SO$_4$ concentration, For "wet acid" at 361 K [83, 84]. According to this figure AISI type 316 stainless steel can be used for only a short time (if at all) under these conditions, with corrosion rates of 1 mm y^{-1}. The decrease in corrosion rate in 85% H$_3$PO$_4$ as a function of coper content is shown for AISI type 316 L stainless steel (Figure 30). This shows that a copper content of about 0.2% in the steel reduces the corrosion rate from more 40 mm y^{-1} to less than 15 mm y^{-1}. Although no resistance results, the trend is detectable [83, 84].

During storage and transportation of phosphoric acid, it is absolutely essential to ensure that this is not too warm when the corresponding tanks are filled, since hot acid is more corrosive than cold. It is important to know the precise composition of the crude acid, so that its corrosiveness can be estimated.

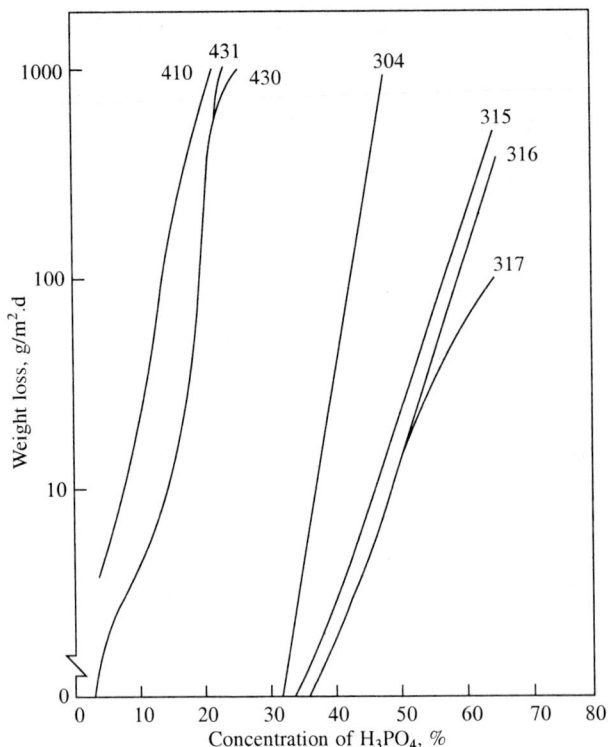

Fig. 25   Corrosion rates of various stainless steels in phosphoric acid solutions at their boiling points [18].

AISI type 317 L stainless steel is used in tanker construction for transportation of superphosphoric acid. The austenitic Cr-Ni-Mo steels has become a standard material for transportation of phosphoric acid, a minimum content of 2.5%, Mo is being prescribed [85].

Anodic protection on Cr-Ni and Cr-Ni-Mo austenitic stainless steels can be used in phosphoric acid, only if they are passive in the medium. Under these conditions, corrosion proceeds uniformly, with corrosion rates of far less than 0.1 mm y^{-1}. These conditions are disturbed in phosphoric acid containing fluoride and/or chloride. In special cases, pitting corrosion may occur and proceed far more vigorously than uniform corrosion due to local corrosion current. The current/potential curves provide information on the position of free potential (corrosion without polarization). The anodic protection to austenitic stainless steels can be provided by impressing a potential in the passive region. It is brought into a region in which it corrodes at a rate far less than 0.1 mm y^{-1} under a low residual current. The applicability of the method depends on large number of factors, including the size and

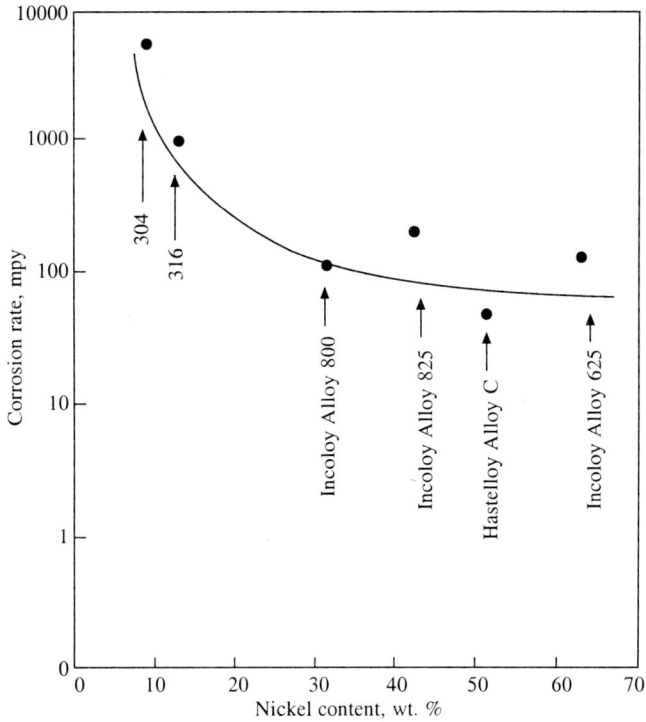

Fig. 26. Effect of nickel content on the corrosion rate of austenitic stainless steels and higher alloys in boiling 85% H_2PO_4 solution.

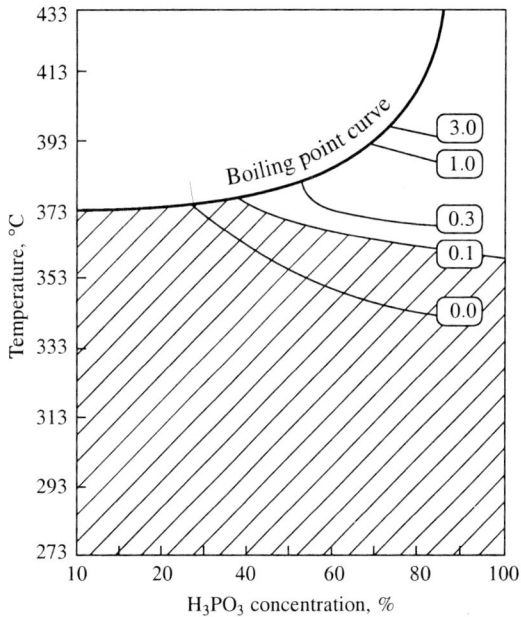

Fig. 27. Isocorrosion curves (mm y^{-1}) of AISI 316 in pure phosphoric acid [80].

shape of the component to be protected. [80, 85]. In a very corrosive phosphoric acid with 54% P_2O_5, $4\,SO_4^{2-}$, 1.05 F^-, 0.0167 Cl^-, 0.27 Fe_2O_3, 0.17 Al_2O_3, 0.1 SiO_2, 0.2 CaO, 0.7 MgO, corrosion studies on AISI type 316 L stainless steel at room temperature showed corrosion rates of more than 0.1 mm y^{-1} with anodic protection [81, 86].

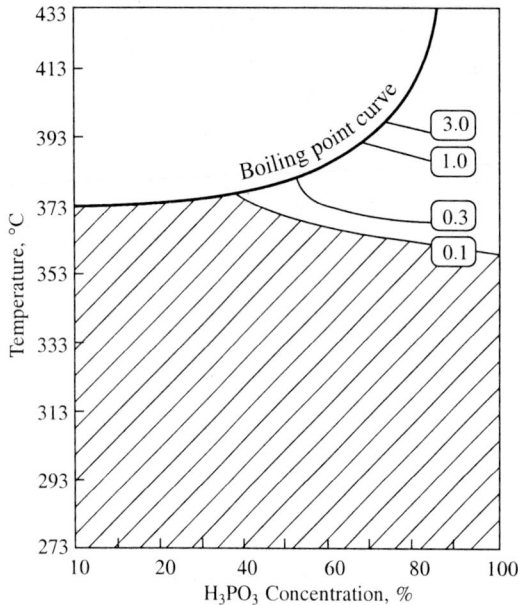

Fig. 28. Isocorrosion curves (mm y^{-1}) of CrNiMo steels of ASTM type A240 in pure phosphoric acid [80].

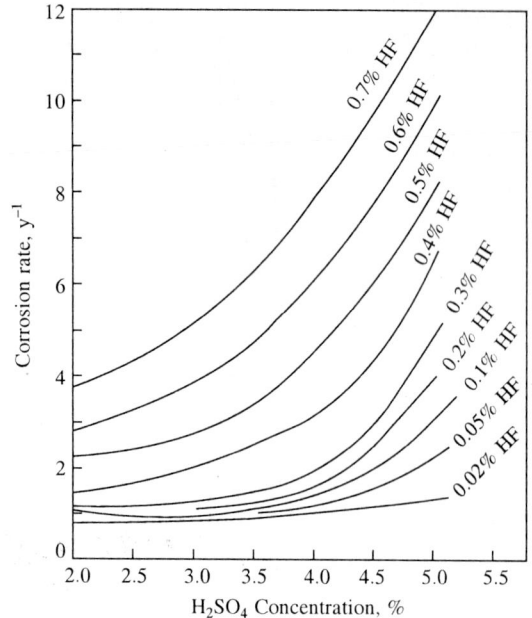

Fig. 29. Corrosion rates of AISI 316 L in 35% H_3PO_4 (25% P_2O_5), with 1.5% F as H_2Si_6, 0.02–0.7% HF and 2–5% H_2SO_4 [83, 84].

General Corrosion of Austenitic Stainless Steels in Alkalies

Alkalies dissove in water with the formation of OH^- ions. One of the most important technical media in this category are aqueous solutions of alkaline hydroxides, such as sodium hydroxide (caustic), caustic potash (KOH) and ammonium hydroxide. Alkaline conditioning of aqueous media is often used technically for means of corrosion protection in heating and steam-producing systems. But with increasing concentration of alkali in solution, up to pH 14 and above, the corrosivity of steels increases. Besides increased uniform corrosion, caustic embrittlement can occur. The common alkalies such as caustic soda (NaOH) and caustic potash (KOH) are not particularly corrosive and can be handled in steel in most applications where contaminations is not a problem.

Austenitic stainless steels exhibit active-passive behavior in sodium or potassium hydroxide solutions, with the active condition being developed with increasing concentration and temperature. AISI types 304 and 316 stainless steels in sodium hydroxide at concentrations up to 50% and temperature at 323 K exhibit low uniform corrosion rates. The passive-active transition for AISI type 304 occurs in 40% NaOH at 353 K and in 50% NaOH at 343 K [87].

Standard 18-8 steels are resistant to dilute sodium hydroxide (below 10%), but are noticeably attacked in 50% solution at temperatures above 340 K and rapidly dissolved above 365 K [88]. The corrosion rate of AISI type 304 stainless steel can be kept below 0.01 mm y^{-1} in 10 mol · L^{-1} sodium hydroxide solution at 323 K and 343 K by cathodic protection at a potential of < –0.8 V (SCE) [89]. As shown in Figure 31, addition of vanadium and especially of cerium (up to 3%) reduce corrosion rate of 12 Kh 118N10T (Fe-0.12C-18Cr-10Ni, Ti-stabilized) steel in boiling 30% and 55% sodium hydroxide solution. The addition of titanium nitride (2.5%) to the steel accelerates corrosion in the 30% sodium hydroxide solution. Under identical test conditions in 50% NaOH solution a significant decrease in brought about [90].

In NaOH solution, containing 140 and 650 gL^{-1} NaOH, the corrosion rates of AISI type 321 stainless steel at 368 K are 0.002 g.m^{-2}. h^{-1} (0.043 mm y^{-1}) respectively. If the steel comes into contact with copper (ratio of areas 1 to 1 or 1 to 10 respectively), the corrosion rates in the more dilute NaOH solution are 0.04 and 0.15 mm y^{-1} and in the concentrated solution (650 g L^{-1} NaOH) about 0.24 and 0.40 mm y^{-1}. The corresponding corrosion data for copper in the same order are: 0.85 or 1.36, and 0.036 or 0.068 mm y^{-1} [91].

The corrosion behavior of chromium-nickel austenitic stainless steels in potassium hydroxide solutions and in water containing potassium hydroxide is of industrial interest since these materials are used for high-pressure boilers in power plant. In oxygen-free water containing KOH at 543 to 533 K, Fe-18 Cr-9 Ni-0.8 Mn-0.5 Si steel exhibit only slight uniform corrosion without stress corrosion cracking after 440 h [92]. Although the chromium-nickel steel are resistant to KOH just like other steels, they are unsuitable for prolonged use even though the rate of corrosion decreases as the duration of exposure increases. In pressurized water containing KOH, iron is preferentially dissolved due to the solubility of magnetite from high-alloy austenitic materials. The rate of corrosion of these steels depends on the concentration of free oxygen in the water containing KOH [93].

Austenitic chromium-nickel steels are completely resistant to gaseous and liquid ammonia; corrosion rate is less than 0.05 mm y^{-1} in gaseous, pure ammonia up to 589 K. Above this temperature corrosion rate increases and exceeds more than 1.27 mm y^{-1} at 773 K [94].

Aqueous, dilute and concentrated ammonium hydroxide solutions attack austenitic Cr-Ni and Cr-Ni-Mo steels only slightly less than 0.05 mm y^{-1} at room temperature (297 K). At boiling temperature (and/or at 373 K), however, the material losses increase to 0.5 mm y^{-1} [2].

The austenitic Cr-Ni and Cr-Ni-Mo steels exposed to ammonia and ammonium hydroxide often fail after a short period of use when carbon dioxide content in the mixtures of CO$_2$, NH$_3$ and NH$_4$OH exceeds 5 to 10% [95]. The addition of oxygen between 330 and 500 ppm is recommended as a remedy for such failures. The effects of such additions can be seen from the example in Table 10 [96].

Severe corrosion damage of austenitic Cr-Ni and Cr-Ni-Mo steels is attributed to marked local ammonia enrichment, which causes destruction of passive layer followed by dissolution of the steel [97].

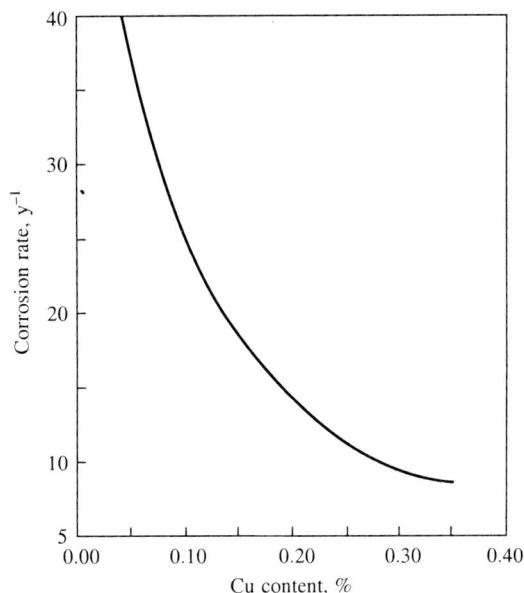

Fig. 30. Influence of copper content on the corrosion rate of AISI 316 L in 85% boiling H$_3$PO$_4$ [84].

Table 9 Influence of halogens on the corrosion rate behavior of the austenitic Cr-Ni steel AISI 304 and the Cr-Ni-Mo steel AISI 316 in phosphoric acid with or without addition of halogens (40 mmol L^{-1}) [77]

Type of Steel	Halogens	Temperature, K	30	H_3PO_4 Concentration, %		
				50	70	85
				Corrosion Rate, mm y^{-1}		
AISI304	–	323	0.023	0.005	0.014	0.023
	F	323	0.023	0.005	1.000	0.280
	Cl	323	0.023	0.005	3.140	24.000
	Br	323	0.023	0.005	0.070	0.046
	–	353	0.023	0.009	0.540	0.070
	F	353	0.302	1.160	1.023	99.000
	Cl	353	0.046	0.009	14.000	145.000
	Br	353	0.023	0.009	0.511	0.600
	–	373	0.093	0.023	2.790	254.000
	F	373	1.750	6.280	4.510	238.000
	Cl	373	0.070	15.020	54.400	400.000
	Br	373	0.070	0.023	1.770	1.260
AISI 316	–	323	0.005	0.005	0.009	0.009
	F	323	0.005	0.005	0.023	0.023
	–	323	0.005	0.005	0.116	0.023
	Br	323	0.005	0.005	0.023	0.014
	–	353	0.009	0.023	0.093	0.600
	F	353	0.139	0.420	0.370	0.190
	–	353	0.023	0.023	0.560	0.370
	Br	353	0.023	0.023	0.116	0.023
	–	373	0.400	0.070	1.813	0.450
	F	373	1.140	1.420	1.930	3.240
	Cl	373	0.139	0.210	2.650	1.170
	Br	373	0.116	0.139	1.540	0.046

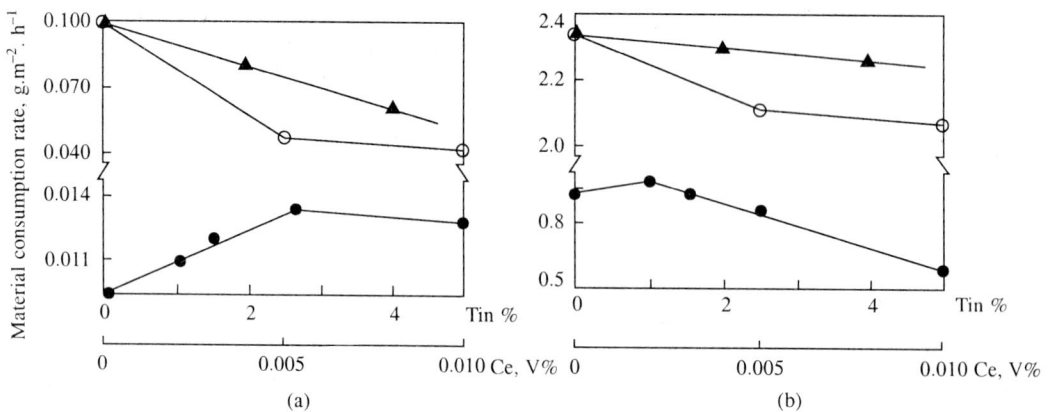

Fig. 31. The influence of TiN (●), Ce(O) and V (▲) on the corrosion rate of AISI type 321 in boiling 30 (a) and 50% (b) sodium hydroxide solution [92].

Table 10 Influence of oxygen on the corrosion rate of austenitic steels in NH_3 plants [96]

Type of Steel	Point of use in NH_3 plant	Corrosion Rate, mm y^{-1}	
		Without oxygen	With oxygen
AISI 304	Top tray of degassing column	4.1	0.2
AISI 316 L	Top tray of degassing column	1.4	0.18
AISI 319 L	Top tray of degassing column	1.2	–

General Corrosion of Austenitic Steels in Sea Water

Seawater contains about 3.4% of salt and is slightly alkaline with pH of 8. Corrosion in seawater is dependent mainly on the salt content (which increases the electrical conductivity) and its oxygen content. A number of variables can influence and complicate the course of corrosion in different ways. The effect of chloride ion is predominant; differences in the aggressiveness of seawater from different geographical locations can be attributed to different salt contents. Sulfates exert a smaller influence, while bicarbonates hinder corrosion. Other variables are the concentration of ions, the temperature and the rate of flow. In addition, living organisms (plants and microorganisms) may have a direct or indirect influence. In the oceans the normal water temperature varies between 277 and 289 K but in tropical regions it usually exceeds 293 K and can attain about 303 K.

Some experience with austenitic chromium-nickel steels shows that they are generally not considered as marine materials. But cast alloy CF3 (19Cr-10Ni) has earned its place as a standard propeller, and materials for workboat, controllable pitch propellers, and bow thrusters, while AISI type 304 stainless steel, for example, is used for wire rope despite of its rather shorter service life. AISI type 304 stainless steel fasteners give reasonably good service when used to fasten steel or aluminium. [98]. Tubes of AISI type 304 and 304 L stainless steels are unsuitable for condenser tube service in seawater due to pitting and crevice corrosion. Nitrogen improved the corrosion resistance of austenitic Cr-Ni steels with low carbon in seawater [99].

Investigations on Type AISI type 304 stainless steel showed that chlorination (1 mg L^{-1}) for controlling biofouling in sea water flowing at the rate of 0.1 m s^{-1} did not remarkably change the uniform corrosion rates but lead to an increase in localized corrosion [100]. Table 11 shows the corrosion behavior of some austenitic Cr-Ni steels in slowly moving seawater [101].

Austenitic stainless steels display excellent resistance to marine atmosphere; although sometime the steels become tinted if not polished, lacquered or waxed. If the surface is kept clean (and polished), further protection is not necessary. A thorough comparison (in terms of weight loss of component/year) has been made for fasteners of Cr-Ni austenitic steels, Ni-Fe-C alloys and constructional steels with or without surface protection in sea and industrial atmosphere (Cleveland, Ohio) over 5-years period [102].

Atmospheric studies made in Panama Canal zone included the AISI types 410, 316, 321, 302, 203 and 17% Cr steels. Both inland and marine atmosphere in the tropics were practically non-corrosive to all the stainless steels. The highest loss was reported for AISI type 410 stainless steel at the marine site, amounting to 0.08 g dm^{-2} in the first year and only 0.11 g dm^{-2} after 8 years. Others report for this alloy a loss of 0.11 g dm^{-2} in 540 days in marine atmosphere and 0.25 g in 600 days in an industrial atmosphere. Austenitic steels suffered a loss of only 0.01 g dm^{-2} in the canal zone and no pitting was reported [103, 104].

In North-Sea atmospheres an austenitic 18Cr-10Ni-3 Mo steel showed slight (~0.1 %) surface corrosion; the usual 18–8 austenitic steels with 0.5 to 10% of the surface attack were unsuitable. In tropical sea water (Panama Canal zone) the austenitic Cr-Ni steels after 4 years (and up to 8 years) showed an approximately linear corrosion rate of 11 to 12.5 μm y^{-1} when fully immersed (comparing with 39 μm y^{-1} for 13% Cr steels). Corrosion in the tidal zone was less, settling down to a linear rate of ~ 2.5 μm y^{-1} for AISI type 302 stainless steel and 0.25 μm y^{-1} for AISI type 316 [102]. In Dead Sea water (alternating between 289 K and 313 K) corrosion occurs at the rate of 1.5 μm y^{-1} (with pitting) compared with 0.55 μm y^{-1} in Mediterranean sea [105].

Table 11   Corrosion rate of some austenitic stainless steels in slow moving seawater [101]

Type of Steel (AISI)	Test Duration, d	Weight Loss, g	Corrosion Rate, μm y^{-1}
302	643	24	9.6
304	320	12	–
304	365	6	12.6
347	755	30	10.2
321	944	–	–
316	365	2.7	5.7
316	730	3.9	8.25
316	1255	3	–
316(2.4MO)	2773	8	0.65
316(3.18MO)	3164	3	0.25
307	1075	5	–
305	198	19	–
308	755	28	9.5
309	320	11	–
310	320	4	–
325	106	60	–
329	106	7	–

In Black Sea water, the austenitic AISI type 347 stainless steel and an experimental alloy (18Cr, 6Ni, 1.2Cu, 0.82Nb, 0.30N) performed better (corrosion rate 2.3 to 2.5 μm y^{-1}), the austenitic 18Cr-8Ni steel undergoing a loss of 34 to 40 μm y^{-1} and pitting [106].

A summary of published corrosion data of austenitic stainless steels is given in Table 12. Total protection from uniform corrosion in seawater for austenitic stainless steels can be achieved by cathodic protection at – 0.8 V (Ag/AgCl electrode); in such cases crevice corrosion is suppressed when stainless steels are in contact with carbon steel [107]. Although protection by contact with rolled steel is practicable, the use of strongly anodic materials (Zn, Al) in seawater is not recommended. In quiet seawater, crevice corrosion of AISI type 304 and 316 stainless steels is completely inhibited by steel and Al anodes. The same is true for stainless steel 2Cr9Mn6Ni [108].

The AISI types 316, 316 L, CF8M (CrNiMo 19.5 10.5 2.5), and CF3M (CrNiMo 19 11 2.5) stainless steels are often used for marine and marine related services. Cleanliness and weld quality are as critical to successful performance as increase or decrease chloride ion concentration. AISI type 316 or AISI type 316 L stainless steel tubes are unsuitable for condenser tube service in seawater due to pitting and crevice corrosion problem. They were found resistant to impingement attack under the

Table 12 Corrosion rates of some austenitic stainless steels under various sea water conditions

Type of Steel	Condition		Corrosion Rate, $\mu m\ y^{-1}$	Literature
304	316	K; $1.85ms^{-1}$; 1.45Ta; aerated seawater	< 2.5	[105]
	300	K; pH 8.2; seawater, 8.5 ppm O_2(Crevice)	75	
	353	K; pH 8.2; conc.sea water 59.3g L^{-1}, 7–8 ppm O_2	2.5	
	355	K; deaerated seawater (0.02 ppm O_2) 1.5 m s^{-1}	15	
	363.5	K; deaerated seawater(0.02 ppm O_2) 1.5 ms^{-1}	10	
	372	K; deaerated seawater (0.02 ppm O_2) 1.5 ms^{-1}	8	
	386	K; deaerated seawater (0.02 ppm O_2) 1.5 ms^{-1}	13	
	433	K; desalination plant; pH 7; 1.5 ms^{-1}; deaerated	2.5	
	355	K; desalination plant; pH 7; 1.5 ms^{-1}; trace O_2	225	
	416	K; desalination plant; pH 6.2; 1.5 ms^{-1}; 15 ppm O_2	300	
316	seawater, high velocity (43 ms^{-1})			
	316	K; 1.85 ms^{-1}; 1.45Ta; aerated seawater;		
	355	K; deaerated seawater (0.02 ms^{-1}) pH 7; 1.5 ms^{-1}	<2.5	
	363.5	K; deaerated seawater (0.02 ms^{-1}) pH 7; 1.5 ms^{-1}	5	
	372	K; deaerated seawater (0.02 ms^{-1}) pH 7; 1.5 ms $^{-1}$	5	
	386	K; deaerated seawater (0.02 ms $^{-1}$) pH 7; 1.5 ms^{-1}	3	
	433	K; desalination plant; pH 7; 1.5 ms^{-1}; deaerated	5	
	355	K; desalination plant; pH 6.7; 1.5 ms^{-1}; deaerated	18	
	416	K; desalination plant; pH 6.2; 1.5 ms $^{-1}$; deaerated	325	
302	313/289 K, Mediterranean water		0.6	[105]
	313/289 K, Dead Sea water			
302	Panama Canal zone, linear corr. rate after 4 years		12.5	[103]
316	Panama Canal zone, linear corr. rate after 4 years		11.0	
347	Black Sea		2.3–2.5	
304	seawater, E coast USA (Kure Beach) RT, ± motion		2.5 – 5.0	
316	seawater, E coast USA (Kure Beach) RT, ± motion		2.5 – 5.0	
304	Curapio. West Indies; 300–305 K; 20g L^{-1}; PH 6.6		2.5	
316	Curapio. West Indies; 300–305 K; 20g L^{-1}; PH 6.6		<2.5	
304	seawater evaporator, I stage; pH 8.1; 360.5 K; 6.07% salts		3.3	
316	seawater evaporator, I stage; pH 8.1; 360.5 K; 6.07% salts		2.8	
304	seawater evaporator, 2 stage; pH 8.1; 352.5 K; 6.07% salts		4.5	
316	Seawater evaporator, 2 stage; pH 8.1; 352.5 K; 6.07% salt		7.8	
304	Seawater evaporator, 3 stage; pH 8.1; 352.5 k; 6.07% salts		2.3	
316	seawater evaporator, 3 stage; pH 8.1; 352.5 K; 6.07% salts		5.3	[106]
316	salt solution concentration (14 NaCl; 3.4 $MgCl_2$) pH 5.4–6.4; 339 K, without aeration or motion		2.5 – 5.0	
304	hot seawater; moving; 450 K; 30 days		2600	
316	hot seawater; moving; 450 K; 30 days		1.3	
347	hot seawater; moving; 450 K; 30 days		2.7	

condition of the jet impingement using flowing seawater and were superior to conventional non-ferrous condenser tubes in this respect [109].

Figure 32 enables comparison to be made of typical values of the rates of corrosion of materials most widely used as condensers. It should be noted, however, that in the case of Ni, CuNi and stainless steels, pitting corrosion must also be taken into consideration [110]. Increasing temperature is accompanied

□	Stainless steel type 316, stainless steel Type 304, NiCr Alloys
	NiCu 30 Fe
	Ni
	CuNi Alloy 70 30 with 0.5% Fe
	CuNi Alloy 90 10 with 1.5% Fe
	Cu
	Admiralty brass
	Al brass
	Sn bronze, NiAl bronze, N(A)Mn bronze
	Mn bronze
	Dezinofication possible
	Austenitic cast iron
	C steel

0   1   2.5   5   10   25   50   100   250

Typical average corrosion rate, $\mu m\ y^{-1}$

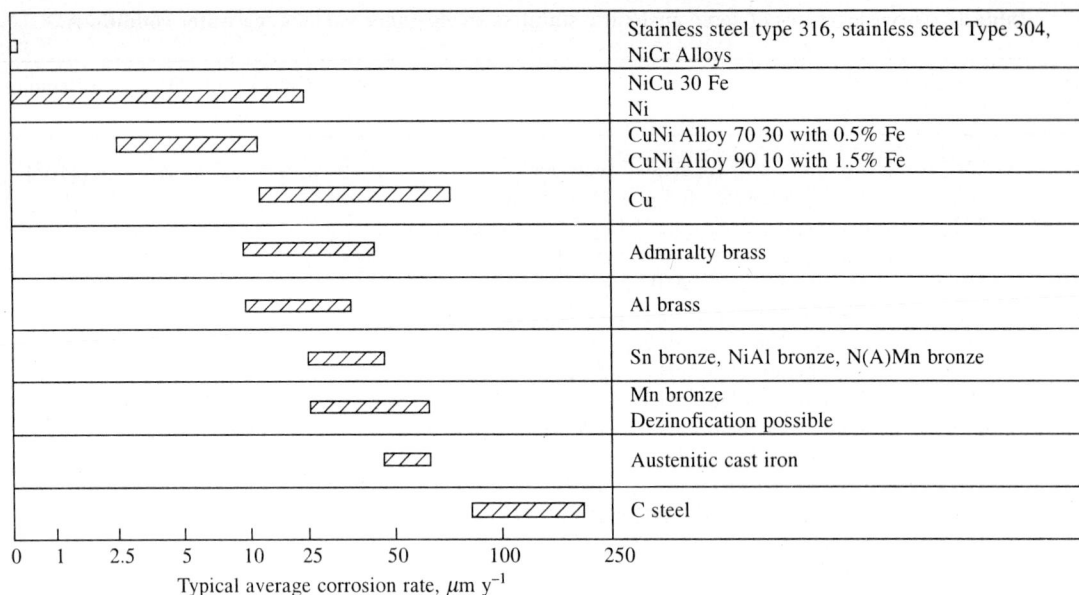

Fig. 32. Comparative range of corrosion rates of austenitic steels and other alloys in seawater [110].

by accelerated attack. In open sea, the corrosion rate of steel is about 50% higher in summer (300 to 305 K) than in winter (277.5 to 280 K). Similarly, in tropical regions, the rate of corrosion is higher than in polar areas.

SUMMARY

The uniform corrosion behavior of various austenitic stainless steels in different corrosive environments of industrial importance like in sulfuric acid, nitric acid, phosphoric acid, hydrochloric acid, seawater and alkalies are discussed. The metallurgical factors and the use of chemical inhibitors that improve the uniform corrosion of austenitic stainless steels in different corrosive media are also mentioned. Although extensive data are available on the corrosion rate of austenitic stainless steels, more information on the uniform corrosion behavior on long-term exposure need be generated. The influence of physical parameters such as temperature, flow rates and the performance of the newer austenitic alloys, and their composition effects, on the corrosion rate are still under investigation.

REFERENCES

1. Fontana, M.G., *Corrosion Engineering,* 3rd Ed, McGraw-Hill, New York, 1987.
2. Nelson, G.A., *Corrosion Data Survey,* 1960. (updated and expanded version published by NACE, Houston, TX in 1974).
3. Polar, J.P., *A Guide to Corrosion Resistance,* Climax Molybdenum, New York, 1961.
4. Uhlig, H.H., *Corrosion Handbook,* Wiley, New York, 1984.
5. Rabald, E., *Corrosion Guide,* Elseiver, New York, 1951.
6. Wagner, C and Traud, W., *Zeit fuer Electrochem,* Vol. 44, 1938, p. 391.
7. Sedriks, A.J., *Corrosion of Stainless Steels.,* John Wiley & Sons, Inc., 1996.

8. Morris, P.E. and R.C. Scarberry, *Corrosion,* Vol. 28, 1972, p, 444.
9. Evans, S and Koehlet, E.L., *J. Elecrochem. Soc.,* Vol. 108, p. 509, 1961.
10. Stern, I.M and Geary, A.L., *J. Elecrochem. Soc,* Vol. 104, 1956, p. 56
11. Stern, I.M and Weisert, E.D., *Proceedings of ASTM,* Vol. 59, 1959, p. 1280.
12. Leroy, R.L., *Corrosion,* Vol. 29, 1973, P. 272.
13. Hladky, K, Callow, L.M. and Dawson, J.L., *British Corrosion Journal,* Vol. 15, 1980, P. 20.
14. Randles, J.E.B., *Discussions of Faraday Society,* Vol. 1, 1947, p. 11.
15. Liening, E.L., *Electrochemical Corosion Testing Techniques in Process Industries Corrosion.,* B.J. Moniz and W.I. Pollock, Ed., NACE, 1986.
16. Cottis, R.A., *Corrosion,* Vol. 57, No. 3, 2001, p. 265.
17. Dean, S.W., *In-Service Monitoring,* ASM Handbook, Vol. 13, originally published as Metal Handbook, Vol. 13, 9th ed), ASM International, p. 197, 1987.
18. Truman, J.E., in Corrosion, Metal/Environment Reactions, Shreir, L.L., eds, Vol. 1, Butterworth-Heinemann Ltd, 1994.
19. Locke, C.E., *Metal Handbook,* 9th ed., Vol. 16. No. 11, 1977, p. 16.
20. Szmanski, W.A., *Mater Performance,* Vol. 16, No. 11, 1977.
21. See Ref [18], p. 3.52.
22. Schillmoler, C.M, *"Selection and Performance of Sainless Steels and Other Nickel-Bearing Alloys in Sulfuric Acid"* NiDi Technical Series N. 10057, Nickel Development Institute, Toronto, 1990.
23. Yau, Y.H., and Streicher, M.A., *Corrosion,* Vol, 47, No. 5, 1991, P. 352.
24. Sedriks, A., *Corrosion,* Vol. 21, No. 7, 1986, p.376.
25. Scharfstein, L.R., *Corrosion,* Vol, 21, No. 8, 1965, p. 254.
26. Renner, M., *Werst Korros,* Vol. 38, 1987, p. 191.
27. Kuron, D., Paulekat, F., Gräfen, H. and Horn, E. M., *Werst Korros,* Vol. 36, No. 11, 1985, p. 489.
28. Brubaker, S.K., *Metal Handbook,* 9th Ed, Vol. 13, ASM International, 1987, p.1151.
29. Omashov, N.D., *Chem. Tech.,* Vol. 30, No. 1, 1987, p. 6.
30. Anonymous. *Chem Eng,* Vol. 67, No. 17, 1960, p. 58.
31. Polar, J.P., *"A guide to Corrosion Resistance",* Climax Molybdenum Co., New York, 1961, p. 121.
32. Schmid, G., Maurer, E. and Steinhausen, H., *Metalloberfläche,* Vol. 10, 1956, p. 289.
33. Scarberry, R.C., Graver, D.L., and Stephen, C.D., *Materials Protection,* Vol. 6, No. 6, 1967, p. 54.
34. Ertal, H. and Horner, H.L., *Werst Korros,* Vol. 23, 1972, p. 9.
35. General Aniline and Film Corp., New York., *French Patent* 139 9173 (14.5. 1965).
36. Mc William, J.A., *Materials of Construction in the Chemical Industry,* Conference at Birmingham, Soc. of Chemical Industry, London, 1950.
37. Batravok, W.P., *Corrosion of Metallic Materials,* VEB-Verlag Technik, Berlin, 1954.
38. Rajagopalan, K.S., Subramanyan, N. and Sundaram, M., *Corrosion Inhibitor,* Ind. Pat. 107 414, Council of Scientific and Industrial Research, (29.3. 1969).
39. Anonymous., *Aluminium Pocket Edition,* 12th Ed., Aluminium-Verlag GmbH, D-4000 Düsseldorf, 1963, p. 167.
40. Galvalle, J.R., de Waxler, S.B. and Gardiazabalal. S.B., *Corrosion,* Vol. 31, No. 10, 1975, p. 352.
41. Nielsen, N.A., *Werk Korros,* Vol. 10, 1959, p. 429.
42. *Product Information,* Cronifer®, Crofer®, 1971, p., 28. (VDM Vereinigte Deutsche Metallwerke AG, D-5990, Altena Germany).
43. Gurry, R.W., *Ind Eng Chem Prod Res Develop,* 10, 1971, p. 112.
44. Carassiti, V., Trabenelli, G. and Zucchi, F. *Ann Univ Ferrara,* Spez. 5, Suppl. 4, 1966, p. 417.
45. Anonymous. "Evaluation of Inhibitors used to prevent Corrosion of Metal by Acid Cleaning Method, *NACE Technical Group* T-8., *Mater Protection,* Vol. 1, No. 5, 1962, p. 107.
46. Ertel, H., *Chem Ing Tech,* Vol. 38, 1966, p. 51.
47. Mott, N.S., *Chem Eng Progr,* Vol. 47, No. 11, 1951, p. 592.
48. Product Information, *"Chemical Resistance of the Remanit® – steels"* No. 1124 736/73, p. 22, (Deutsche Edelstahlwerke AG, D-4150, Krefeld, Germany).

49. Friend, W.Z. and LaQue, I.L. *Ind Eng Chem,* Vol. 44, No. 5, 1952, p. 965.
50. Golden, L.B., Lane, I.R. and Achermann, W.L. *Ind Eng Chem,* Vol. 44, 1952, p. 1930.
51. Product Information, "*Sandvik*® rust, acid and heat resistant pipes" No. TY-300 V, 1967, p. 27. (Sandvik Steel, Sandviken, Sweden).
52. Mott, N.S., *Chem Eng Progr,* Vol. 50, 1954, p. 324.
53. Product Information, "*Chart of resistance*", Harzer Apparatewrke AG, D-3205, Bornum Hartz, Germany).
54. Gindin, L.G., Dmitrenko, V.E., Zherdeva, T.I., Smirenkina, I.P. and Fedorov, V.V. *Russian J* Phys Chem, Vol. 45, 1971, p. 240.
55. Kügler, A. "The selection of high-grade steels in aggressive environment", *VDIZ* Vol. 119, 1977, p. 411.
56. Donat, H. and Schäfer, K. Schweissen Schneiden, Vol. 27, 1975, p. 343.
57. Herbsleb, G., Jäkel, U. and Schwaab, P. *Werskst Korros,* Vol. 41, No. 4, 1990, p. 343.
58. Denhard Jr, E.E. "*Austenitic stainless steel combining strength and resistance to intergranular corrosion*", US Pat 3 645 725 (Feb. 1972).
59. Kobayashi, M., Fujiyama, S., Araya. Y., Wada, S. and Sunayama, Y. *Nippon Sutenreso Giho,* No. 12, 1076, p. 1.
60. Kabayashi, M., Miki, M. and Ohkubo. K, *Bull Jpn Inst Met,* Vol. 22, 1983, p. 320.
61. Slate, S.C. and Maness, R.F. *Mater Performance,* Vol. 17, No. 6, 1978, p. 13.
62. Hahin, C., Stoss, R.M., Nelson, B.H and Reucroft, P.J. *Corrosion,* Vol. 32, No. 6, 1976, p. 229.
63. Hahin, C. *Corrosion,* Vol. 38, No. 2, 1982, p. 116.
64. Chung, P. and Szklarska-Smialowska, S. *Corrosion,* Vol. 37, No. 1, 1981, p. 39.
65. Brown, M.H, *Corrosion,* Vol. 30, No. 1, 1974, p. 1.
66. Robinson, F.P.A. and Scurr, W.G. *Corrosion,* Vol. 33, No. 11, 1977, p. 408.
67. Briant, C.L. *Corrosion,* Vol. 36, No. 9, 1980, p. 497.
68. Kamachi Mudali, U., Dayal, R.K. and Gananamoorthy, J.B. *J. Nucl Mat,* Vol. 203, 1993 p. 73).
69. Holtser, M., *Werst Korros,* Vol. 41, No. 1, 1990, p. 25.
70. Casarini, G., Colonna, C. and Somga, T. *Metall Ital,* Vol. 62, 1970, p. 183.
71. Schillmoller, C.M. "Alloy Selection in Wet-process Phosphoric Acid Plants," *NiDi Technical Series No. 10015,* Nickel Development Institute, Toronto, 1988.
72. McDowell, D.W., *Chem Eng,* Vol. 82, No. 18, 1975, p. 121.
73. Product Information. "*Corrosion resistance of stainless steels and nickel alloys in phosphoric acid*", (Vereinigte Deutsche Metallwerke Ag, Duisburg, Germany, 1988).
74. Anonymous, *Corrosion Data Survey,* Metal Section, 6th Ed., NACE Houston (Taxas/USA), 1985, p. 95.
75. Anonymous, *Werstoffe und Ihre Veredlung,* Vol. 1, 1979, p. 24.
76. Guenbour, A., Faucheu, J., Ben Bachir, A., Dabosi, F. and Bui, N. *Br Corros J,* Vol. 23, No. 4, 1988, 234.
77. Alon, A., Yahalom, J. and Schnorr, M., *Corrosion,* Vol. 31, No. 5, 1975, p. 325.
78. Product Information., "*ABC of Steels Corrosion*" 2nd Eds., 1966, p. 6 (Mannesmann AG, Düsseldorf, Germany).
79. Rabald, E., *Corrosion Guide,* 2nd Eds., Elseiver, New York, 1968, p. 556.
80. Berg, F.F., "*Corrosion Diagram*" 2nd Eds., 1969, p. 2, (VDI-Verlag GmbH, Düsseldorf, Germany).
81. Linder, B., "Anodic Protection of Stainless Steels in Phosphoric Acid containing halide ions", *The Institute of Corrosion Science and Tech,* Vol. 1, 1984, p. 73 (Birmingham, UK).
82. Süry, P. *Material Und Technik,* Vol. 8 No. 4, 1980, p. 163.
83. Product Information, "Corrosion Resistance Material Against Phosphoric Acid and Phosphates" 1st Eds., No. 61, 1970, p. 1, (International Nickel Deutschland GmbH, Düsseldorf,Germany).
84. Product Information. "*Corrosion Resistance of Nickel Containing Alloys in Phosphoric Acid*" P. 1, The International Nickel Company, Inc., New York, USA.
85. Heurling, K., *Blech-Rohre-Profile,* Vol. 28, No. 8, 1981, p. 481.
86. Linder, B., *Industrial Corrosion,* Vol. 5, No. 3, 1987, p. 12.
87. Schillmoller, C.M. "Alloy Selection for Caustic Soda Service" NiDi Technical Series No. 10019, Nickel Development Institute, Toronto, 1988.
88. Lew, S. *Anti-Corrosion,* Vol. 16, No. 7, 1969, p. 17.
89. Nakanishi, K. and Ohtsuka, H. *J Metal Finish Soc Japan,* Vol. 34, 1983, p. 171.

90. Lazebnov, P.P., Savonov, Yu, N. and Alkeksandrov, A.G. *Avtom* Suarka USSR, No. 8, 1981, p. 69.
91. Babkina, V. Yu., Chuba, E.G., Vasilieva, I.K., Gapunina, O.V. and Kogan, E.I., *Khim Neft Mashinostr,* No. 2, 1978, p. 22.
92. Wankyl, J.N. and Jones, D. *J. Nucl Mater,* Vol. 2, 1959, p. 154.
93. Murukami, K. and Murukami, Y. Japan Patent: 77 56, 030 (cl.c23f7/04), 09-5-1977.
94. Nordon, R.B. and Hughes, R.V. *"Chem Eng Materials of Construction"*, Report No: 218, 1962, p. 199.
95. Perelman, L.A., Strinin, A.F. and Gendelman, A.B. *Kim Prom,* Vol. 47, No. 9, 1971, p. 688.
96. Hong, D.J., Chung-ju. And Johnson, E.R. *US Patent:* 3 488 293V (6.1.1970).
97. Van der Horst, J.M.A., *Z Werstofftechnik,* Vol. 7, No. 11, 1976,
98. Tuthill, A.H. *Mater Performance,* Vol. 27, No. 7, 1988, p. 47.
99. Baker, D.W.C., Heaton, W.E. and Patient, B.C. *Corros Sci,* Vol. 12, No. 3, 1972, p. 247.
100. Thomas, E.D., Lucas, K.E., Peterson, M.H. and Christian, D.K. *Matter Performance,* Vol. 27, No. 7, 1988, p. 36.
101. Product Information: Bulletin No. 34, 1967 (International Nickel Deutschland GmbH, D-4000, Düsseldorf, Germany).
102. May, T.P., Holmberg, E.G. and Hinde, J. *Dechema Monographic,* Vol. 47, 1962, p. 253.
103. Alexandar, A.L., Southwell, C.R. and Forgeson, B.W, Corrosion, Vol. 17, No. 7, 1961, p. 345t.
104. Binder, N.O, "Corrosion of Metals", *Am. Soc. Metals,* Cleveland., 1964 (USA).
105. Swales, G.W., *S.C.I Monograph* No. 10, p. 203.
106. White, J.H., Yaniv, A.E. and Scick, H, *Corros* Sci, Vol. 6, 1966, p. 447.
107. Huyghe, G.E, *Ingeniur,* Vol. 68, 1956, p. 37.
108. May, T.P. and Weldon, B.A *"Copper Nickel Alloys for Service in Seawater"* Bulletin No: 2981, 1964. (International Nickel Deustschland GmbH, D-4000, Düsseldorf, Germany).
109. Anonymous, *"Material Seawater Plants"* 1st Eds., No: 54, 1968 (Internatinal Nickel Deutschland GmbH, D-4000, Düsseldorf, Germany).
110. Janssen, K. and Moser, R., *Werst Korros,* Vol. 15, No. 10, 1964, p. 804.

3. Pitting Corrosion of Austenitic Stainless Steels and Their Weldments

U. Kamachi Mudali and M.G. Pujar[1]

Abstract Pitting corrosion is a major problem associated with the application of austenitic stainless steels in industries. The naturally formed protective passive film on stainless steels is damaged by halide ions leading to the formation of pits. Pits provide sites for cracks to initiate and propagate, and thus reduce the useful life of engineering components in service. Passivity and its significance, influence of alloy composition, microstructure, cold working, grain size etc., and the parameters of environments are discussed in detail with respect to pitting corrosion. Influence of welding and associated microstructural changes on pitting corrosion are also discussed. Various morphologies of pits are highlighted.

Key Words Pitting corrosion, passivity, pitting mechanism, alloy composition, cold working, grain size, welding, microstructural changes, heat input, morphology of pits.

1. GENERAL INTRODUCTION

Materials chosen for engineering applications are expected to remain with the designed strength and corrosion resistance throughout their life. However, being thermodynamically unstable, materials react with the environment and depending on the corrosivity of the environment they either corrode severely or corrode at a very low rate with the formation of a passive film [1]. The destruction of this passive film either uniformly or at a localized spot leads to further corrosion and ultimately the materials fail to perform with the expected strength and corrosion resistance. Hence, the probability of breakdown of the passive film leading to "localized corrosion", namely, pitting, crevice corrosion, stress corrosion cracking, intergranular corrosion and corrosion fatigue, has a significant role to play in the service life of engineering components.

Pitting corrosion refers to the formation of microscopic holes/cavities on the surface of metals/ alloys either due to direct corrosion of heterogeneities present on the surface or due to the localised damage caused to a protective passive film present on the surface. It is well known that pitting corrosion occurs on passivated surfaces that are protected by a thin self-healing, tenacious, stable oxide layer. The thickness of this layer ranges from a monolayer to a few tens of angstroms. Under the influence of aggressive anions such as halides, localised attack takes place on the passivated surface

[1]Scientific Officers, Corrosion Science and Technology Division, Indira Gandhi Centre for Atomic Research, Kalpakkam-603 102, India.

causing the formation of hemispherical and other shapes of polygonal holes called pits. These pits could grow to diameters of 1 mm and above, and often they are filled with solid reaction products. Pitting corrosion causes severe reduction in the service life of many engineering components either by directly damaging the structure or by providing sites at the bottom of the pit for the stress corrosion or fatigue cracks to initiate and propagate. It was estimated that pitting corrosion caused 11% of the failure of components in chemical industries [2]. This indicates the importance of pitting corrosion and the necessity to understand the pitting mechanisms.

Austenitic stainless steels form a thin, protective passive film consisting of mixed oxides of chromium and iron which provides excellent corrosion resistance in a variety of chemical environments. This passive film is self-healing in nature, hence whenever a damage is caused to it immediately the surface is covered with a new layer of the film. However, under certain environmental conditions, particularly in the presence of halide ions, this films is damaged at weak sites either by adsorption and penetration or by penetration and migration of aggressive halide ions. Such weak sites are locations where inclusions, second phase precipitates, grain boundaries, slip steps, segregated interfaces etc. are present on the surface. The aggressive anions react with the metal atoms at the film/substrate interface to form metal chlorides, which hydrolyse to give metal hydroxide, but at the cost of increasing the acidity at the reaction site. The decrease in pH at such sites enhances the metal dissolution, and electrons released during dissolution are consumed by the cathodic reaction occurring on passive film present farther from the pit sites. Thus, the events like ingress of aggressive halide ions into the passive film, metal dissolution at passive film/substrate interface, hydroxide formation producing repassivation, acidity increase at pit site and further metal dissolution, and cathodic reduction at passive film surface outside the pit site happen sequentially leading to an autocatalytic situation enhancing the growth of pits. Figure 1 shows the schematic of the events happening during pitting corrosion.

2. PASSIVITY AND ITS SIGNIFICANCE

2.1 What is Passivity?

According to the American Society for Testing and Materials (ASTM G 15–83) [3], passivity is defined as the state of a metal surface characterized by low corrosion rates in a potential region that is strongly oxidising for the metal. Passivity is classified into two types. Type (1) a metal active in the emf (electromotive force) series is passive when its electrochemical behaviour becomes that of a metal noble in the emf series with noble potential and low corrosion rate (examples are Ni, Cr, Ti, Fe and their respective alloys), and Type (2) a metal is passive while still at an active potential and exhibits a low corrosion rate (examples are lead in sulphuric acid and iron in inhibited pickling solution). Type (1) passivity can be explained on the basis of a polarization curve as shown in Fig. 2. In general, a metal becomes passive when the potential is increased in the positive or anodic direction after reaching a potential where the current (rate of anodic dissolution) sharply decreases to a value less than that observed at a less anodic potential called *Flade potential*. The reduction in the rate of dissolution is attributed to the formation of the passive oxide film on the surface at that potential. After the passive region where the anodic current is low, a sharp increase in the current is observed at nobler (higher) potentials leading to the transpassive dissolution of the oxide film, or to the pitting attack.

The phenomenon of passivity is exhibited by most of the metals and their alloys, with the possible exception of gold and platinum. Schonbein coined the word "passivity" in the year 1836, and the first

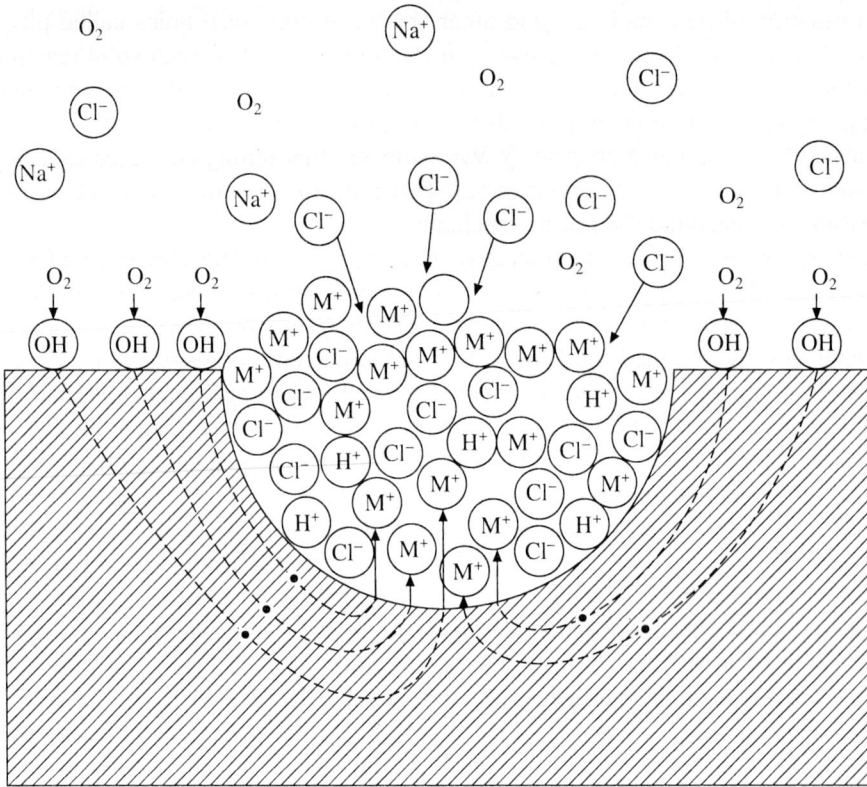

Fig. 1. Schematic of the various events occurring during the growth of pits [2].

metal found to exhibit the phenomenon of passivity was iron [1]. Michael Faraday referred to the passivity of iron as "*this very beautiful and important case of voltaic condition presented to us by the metal iron*" [1]. According to Uhlig [1], three scientists of the eighteenth century, Lomonosov (USSR, 1738), Wenzel (Germany, 1782) and Keir (England, 1790) had observed that a highly reactive surface of iron became surprisingly nonreactive after immersion in concentrated nitric acid. Today, a number of theories and models are available to explain the concept of passivity of different metals and alloys.

2.2   Breakdown of Passivity

Breakdown of passivity of various metals and their alloys occurs in the presence of aggressive ions, particularly halide ions. Chloride ions are reported

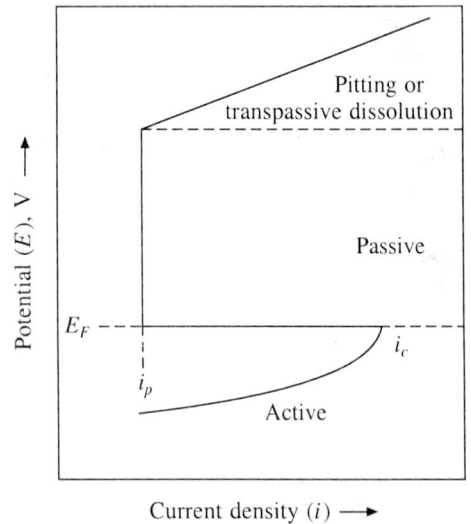

Fig. 2.   Polarisation curve showing various zones of corrosion events including passivity [2].

to cause severe damage to the passive film, and their role has been extensively examined. Depending on the electrode potential, environment and the inhomogeneities at the metal surface, the breakdown of the film results in either general corrosion or localized corrosion. Chemical as well as mechanical breakdown of the film also occurs depending on the environmental parameters and the nature of the film formed. Many theories and models are available to explain breakdown of the passive film [4-8]. In considering the initiation of pitting attack or crevice corrosion subsequent to the breakdown of the passive film, the following points are important for consideration: (1) presence of aggressive ions, (2) exceeding of a critical potential, (3) induction time for the breakdown to occur and (4) presence of susceptible sites at the metal surface. Three main mechanisms are discussed in the literature [6] for the breakdown of the films, namely, penetration mechanism, adsorption mechanism and film-breaking mechanism, and are schematically shown in Fig. 3.

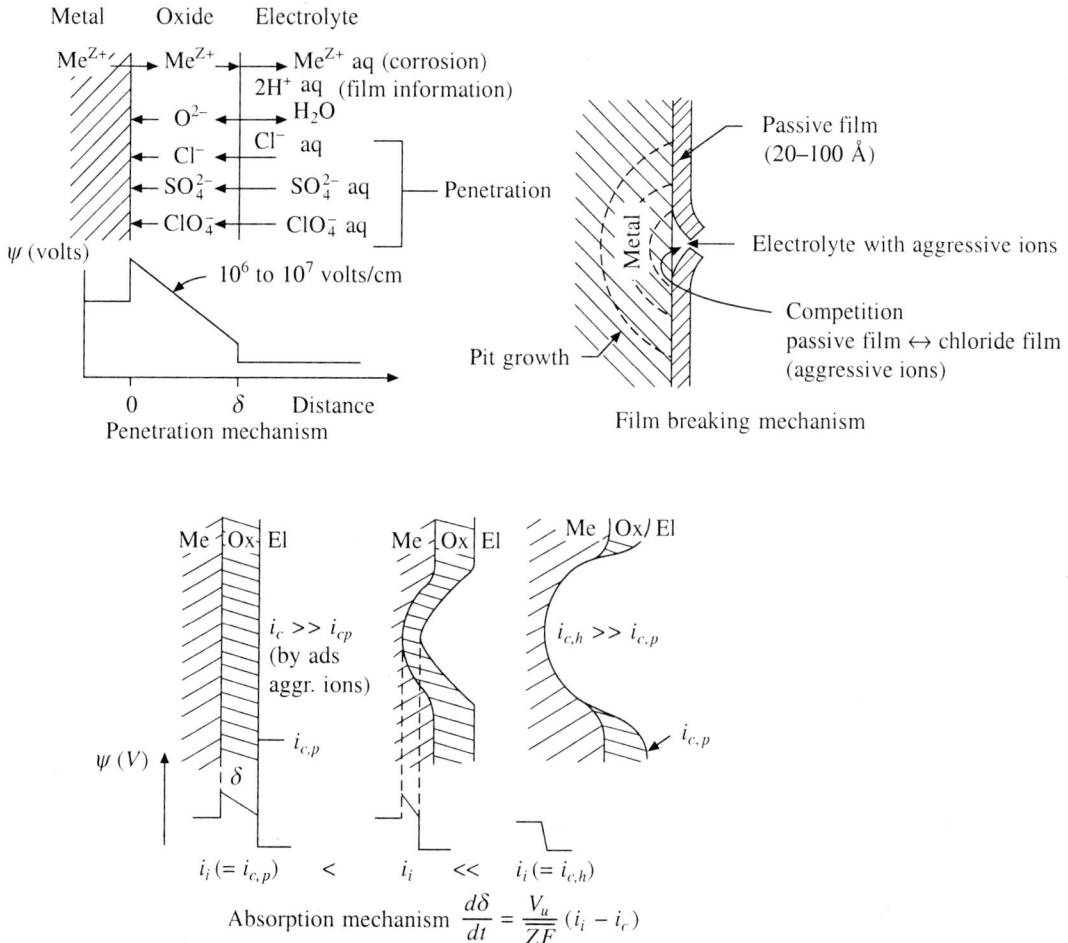

Fig. 3.   Schematic representation of various mechanisms of pitting attack [6].

In the adsorption mechanism [8], the passive film is considered to be an adsorbed film on the surface of the metal, probably made of a monolayer of oxygen. When a strongly adsorbing aggressive anion is added, it displaces oxygen from the passive film. Once the aggressive anion is adsorbed on the surface, the breakdown process is initiated because the bonding of the metal ions to the metal lattice is weakened. When selective adsorption of the inhibiting ions takes place compared to the aggressive anions, the breakdown is either stopped or slowed down. Also, adsorption of the aggressive anions at the surface leads to the formation of a surface complex, which dissolves, into the electrolyte. This leads to the thinning down of the oxide layer at a localized spot, and at those spots a higher electric field strength ($> 10^6$ V/cm) is produced. Hence more number of metal ions migrate within the film and move to the film/electrolyte interface. This leads to a situation of catalytically increased dissolution of metal ions by forming complexes with the aggressive anions.

In the penetration mechanism, the aggressive anions migrate through the film under high electric field strength and the breakdown is completed once they reach the metal/film interface. This requires an induction time for the anions to migrate and initiate the breakdown process. The anions can also penetrate through a lattice, via defects or through some ion exchange process. The exchange of aggressive anions for the O^{2-} or OH^- ions can create anion vacancies that will further enhance the migration of the aggressive anions to the film/metal interface. The breakdown has also been attributed to the pile-up of cation vacancies or metal holes at the metal/film interface during the passive film growth process [9]. The accumulation of the vacancies results in the formation of microvoids at the metal/film interface. The growth of the microvoids to a critical size requires an incubation period, and thereafter the passive film collapses at these localized sites leading to breakdown. As a corroboration of this mechanism, an experimental observation of semilograthmic dependence of pitting potential to chloride ion concentration was obtained.

In the film-breaking mechanism, the mechanical breakdown of the film due to various types of stresses present in the film leads to the exposure of the bare metal surface to the electrolyte. The direct access of the aggressive anions to the bare metal surface causes severe localized corrosion. Stress in the passive oxide film can arise for several reasons: (a) interfacial tension of the film (b) electrostriction pressure resulting from the presence of a high electric field ($> 10^6$ V/cm) across the film, (c) internal stress caused by the volume ratio of the film and the metal, being compressive if anions are more mobile than cations in the film, (d) internal stress due to partial hydration or dehydration of the film, and (e) local stress caused by impurities such as inclusions, or by flaws, pores and microcracks. Thus, stress in the film is dependent on many factors such as metal surface condition, impurity, film-formation condition and film-growth mechanism. Sato [10] attributed the breakdown of the film to the presence of high electric field strength and surface tension. The specific adsorption of anions leads to the lowering of the surface tension and the subsequent reduction in the critical thickness of the film of breakdown. This was experimentally supported by studying the breakdown of the film on stainless steels in a solution containing various concentrations of halide ions.

In general, the breakdown of the passive film results in either pitting or transpassive dissolution. Pitting occurs in the presence of aggressive anions through various stages, viz. (1) processes leading to the breakdown of passivity, (2) early stages of pit growth, and (3) pit propagation and (4) repassivation of pits. Stage (1) refers to the onset of destruction of passive film where no visible pitting occurs: stage (2) refers to the appearance of pits and the resultant monotonic increase in the anodic current; stage (3) refers to the growth of pits and the formation of salt films at the pit site and stage (4) refers to the drop in the anodic current with the stabilization of the passive films at the pit sites by the addition of inhibitors or by suitably alloying of the metal. Many review articles [1!–13] on pitting corrosion are available, and the present paper is restricted to pitting corrosion of austenitic stainless steels and their weldments.

2.3   Theories of Pitting Attack

In general, most of the investigators recognise two stages of pitting attack namely, the nucleation of pits on the passivated metal surface and the growth of pits. A number of published papers available in the literature explain the theories and the concepts involved in pitting [14–21]. Several different mechanisms for the initiation and growth of pits have been proposed. It is agreed in all the mechanisms that pits propagate as a result of the development and maintenance of an aggressive local environment. As far as the nucleation of pits is concerned, several factors such as inhomogeneities in the metal, cracking and slow healing of the film, development of a critical acidity in the microscopic flaws, defect transport in the passive film, and chloride adsorption or incorporation into localised areas, are considered. When a pit has been initiated and its growth has reached a steady state, it has come to the propagation stage. Irrespective of the mode of pitting, certain characteristic conditions exist within the pit. The potential and pH conditions existing during the pitting of iron in a chloride solution are explained [15] in a potential-pH diagram shown in Fig. 4.

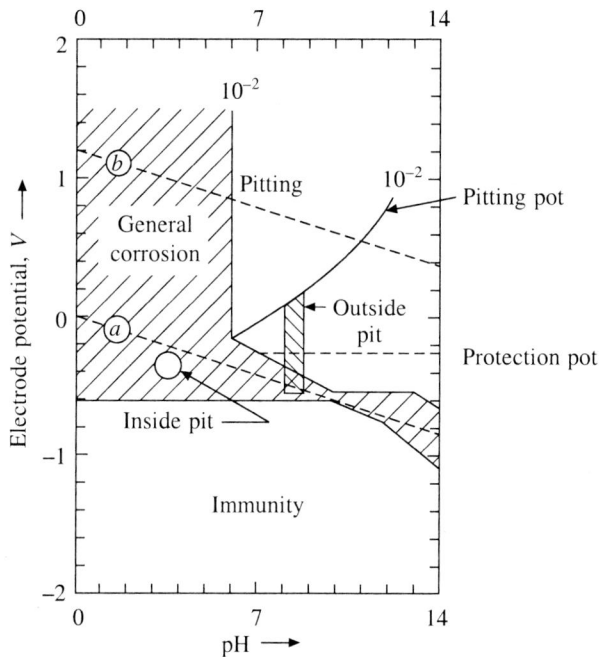

Fig. 4.   Potential and pH conditions existing during the pitting of iron in a chloride solution.

The diagram shows that with sufficient access for oxygen outside the pit the potential may rise above the pitting potential, so that pitting is initiated. The pit solution has a lower pH value due to the hydrolysis of the dissolved metal ions, and under normal conditions the pit solution might have a pH value of the order of 0–1. The pit solution also has a higher concentration of chloride ions due to the migration of chloride ions into the pit. In practice, the chloride ion content may rise to as high a value as 5 M. In the presence of a salt layer developed at very high potentials, pits formed are partly spherical and have bright and smooth surfaces. The pits formed at lower potentials just above the pitting potential have flat walls corresponding to crystallographic planes. Also, it has been reported [11, 22] that hydrogen evolution may take place during the propagation stage especially within the pit where the pH is low. The overall mechanism put forward in the available literature can be described as follows:

1. adsorption and/or penetration of chloride ions on or in to the passive film,
2. formation of chlorides replacing oxyhydroxides in the film, under favourable thermodynamic conditions,
3. breakdown of the film by electrostriction, if the film suffers a sufficient decrease in its surface tension due to the incorporation of chloride ions,
4. accumulation of cation vacancies and formation of pores in the film or at the metal/film interface, and
5. instability of the passive film due to perturbations of the potential distribution at the film/ electrolyte interface due to local acidification.

All the above mechanisms imply an accentuating effect on the total potential difference and the chloride ion concentration. Depending on the experimental conditions, the material and its film-forming tendency, one or more of the above factors will dominate the mechanism and will be the rate determining step in the initiation and growth of the pits.

3. INFLUENCE OF METALLURGICAL VARIABLES ON PITTING CORROSION OF AUSTENITIC STAINLESS STEELS AND THEIR WELDMENTS

3.1   Microstructure and Thermal Ageing

The sites which are prone to pitting corrosion are grain boundaries, inclusions, second phase precipitates, mechanical scratches, slip steps emerging at surfaces and other heterogeneities existing on the surface. Pits are primarily formed at grain boundaries (Fig. 5), probably due to precipitation of complex carbides, which deplete this region from chromium and other alloying elements. Kamachi Mudali et al [23] observed that the presence of sensitised microstructure at grain boundaries led to the initiation and growth of pits for nitrogen-bearing austenitic stainless steels. A detailed investigation on the role of sensitised microstructure indicated that pitting corrosion preceded intergranular corrosion along the grain boundaries, as it was very sensitive to the chromium-depleted zones. Time-temperature-sensitisation-pitting (TTSP) diagram developed for types 304LN and 316LN SS delineating the regions prone to intergranular corrosion and pitting corrosion are shown in Fig. 6. Szklarska-Smialowska and Janik-Czachor [24] observed in a 13 Cr-Fe alloy that pits are initiated at grain boundaries, and that during their growth, pits belonging to different grains are developed at different rates. In the case of commercial stainless steels a large number of metallurgical factors, namely, alloying elements in the solid solution, second phases like, sigma, chi, manganese sulphide, carbides etc. have a significant influence on passivity and pitting attack.

3.2   Inclusions

A number of papers reporting the initiation of localized corrosion at structural heterogeneities are available [25–28]. For example pitting is often associated with manganese sulphide inclusions present in commercial stainless steels. It was suggested [26] that owing to the high conductivity of sulphide inclusions compared to the surrounding oxide film, chloride ions are adsorbed on the surface of

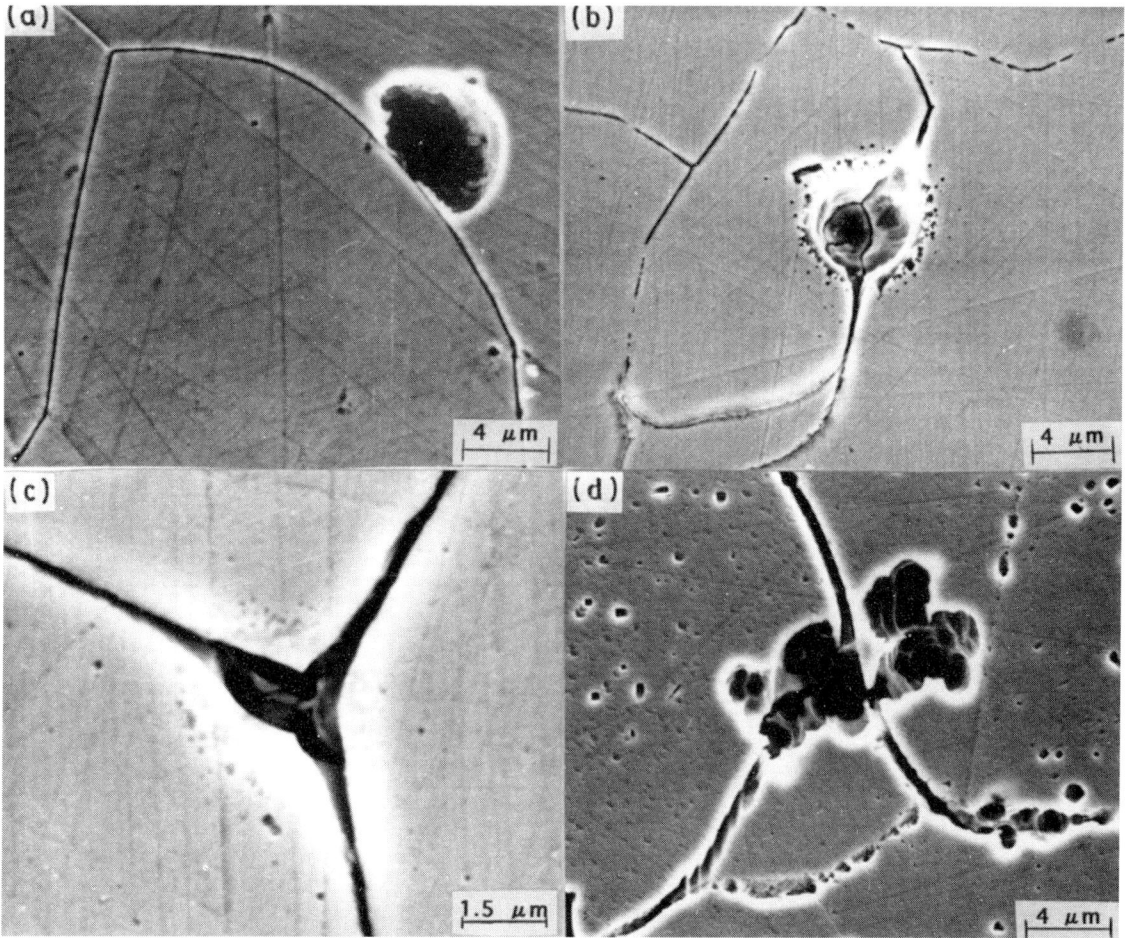

Fig. 5. Pitting attack at grain boundaries in austenitic stainless steels [23].

sulphide inclusions. Adsorbed chloride ions facilitate the anodic dissolution of sulphide inclusions. Also, the corrosion potential of a passive stainless steel surface in an aqueous chloride solution is generally between 0 and + 200 mV (SHE), and in this potential range the sulphide is thermodynamically unstable and will tend to dissolve as per the potential-pH diagram of $MnS-H_2O-Cl^-$ system [25]. According to Eklund [27], when the sulphide dissolves, a virgin metal surface is exposed to the environment. Depending on the dissolution and mass transfer conditions existing at such locations pitting attack will continue to grow. Three different stages are suggested while considering pitting attack at inclusions: (i) dissolution of active inclusion and initiation of microcavities, (ii) agglomeration of chloride ions at such microcavities (incubation stage) and (iii) initiation and growth of pits at such sites. The remedies to remove these deleterious inclusions include, reducing the bulk sulphur content, laser surface treatment for melting such inclusions [28, 29], reducing the Mn content during alloy making and adding rare earth metals (REMs) like cerium to reduce the probability of formation of MnS inclusions [26]. It was reported recently [28] that reducing the bulk sulphur content itself does not control pitting attack, rather it was the size and distribution of inclusions which significantly

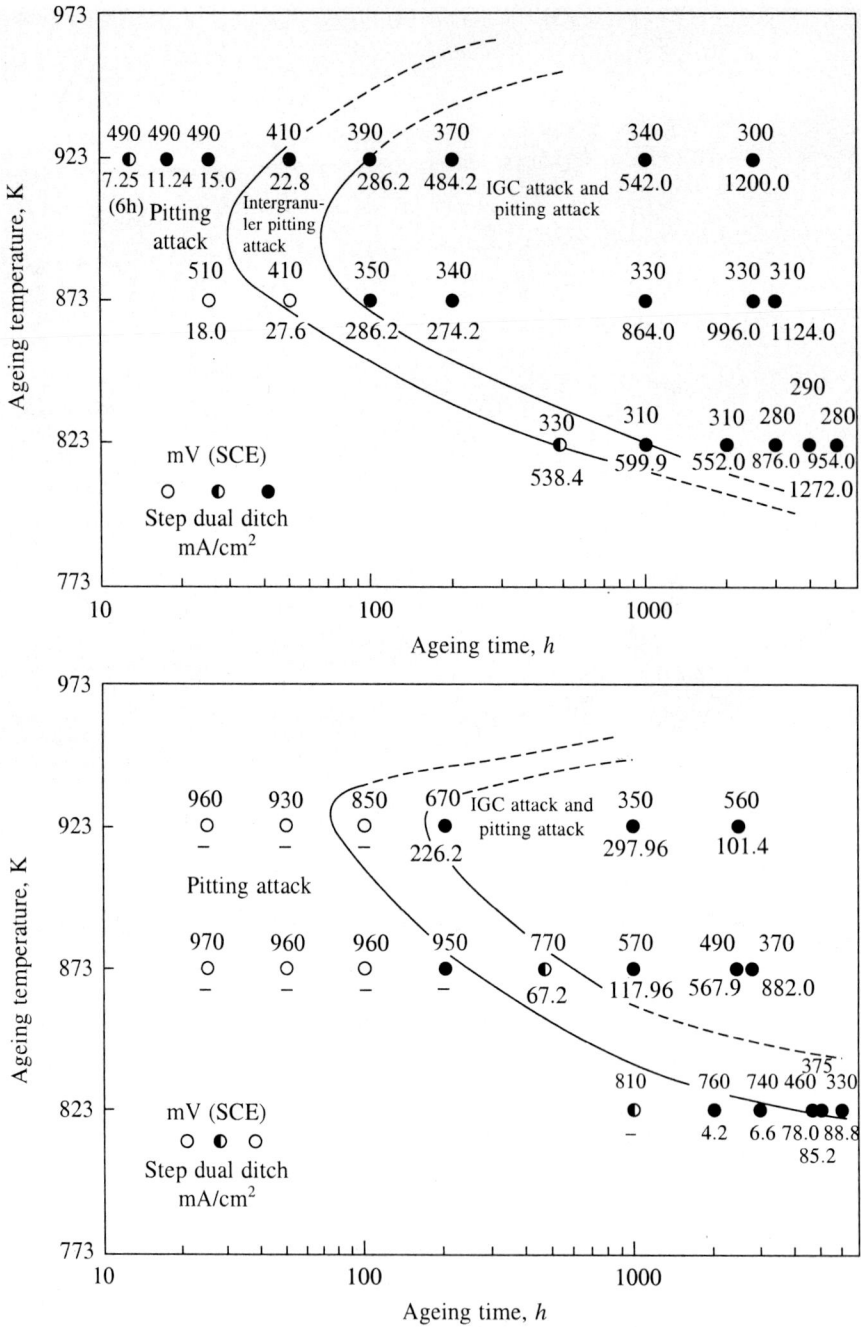

Fig. 6. Time-temperature-sensitisation-pitting (TTSP) diagrams developed for both types 304LN and 316LN SS [23].

affected the pitting initiation. For a given type of inclusion, spheroidised particles are less susceptible to corrosion than elongated and plastically deformed ones. Thus, between 0.008 μm and 0.5 μm (if spheroidal), individual sulphide particles were found to be too small to initiate pitting. It was also noticed that even a bulk content of 0.07% sulphur did not induce pitting attack when the surface was treated with a laser beam, as the treatment produced sulphide inclusions of smaller sizes. Improvement in pitting resistance after laser surface melting of type 316 SS has been reported by U. Kamachi Mudali et al [29]. Iron sulphide (FeS), is a good electronic conductor with low hydrogen overvoltage, and that the presence of FeS inclusions enhanced the pitting probability. It was also observed that other than sulphide inclusions, oxides, carbides, silicates and oxy-sulphides also act as sites for pitting attack [2]. Using electron microprobe analysis, corrosion pits were predominantly found to initiate at inclusions of oxides, sulphides and oxysulphides. Studies carried out in the authors' laboratory indicated [30] that longitudinal stringers of MnS inclusions provided site for pitting to take place while globular oxide inclusions provided interfaces with matrix for initiation of corrosion attack (Fig. 7).

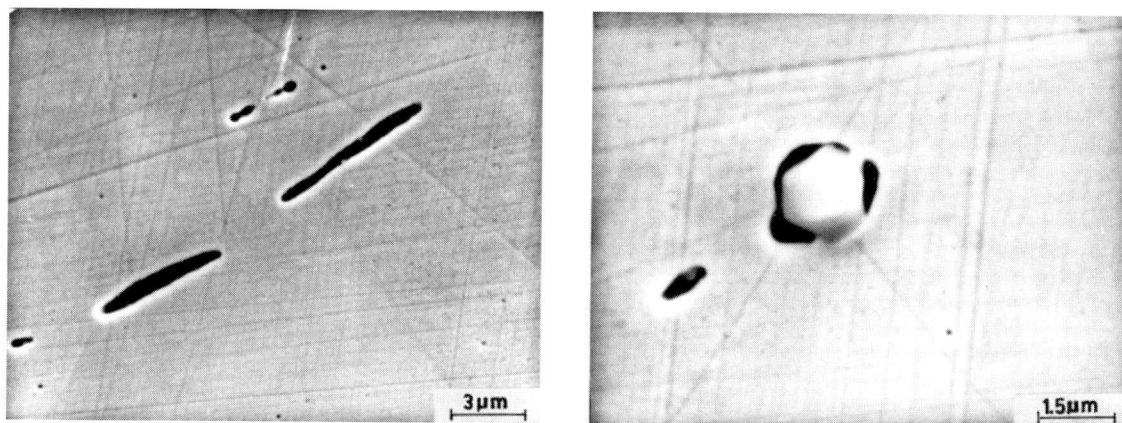

Fig. 7. **Longitudinal stringers of MnS inclusions and globular oxide inclusions provide sites for pitting attack to take place [30].**

These inclusions are dislodged from the site and pits continue to grow there with time. Presence of sigma in a type 317L stainless steel shifted the pitting potential in the active direction [2]. Alpha prime and martensite phases formed in austenitic stainless steels enhanced the pitting tendency due to their solubility in acidic solutions. A decrease in the pitting resistance is reported for austenitic stainless steel weld metals aged at higher temperatures, and is attributed to the formation of sigma and carbide precipitates [31]. Delta-ferrite in austenitic stainless steels and their weld metals is also reported to be detrimental to the pitting resistance [32, 33]. Pitting attack was found at the delta-ferrite/austenite interface, and the pits grew into austenite matrix. Dundas and Bond [34] reported that the delta-ferrite formed on heat-treating Fe-18% Cr-10% Ni-2.5% Mo-0.16% N alloy decreased the pitting resistance. The main reason for pitting attack at delta-ferrite/austenite, sigma/austenite and carbide/austenite interfaces is the depletion of major alloying elements like Cr and Mo in the austenite phase due to the formation of the second phases that are rich in these elements.

3.3 Alloy Composition

3.3.1 Chromium, Nickel and Molybdenum

The chemical composition of the alloy plays a major role in affecting the pitting resistance. In stainless steels, chromium, molybdenum and to a lesser extent nickel are the main alloying elements required to improve the pitting resistance. The other alloying elements also have a significant role in affecting the pitting resistance as shown in Fig. 8 [2]. Increasing chromium content enhanced the stability of the passive film against pitting attack, while increasing nickel content delayed the decrease in pH during pit growth by neutralising the solution in the pit. Molybdenum is specially added to improve the localized corrosion resistance and mechanical properties at elevated temperatures. Tomashov et al [35] measured the depth and width of the pits on 18Cr-14Ni and 18Cr-14Ni-2.5Mo stainless steels. They found that for the steel containing Mo, both the diameter and the depth of the pits grew more slowly than for the Mo-free steel. Though several investigators [36, 37] reported beneficial effect of Mo in improving pitting resistance, the exact mechanism by which Mo improved the pitting resistance is still not clear.

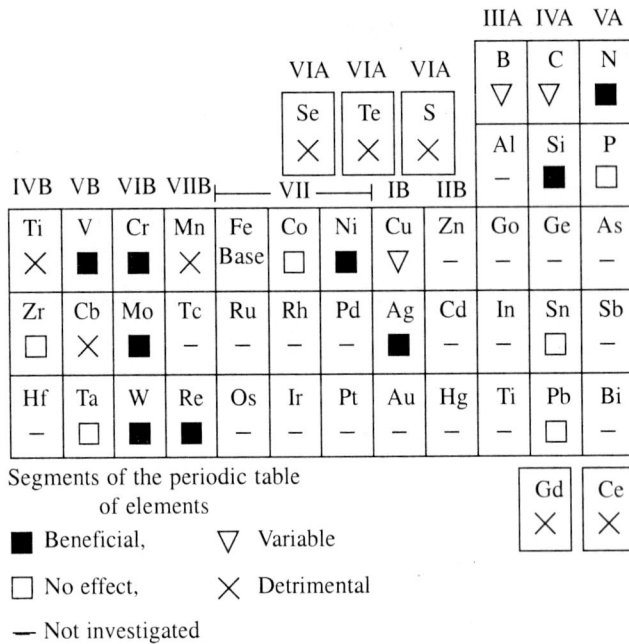

					IIIA	IVA	VA
	VIA	VIA	VIA		B ∇	C ∇	N ■
	Se ×	Te ×	S ×		Al −	Si ■	P □

IVB	VB	VIB	VIIB	VII		IB	IIB				
Ti ×	V ■	Cr ■	Mn ×	Fe Base	Co □	Ni ■	Cu ∇	Zn −	Go −	Ge −	As −
Zr □	Cb ×	Mo ■	Tc −	Ru −	Rh −	Pd −	Ag ■	Cd −	In −	Sn □	Sb −
Hf −	Ta □	W ■	Re ■	Os −	Ir −	Pt −	Au −	Hg −	Ti −	Pb □	Bi −

Segments of the periodic table of elements

■ Beneficial,   ∇ Variable

□ No effect,   × Detrimental

— Not investigated

Gd ×	Ce ×

Fig. 8.   Influence of various alloying elements on the pitting corrosion resistance [2].

3.3.2 Titanium

Small additions of Ti are beneficial in improving the pitting resistance [2, 25, 38]; but the effect is minimum at higher concentrations (1.8% Ti) due to the formation of new phases which are susceptible to pitting attack [39]. Studies carried out at the authors' laboratory [40] indicated that the increase in titanium addition from 0.21% to 0.42% to 15Cr-15Ni-2.5Mo stainless steel increased the pitting corrosion resistance. It was postulated that titanium was incorporated into the passive film and strengthened it against pitting attack by chloride ions. When these alloys were investigated for their pitting corrosion resistance in a simulated body fluid environment, significant improvement was noticed in the pitting corrosion resistance [41, 42].

3.3.3 Nitrogen

Nitrogen is considered an important alloying addition to austenitic stainless steels, in terms of corrosion resistance as it promotes passivity, widens the passive range in which pitting is less probable [43–47]. The addition of nitrogen has been reported to improve the pitting corrosion resistance of austenitic stainless steels (Fig. 9) and their weldments (Fig. 10). A particularly interesting development has been

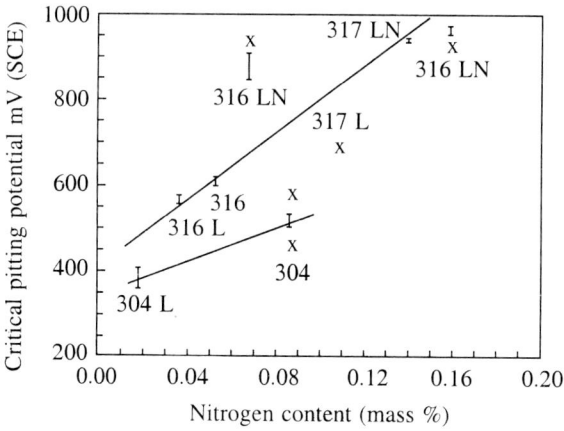

Fig. 9. Influence of nitrogen addition on the pitting corrosion resistance of austenitic stainless steels [30].

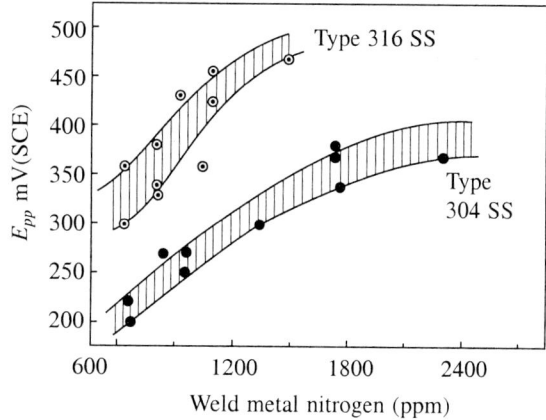

Fig. 10. Pitting corrosion reistance of nitrogen-added austenitic weld metals [43].

the identification of the synergistic effect of Mo and N addition on the pitting corrosion resistance [48, 49]. By increasing the Mo and N contents to 6 wt.% and 0.45 wt.% respectively, stainless steels with outstanding pitting corrosion resistance have been developed. Significant increase in the breakdown potential was found in aerated acidic carbonate solutions, for a number of alloys with increase in nitrogen content [48]. For a steel with composition 22Cr-3Mo-0.435N the breakdown potential was + 940 mV (SCE) whereas it was + 20 mV (SCE) for an alloy of 16Cr-3Mo-0.034N composition.

Investigations at the authors' laboratory [30, 44, 46] indicated that the addition of nitrogen significantly improved the pitting corrosion resistance in acidic and neutral chloride media up to a temperature of 338 K (Fig. 11). Based on extensive studies it was proposed that the formation of nitrate compounds at the pit site was responsible for enhanced repassivation tendency of nitrogen-containing alloys [43, 46]. This "localised" nitrate inhibition theory was supported by the

Fig. 11. Influence of nitrogen addition on pitting corrosion resistance at various temperatures [46].

identification of nitrate species at the pit bottom using microlaser Raman spectroscopy [50]. The increase in pH at the pit site due to the presence of nitrogen was also supported by the verification of

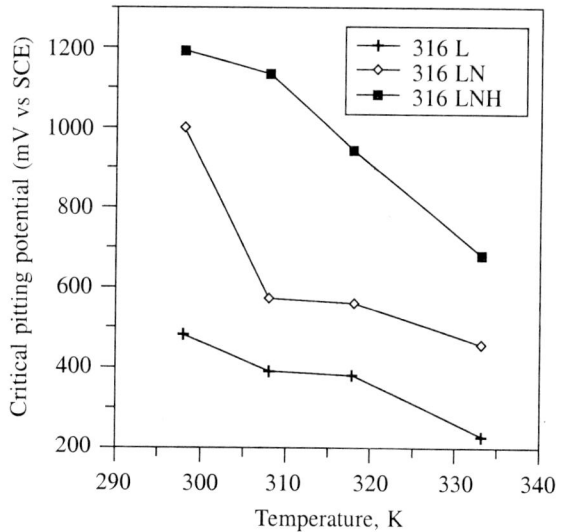

pH rise at the pit site during pitting attack of high-nitrogen stainless steels using scanning electrochemical microscopy and electrochemical scanning tunneling microscopy [51, 52].

Ohta et al [53] reported that for an alloy 20Cr-10Ni-0.7N, pitting potential of +900 mV (SCE) was obtained in a neutral chloride solution compared to a value of + 180 mV for a type 304 SS alloy. Speidel [54] found a chromium equivalent of 30 for nitrogen through critical pitting temperature measurements in an aqueous solution of 6% ferric chloride for pitting resistance. Figure 12 shows the plot of pitting resistance equivalent (PRE) of %Cr + 3.3 x%Mo + 30 x%N versus critical pitting temperature, which indicated that with the increase in nitrogen, the pitting temperature increases. Suutala and Kurkela [55] reported that the pitting corrosion resistance of austenitic stainless steels was primarily controlled by the amounts of Cr, Mo and N. A commonly used pitting resistant equivalent is %Cr + 3.3 X%Mo + 13 or 16 or 30 x%N. Mudali et al [43] reported a nitrogen equivalent of the molybdenum present in type 316

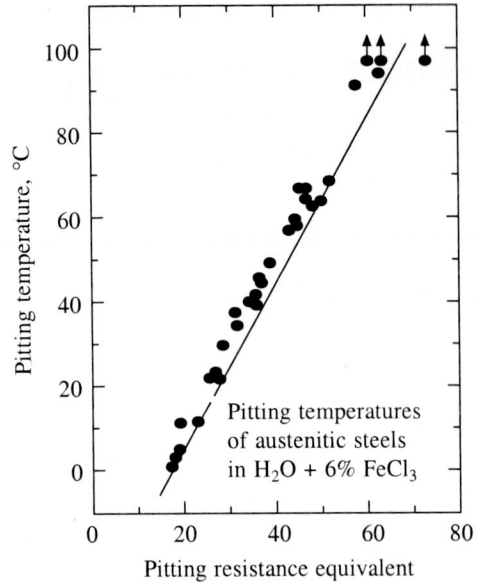

Fig. 12.   Pitting resistance equivalent versus critical pitting temperature [54].

SS weld metal as 530 ppm of nitrogen per weight percent of molybdenum, and an inter-relationship between the critical pitting potential and nitrogen content as $E_{pp} = 0.126$ [N/ppm] + 135. Ogawa et al [56] reported that addition of nitrogen to the austenitic weld metals significantly improved the pitting resistance irrespective of the presence of molybdenum. They attributed the combined effect of Cr, Mo and N in improving the pitting resistance.

3.3.4   Other elements

Tomashov and Chernova [57] have studied the influence of alloying elements Mo, Si, V, Re, W, Ce, Nb, Zr and Ta in 18Cr-14Ni steel. They found that Mo, Si, V and Re increased the pitting resistance, and explained it on the basis of increased stability of the passive film. Uhlig also noticed an increased resistance to pitting attack with the addition of Re [58]. The elements B, C, and Cu have been reported to have variable effect on the pitting resistance [2]. Boron is beneficial when in solid solution, but detrimental when precipitated as borides at the grain boundaries. Similarly, carbon has no effect on the pitting resistance when it is present in the solid solution, but is detrimental when it forms carbides [2]. Mudali et al reported [59] that the presence of Cu in a 17–4 PH SS had variable effects on pitting resistance depending on the microstructure produced after thermal ageing treatments. Earlier studies [60] using copper additions to type 301 SS found a beneficial effect at 0.2% and 0.5% Cu, and no effect at 1% and 1.9% Cu. Further studies [61] showed that a 1.5% Cu addition was detrimental and that additions of 0.8% or less, produced borderline behaviour characterised by the formation of small pits and their repassivation. Wilde [62] reported that increasing additions of Si from 1.01% to 4.45% increased the critical breakdown potential for 18Cr-8Ni alloy indicating the increasing resistance to pit initiation. Addition of rare earth elements (REMs) like Se and Te was reported to be detrimental due to the formation of chromium tellurides and selenides as precipitates [63].

3.4 Cold Working

It is well known that the cold working of austenitic stainless steels is necessary for the fabrication of components, and cold working is also intentionally introduced for the core components of the fast breeder reactors for improving their irradiation resistance. Cold working at higher levels introduces strain-induced martensite and residual stresses on the surface, which significantly decreased the localised corrosion resistance by increasing the number of active anodic sites on the surface. Austenitic stainless steels generally have relatively low stacking-fault energies (<100 mJ/m^2) and cold working them introduces a large amount of deformation faults including dislocations. This affects significantly the diffusion kinetics of the alloying elements and the defect structure depending on the cold working level. These changes in the microstructure markedly reduces the general and pitting corrosion resistance by increasing the number of anodic defect sites in the passive film [64, 65]. The martensite phase produced after cold working does not change the pitting potential; but with increase in the degree of cold work the number of pits increased [46]. Recent investigations at author's laboratory [66] showed that the effect of cold working on pitting corrosion resistance depended on the nitrogen content.

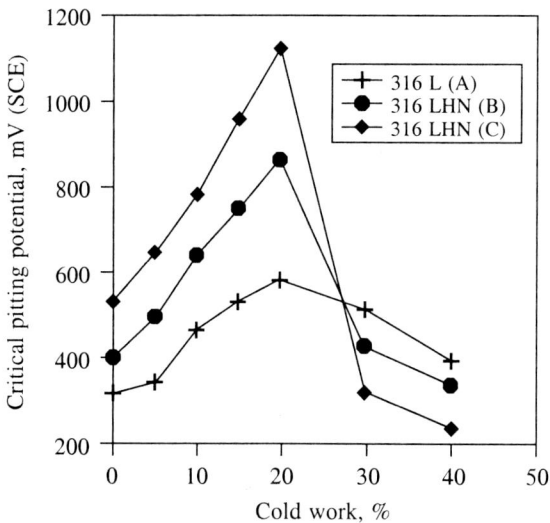

Fig. 13. Pitting corrosion resistance of cold worked Nitrogen-containing alloys [66].

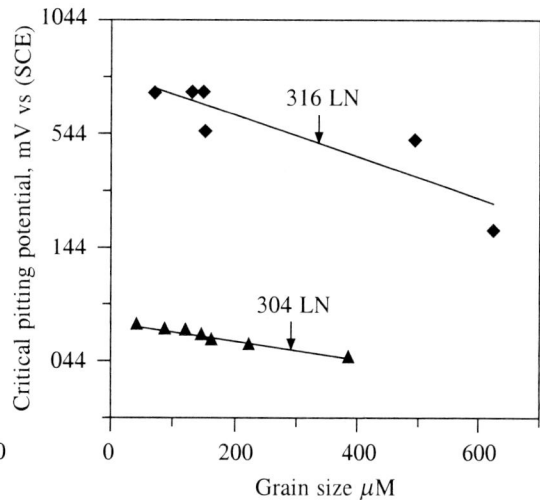

Fig. 14. Pitting corrosion resistance of alloys with various grain size [46].

Specimens of cold worked type 316L SS containing various nitrogen contents showed that the pitting resistance in neutral chloride medium increased up to 20% cold work while at 30% and 40% cold work pitting resistance drastically decreased. The substructure developed showed drastic changes from twins to micro deformation bands depending on the nitrogen content which influenced the pitting corrosion resistance (Fig. 13).

3.5 Grain Size

An increase in breakdown potential of Cr-Ni steel has been reported [67] with an increase in grain size from 0.005 mm to 0.07 mm. This grain size effect has been attributed to lesser grain boundaries and heterogeneous inclusions. Further, increase in grain size beyond 0.07 mm was found to show constant breakdown potentials and the number of pits decreased. Potentiodynamic anodic polarisation studies

[46] were carried out in acidic chloride medium for type 304LN SS (0.086% N) and type 316LN SS (0.07% N) specimens with varying grain sizes. Specimens with different grain sizes ranging from 40 μm to 380 μm for type 304L SS and from 70 μm to 570 μm for type 316L SS were prepared. The studies indicated that as the grain size increased the pitting resistance deteriorated for both alloys (Fig. 14). SEM observation of the pitted specimens indicated that at lower grain sizes, deep and stable pitting attack was seen whereas at higher grain sizes a large number of shallow pits were present. Due to the large sized grains the reduction in the grain boundary area would have increased the concentration of segregated impurities. This could have weakened the passive films at such grain boundary areas leading to decrease in the pitting resistance. Saturation of pitting potentials in the beginning could be due to the redistribution of such impurities over a large grain boundary area with smaller grains.

4.   EFFECTS OF EXTERNAL FACTORS ON PITTING CORROSION

4.1   Composition of the Electrolyte

Halide ions, especially chloride ions, are reported [6] to be the major species causing severe pitting attack for stainless steels. It is well known that pitting depends not only on the concentration of the halide ions present in the solution, but also on the concentration of the other ions as well. The chemical nature of the non-halide ions can affect the value of the critical pitting potential, the induction time for the stable pit formation and the number of pits. Among metal ions cupric, ferric, and mercuric ions in chloride solutions are considered to be aggressive. Among anions that reduce the tendency to pitting in chloride solutions are SO_4^{2-}, OH^-, ClO_3^-, CO_3^{2-}, CrO_4^{2-} and NO_3^-. The inhibiting tendency of these ions depend on their own concentrations as well as the concentration of the chloride ions in the solution. It was found [44, 68–70] that with nitrate addition the potential was shifted further in the noble direction, though it was also observed that nitrate ions did not have any influence on pitting potential. The inhibition efficiency of the ions for 18Cr-8Ni stainless steel was found to decrease in the order $OH^- > NO_3^- > CH_3COO^- > SO_4^{2-} > ClO_4^-$. Studies on the effect of pH revealed very little effect on the pitting potential in the pH range of 1.6 to 10 [2, 71]. This may be due to the fact that the bulk acidity has little role on the acidity at the pit site. However, in alkaline solutions, as the pH was increased the pitting potential increased in the noble direction indicating the inhibiting effect of hydroxyl ions. In general, with increasing temperature of the electrolyte the pitting potential decreases. Szklarska-Smialowska [71] and Forchhammer and Engell [72] reported a linear relationship between the pitting potential and the temperature. Forchhammer also noted a decrease in the pitting potential with the increase in temperature; however, it attained a stable value at higher temperatures for an 18Cr-10Ni-0.3Mo alloy. Increasing adsorption of chloride ions on the passive films and the temperature induced modifications of the passive films contributed to the decrease in pitting resistance. As mentioned in section 3.3 (Fig. 11), studies carried out at the author's laboratory showed that the pitting resistance decreased as the temperature of the neutral chloride medium increased. Another interesting parameter that affects the pitting potential is the extent of dissolved gases present in the electrolyte solution. Considerable difference in the pitting potentials [2] can be observed in the presence of different dissolved gases. A higher pitting potential was observed when a high concentration of oxygen was present in the solution.

4.2 Surface Finish and Potential Scan Rate

The surface finish of the test specimens also plays a major role in determining the pitting potential of stainless steels. It has been reported [2] that the resistance to pitting changes with the degree of mechanical polishing; electropolishing or etching leads to decrease in resistance to pitting. The reason being mechanical polishing covers grain boundaries and inclusions with the flowed metal. This correlates well with the observation that pitting initiates at grain boundaries and inclusions. The oxide film formation for a given passivation treatment or heat treatment also significantly affected the pitting resistance. The beneficial effect of the passivation treatment on pitting was attributed to the dissolution of heterogeneities like inclusions from the surface. The rate of change of potential and the initial potential are of considerable importance in the determination of pitting potential of stainless steels. Many investigators [2, 44] have reported that there was an induction time that varied with the concentration of the chloride ions and potential at which the test was conducted. This is the reason for the difference in the critical pitting potentials observed in a potentiodynamic polarisation test with the increase in the scan rate. In general, the faster the scan rate, the greater will be the potential difference across the film solution interface, and this would favour breakdown of the film. However at slow scan rates this potential difference remains relatively constant and therefore the best procedure is to use slow scan rates for potentiodynamic polarisation tests.

5. MORPHOLOGY OF PITS

In recent years, much attention has been given to the morphology of pits and the conditions leading to the formation of pits of different shapes. Figure 15 shows the variation in the cross-sectional shapes of pits observed during pitting attack [2]. The shapes of pits depend upon the metal/alloy and environment conditions under consideration during the test. Brauns and Schwenk [73] observed anisotropic dissolution of 18Cr-10Ni stainless steel in solutions containing chloride ions. They have found flat walled pits mostly in rectangular and square shapes at low potentials and at low current densities. However at more noble potentials, that is, at high current densities within the pit, the attack was isotropic and the conditions within the pit were similar to those occurring during electropolishing. Polyhedron shaped pits composing of most closely packed {110} and less closely packed {100} crystal planes were reported for a 16Cr-Fe single crystal in an aqueous solution containing chloride ions near breakdown potential [74]. Though a number of shapes are reported, it is still not clear whether there is any correlation between the shape and other parameters of pitting. However, it should be realised that the

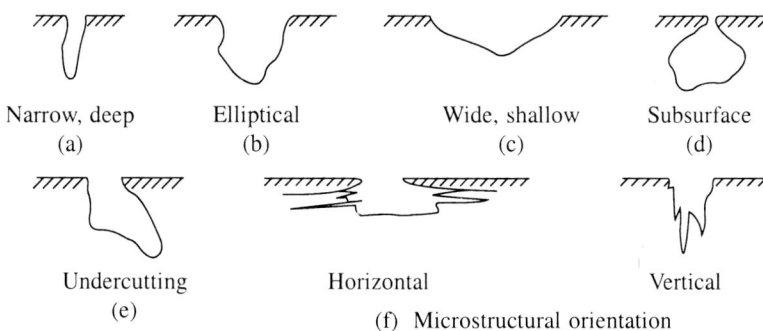

Narrow, deep (a) Elliptical (b) Wide, shallow (c) Subsurface (d)

Undercutting (e) Horizontal Vertical

(f) Microstructural orientation

Fig. 15. Variations in the cross-sectional shapes of pits observed during pitting attack [2].

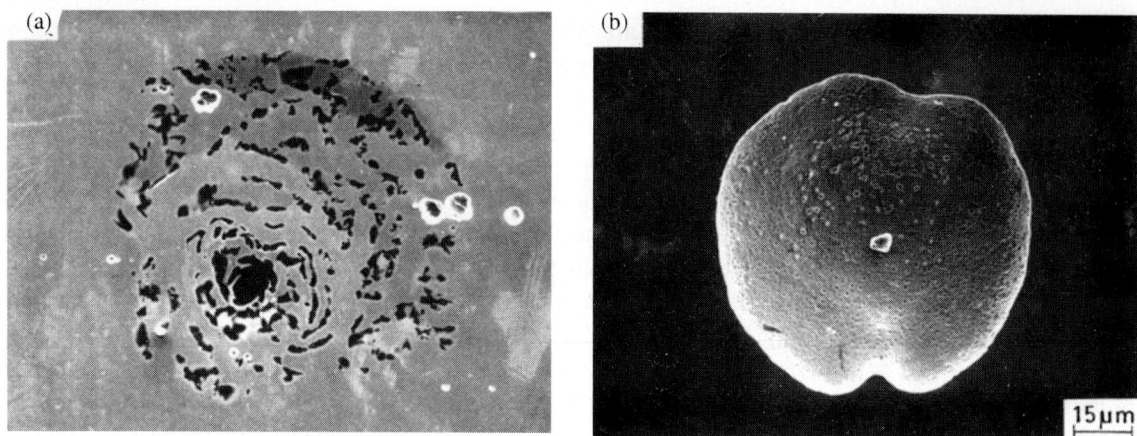

Fig. 16. Pits with (a) lace-like cover over the top and (b) flat-walled opening [30].

conditions existing both inside the pit and the bulk affect the shape of the pits. The formation of pits covered with a lace-like layer (Fig. 16a) indicated that the passive film is scarcely soluble, both in the bulk electrolyte and that within the pit. A flat-walled pit (Fig. 16b) indicated that there is no ohmic layer within the pit which would maintain equal current density at all points of the pit surface. The hemispherical or partly spherical pits suggested that within the pit there probably occur processes similar to those accompanying electropolishing.

6. PITTING CORROSION OF AUSTENITIC STAINLESS STEEL WELDS

6.1 Welding of Austenitic Stainless Steels

Welding of austenitic stainless steels usually results in a weld metal with a dendritic and inhomogeneous microstructure having a small amount of delta-ferrite, $M_{23}C_6$ carbides, sigma etc., and significant segregation of major alloying elements at the phase interfaces. Delta-ferrite, rich in Cr and Mo, formed during the welding of austenitic stainless steels is required up to a limit of about 10 ferrite number (FN) in order to avoid hot cracking and microfissuring of the welded components [75, 76]. The presence of delta-ferrite leads to preferential corrosion attack in the weld metal in certain environments. Pits have been shown [43] to nucleate preferentially, depending on the alloy composition, either at the austenite/delta-ferrite interfaces or inside the dendrite cores of austenite. The ageing of weld deposits, either during stress-relieving operations or during exposure to high temperature in service, leads to the formation of a complex precipitate microstructure. Delta-ferrite transforms into sigma, carbide, R-phase, chi-phase etc. during exposure to high temperatures in service [77, 78]. The inferior corrosion and mechanical properties of the welded components in comparison with the base metal are due to the preferential corrosion attack at the alloy-depleted regions, segregated interfaces, dendritic cores, and austenite/delta-ferrite and other secondary precipitate interfaces in the weld metals [79–83]. Selective corrosion failures at the weld regions have been reported in the nuclear fuel reprocessing plants, pulp and paper industries and urea plants [84, 85].

6.2 Solidification Behaviour of Austenitic Stainless Steel Weld Metals

Most of the commercial stainless steels contain about 70% Fe and are often discussed on the basis of constant Fe vertical section as shown in Fig. 17 which exhibit eutectic behaviour [86].

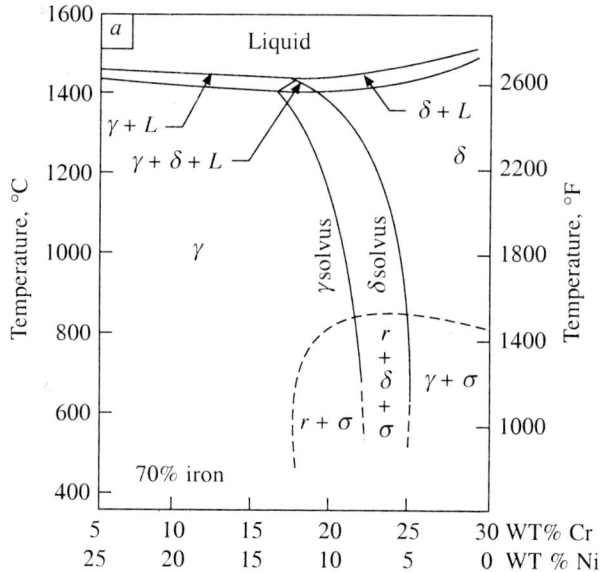

Fig. 17.   Vertical sections of Fe-Ni-Cr ternary diagram at constant Fe content [86].

It is documented [87–89] that welds can solidify with the primary phase solidifying as either ferrite or austenite, depending on which side of the peritectic-eutectic liquidus contains the nominal composition. For primary ferrite solidification, a large fraction of the primary ferrite would transform to austenite during cooling. The compositions on the Ni-rich side of the eutectic trough will solidify as austenite. If the composition is sufficiently far from the eutectic trough, solidification will occur as complete austenite, which contains Cr-enriched and to a lesser extent Ni-enriched cell boundaries resulting from microsegregation. As the compositions approach the eutectic trough some ferrite can form during the last stages of solidification by the eutectic reaction, which is confined to the solidification cell boundaries. The earlier concept of simultaneous formation of ferrite and austenite [90] was strongly refuted by Fredriksson [91] who proved that alloys could solidify as primary ferrite or primary austenite, and still could contain a rather small amount of ferrite at room temperature. He proved the fact that each dendrite solidified as primary ferrite but upon cooling through δ/γ two phase region, transformed to austenite except for thin rods or stringers of ferrite along the core of the original dendrite. Most of the ferrite formed initially transformed to austenite through a solid-state transformation reaction. One way of representing the complexities of solidification in a simplified ways is through plots of chromium equivalent (Cr_{eq}) vs nickel equivalent (Ni_{eq}) [92]. The line separating predominantly austenitic from predominantly ferritic solidification has a slope of 0.68 (Cr_{eq}/Ni_{eq} = 1.5) [93, 94]. Suuatala [95], DeLong [96] and WRC-92 [97] have given different formulae to calculate the Cr_{eq}/Ni_{eq} values using the chemical composition of the weld metals as follows:

(a) Suuatala's $\dfrac{Cr_{eq}}{Ni_{eq}} = \dfrac{Cr + 1.37Mo + 1.5Si + 2Nb + 3Ti}{Ni + 0.31Mn + 22C + 14.2N + Cu}$

(b) DeLong's $\dfrac{Cr_{eq}}{Ni_{eq}} = \dfrac{Cr + Mo + 1.5Si + 0.5Nb}{Ni + 0.5Mn + 30(C + N)}$

(c) WRC-92 $\dfrac{Cr_{eq}}{Ni_{eq}} = \dfrac{Cr + Mo + 0.7Nb}{Ni + 35C + 20N + 0.25Cu}$

These formulae were based originally on Schaefflier's concept [98] of $Cr_{eq}(Cr_{eq} = Cr + Mo + Nb)$ and $Ni_{eq}(Ni_{eq} = Ni + 0.5Mn + 30C)$. In order to predict the delta-ferrite content of the austenitic stainless steel weld metal from its chemical composition, Schaeffler proposed a diagram where Ni_{eq} is plotted against the Cr_{eq} for a given weld metal [98]. This diagram depicts various phase fields like austenite, martensite and up to 100% delta-ferrite (Fig. 18). This diagram was later modified by DeLong [96] and Kotecki and Siewart [97] by taking into consideration the roles of N, Cu and Ti.

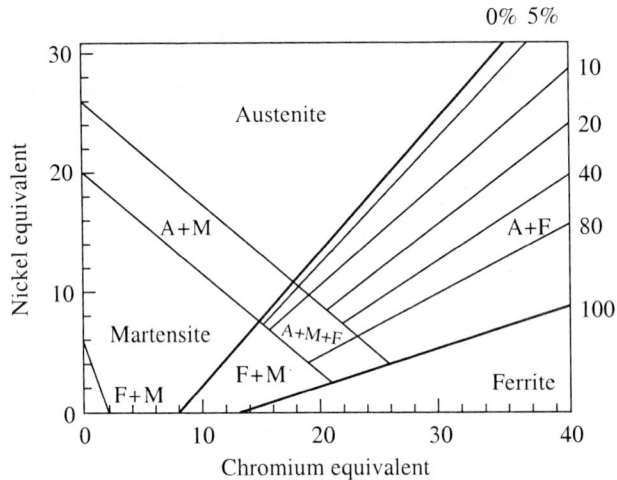

Fig. 18.   Schaeffler's diagram to predict the weld metal delta-ferrite content [98].

6.3   Microsegregation During Solidification

Microsegregation during weld solidification is an extremely important phenomenon. It occurs over distances of the order of the cell or dendrite spacing and can lead to large variations in alloying elements or impurity concentrations between the solidification cell or dendrite core. In welds and castings cooling rates are usually very high or the diffusivities in the solid are too slow for the solid to be compositionally homogeneous. This leads to the enrichment of the solid with the higher amounts of solute. Consequently, the weld metal microstructure is considerably different from those of the base metal [99]. Depending somewhat on the solidification parameters, but primarily on the diffusivities, solid-state diffusion can greatly reduce the degree of microsegregation during solidification and cooling. The amount of solid-state diffusion can be very different for different alloying elements and alloys systems.

Segregation of alloying elements at the δ/γ boundaries would depend on the austenite stability and heat input. Austenite stability increases for faster cooling rates and lower heat inputs. As the heat input increases cooling rate decreases and sufficient time is available for the ferrite to grow. During primary solidification mode, the ferrite formed is rich in ferrite formers, hence the austenite formers are rejected into the liquid adjacent to the solid/liquid interface. Subsequent rapid cooling to room temperature most of the ferrite formed transforms to austenite by solid state transformation mechanism and only the primary ferrite stringers rich in ferrite formers are retained in the final weld microstructure. This phenomenon is called coring. The diffusivities of Cr and Ni are much higher in ferrite than in austenite, leading to the very uniform structure within ferrite compared to the austenite solidification [100, 101]. Experimental evidence towards the existence of microsegregation of Cr, Mo, Mn, Ni and other impurity elements has been provided by a number of investigators [102–106]. It was established [94] that the segregation ratios of the main alloying elements steeply changed at $Cr_{eq}/Ni_{eq} \cong 1.5$, which corresponds well to the transition from primary austenite to ferritic solidification. Studies on 316 stainless steel Gas Tungsten Arc (GTA) welds revealed that both Mo and Cr segregated at the δ/γ during solidification and cooling [107]. The extent of microsegregation was much higher for Mo than Cr in the welds where solidification mode was primary austenite compared to the welds where ferrite was primary mode of solidification. Similar effects of microsegregation of Cr and Mo were also observed [81] for AISI 316L GTA welds where Mo concentration levels at the δ/γ boundary were found to be almost double the level of Mo present in the parent metal (2–5 wt.%) for welds when the mode of solidification was primary austenite. Higher concentration of segregated Mo were reported [102, 104, 105] (2.5 times) compared to Cr (1.2 times) in both ferritic as well as austenitic mode of solidification at the δ/γ boundaries. Apart from these major alloying elements, minor elements like sulphur, phosphorus, boron, niobium, titanium and silicon get segregated at the δ/γ boundaries. These elements strongly partition to the liquid, possess low solubilities in solidified steel and form low-melting eutectics with iron, chromium or nickel. Sulphur is known to be an undesirable impurity in welding of stainless steels due to the formation of low-melting sulphide films along the interdendritic and grain boundary regions. Sulphur is strongly rejected into the liquid during solidification of austenite, rapidly lowering the melting point of the interdendritic liquid forming low-melting eutectics even at very low concentration (< 0.005%) [108]. On the other hand, δ-ferrite shows higher solubility for elements like S, P, Si and Nb. Like sulphur, P forms low-melting eutectics with iron, chromium and nickel. The segregation tendency remains high due to the wide solid-liquid range and low eutectic temperatures (1373 K) [109]. Investigations showed that the segregation of Cr, Mo and N in a number of austenitic stainless steel weld metals with chemical composition in the range of Cr (17 to 28%), Ni(6 to 60%), Mo (0 to 9%) and N (0 to 0.37%) revealed that the segregation was greatest for N followed by Mo and Cr, the segregation being maximum at $Ni_{eq}/Cr_{eq} = 0.7$ (Suuatal's formula) [110]. Increasing heat input leads to a small variation in the chemical composition of the weld metal. It is reported that increase in the heat input resulted in the reduction of chromium and molybdenum and the increase in the carbon concentration of the 316L weld metals prepared by submerged arc welding [111]. These weld metals prepared at 1.73 kJ/mm, 2.58 kJ/mm, 4.08 kJ/mm and 5.77 kJ/mm heat inputs showed progressive enrichment of ferrite-formers (e.g. Cr and Mo) and impoverishment of austenite-formers (e.g. Ni) in the delta-ferrite phase with increasing heat input (Fig. 19). Since, all these weld metals had Cr_{eq}/Ni_{eq} > 1.55, the mode of solidification was primary ferrite thereby conforming the fact that elemental segregation is equally prominent in these weld metals [111].

6.4   Effect of Solidification Mode and Impurities on Pitting Corrosion

A number of authors have proposed that sulphides and complex oxysulphides are preferred sites for pit initiation in wrought steels [26]. However, the inferior pitting resistance of the weld metal is more complex because sulphides tend to dissolve and spheroidize during the thermal cycle of the welding which would promote pitting resistance [112, 113]. Pit initiation takes place at the last interdendritic regions to solidify in austenitic-ferritic welds and at the delta-ferrite/austenite phase interface in ferritic-austenitic welds. These regions match those of sulphide precipitation during the solidification of austenitic stainless steels, as reported earlier [114]. The most important factor controlling the location of pit initiation might be the distribution of sulphur as either sulphides or elemental sulphur segregate in the weld microstructure [115]. A study on [115] the pitting corrosion resistance of the

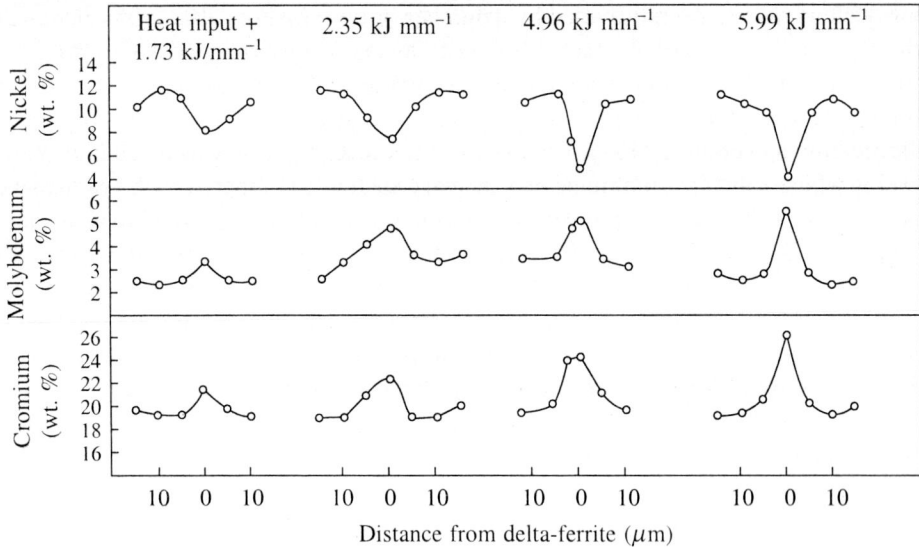

Fig. 19.   Segregation of elements at the delta-ferrite boundary at different heat inputs [111].

GTAW welds of 26 commercial heats of AISI 316 stainless steels with 0.001% to 0.02% sulphur with phosphorus content of 0.006% using ferric chloride immersion test (ASTM G 48–76) showed that the weight loss (Fig. 20 and 21) in the primary austenitically solidifying welds (AF) was approximately 20 g/m²h, independent of the sulphur content, whereas the weight loss in the primary ferritically solidifying welds (FA) was markedly influenced by the sulphur content.

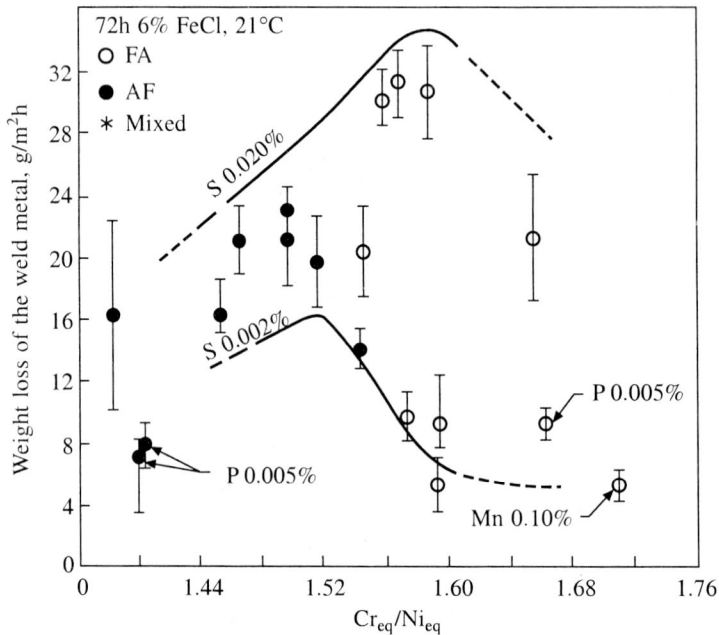

Fig. 20.   Effect of solidification mode on the weight loss of welds in the $FeCl_3$ test [115].

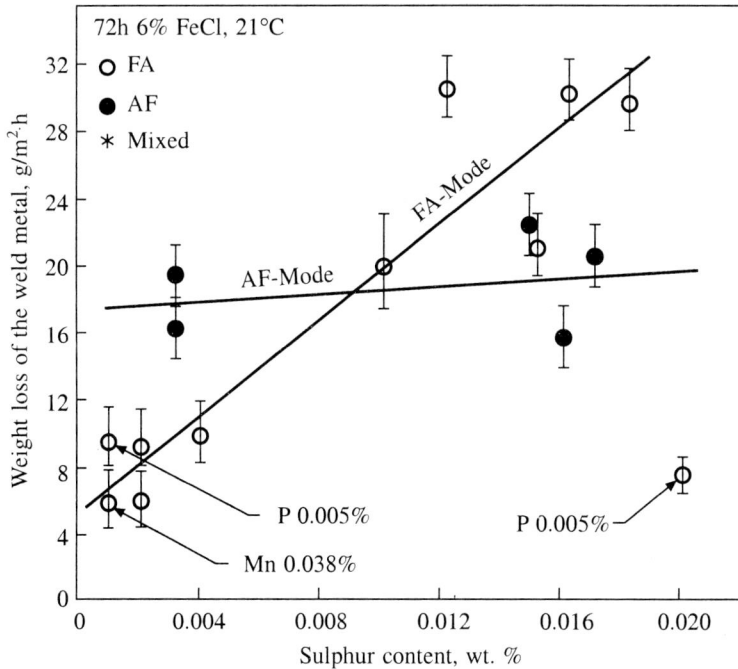

Fig. 21. Effect of sulphur content on weight loss of welds solidifying in different mode [115].

These results showed that the effect of sulphur seemed to be at its strongest just beyond the shift of the solidification mode from primary austenitic to primary ferritic, i.e., $1.52 < Cr_{eq}/Ni_{eq} < 1.60$. It was observed that the pit density values for the welds solidified in the FA mode were higher suggesting higher pit initiation sites. This fact conclusively proved that pit initiation in weld metals was controlled by distribution of impurities during solidification and the change in the mode of solidification influenced the segregation of impurity elements across δ/γ boundaries thereby accentuating pitting attack at the δ/γ interface. In the light of these observations the earlier view [116] that any type of ferrite in austenitic stainless steel weld metals was susceptible to pitting attack was refuted.

Pits also tended to grow along solute-depleted zones in the microstructure caused by microsegregation which is considerably stronger is austenitic-ferritic (AF) solidification than in the ferritic-austenitic (FA) mode [117]. Garner [79] studied the pitting corrosion resistance of the commercial 316L and 317L autogenous welds in an acidic 10% $FeCl_3$ (pH $\cong$ 1.0) solution. Electron microprobe measurements of Cr and Mo concentrations in weld metal microstructures were made. The Cr and Mo concentrations at different locations in the weld microstructure are listed in Table 1. It is evident that Cr and Mo segregated in the interdendritic regions of the weld metal leaving the dendrite centers depleted of these two elements. This effect is called as coring. This effect is more pronounced in primary austenitically solidified weld metals where dendrite cell centers are depleted in Cr and Mo compared to parent metal; the last-to-solidify interdendritic regions are enriched in Cr and Mo. Microscopic observations of pits [79] in the early stages of initiation in unannealed and annealed weld metal, revealed the preferential pitting corrosion attack of the dendrite centers. This established the fact that preferential pitting attack on the cored austenite dendrites signified that influence of ferrite was less important in presence of oxidizing acid chlorides and when the cathodic reaction was not rate controlling [79, 82].

Table 1   Microprobe analysis of the weld metals [79]

Specimen		316L		317L		28V3	
		Cr	Mo	Cr	Mo	Cr	Mo
Parent Metal		16.3	2.8	18.4	3.2	15.9	5.0
Weld Metal: As-welded	Dendrite Centre	14.3	1.8	14.2	2.0	14.7	3.1
	Interdendritic Phase	20.1	5.7	24.0	6.6	18.0	9.8
Weld Metal+ 1 hr at 900°C	Dendite Centre	14.8	2.4	17.6	2.3	14.1	2.8
	Interdendritic Phase	21.7	10.5	25.2	9.3	18.9	14.1

Initiation of pitting corrosion attack at the austenite cell centres of the as-deposited 316 GTA weld metal is shown in Fig. 22(b) [43]. Pit initiation studies on 904MS stainless steel weld metals (Cr 19.9%, Ni 24.8%, Mo 4.66%, P 0.023%, S 0.011% and delta-ferrite <0.1) showed that, the most important pit initiation sites were non-metallic inclusions (MnS or $MnS.SiO_2$) at the dendrite centers [105]. The pitting attack then propagated preferentially along the dendrite centers, where Cr and Mo content was lower than the mean value for the weld metal as a whole. It was thought that the type, number and size of the non-metallic inclusion exerted a secondary influence on the weld metal pitting resistance [105]. Pit initiation process on 304L autogenous welds in 0.33M $FeCl_3$ solution revealed that the most susceptible sites were δ/γ interface boundary in the primary ferritic solidification mode [118]. Pit initiation at these sites was related to the sulphur and phosphorus segregation, which increased the dissolution tendency of the passive film over the interface. The sulphur impurity would also tend to inhibit the repassivation kinetics at the δ/γ interface.

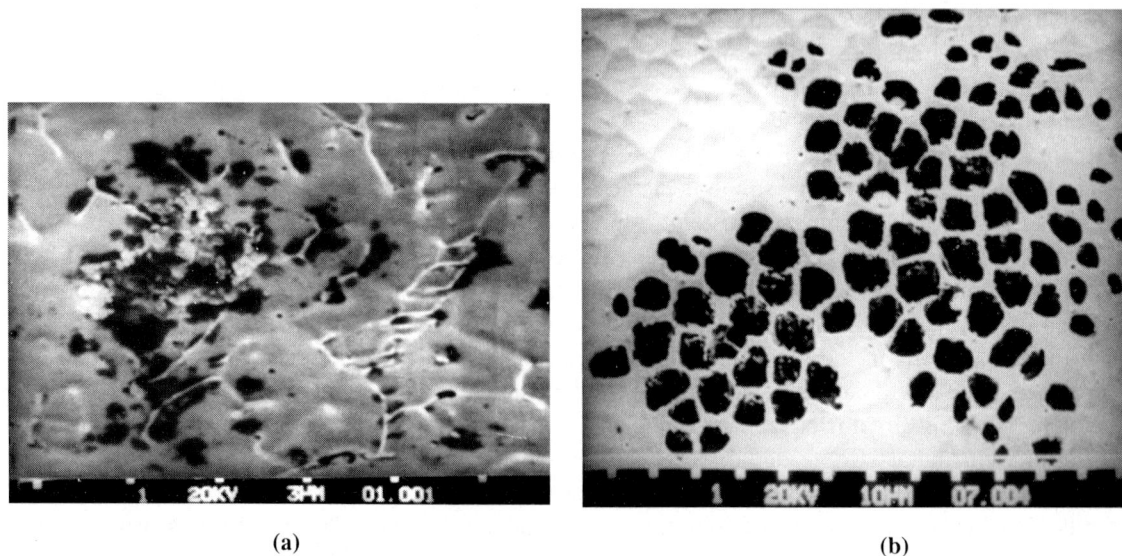

(a)   (b)

Fig. 22. (a) Initiation of pitting corrosion attack at the δ/γ interface of a 316L weld metal [111] and (b) Pits initiated at the austenite cell centers [43].

A submerged arc welded (SAW) specimen of 316L weld metal shows (Fig. 22(a)) the pitting attack initiated at the δ/γ interface in acidic chloride medium [111]. Transformation of delta-ferrite in 316L SAW specimen to sigma phase (white block) and secondary austenite (dark areas) by thermal ageing is shown in Fig. 23(a). Pitting corrosion attack was found to have initiated at the depleted secondary austenite leaving sigma intact (Fig. 23(b)). Delta-ferrite and other intermediate phases such as σ or χ phase which are Cr- and Mo-rich, are not usually attacked under mildly or moderately oxidizing conditions. However, these transformation products of delta-ferrite are attacked in presence of reducing acids, such as HCl, or highly oxidizing acids, such as HNO_3 [32, 119].

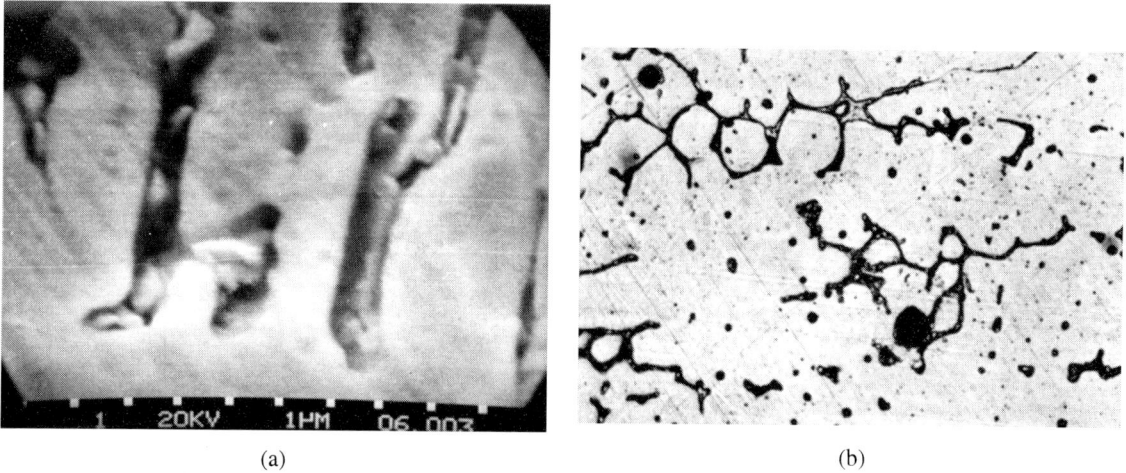

| (a) | (b) |

Fig. 23.   (a) Transformed delta-ferrite to sigma phase and (b) Pit initiation attack at the secondary austenite near sigma phase for the same specimen [111].

6.5   Effect of Heat Input

Increase in heat input increases the cooling time resulting in the coarser microstructure. A more precise correlation between pitting resistance and coarseness could be explained by fact that higher heat input welds solidified with more grains in which primary dendrites lie parallel to the surface of the weld [32]. It was found that higher average dendrite section-lengths were found to have lower pitting resistance [120]. But it has been observed that, decrease in pitting resistance could also be due to the increased segregation of ferrite-formers to delta-ferrite resulting in the impoverishment of austenite in these elements as well as the precipitation of $M_{23}C_6$ carbides or sigma phase particles along the austenite/ferrite interface. This is exemplified in Fig. 24 where the pitting resistance of 316L weld metal prepared by submerged arc welding is shown in as-welded and aged (1073 K/10 h) conditions. The decrease in E_{pp} values of the aged specimens was attributed to the formation of sigma phase particles at the austenite/ferrite interface [80, 111]. In order to mitigate the deleterious effects of the increasing heat input Mudali et al [43] suggested the addition of nitrogen gas through the shielding gas. The investigation clearly showed that the increase in the volume percent nitrogen in the shielding gas of 304 autogenous TIG weld metals substantially raised the E_{pp} values (Fig. 25) [43].

Areas, which contain these long, broad segregated dendrites, are very susceptible to pitting attack due to the effect of coring. The preferential pitting attack on the surface-lying primary dendrites rather

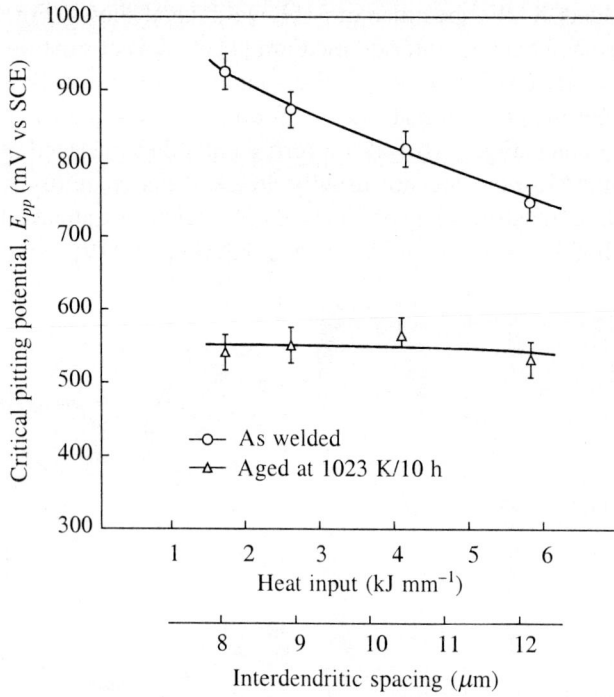

Fig. 24.   Effect of heat input on the E_{pp} values of the 316L SAW weld metal in as-welded and aged condition [80].

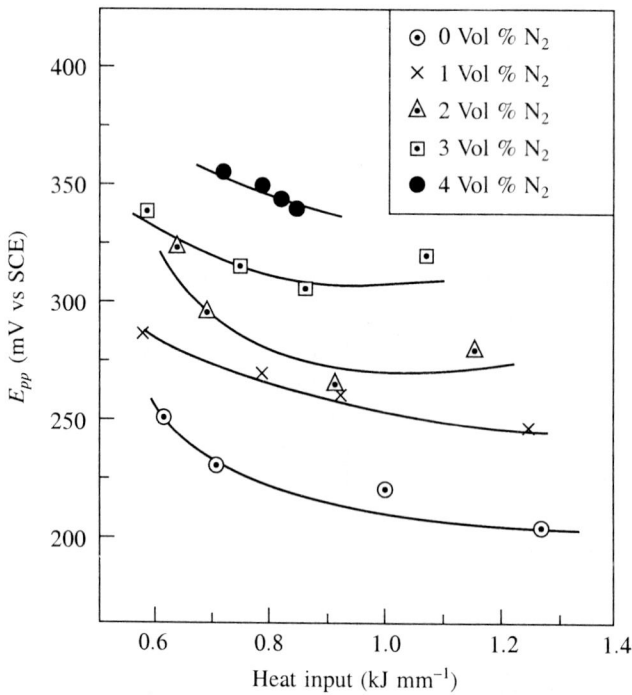

Fig. 25.   Effect of addition of nitrogen through shielding gas on E_{pp} values of 304 weld metal [81].

than in adjacent grains or in the base metal is shown for 317L autogenous weld in a potentiostatic scratch test (0.6N NaCl, pH 3 at 40°C) (Fig. 26) [120].

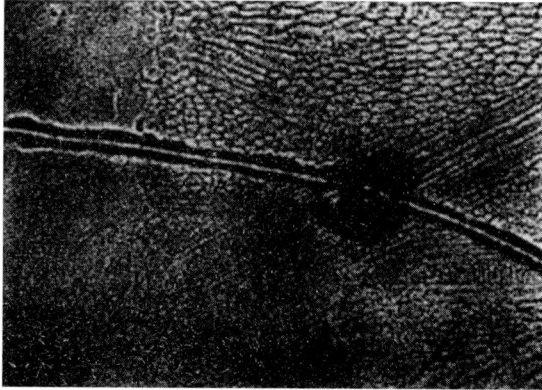

Fig. 26. Preferential pitting attack on the surface-lying primary dendrites [120].

Long primary dendrite structures are most prone to attack probably because the alloy-depleted initiation point is not entirely surrounded by higher alloy material. These adversely oriented grains, then provide larger scale initiation sites for further pit propagation, and thus lower the pitting resistance of autogenous welds.

6.6 Effect of Chemical Composition

Systematic work on the effect of addition of alloying elements on weld metal pitting behaviour was carried out wherein emphasis was placed on the beneficial effect of addition of Mo and N to the weld metal [79]. Addition of Mo led to the increase in critical pitting temperature (CPT) (when tested in 10% $FeCl_3$ solution) as well as pitting potentials but decreased after autogenous GTAW welding as well as welding at higher heat inputs forming coarse delta-ferrite [120]. Mudali et al [43, 81] investigated the effect of nitrogen, added through shielding gas, on pitting corrosion of type 304 and type 316 weld metals and concluded that added N improved pitting resistance and also it helped in offsetting the reduction in pitting potentials that could be caused by increasing heat input. These authors contended that nitrogen formed NH_4^+ ions and other inhibiting compounds that helped in repassivating the pits. According to the earlier results [55] on type 316 stainless steel weld metals with different Cr, Mo and N contents, that pitting corrosion resistance was mainly controlled by Cr, Mo and N in the weld metal and that the Mo seemed to exert stronger positive effect on the pitting resistance of the base metal than the weld

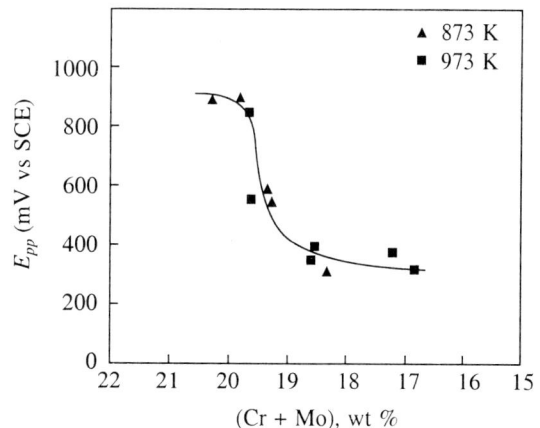

Fig. 27. Effect of total Cr + Mo concentration of the matrix of 316L weld metal on E_{pp} values [31].

metal [56]. It was observed that the total concentration of Cr and Mo present in the matrix was directly correlated with the pitting resistance. Weld metal specimens prepared from 316L by GTAW welding were thermally aged at 873 K and 973 K for different time durations and their pitting resistance studied in acidic chloride solution potentiodynamically. The E_{pp} values of these specimens were plotted as a function of total concentration of Cr + Mo of the matrix revealing the drastic drop in E_{pp} values were directly related to the decrease in total Cr + Mo concentration of the matrix (Fig. 27) [31]. The contribution of Cr, Mo and N in raising the pitting resistance of the base and weld metals was obtained in the form of two different equations signifying the pitting corrosion resistance of base and weld metals as [55]:

$$CPT \text{ (sheet)} = -45 + 2.48Cr + 8.10Mo + 32.5N \text{ and}$$

$$CPT \text{ (weld)} = -68 + 3.65Cr + 4.76Mo + 17.8N$$

The coefficients of Mo and that of N in the equation for the weld metal were nearly half of that of the base metal. Because nitrogen segregates into dendrite cores (unlike Mo), N may have mitigated the negative effects of Mo segregation during solidification.

However, since the differential segregation reduced any synergistic action between the two elements, N appeared to have a less marked effect in weld metal than in parent steel. A new pitting index known as "Modified Pitting Index (MPI)" was formulated [110] for high alloy austenitic stainless steel weld metals, where Ni_{eq}/Cr_{eq} factor was included considering the fact that the elements like Mo and N interacted together to give synergistic effect in raising pitting resistance and is given by ($FeCl_3$, CPT),

$$MPI(1) = 6Mo + (Cr + 1.9Mo)Ni_{eq}/Cr_{eq} \text{ and using potentiodynamic } E_b \text{ values,}$$

$$MPI(2) = Cr + 3.5Mo - 13N + (1.7Mo + 23N)Ni_{eq}/Cr_{eq}$$

6.7   Determination of Pitting Corrosion Resistance

(a) Determination of Critical Pitting Temperature (CPT): The CPT is defined as the highest temperature at which the specimen does not exhibit persistent pitting. A test commonly used to determine the CPT is standardized and is available in ASTM G48–92 [122]. In this test, coupons from the base steels and weld pads are immersed in 6% $FeCl_3.6H_2O$ solution in individual flasks housed in a constant temperature bath. After exposure for 72 h at an appropriate temperature, the coupons are removed and examined for pitting with an x20 eyepiece. The procedure to evaluate the pits on the surface is given in ASTM G46–94 [123]. The coupons are prepared again to a 120-grit finish and immersed in a fresh solution at a temperature 2.5°C higher. This process is continued until persistent pits are established. The location of pits is documented. Edge and crevice corrosion attack, when present are ignored. Duplicate samples are tested and CPTs are generally reported within 2.5°C. The lower values are used to describe the material behaviour.

(b) Determination of breakdown potential (E_b): In order to determine the E_b values potentiodynamic testing is used as described in ASTM G61–86 with polarization of samples from their rest potential at 1 mV/s in an anodic direction until pitting occurred. Weld metal or base steel specimens are mounted in an epoxy resin with one face (1 cm^2 surface area) exposed. Electrical connection is made to the potentiostat through a stainless steel wire spot-welded to an unexposed surface of the specimen. The environment is aerated 3% NaCl solution at room temperature. The current response is continuously monitored. After the current density exceeds 1 mA/cm^2, the potential scan is reversed. Samples which had pitted described a hysteresis loop. Those, which do not pit show the reversible oxygen evolution reaction (Fig. 28) [110].

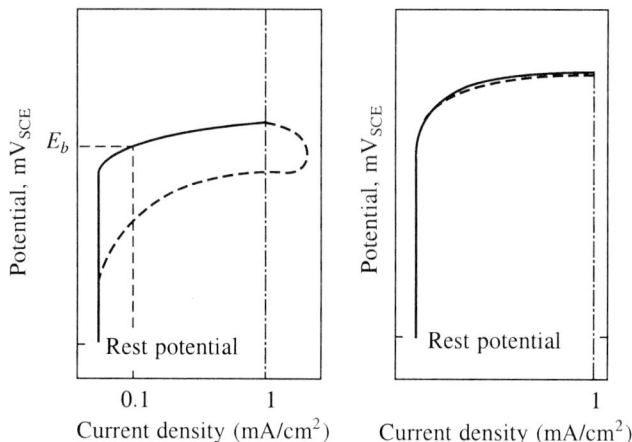

Fig. 28. Technique to determine the breakdown potential E_b [110].

The breakdown potential, E_b was measured at 0.1 mA/cm^2, representing a condition where pits grew rapidly, Duplicate of triplicate scans are carried out and results from samples showing crevice attack are ignored. Breakdown potentials are generally within about 30 mV. Marshall and Gooch [110] on their studies on high alloyed austenitic stainless steel welds showed that there was a reasonable, although not necessarily linear, correlation between E_b and CPT (Fig. 29).

(c) Determination of Pit Initiation Site: Scanning Transmission Electron Microscopy (STEM) was employed to determine the pit initiation site. Transmission foils are prepared by mechanically grinding thin sheets of material on 600 grit paper to 0.127 mm. Discs of 3 mm diameter were cut and thinned electrochemically in a jet polishing apparatus utilizing a 9:1 solution of methanol/perchloric acid −313 K with an applied voltage of 25 V. Initially, regions are documented where no electropolish pits are observed. Pits were initiated in foils according to the following procedure. The edge of the specimen was spot welded to a stainless steel wire. In order to obtain pit initiation only on the polished foil, the wire and spot-welded area are prepassivated in 50%

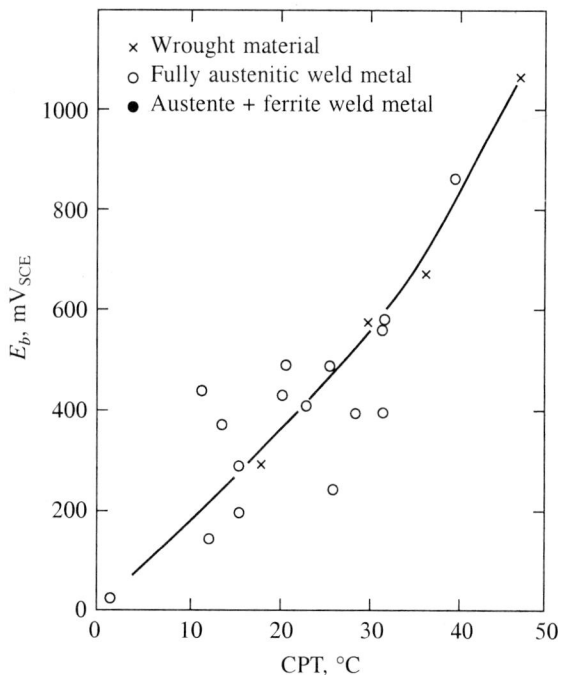

Fig. 29. Correlation between CPT and E_b for different specimens [110].

HNO$_3$ at 323 K for 30 min. The specimen was placed in a glass test cell, which contained deaerated 1N NaCl solution, adjusted to pH 4, at a constant temperature of 295 K. Anodic polarization from the corrosion potential at a rate of 600 mV/h leads to pitting of the foil. Polarisation is arrested when a current of 100 μA is reached and the potential is held constant at this level for 5 min. The specimen is then separated from the wire about 1.5 mm from the spot weld by exploding the wire with a high energy setting on a spot welding unit. The specimen is placed in the STEM microscope, and regions, which were well documented in the as-electropolished condition, were relocated and examined for pit initiation sites [118].

7. SUMMARY

Understanding pitting corrosion is very important as it causes several failures in many industries. The theories and mechanisms proposed so far on the pitting corrosion aspects directed attention on two important aspects, namely, structural heterogenieties and environmental parameters. However, more insight is required on the interfacial reactions and phenomena occurring at the electrode-electrolyte interface to understand and control the pitting events. This necessitates the study by finer techniques on the interfacial reactions promoting pitting corrosion. With the development of high corrosion resistant materials the improvements in pitting corrosion resistance can be demanded for several applications

REFERENCES

1. Passivity of Metals, Eds. R.P. Frankenthal and J. Kruger, Corrosion Monograph Series, The Electrochemical Society, USA, 1978.
2. Corrosion of Stainless Steels, A.J. Sedriks, John-Wiley and Sons, New York, USA, 1979.
3. ASTM G15–89A: Standard Terminology Relating to Corrosion and Corrosion Testing, Annual Book of ASTM Standards, Vol. 03.02, ASTM, Philadelphia, USA, 1990, pp. 84–87.
4. Kruger, J., In: Passivity and Its Breakdown on Iron and Iron-Base Alloys, NACE, Houston, USA, 1976, pp. 91–98.
5. J. Kruger, International Metals Review, 33 (1988), pp. 113–130.
6. H.-H Strehblow, Werkstoffe and Korrosion, 35 (1984), pp. 437–448.
7. T.P. Hoar and W.R. Jacob, Nature, 216 (1967), p. 1299.
8. Y. Kolotyrkin, Corrosion, 19 (1963), pp. 261–268.
9. L.F. Lin, C.Y. Chao and D.D. Macdonald, J. of the Electrochemical Society, 128 (1981), p. 1194.
10. N. Sato, Electrochimica Acta, 16 (1971), pp. 1683–1692.
11. J. Tousek, Theoretical Aspects of the Localised Corrosion of Metals, Trans Tech Publications, Switzerland, 1985.
12. K.J. Vetter and H.-H. Strehblow, in Localised Corrosion, eds. R.W. Stahle, B.F. Brown, J, Kruger and A. Agrawal, NACE, Houston, USA, 1974, pp. 240–251.
13. Z. Szklarska-Smialowska, Pitting Corrosion of Metals, NACE, Houston, USA, 1986.
14. Z. Szklarska-Smialowska, Corrosion, 27 (1971), pp. 223–233.
15. E. Mattsson, British Corrosion Journal, 13 (1978), pp. 5–12.
16. H. Bohni, in corrosion in Power Generating Equipment, eds. M,O. Speidel and A. Atrens, Plenum Press, New York, 1984, p. 85.
17. A.J. Sedriks, in Proceedings of the International Conference on Stainless Steels 87, The Institute of Metals, London, UK, 1988, pp. 127–137.
18. Susan Szklarska-Smialowska, British Corrosion Journal, 10 (1975), pp. 11–16.
19. N. Sato, Transactions of the SAEST, 34 (1999), pp. 81–87.

20. A.P. Bond, in Localised Corrosion – Cause of Metal Failure, ASTM STP 516, American Society for Testing Materials, Philadelphia, USA, 1972, pp. 250–261.

21. N. Sato, Corrosion Science, 37 (1995), p. 1947.

22. A.A. Seys, M.J. Brabers and A.A. van Haute, Corrosion, 30 (1974), p. 47.

23. U.U. Kamachi Mudali, R.K. Dayal, J.B. Gnanamoorthy and P. Rodriguez, Iron and Steel Institute of Japan International, 36 (1996), pp. 799–806.

24. Z. Szklarska-Smialowska and M. Janik-Czachor, Corrosion Science, 7 (1967), p. 65.

25. A.J. Sedriks, International Metals Reviews, 28 (1983), pp. 295–307.

26. Z. Szklarska-Smialowska and E. Lunarska, Werkstoffe and Korrosion, 32 (1981), p. 478.

27. Eklund, N.D. Tomashow, G.P. Tchernova and N. Markova, Corrosion, 20 (1964), p. 166t.

28. J. Stewart and D.E. Williams, Corrosion Science, 33 (1992), pp. 457–474.

29. U. Kamachi Mudali, R.K. Dayal, J.B. Gnanamoorthy, S.M. Kanetkar and S.B. Ogale, Materials Transactions JIM, 31 (1991), pp. 845–853.

30. U. Kamachi Mudali, R.K. Dayal, J.B. Gnanamoorthy and P. Rodriguez, Materials Transactions JIM, 37 (1996), pp. 1568–1573.

31. T.P.S. Gill, J.B. Gnanamoorthy and K.A. Padmanabhan, Corrosion, 43 (1987), p. 208.

32. A. Garner, Materials Performance, 21 (1982), p. 9.

33. K.E. Pinnow and A. Moskowitz, Welding Journal, 49 (1970), p. 278s.

34. H.J. Dundas and A.P. Bond, Paper presented at NACE Corrosion 75, preprint No. 159, 1975.

35. N.D. Tomoshav, G.P. Chernova and O.N. Markova, Zashchita Metallov, 7(1971), p. 104.

36. J.B. Lumsden, in Passivity of Metals, Eds. R.P. Frankenthal and J. Kruger, Corrosion Monograph Series, The Electrochemical Society, USA, 1978, pp. 730–739.

37. K. Sugimoto and Y. Sawada, Corrosion Science, 17 (1977), p. 425.

38. E.A. Lizlovs and A.P. Bond, J. of Electrochemical Society, 116 (1969), p. 574.

39. T. Sydberger, Werkstoffe and Korrosion, 32 (1981), p. 179.

40. U. Kamachi Mudali, R.K. Dayal, S. Venkadesan and J.B. Gnanamoorthy, Metals, Materials and Processes, 8 (1996), pp. 139–146.

41. T. Sundararajan, U. Kamachi Mudali, K.G.M. Nair, S. Rajeswari and M. Subbaiyan, Werkstoffe and Korrosion, 50 (1999), pp. 344–349.

42. K. Arumugam, Ph.D. Thesis, University of Madras, 1998.

43. U. Kamachi Mudali, R.K. Dayal, T.P.S. Gill and J.B. Gnanamoorthy, Werkstoffe and Korrosion, 37 (1986), pp. 637–643.

44. U. Kamachi Mudali, R.K. Dayal, J.B. Gnanamoorthy and P. Rodriguez, Transactions of the IIM, 50 (1997), pp. 37–47.

45. U. Kamachi Mudali, R.K. Dayal, J.B. Gnanamoorthy and P. Rodriguez, Metallurgical Transactions A, 27 (1996), pp. 2881–2887.

46. U. Kamachi Mudali, S. Ningshen, A.K. Tyagi and R.K. Dayal, Materials Science Forum, 318–320 (1999), pp. 495–502.

47. U. Kamachi Mudali and R.K. Dayal, J. Materials Science, 35 (2000), pp. 1799–1803.

48. J.E. Truman, M.J. Coleman and K.R. Pirt, British Corrosion Journal, 12 (1977), p. 236.

49. R.C. Newman, Y.C. Lu, R. Bandy and C.R. Clayton, in Proceedings of the Ninth International Congress on Metallic Corrosion, National Research Council, Ottawa, Canada, 4 (1984), p. 394.

50. H. Ohno, H. Tanabe, A. Sakai and T. Misawa, Zairyo-to-Kankyo, 47 (1998), pp. 584–590.

51. U. Kamachi Mudali, Corrosion of High Nitrogen Steels – Effects of Nitrogen Addition on Passivation Kinetics, Composition of Passive Films and Pitting Corrosion in Fe-N Model Alloys, Max-Planck Institute for Iron Research, Dusseldorf, FRG, 1994.

52. H. Tanabe, U. Kamachi Mudali, K. Togashi and T. Misawa, Journal of Materials Science Letters, 17 (1998), pp. 551–553.

53. Y. Ohta, Kiyoshi, T. Yoshida and I. Takahashi, Transactions of Iron and Steel Institute of Japan, 27 (1987), p. B-115.

54. M.O. Speidel, Proceedings of the International Conference on Stainless Steels 87, The Institute of Metals, London, UK, 1998, pp. 247–252.

55. N. Suutala and M. Kurkela, Proceedings of the International Conference on Stainless Steels 84, The Institute of Metals, London, UK, 1984, pp. 240–247.
56. T. Ogawa, S. Aoki, T. Sakamoto and T. Zaizen, Welding Journal, 61 (1982), p. 139s.
57. N.D. Tomoshav, G.P. Chernova and O.N. Markova, Corrosion, 20 (1964), p. 166t.
58. H. Bohni and H.H. Uhlig, Corrosion Science, 9 (1969), p. 353.
59. U. Kamachi Mudali, A.K. Bhaduri and J.B. Gnanamoorthy, Materials Science and Technology, 42 (1990), pp. 475–481.
60. A. Moskowitz et. al., ASTM STP 418, American Society for Testing Materials, Philadelphia, USA, 1967, p. 3.
61. H.J. Dundas and A.P. Bond, Paper presented at NACE Corrosion 81, Paper No. 122, 1981.
62. B.E. Wilde, Corrosion, 42 (1986), p. 147.
63. G.S. Eklund, Scandinavian Journal of Metallurgy, 6 (1977), pp. 196–201.
64. K. Elayaperumal, P.K. De and J. Balachandra, Corrosion, 28 (1972), pp. 269–273.
65. U. Kamachi Mudali, N. Parvathavarthini, R.K. Dayal and J.B. Gnanamoorthy, Transactions of the Indian Institute of Metals, 41 (1988), pp. 35–40.
66. U. Kamachi Muddle, S. Ningshen, P. Shankar, K. Sanjay Rai and R.K. Dayal, Proceedings of the International Symposium on Electrochemical Methods in Corrosion Research (EMCR 2000), Budapest (Hungary), 2000, paper no. 125.
67. N.D. Tomoshav, G.P. Chernova and O.N. Markova, Zashchita Metallov, 6 (1970), p. 21.
68. H.H. Uhlig and J.R. Gilman, Corrosion, 20 (1964) 289t.
69. R.C. Newman and M.A.A. Ajjawi, Corrosion Science, 26 (1986) 1057.
70. I.L. Rosenfeld, Corrosion Inhibitors, McGraw-Hill International Book Company, New York, USA (1981), p. 27 and p. 160.
71. Z. Szklarska-Smialowska, Corrosion, 27 (1971), pp. 223–233.
72. P. Forchhammer and H.J. Engell, Werkstoffe and Korrosion, 20 (1969), p. 1.
73. E. Brauns and W. Schwenk, Arch. Eisenhuttenw, 32 (1961), p. 387.
74. M. Janik-Czachor and Z. Szklarska-Smialowska, Corrosion Science, 8 (1968), p. 215.
75. T.G. Gooch, The Welding Institute Research Bulletin, USA, 15 (1974), p. 183.
76. P. Rodriguez, D.K. Bhattacharya and S.L. Mannan, Paper presented at Indian Institute of Welding Seminar, Bombay, India, 1976.
77. T.P.S. Gill, M. Vijayalakshmi, J.B. Gnanamoorthy and K.A. Padmanabhan, Welding Journal, 66 (1986), p. 122-s.
78. R.A. Farrar, Journal of Materials Science, 20 (1985), p. 4215.
79. A. Garner, Corrosion, 35 (1979), p. 108.
80. U. Kamachi Mudali, T.P.S. Gill, R.K. Dayal and J.B. Gnanamoorthy, Proceedings of the International Welding Conference IWC-87, New Delhi, 1987, pp. 465–476.
81. U. Kamachi Mudali, R.K. Dayal, T.P.S. Gill and J.B. Gnanamoorthy, Corrosion, 46 (1990), pp. 454–460.
82. M.G. Pujar, U. Kamachi Mudali, R.K. Dayal and T.P.S. Gill, Corrosion, 48 (1992), pp. 579–586.
83. U. Kamachi Mudali, M.G. Pujar and R.K. Dayal, Materials Science and Technology, 16 (2000), pp. 393–398.
84. A. Garner, Metals Progress, 127 (1985), p. 31.
85. M.E. Carruthers, Welding Journal, 35 (1959), p. 259-s.
86. J.W. Pugh and J.D. Nisbet, Trans. AIME, 188 (1950) 268.
87. Y. Arata, F. Matsuda and S. Saruwatari, Trans. JWRI 3 (1974) 79.
88. Y. Arata, F. Matsuda and S. Katayama, Trans. JWRI 6 (1977) 105.
89. H. Their, DVS-Ber. 41 (1976) 100.
90. F.C. Hull, Weld. J. 46 (1967) 399-s.
91. H. Fredriksson, Metall. Trans. 3 (1972) 2989.
92. D.L. Olson, Weld. J. 64 (1985) 281-s.
93. N. Suuatala, T. Takalo and T. Moisio, Metall. Trans. 11A (1980) 717.
94. N. Suuatala and T. Moisio, Solidification technology in the foundry and casthouse, The Metals Society, London, pp. 310–314 (1983).
95. N. Suuatala, Met. Trans. 14A (1983) 191.

96. W.T. DeLong, Weld. J. 53 (1974) 273-s.
97. D.J. Kotecki and T.A. Siewert, Weld. J. 71 (1992) 171-s.
98. A. Schaeffler, Weld. J. 26 (1947) 1.
99. J.A. Brooks, Weldability of Materials, Ed. R.A. Patterson and K.W. Mahin, ASM International, Ohio, p. 41-47 (1990).
100. J.A. Brooks, M.I. Baskes and F.A. Greulich, Metall. Trans. A in press.
101. J.A. Brooks and M.I. Baskes, Second Intl. Conf. On Trends in welding research, Ed. S.A. David, American Society for Metals, Materials Park, OH, pp. 153–158 (1989).
102. O. Hammar and U. Svenson, Solidification and Casting of Metals, The Metals Soc., London, pp. 401–410 (1979).
103. A.J. Gouch and H. Muir, Metal Construction, 150 (1981).
104. R.A. Farrar, Stainless Steels 1984, The Institute of Metals, London, pp. 336–342 (1985).
105. M. Lindenmo, Stainless Steels 1984, The Institute of Metals, London, pp. 262–270 (1985).
106. M. Liljas, B. Holmerg and A. Ulander, Stainless Steels 1984, The Institute of Metals, London, pp. 323–329 (1985).
107. M.J. Cieslak and W.F. Savage, Weld. J. (1981) 131-s.
108. F. Matsuda, S. Katayama and Y. Arata, Trans. JWRI 10 (1981) 201.
109. J.A. Brooks, Weld. J. 53 (1974) 517-s.
110. P.I. Marshall and T.G. Gooch, Corrosion, 49 (1993) 514.
111. T.P.S. Gill, U. Kamachi Mudali, V. Seetharaman and J.B. Gnanamoorthy, Corrosion 44 (1988) 511.
112. B.E. Wilde, J.S. Armijo, Corrosion, 23 (1967) 208.
113. V. Scotto, G. Ventura and E. Traverso, Corros. Sci., 19 (1979) 237.
114. J. Stjerndahl and C. Dacker, Scand. J. Met., 9 (1980) 217.
115. A.I. Grekula, V.P. Kujanpaa and L.P. Karjalainen, Corrosion 40 (1984) 569.
116. W.R. Cieslak, D.J. Duquette and W.F. Savage, Proc. Conf. Sponsored by the Joining Div. ASM., Ed. S.A. David, American Society for Metals, Ohio, pp. 361 (1981).
117. N. Suuatala, "Solidification Studies on Austenitic Stainless Steels", Acta Universitatis Ouluensis C23, Metallurgica 3., University of Oulu, Finland, 1982.
118. P.E. Manning, C.E. Lyman and D.J. Duquette, Corrosion 36 (1980) 246.
119. R.J. Brigham and E.W. Tozer, Can. Metall. Q. 16 (1977) 48.
120. A. Garner, Welding J., (1983) 27.
121. Standard test methods for Pitting and Crevice Corrosion Resistance of Stainless Steels and Related Alloys by Use of Ferric Chloride Solution, Annual Book of ASTM Standards, Vol. 03.02, ASTM, Philadelphia, PA, 1995. pp. 174–179.
122. Standard Guide of Examination and Evaluation of Pitting Corrosion, Annual Book of ASTM Standards, Vol. 03.02, ASTM, Philadelphia, PA, 1995. pp. 162–168.

4. Crevice Corrosion of Stainless Steel

R.K. Dayal[1]

Abstract Crevice corrosion is a serious problem for stainless steel exposed in corrosive environment. This article describes the corrosion mechanisms, its similarities with other types of localized corrosion, various test methods, factors influencing the corrosion process and different protection methods. Particular emphasis has been placed on the crevice corrosion of austenitic stainless steels, since these are used as major structural materials for nuclear power plants.
Key Words Crevice corrosion, austenitic stainless steels, cold work, grain size, texture, sensitization.

INTRODUCTION

A form of localised attack that occurs at shielded areas on the metal surface exposed to certain specific environments is known as Crevice Corrosion. These shielded areas or crevices could be formed on riveted connections, gasket fittings, spot-welded lap joints, porous welds, coiled or stacked sheets of metals, marine or debris deposits. These sites are often unavoidable in component design or may arise during exposure period. This type of corrosion is a major practical problem affecting the stainless steels since its occurrence is unpredictable in chloride containing environments e.g. sea water, brackish water and many other environments used in chemical and power industries. The crevice of a few micrometer width is quite susceptible to this type of corrosion and can lead to various types of failures [1–8]. This article describes the corrosion mechanisms, its similarities with other types of localised corrosion, various test methods, factors influencing the corrosion process and different protection methods. Particular emphasis has been placed on the crevice corrosion of austenitic stainless.steels, since these are used as major structural materials for nuclear power plants.

MECHANISM OF CREVICE CORROSION

Extensive investigations have been carried out towards understanding the mechanism of crevice corrosion and reviews on this aspect have been published [2, 9, 10].

All the metals, which are ordinarily in a passive state, are prone to crevice corrosion. Principally, the alloys whose corrosion rate is controlled by the anodic process are susceptible to this type of corrosion. As a result of the small volume of the crevice, the poor access of electrolyte to it and the

[1]Aqueous Corrosion and Surface Studies Section, Corrosion Science and Technology Division, Materials Characterisation Group, Metallurgy and Materials Group, Indira Gandhi Centre for Atomic Research, Kalpakkam-603 102, India.

slow removal of reaction products, certain changes take place in the composition and the nature of the corrosion medium in the crevice. Such changes may be manifested as a reduction in the concentration of the oxidiser or passivator or as acidification of the electrolyte as a result of hydrolysis.

1. Change in Oxygen Concentration in Crevices

In neutral electrolytes oxygen is the principal cathodic depolariser. The oxygen in the crevice is consumed quite rapidly either in the cathodic reaction or in passivation. This inhomogeneity in oxygen concentration leads to the formation of concentration cell. Schafer et al [11, 12] suggested that crevice corrosion is a special case of differential aeration. However, the change in pH of the medium seems to be of greater importance than the change in the oxygen concentration. It follows that differential aeration while changing markedly the potential of metals, is incapable of effecting a substantial change in the rate of anodic process in passive alloys resulting in the severity of the corrosion attack. Therefore, there must be other factors responsible for the intensive corrosion of metals in the crevice.

2. Acidification of the Medium Due to Hydrolysis

When corrosion proceeds inside the crevice, one of processes is the hydrolysis of. the corrosion products. Peterson et al [13] have established that during crevice corrosion of type 304 stainless steel in sea water, a strongly acidic electrolyte (pH $\approx$ 1.2–2) exists within the crevice and the corrosion product contained a substantial amount of Fe^{+2} ions and only traces of Fe^{+3} ions. Considering the thermodynamics of the possible reactions it was concluded that such a low value of pH could be due to the hydrolysis of chromic ions in the crevice. The lowering of pH has also been reported by Bates [14] for 304 stainless steels, by Marek and Hochman [15] for 316 stainless steels in neutral chloride environment and also by other investigators [16–18].

3. Acidification of the Medium Due to Anodic Polarisation

Acidification of the corrosive medium may also take place as a result of anodic polarisation, which apparently is not accompanied by hydrolysis.

$$M + H_2O \rightarrow MO + 2H^+ + 2e$$

$$M + 2H_2O \rightarrow M(OH)_2 + 2H^+ + 2e$$

$$M + 2Cl^- \rightarrow MCl_2 + H_2O \rightarrow MO + 2H^+ + 2Cl^-$$

It can be seen that irrespective of the adopted scheme of anodic oxidation, it invariably leads to acidification of the medium.

4. Influence of pH on Corrosion

The influence of concentration of hydrogen ions on the anodic behaviour of stainless steels has been reported [2]. It was shown on 17% Cr steel that activation of the steel in 0.5 N NaCl at room temperature already sets in when the pH drops to 1.1 to 0.34. As the pH of the electrolyte dropped, the passivation current rises.

In view of the fact that the electrolyte in the crevice is acidified, many investigators believe that the possible onset of crevice corrosion and the rate of its propagation depend upon the behaviour of metal in an active state. For this reason efforts were made to investigate the anodic behaviour of metals in active state. The critical passivation current serves as the criterion for the ranking of alloys towards crevice corrosion resistance.

5. Change in Potential of Metals in Crevices
A change in the kinetics of electrochemical reactions arising either as a result of lower concentration of oxygen or of higher concentration of hydrogen ions, brings a change in the open circuit potential of metals in crevices. As the concentration of oxygen diminishes the potential of most of metals is shifted to a more active value [2]. This effect is greatly dependent upon the pH of the bulk solution.

6. Formation of Active-Passive Cell
Since the potential of the metal in the crevice differs considerably from the potential of the metal freely exposed to the electrolyte, favourable conditions are created for the formation of a macro-galvanic couple in which the metal in the crevice acts as an anode. Under these conditions quite powerful corrosion couples function on different metals. There exists an incubation period during which the cell current is very small (~1 $\mu A/cm^2$) and after which a high current (~400 $\mu A/cm^2$) is observed. Differential aeration by itself is incapable of developing a considerable electromotive force in the cell. During the incubation period the concentration of hydrogen ions rises in the crevice as a result of anodic polarisation and when a critical magnitude is reached, the alloy passes from passive to active state. Hence the differential aeration is only the initial cause and the basic reason for destruction of the metal in the crevice is the development of active-passive macro-couples and the acidification of the medium in the crevice [2, 12, 14, 19].

7. Unified Crevice Corrosion Mechanism
Based on the knowledge of the influence of various factors described above a unified crevice corrosion mechanism has been proposed [20] and is described below:

Initially the anodic dissolution ($M \rightarrow M^+ + e$) and cathodic reduction ($O_2 + 2H_2O + 4e \rightarrow 4OH^-$) processes occur uniformly over the entire metal surface, including the crevice exterior (Fig. 1). The oxygen in the shielded crevice area is consumed after some incubation period, but the decrease in cathodic reaction rate is negligible because of the small area involved. Consequently, the corrosion of the metal inside and outside the crevice continues at the same rate. With the cessation of the cathodic

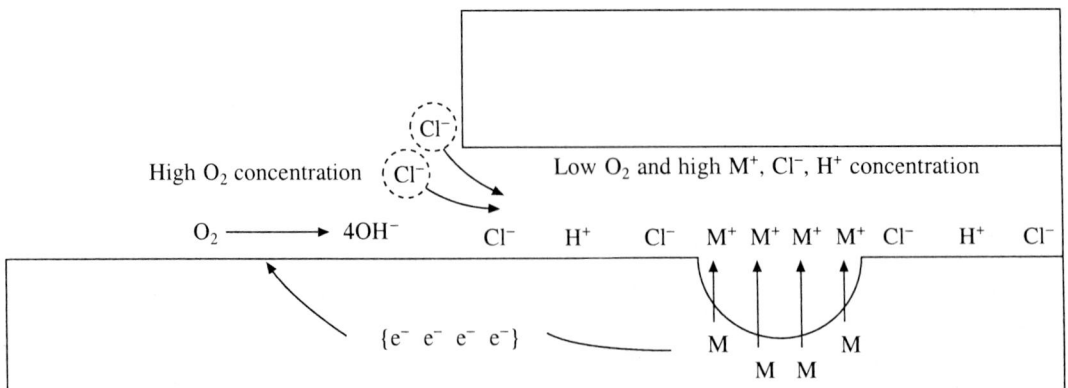

Fig 1.   Crevice corrosion mechanism

hydroxide producing reactions however, the migration of negative ions (e.g. chlorides) into the crevice area is required to maintain charge balance. The resulting metal chloride hydrolyses to insoluble metal hydroxides and hydrochloric acid, which results in the progressive acidification of the crevice. Both the chloride ions and low pH accelerate crevice corrosion in a manner similar to autocatalytic pitting while reduction reaction cathodically protects the exterior surface.

SIMILARITIES WITH OTHER FORMS OF LOCALISED CORROSION

Brown [21] has proposed an "occluded corrosion cell" concept in which local acidification by hydrolysis is a common feature of all types of localised corrosion attack such as crevice corrosion, filiform corrosion, pitting, stress corrosion cracking, intergranular attack. In all these types of corrosion attack is confined to only a local site which becomes anodic and the outside area becomes cathodic.

With the help of electrochemical techniques Bombara [22] has shown an electrochemical and geometrical analogy between stress corrosion cracking and crevice corrosion. Initiation of stress corrosion cracking is normally observed from deep narrow pits that appear to be similar to crevices.

Schafer et al [12] proposed a similarity in the mechanism of crevice corrosion and pitting. In both the cases differential aeration or inhibitor depletion is decisive in determining the locations at which corrosion is initiated. Pitting is considered to be merely a limiting case of crevice corrosion, the difference being that macroscopic geometrical factors determine the initiation sites of crevice corrosion while random microscopic factors usually determine the sites of initiation of pitting corrosion.

The initiation sites for local corrosion attack are inclusions in stainless steels [23–25]. It has been found that predominant initiation sites are complex sulphide/oxide inclusions in the steel. Dissolution of sulphide shell around an oxide particle results in narrow crevices between the oxide and the metal. From this active site the propagation of pits can be considered as one kind of crevice corrosion. Nearly close pits on stainless steels act as crevices and retard the diffusion of passivators [26]. Uhlig [27] has presented the distinguishing characteristics of pitting and crevice corrosion. Pitting of a passive metal initiates at a well defined potential at which chloride ions on favoured sites successfully displace oxygen of the passive film and form anodic areas whereas crevice corrosion occurs because of changes in environment which lead to a loss in passivity.

METHODS FOR ASSESSING CREVICE CORROSION

Various laboratories have developed different crevice corrosion tests. The procedure and specimen design is unique to the originating laboratory and their design is based on the nature of the specific investigation. All these methods can be classified into two main groups: (1) direct immersion tests and (2) electrochemical tests.

1. Direct Immersion Tests
In these tests a part of the metal surface is covered by a non-metal to produce a crevice at the contact surface and the assembly is put in the corrosive environment for a certain length of time so as to simulate the service conditions. Various types of such assemblies (Fig. 2) have been described in literature [13, 28–37]. A few tests have been included as ASTM standards [35] and are described below:

(i) Ferric Chloride (ASTM G-48B) Test
In this maximum corrosion depth and or mass loss is measured after immersion for a set period of 72

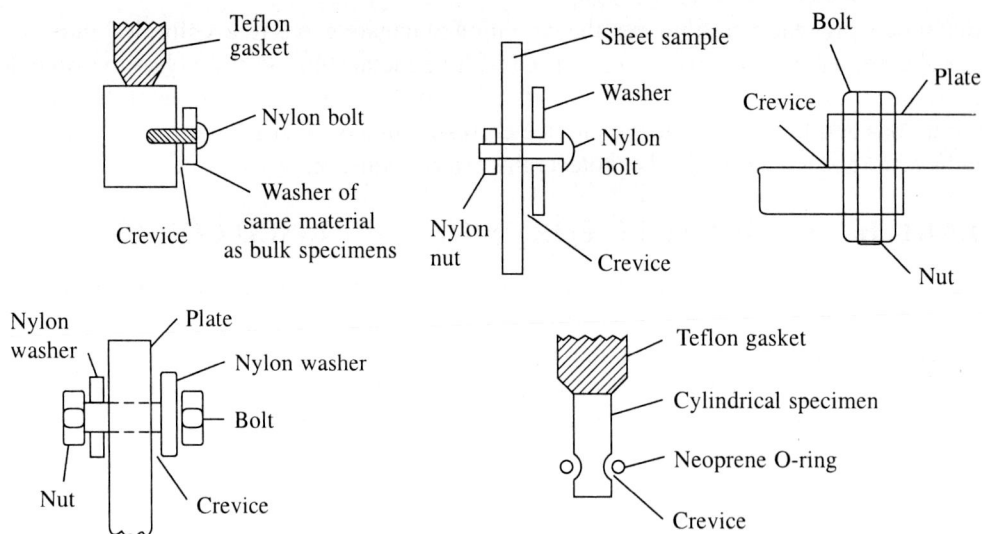

Fig. 2.   Various forms of crevice assemblies

hours at a set temperature as established by user. Two plastic blocks held in a place on specimen faces with rubber bands form crevices.

(ii) Multiple Crevice Assembly (MCA) (ASTM G-78) Test

In this test a large number of crevices are introduced. The multiple crevices are formed by two plastic serrated washer bolted to the specimen faces through a hole drilled in the specimen. MCA's typically have 20 contact points on each washer. The torque applied to the assembly determines the crevice gap. The assembly is immersed in the solution for a set period typically 30 days at ambient temperature. The mass loss, maximum crevice depth and measurement of maximum depth are suggested as evaluation techniques. Statistical treatment (e.g. counting the number and depth of attacked sites) is possible for minimum three tests.

(iii) Critical Crevice Temperature Test (ASTM G-48D)

Specimen with a crevice former is immersed in a chloride bearing solution at a set temperature for a set period of 72 h. The sample is examined for sign of attack after the test. If no attack is observed, new sample is immersed again at an incrementally higher temperature in a fresh solution. The process is repeated until attack is observed at a particular temperature, which is termed as Critical Crevice Temperature (CCT) and is used for ranking of various alloys [36, 38].

2. Electrochemical Tests

Direct immersion tests mostly provide data for comparing the susceptibility of metal to crevice corrosion in a corrosive environment whereas the electrochemical tests, besides comparing the crevice corrosion susceptibility provide necessary data for understanding the mechanism of corrosion. Many investigators have used specially designed assemblies for the electrochemical studies of the crevice corrosion [14–16, 39–45].

(i) Critical Crevice Potential

A concept of critical crevice potential analogues to Critical pitting potential has been used to study the influence of chemical and metallurgical factors on crevice corrosion [4, 45]. In this specimen with fixed crevice geometry (Fig. 3) is potentiodynamically tested in a given solution. The critical crevice potential (E_{cc}) is determined above which monotonous increase in current is observed (Fig. 4) and crevice corrosion takes place under the crevice former. The E_{cc} has been found to be independent of potential scan rate in the range 5–100 mV/min [45].

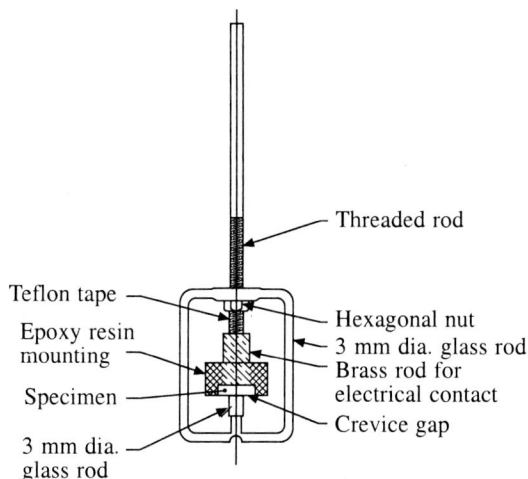

Fig. 3. Crevice assembly for electrochemical test.

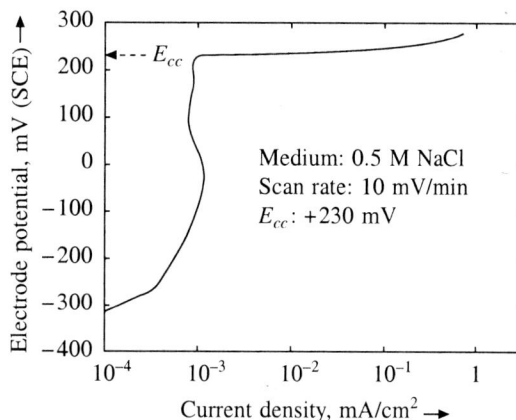

Fig. 4. Polarisation curve in presence of crevice.

(ii) Crevice Repassivation Potential

In this technique a cylindrical specimen with a tapered PTFE collar is polarised to a high noble potential to initiate crevice corrosion. After a certain current density is reached, the potential is decreased and the potential at which the material is repassivated is determined for crevice corrosion evaluation. With some modification in cyclic polarisation test, a parameter, Crevice Repassivation Potential has been proposed. This is the potential below which crevice corrosion does not propagate. It is recognised to be an electrochemical parameter for repassivation of crevice corrosion [46].

The crevice corrosion resistance of various alloys can also be evaluated by conducting cyclic potentiodynamic polarisation test on samples without crevice and estimating hysterisis loop area (ASTM G-61).

INFLUENCE OF VARIOUS FACTORS ON CREVICE CORROSION

Figure 5 illustrates a number of possible factors that can influence the initiation of crevice corrosion attack [47]. In the following sections some of these factors have been discussed.

1. Crevice Geometry

The crevice geometry includes the gap, depth, bulk to crevice area ratio, number of crevices and type of crevices (metal or non-metal). In practical situations, the area ratio is likely to be high and therefore the initiation of crevice corrosion is generally controlled by the passive current. If a number of crevices are present on one piece of equipment then, depending on their proximity some interaction by way of cathodic protection might be expected once one crevice has broken down.

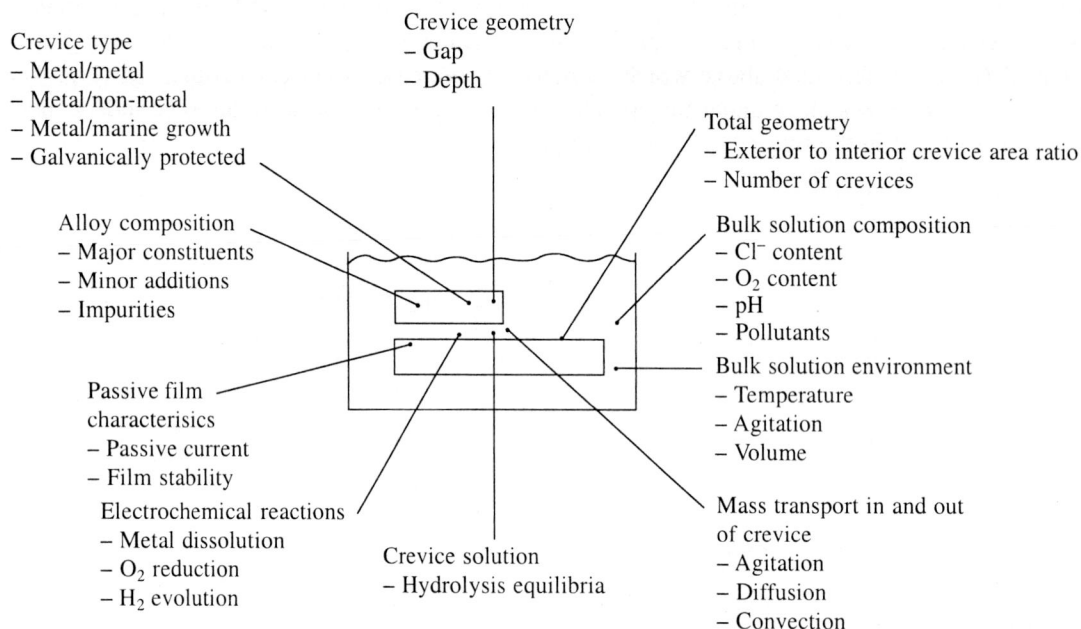

Crevice type
– Metal/metal
– Metal/non-metal
– Metal/marine growth
– Galvanically protected

Crevice geometry
– Gap
– Depth

Total geometry
– Exterior to interior crevice area ratio
– Number of crevices

Alloy composition
– Major constituents
– Minor additions
– Impurities

Bulk solution composition
– Cl⁻ content
– O₂ content
– pH
– Pollutants

Passive film
characterisics
– Passive current
– Film stability

Bulk solution environment
– Temperature
– Agitation
– Volume

Electrochemical reactions
– Metal dissolution
– O₂ reduction
– H₂ evolution

Crevice solution
– Hydrolysis equilibria

Mass transport in and out
of crevice
– Agitation
– Diffusion
– Convection

Fig. 5.   Various factors influencing crevice corrosion.

Wide scatter of data in crevice corrosion studies results from the inability to reproduce the geometry of a tight crevice. With all other parameters held constant it is predicted [48], using a mathematical model, that a difference in crevice gap of 0.01 µm or less could be the determining factor controlling the occurrence of the attack. Variation in the degree of tightening could obviously affect an entire range of gaps; likewise, irregularities or variation in specimen surface roughness could also affect the true gap. For each crevice gap a minimum crevice depth exists to satisfy the condition for breakdown. Increasing the depth and decreasing the gap are detrimental [47–49] in combating the crevice corrosion.

2. Alloying Elements

Major alloying elements like Ni, Cr, Mo , N increase the resistance to crevice corrosion. The addition of Mo in stainless steel has been shown to cause a large increase in resistance to crevice corrosion [50, 51]. Increasing the Cr content from 20 to 30% while maintaining the Mo content at 3.5% and Ni content at 22% has also shown a large increase in crevice corrosion resistance [52–54]. Si, N and Cu when present in Mo containing stainless steel have beneficial effect on the crevice corrosion resistance in sea water [10, 55].

The addition of S is highly detrimental to austenitic stainless steel. Eklund [24] using scanning electron microscopy has shown that crevice corrosion initiates from sulphide inclusions present under the crevice. No effect of Nb and Ti has been found during a sea water test. Pd and Rh were found to be detrimental in Fe-Cr-Mo alloys in ferric chloride solution [10].

3. Surface Composition and Surface Treatment

The surface conditions and composition of stainless steel plays an important role in the initiation of crevice corrosion. A direct relationship between surface enrichment of chromium and the resistance against crevice corrosion has been found which is independent of the presence of non-metallic inclusions [56]. The electropolished surface [57] showed least effect on crevice corrosion. MnS inclusions have been shown to be the initiation sites for the crevice corrosion [24, 55, 58]. Pickling and passivation treatment removes surface inclusions, which act as crevice sites. Pickling has also been demonstrated to improve crevice corrosion resistance by removal of Cr depleted layers formed during annealing treatment [54]. Surface modification of stainless steels by Ce ions can also make the surface more resistance against crevice corrosion [10, 59].

4. Microstructure

There have been limited evaluations of crevice corrosion resistance of stainless steels which have examined the effect of microstructure. The studies on duplex stainless steels have identified austenitic/ferrite boundaries as susceptible sites for initiation and propagation of crevice attack [10]. Studies by Rowland [60] suggest that precipitation hardening may have a slightly detrimental effect on the crevice corrosion resistance. Sigma and chi phases have been shown to be detrimental to various types of stainless steels [61, 62]. It has been reported that the martensite formed by cold working of type 304 stainless steels is detrimental to crevice corrosion [63].

Influence of microstructure on the crevice corrosion behaviour of austenitic steels (AISI types 304, 310 and 316) was investigated in detail [4]. The grain size, carbide precipitation, cold work and texture in the material have been found to influence the crevice corrosion resistance significantly. The crevice corrosion resistance was evaluated in a 0.5N NaCl solution using potentiodynamic method for a fixed crevice geometry and represented in terms of a critical crevice potential E_{cc}.

1. Influence of Grain Size

The resistance to crevice corrosion decreases with a decrease in the average grain diameter d following a linear relationship between E_{cc} and $d^{-1/2}$ (Fig. 6) [64]. The inverse grain size dependence for the E_{cc} was also found to be appropriate [65]. The grain size influence is explained due to the presence of

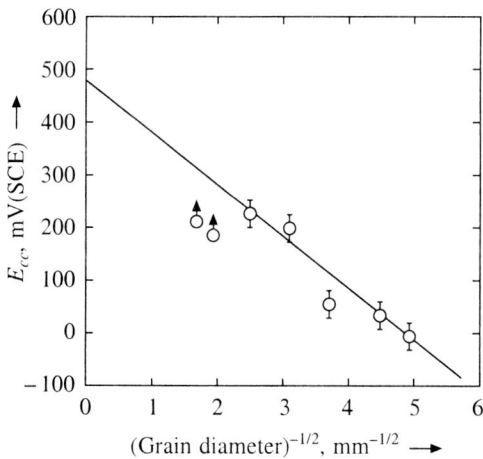

Fig. 6.   Grain size effect on crevice corrosion.

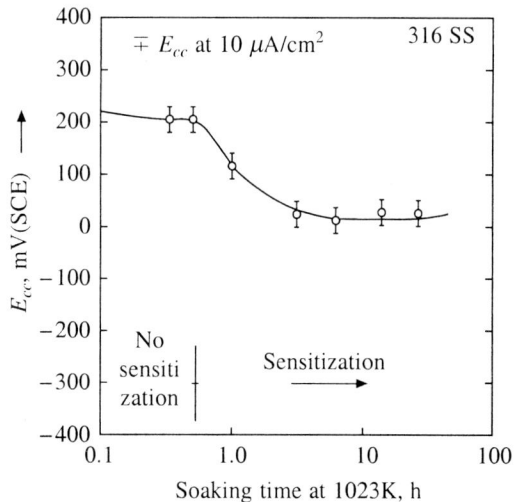

Fig. 7.   Sensitization effect on crevice corrosion.

large grain boundary length per unit area for smaller grain size material where crevice corrosion attack initiates.

2. Influence of Carbide Precipitation (Sensitization)
The sensitised specimens shows increased susceptibility to crevice corrosion and exhibits intergranular attack inside the crevice (Fig. 7) [64]. Due to the presence of more numerous active sites resulting from the chromium depletion in sensitized material, the crevice corrosion resistance is lowered. At higher soaking time, as the sensitization is complete, no further lowering of the crevice corrosion resistance is observed.

3. Influence of cold work
The crevice corrosion resistance decreases with increase in cold work up to about 15% and any further increase in cold work has negligible effect (Fig. 8) [66]. The increase in the number of defect sites and change in the texture produced by cold work markedly influence the crevice corrosion behaviour.

4. Influence of Texture
In cold rolled condition the stainless steel shows more noble E_{cc} values on the rolling surface than on the other two sections (Fig. 8). These anisotropy effects are related to the difference in the crystallography of the planes parallel to the three different cross sections [66].

The crevice corrosion studies were made on types 304, 310 and 316 stainless steels as a function of crystallographic texture [66]. The results have indicated that the resistance to crevice corrosion improves when a less closely packed plane was preferentially oriented parallel to the exposed surface. This was explained due to the formation of an active-passive galvanic cell between crevice and no-crevice areas. If a plane of lower packing density is oriented parallel to the exposed surface, the macrogalvanic couple will have a lower intensity since the difference in the potential between the active state inside the crevice and passive state outside the crevice remains smaller.

As seen in the literature limited data are available regarding the effect of metallurgical variables on crevice corrosion. During the fabrication of a component, the microstructure of the material undergoes considerable changes. Therefore during the design stage itself one should take into account the beneficial or the adverse effect of different metallurgical variables.

Fig. 8. Cold work and texture effect on crevice corrosion.

PROTECTION AGAINST CREVICE CORROSION

Crevice corrosion problem in stainless steels can be avoided by proper design of components and

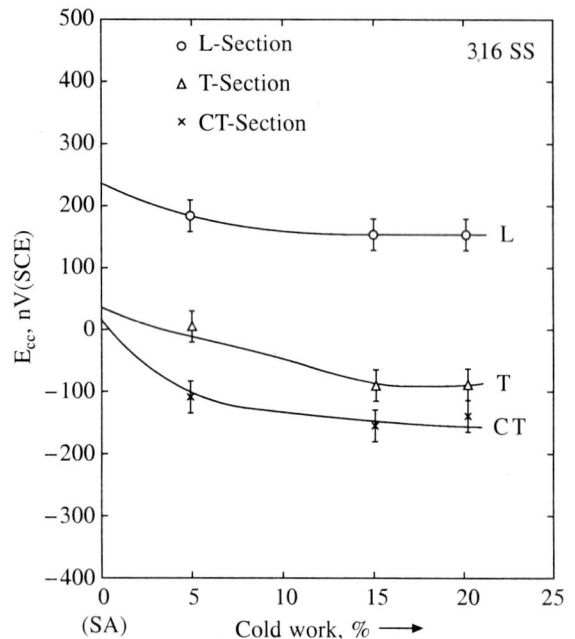

structures so as to avoid crevices. Elimination of crevices is the best remedy for solving crevice corrosion problems. A number of methods for minimising crevice corrosion through proper design have been developed [1, 2, 20, 67]. Welding of joints rather than bolting or revetting them, butt welded joints instead of spot welding lap joints, sealing of crevices, elimination of marine or debris deposit by periodic cleaning, avoid in design that have sharp corners or areas where moisture or debris can collect are the common methods used for preventing crevice corrosion. Avoiding metal to non-metal joints is better as these are usually tighter than metal joints.

Proper material selection is another solution for minimising crevice corrosion. Various stainless steels have been rated in order of merits for their resistance against crevice corrosion. High Mo + high N stainless steels, higher alloys containing Mo are recommended [10] for severe conditions in which crevices are not avoidable.

The crevice corrosion of stainless steels in seawater can be suppressed or minimised by cathodic protection [10]. Zn, Al or C-steel anodes are suitable for preventing crevice corrosion of type 316 stainless steels. Recently a new approach on prevention of crevice corrosion by coupling to more noble materials like graphite has been suggested [68].

REFERENCES

1. Landrum, R.L., Chemical Engineering, 76 (1969) 118.
2. Rosenfeld, I.L., Localised Corrosion, eds. R.S. Staehle, B.F. Brown and J. Kruger, NACE (1974) 373.
3. France, W.D., ASTM STP 516, ASTM (1972) 164.
4. Dayal, R.K., Ph.D. Thesis, IISc., Bangalore, (1984).
5. Dayal, R.K. and Gnanamoorthy, J.B., Corrosion and Maintenance, 4 (1981) 311.
6. Rakesh Kaul, Muralidharan, N.G., Dayal, R.K., Raju, V.R., Gnanamoorthy, J.B., Jayakumar, Kasiviswanathan, K.V., Baldev Raj and Pattu, S.; Engineering Failure Analysis, vol. 2(3) (1995) 165.
7. Shaikh, H., Khatak, H.S., Dayal, R.K., Gnanamoorthy J.B., and Sur, A., Practical Metallography, vol. 28, (1991), 143).
8. Dayal, R.K., International Corrosion Conference (CORCON-97), Dec. 3-6, 1997, Mumbai, 323.
9. Ijsseling F.B., British Corrosion J. 15 (1980) 51.
10. Sedriks, A.J., Corrosion of Stainless Steels, John Wiley and Sons, N.Y. (1996).
11. Schafer G., and Foster, P. K., J. Electrochemical Society, 106 (1959) 467.
12. Schafer, G., Gabriel I.R., and Foster, P.K., J. Electrochemical Society, 107 (1960) 1002.
13. Peterson, M. H., Lennox, T.J. (Jr) and Groover, R.E., Materials Protection and Performance, 9 (1) (1970) 232.
14. Bates, J.B., Corrosion, 29 (1973) 28.
15. Marek M. and Hochman, R.F., Corrosion, 30 (1974) 208.
16. Vernik, E.D., Vernik, E.D. (Jr.), Starr K.K. and Bowers, J.M., Corrosion, 32 (1976) 60.
17. Karlberg G., and Wranglen, G., Corrosion Science, 11 (1971) 499.
18. Lee, Y.H. Takehara, Z. and Yoshizawa, Corrosion Science, 21 (1981) 391.
19. Korovin, U.V.M. and Ulanovskii, Z.B., Corrosion, 22 (1966) 16.
20. Fontana M.G. and Greene, N.D., Corrosion Engineering, McGraw Hill Book Co., N.Y. (1978).
21. Brown, B.F., Corrosion, 26 (1970) 249.
22. Bombara, G., Corrosion Science, 9 (1969) 519.
23. Szklarska-Smialowska, Z., Szummer A., and Janik-Czachor, M., British Corrosion J., 5 (1970) 159.
24. Eklund, G.S., J. Electrochemical Society, 123 (1976) 170.
25. Karlborg G., J. Metals, 3 (1974) 46.
26. Rosenfeld, I.L. and Danilov, I.S., Corrosion Science, 7 (1967) 129.
27. Uhlig, H.H., Materials Protection and Performance, 12 (1973) 42.

28. Oldfield, J.W., International Materials Review, 32(3) (1987) 153.
29. Streicher, M.A., Corrosion, 30 (1974) 77.
30. Jackson, R.P., and Van Rooyen D., ASTM STP 516, ASTM (1972).
31. Vreeland, D.C., Materials Protection and Performance, 9 (8) (1970) 31.
32. Anderson, D.B., ASTM STP 576, ASTM (1976) 231.
33. Degerbeck J., and Gille, I., Corrosion Science, 19 (1979) 1113.
34. Streicher, M.A., Materials Protection and Performance, 22(5) (1983) 37.
35. Annual Book of ASTM Standard, ASTM, vol. 3.02 (1998).
36. Brigham, R.J., Corrosion, 30 (1974) 396.
37. Szklarska-Smialowska, Z., Corrosion Science, 18 (1978) 953.
38. Hibner, E.L., Materials Protection and Performance, 26(3) (1987) 37.
39. France (Jr.), W.D., and Green (Jr.), N.D., Corrosion, 24 (1968) 247.
40. Rosenfeld, I.L., and Marshakov., I.K., Corrosion 20 (1964) 115t.
41. Petersen, M.H., and Lennox (Jr.), T.J., Corrosion, 29(1973) 406.
42. Lizlovs, E.A., J. Electrochemical Society, 117 (1970) 1335.
43. Suzuki, T., and Kitamura, Y., Corrosion, 28 (1972) 1.
44. Jones, D.A., and Green, N.D., Corrosion, 25 (1969) 367.
45. Dayal, R.K., Parvathavarthini, N., and Gnanamoorthy, J.B., British Corrosion J., vol. 18, (1983) 184.
46. Tswijikawa, S., Motoda, S., Suzuki, Y., and Shinohora, T., Corrosion Science, 31 (1990) 441.
47. Oldfield, J.W. and Suttan, W.H., British Corrosion J., 13 (1978) 13.
48. Kain, R.M., Corrosion/81, NACE, Toronto.
49. Oldfield, J.W. and Suttan, W.H., British Corrosion J., 13 (1978) 104.
50. Lizlovs, E.A., J. Electrochemical Society, 118 (1971) 22.
51. Suzuki, T. and Kitamura, Y., Materials Protection and Performance, 16 (10) (1977) 16.
52. Dundas, H.J. and Bond, A.P., Corrosion/81, NACE, Toronto.
53. Oldfield, J.W., ACOM Report 1 (1988), Avesta, Sweden.
54. Grubb, R.E., Procd. Int. Conf. On Stainless Steels, (Tokyo) (1991) 944.
55. Ujiro, T., Yoshioka, K., Hashimoto, O., Kawasaki T., Fuyuki, S., and Amano, S., Procd. Int. Conf. On Stainless Steels, (Tokyo) (1991) 86.
56. Hultquist G. and Leygraf, C., Corrosion 36 (1980) 126.
57. Kain, R.M., NACE Corrosion/91.
58. Lott, S.E. and Alkire, R.C., J. Electrochemical Science, 136 (1989) 973.
59. Lu. Y.C., and Ives, M.B., Corrosion Science, 34 (1993).
60. Rowland, J.C., British Corrosion J. 11 (1976) 195.
61. Lott, S.C. and Alkire, R.C., J. Electrochemical Science, 136 (1989) 3256.
62. Lindsay, P.B., Materials Protection and Performance, 25(12) (1986) 23.
63. Handa, T., Miyata, Y., and Takazawa, H., Proc. 12th Int. Corrosion Congress, vol. 3B (NACE) (1993) 1986.
64. Dayal, R.K., Parvathavarthini, N., Gnanamoorthy, J.B., Rodriguez, P. and Prasad, Y.V.R.K.; Materials Letters, 2, (1984) 248.
65. Dayal, R.K., Parvathavarthini, N., Gnanamoorthy, J.B. and Rodriguez, P., Transactions of the Indian Institute of Metals, 40, (1987) 74.
66. Dayal, R.K., Parvathavarthini, N., Gnanamoorthy, J.B. and Rodriguez, P., Metals, Materials and Processes, 1, (1989) 123.
67. Degerback, J., Chemical Process Engineering, 52(12) (1971) 47.
68. Turnbull, A., Corrosion Science, 40(4/5) (1998) 843.

5. Sensitization and Testing for Intergranular Corrosion

N. Parvathavarthini[1]

Abstract   Sensitization refers to the intergranular precipitation of chromium carbides and the concomitant depletion of chromium in the regions adjacent to the grain boundaries, when austenitic stainless steels are extensively heated or slowly cooled through the temperature range of 1123K to 723K. In the sensitized condition, the material is exceptionally vulnerable to intergranular corrosion (IGC) in certain oxidising environments and high temperature aqueous environments resulting in increased susceptibility to intergranular stress corrosion cracking (IGSCC). The development of sensitized microstructure is controlled by thermodynamics of carbide precipitation and kinetics of chromium diffusion. The degree of sensitization (DOS) can be greatly influenced by several parameters such as chemical composition, cold work, grain size, microstructure and heating/cooling rate. In this chapter, the phenomenon of sensitization is analysed and the recent evaluation test methods and remedies are reviewed. Emphasis is given to the influence of some important metallurgical factors on sensitization kinetics.
Key Words   Austenitic stainless steels, sensitization, intergranular corrosion, intergranular stress corrosion cracking, Electrochemical potentiokinetic reactivation technique, cold work.

INTRODUCTION

Austenitic stainless steels have good combination of mechanical strength, fabricability and general corrosion resistance and hence are extensively used as construction material in chemical, petrochemical, fertilizer and nuclear industries. One of the major problems associated with these steels is their susceptibility to IGC due to sensitization. Numerous failures which have been reported in nuclear and petrochemical industries have been attributed to IGC/IGSCC [1, 2]. When austenitic stainless steels are extensively heated or slowly cooled in the temperature range of 1123K to 723K, chromium rich carbides precipitate along the grain boundaries leading to subsequent chromium depletion in the vicinity of the grain boundaries. This phenomenon is called sensitization. When sensitized austenitic stainless steel is exposed to corrosive environment, chromium depleted zones preferentially dissolve leading to IGC. The solubility of carbon in austenite is about 0.006% at ambient temperature. However, austenitic stainless steels generally contain about 0.05% carbon. Since chromium has high affinity for carbon, there is always strong tendency for carbide formation. During the normal cooling rates encountered during fabrication of stainless steel (e.g., welding, hot working etc.), chromium carbides can be

[1]Scientific Officer, Aqueous Corrosion and Surface Studies Section, Corrosion Science and Technology Division, Materials Characterisation Group, Indira Gandhi Centre for Atomic Research, Kalpakkam-603 102, India.

precipitated, making the steel susceptible to IGC and IGSCC. For this reason, austenitic stainless steels are generally subjected to a solution treatment between 1323K–1423K which brings precipitated carbides as well as most other intermetallic phases back into solution. This condition is maintained by quenching from the solution annealing temperature to ambient temperature which forces the elements responsible for the formation of carbides and inter metallic phases to stay in solid solution by super cooling. The material remains in a condition of metallurgical nonequilibrium. At ambient temperature the diffusivity of most of the elements is very low and hence non equilibrium condition may be maintained for practically infinite period of time. However if they are exposed to the sensitization temperature range, carbides and intermetallic phases precipitate again. Carbide precipitation is a time-temperature dependent phenomena which is determined mainly by carbon diffusion at low temperatures and carbide solubility at high temperatures. In austenitic stainless steels, the precipitated carbides are of $M_{23}C_6$ type which contains 2 to 4 times the amount of chromium than base metal and hence the immediate surroundings of the precipitates are depleted of chromium. These zones can be replenished by chromium diffusion from rest of the matrix. This phenomenon is called 'desensitization' and this is much pronounced only at high temperatures. It is well- known that minimum 12% chromium is required to maintain passivity in stainless steels. Hence although carbides precipitate, the material becomes prone to IGC only and only if chromium depleted region along the grain boundaries are continuous and chromium content is less than 12%. This is illustrated in Fig. 1 [3] where time-temperature combinations leading to initiation of chromium carbide precipitation is indicated by curve (i) and regions prone to IGC is indicated by curve (ii).

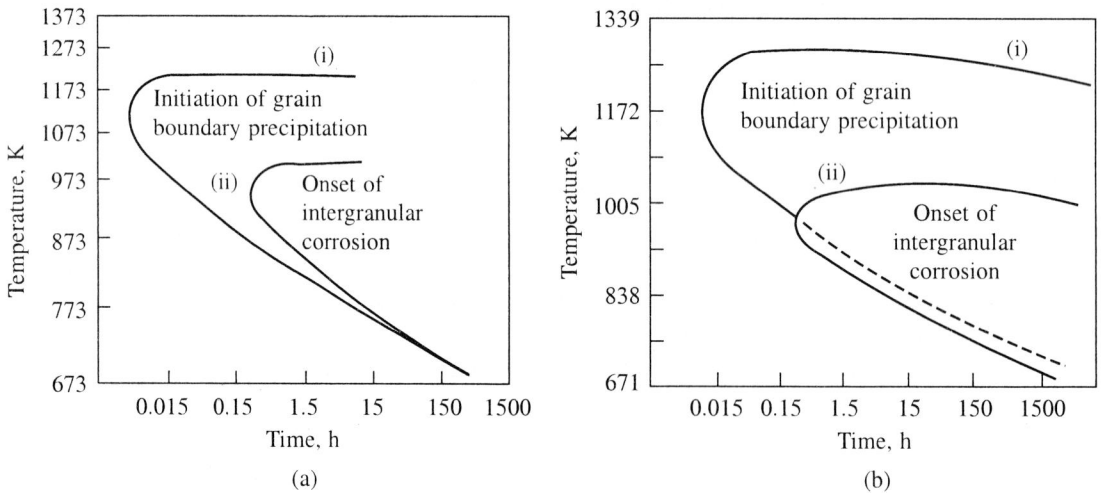

Fig. 1.   Relation between $M_{23}C_6$ precipitation and intergranular corrosion in type 304 SS (intergranular corrosion detected by ASTM A262 practice E test): (a) Alloy containing 0.05% carbon originally quenched from 1523K. (b) Alloy containing 0.038% carbon originally quenched from 1533K [3].

WELD DECAY AND KNIFE LINE ATTACK

Sensitization and the concomitant IGC of the heat affected zone (HAZ) of the weldment is known as weld decay. The HAZ in a weldment is situated in the base metal adjacent to fusion line. It experiences thermal cycles which depend upon the distance from the fusion line. The thermal cycles imposed on HAZ brings about microstructural changes which lead to sensitization in austenitic stainless steel weldments. Other factors like carbon content, plate thickness, joint geometry remaining constant, extent of $M_{23}C_6$ formation during welding depends on the time the HAZ spends in the critical temperature range which in turn depends on the heat input. Sensitized microstructure, in the presence of tensile stress and corrosive medium leads to IGSCC.

Knife line attack (KLA) is a highly localized form of IGC that occurs for only a few grains diameter immediately adjacent to the weld bead in Type 321 and 347 austenitic stainless steels. These stabilized stainless steels contain titanium and niobium respectively which react with carbon and prevents chromium carbide precipitation and chromium depletion. During welding, titanium or niobium carbides dissolve in stainless steel at high temperatures (>1503K) next to the weld bead followed by rapid cooling which retains all carbides in solid solution. Such conditions are more likely in thin welded sheet which allows very rapid cooling. Later weld passes or stress relief annealing in the chromium carbide precipitation range results in conventional sensitization, because the titanium and niobium has not had any opportunity to react with carbon. Thus failure occurs by IGC in a very narrow region next to weld metal.

SENSITIZATION DURING ISOTHERMAL HEATING

Sensitization resulting from isothermal exposures is normally represented by Time-Temperature-Sensitization (TTS) diagram. These diagrams show the durations required for isothermal sensitization at various temperatures and can be used to solve problems such as the choice of conditions of annealing or stress relieving which will not result in sensitization.

SENSITIZATION DURING CONTINUOUS COOLING

Sensitization may also result from cooling through the sensitization temperature range. This is of great practical importance since it is this type of thermal exposure that occur in slow cooling after high temperature annealing or in the cooling of a weld or weld HAZ. TTS diagrams can not be used directly to determine the extent of sensitization that can occur when the material is continuously cooled. The intersection of a superimposed cooling curve with the isothermal TTS diagram will not indicate whether the material is sensitized or not because it does not take into account the effect of the time spent in the different temperature regions. Dayal and Gnanamoorthy have reported a method to predict the extent of sensitization during continuous cooling/heating of the material. The principle of this method is described in Fig. 2(a) [4]. The cooling curve is divided into small segments ΔT from the highest temperature T_H to lowest temperature T_L of the relevant TTS diagram. The time of transit Δt is determined for each segment of the cooling curve and is divided by the sensitization time (t_s) at the mean temperature T of this segment from the TTS diagram. The cumulative fraction α of the resident time in successive segments from T_H to T_L is calculated and is defined as

$$\alpha = \sum_{T_L}^{T_H} \Delta t / t_s \tag{1}$$

Sensitization takes place only when $\alpha \geq 1$.

With the help of Eq. (1), a critical linear cooling rate R to cause sensitization can be calculated and is given by

$$R = \Delta T \sum_{T_L}^{T_H} 1/t_s \qquad (2)$$

It has been confirmed that the value of α is not dependent on the magnitude of ΔT and hence no great advantage can be obtained by taking very small ΔT values. For a given cooling rate, slower than the critical cooling rate, a temperature can be found, up to which α is just equal to 1. This gives one point on the cooling curve. For different cooling curves, the locus of such points determine the continuous cooling sensitization (CCS) diagram. The method to construct CCS diagram is shown in Fig. 2 (b).

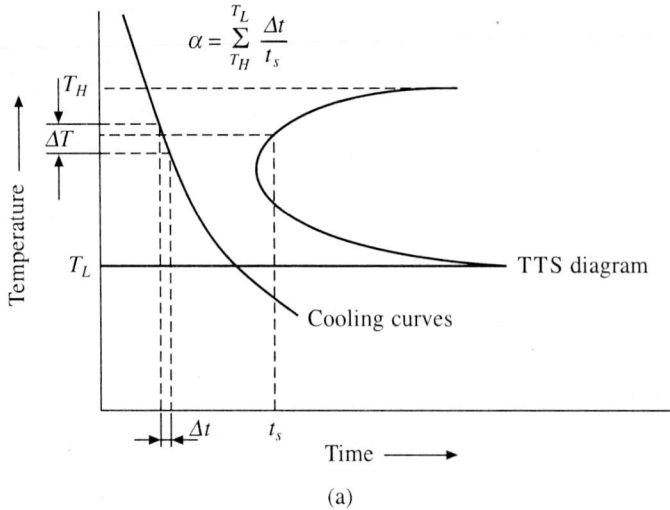

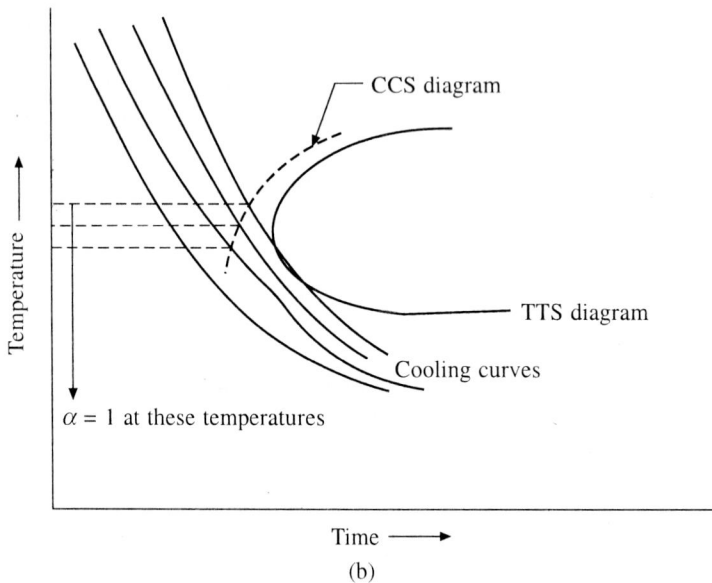

Fig. 2   (a) Principle of the method and (b) Continuous cooling
sensitization diagram obtained from TTS diagram.

Using this method Parvathavarthini et al have calculated critical linear cooling rate for nuclear grade 304, 316 and 316 LN SS, whose compositions are presented in Table 1. The critical cooling rates for these alloys at different levels of cold work (CW) are given in Table 2.

Table 1.   Chemical composition (wt. %)

Elements	304 SS	316 SS	316 (N) SS	316 LN SS
Carbon	0.040	0.054	0.043	0.030
Nitrogen	0.087	0.053	0.075	0.086
Chromium	18.3	16.46	17.18	16.6
Nickel	9.25	12.43	10.23	12.2
Molybdenum	–	2.28	1.85	2.61
Manganese	1.660	1.69	1.54	1.54
Phosphorous	0.023	0.025	0.022	0.024
Sulphur	0.003	0.006	0.005	0.003
Silicon	0.375	0.64	0.585	0.29
Vanadium	–	–	0.061	0.092
Copper	–	–	0.207	0.09
Cobalt	–	–	0.230	–
Boron	–	–	–	0.0012
Iron	balance	balance	balance	balance

Table 2.   Critical linear cooling rate (K/h)

% Cold work	304 SS	316 SS	316 (N) SS	316 LN SS
0	197	365	17	0.43
5	302	710	22	0.54
10	422	765	27	0.73
15	420	515	27	0.76
20	385	815	26	0.93
25	369	790	18	0.97

INFLUENCE OF VARIOUS FACTORS ON KINETICS OF SENSITIZATION

IGC arising due to sensitization depends on various factors like chemical composition, grain size, cold work, heating/cooling rates, heat treating temperature and time. The number of factors and their inter related influence clearly indicate that large number of experiments and statistical studies are necessary to determine reliably the influence of any of these factors.

CHEMICAL COMPOSITION

Sensitization of austenitic SS requires the precipitation of chromium rich carbides along grain boundaries. Hence carbon and chromium are the predominant compositional variables controlling sensitization. By reducing the carbon content in SS, TTS curve is displaced towards longer time because carbon concentration in austenite becomes insufficient to form chromium carbide readily (Fig. 3(a)) [5]. The limit of carbon content for which a steel is not sensitive to IGC is closely connected with the presence of other alloying elements such as chromium, molybdenum, nickel, nitrogen, manganese, boron, silicon as well as titanium and niobium in stabilised steels.

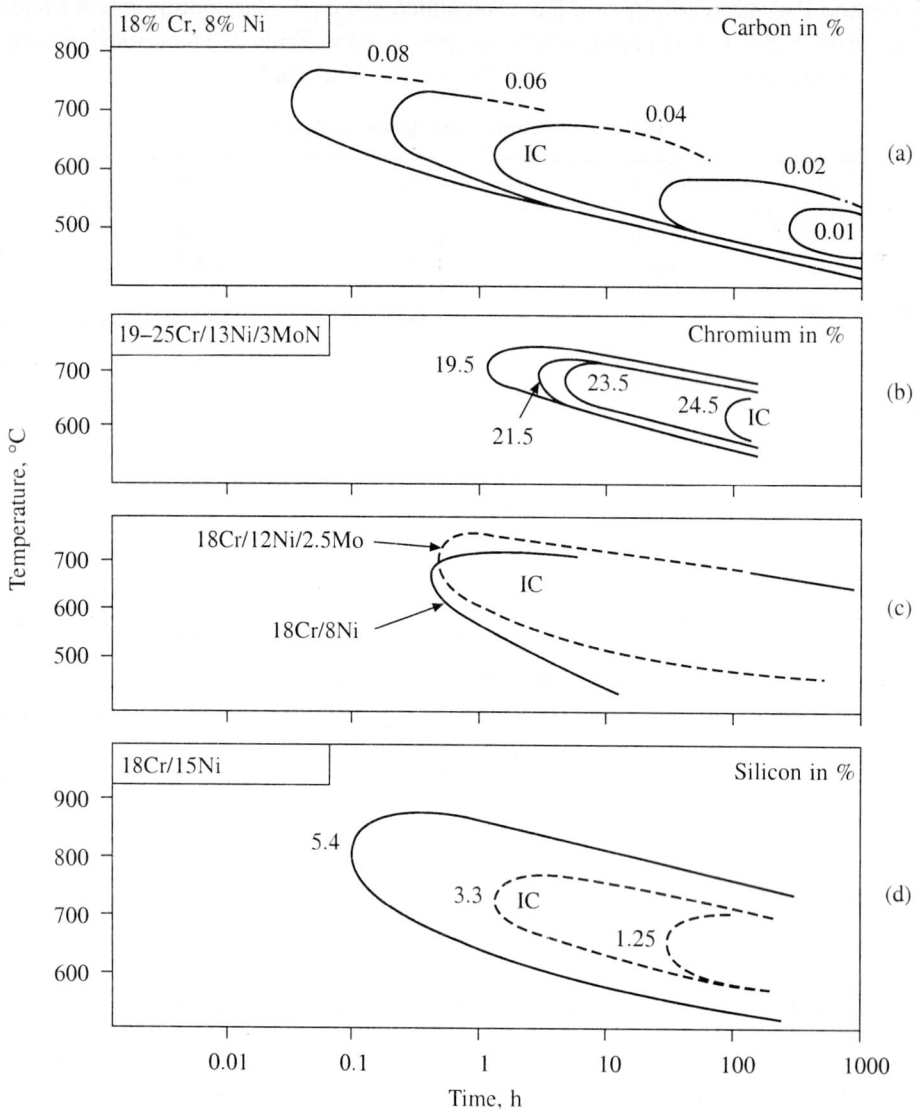

Fig. 3. Influence of alloying additions on $M_{23}C_6$ precipitation and intergranular corrosion determined by ASTM A262 Practice E: **(a) Carbon, (b) Chromium, (c) Molybdenum and (d) Silicon** [5].

Chromium has a pronounced effect on the passivation characteristics of SS. With higher chromium contents, time to reach the resistance limit of chromium depletion at the grain boundaries is shifted to longer time. Higher chromium contents facilitates the diffusion of chromium into the depleted grain boundary area. Fig. 3(b) [5] shows the influence of chromium on the sensitization kinetics [5]. Alloys with higher chromium contents will be more resistant to sensitization.

Nickel is required in austenitic stainless steel to stabilise the austenitic phase and must be increased with increasing chromium concentration. Increasing the bulk nickel content decreases the solubility and increases the diffusivity of carbon. This effect is much more pronounced when nickel content is greater than 20%. It is generally recommended that in 25/20 Cr-Ni steel, carbon content should be less than 0.02% to guarantee resistance to IGC.

Molybdenum reduces the solubility of carbon in austenite. Carbide precipitation is accelerated at higher temperatures whereas at lower temperatures it is slowed down (Fig. 3(c)) [5]. When molybdenum is present, it is also incorporated in $M_{23}C_6$. Therefore in addition to chromium depletion, molybdenum depletion is also revealed. In molybdenum containing chromium-nickel austenitic stainless steels $(Fe, Cr)_{23}C_6$ is precipitated first at 1023 to 1123 K. With prolonged ageing molybdenum is also incorporated as $(Fe, Cr)_{21}Mo_2C_6$ which is finally converted to 'Chi' phase. With increasing molybdenum contents, $M_{23}C_6$ precipitation and IGC becomes increasingly influenced by the precipitation of intermetallic phases.

The influence of manganese is of special importance because in fully austenitic welds this element is added. Manganese reduces the carbon activity and increases its solubility. Carbide precipitation is slowed down and hence it appears to inhibit carbide precipitation [6]. Boron retards the precipitation of chromium carbide but depending upon the heat treatment it promotes IGC [7]. Silicon promotes IGC of high purity and commercial stainless steels [8, 9]. Steels containing molybdenum were found to be much more sensitive to silicon addition. The increased susceptibility to IGC in highly oxidising solution is due to the segregation of silicon to grain boundaries. Fig. 3(d) [5] represents the influence of silicon on the kinetics of sensitization. Apart from $M_{23}C_6$, 'Pi' phase which is a carbonitride is precipitated with increasing silicon contents. $M_{23}C_6$ precipitation is slowed down more and is substituted by 'Pi' phase $(M_{11}(CN)_2)$. The cause for its precipitation seems to be the simultaneous effect of silicon on C and N activity.

One of the alloying additions studied extensively in recent years is nitrogen. Its effect is quite complex and is dependent on the presence of other alloying additions. Nitrogen content up to 0.16 wt% is reported to improve sensitization resistance by retarding the precipitation and growth of $Cr_{23}C_6$ [10]. Parvathavarthini et al have established TTS and CCS diagrams for 316 SS with various amounts of carbon and nitrogen levels (Figs. 4 and 5) using ASTM standard A262 practice A and E test [11–12]. Kamachi Mudali et al have established TTS diagrams for 304LN and 316LN SS using ASTM standard A and EPR test [13]. From these diagrams it can be inferred that as the nitrogen content increases, the time required for sensitization at nose temperature increases from 0.5h (in 316 SS) to as much as 80h in 316 LN SS indicating the beneficial effect of nitrogen. Various views have been proposed by different authors to explain the effect of nitrogen on sensitization kinetics of SS [6, 10, 14–20]. The computation of volume diffusion coefficient of chromium as a function of nitrogen indicates that nitrogen addition decreases the chromium diffusivity thereby retarding the nucleation and growth of carbides [14]. The systematic study using analytical electron microscopy was conducted by Briant et al [6] for 304 SS containing various amounts of nitrogen. They found that chromium concentration near the grain boundary is more and that the volume of the chromium depleted zone (considering both width and depth) decreases as the nitrogen content of the alloy increases. In addition, the solubility of nitrogen in austenite is greater than that of carbon. Another view is that in the presence of nitrogen, the passivation characteristics of the alloy is so superior that higher chromium depletion levels are necessary for IGC [10]. In essence, nitrogen retards $M_{23}C_6$ precipitation by decreasing the diffusivity of Cr [21]. When the concentration of nitrogen is higher than 0.16 wt%, M_2N is formed. However, the kinetics of formation M_2N is very sluggish. Even if it forms the extent of chromium depletion around M_2N will be less compared to that of $M_{23}C_6$ because on mole per solute basis, less chromium is precipitated by nitrogen than by carbon.

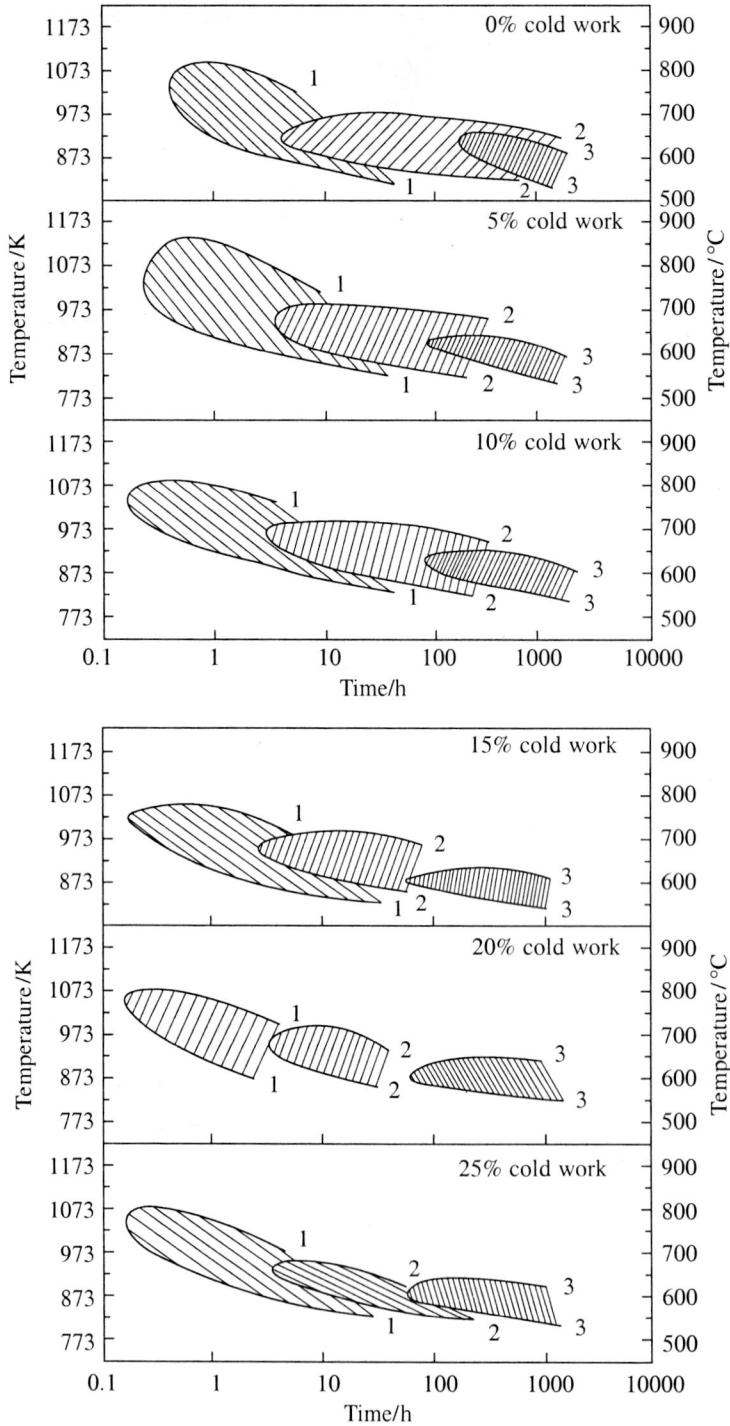

Fig. 4. Time-temperature-sensitization (TTS) diagrams for (1-1) Virgo 14SB (316 SS); (2-2) Nitrogen added 316SS; (3–3) 316 LN SS with different degrees of cold work established as per ASTM A262 practice E (hatched area shows sensitized region).

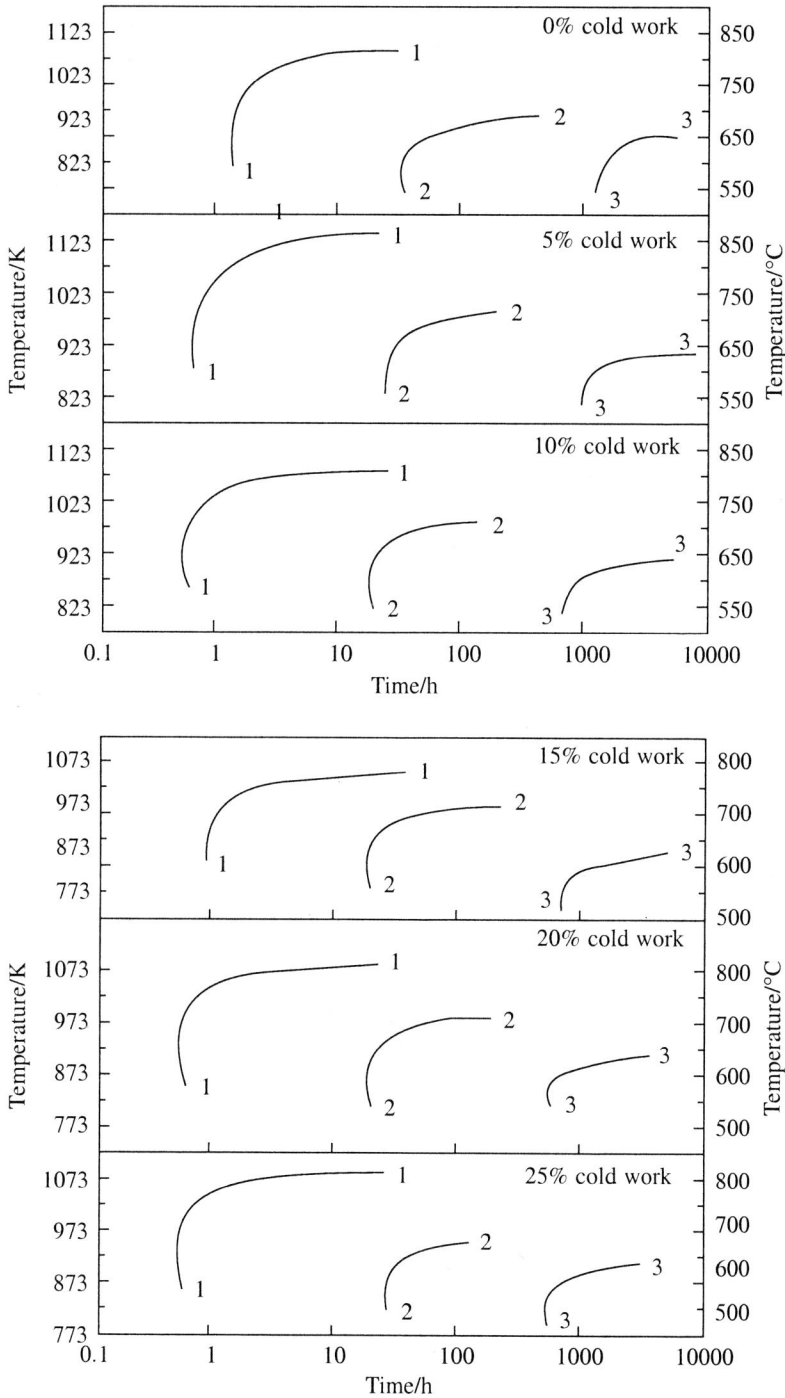

Fig. 5. Continuous cooling sensitization (CCS) diagrams for (1-1) Virgo 14SB (316SS); (2-2) Nitrogen added 316 SS; (3-3) 316 LN SS with different degrees of cold work established as per ASTM A262 practice E test.

The detrimental effect of carbon on sensitization can be reduced by the addition of stabilising elements like titanium and niobium. As explained earlier, these grades are susceptible to KLA.

Based on the numerous data reported in literature by various investigators, several attempts were made to predict time required for sensitization using composition-based correlations [22, 23]. Those elements which have major influence on kinetics of sensitization are given proper weightage and effective chromium content (Cr^{eff}) was calculated. Kain et al [23] have shown that

$$Cr^{eff} = \% \ Cr - 0.18 \ (\%Ni) - 100 \ (\%C) \tag{3}$$

for type 304/304 L stainless steels. They have reported that $Cr^{eff} > 14.0$ ensures material's resistance to IGC in ASTM standard A262 practice C and will be described in detail later. This is only conservative and can be used as a guide rule to reduce the proportion of heats subjected to the time-consuming tests for sensitization detection. For practice E, 304 L with $Cr^{eff} > 13.5$ is shown to be resistant to IGC. The authors of this work have indicated that these should be used with caution and should be supplemented by microstructural studies.

GRAIN SIZE

Precipitation of chromium carbides tends to be localised at the grain boundaries because they are high energy regions. There is some evidence to suggest that increasing grain size increases susceptibility to IGC [24]. This has been explained as being due to the fact that in fine grained material, there is more grain boundary area and therefore less chance for a continuous net work of carbides to form at the grain boundaries. More over, chromium has to diffuse over a less distance in a fine grained material and hence possibility of sensitization is reduced. Desensitization is also faster in fine grained material because it has to diffuse only through a shorter distance for replenishment.

COLD WORK

For numerous applications of stainless steel in nuclear, chemical, petrochemical and fertilizer industry cold working is the final manufacturing operation and the components are subjected to different levels of cold work. Under certain conditions (eg. lower carbon levels), cold work induces martensite formation in type 304 SS which is very dangerous, since diffusion rates of chromium and carbon are higher in martensite while solubility is much less. Martensite has enhanced corrosion rates in certain environments. Martensite formation is favoured by increasing the amount of deformation, decreasing the temperature at which deformation occurs and by decreasing the total content of alloying elements in steel [25]. For instance, in stainless steels with higher levels of Ni (γ stabiliser) this transformation is suppressed and can be prevented totally. Several studies have shown that kinetics of sensitization can be influenced by prior deformation [26–34]. Cold work has been reported to increase the sensitization kinetics at moderate cold work levels (5 to 15%) and decrease the kinetics at higher levels of cold work. Parvathavarthini et al have established TTS and CCS diagram for 316 SS (containing various levels of C and N) for various degrees of cold work ranging from 5 to 25% (Fig. 4) [11, 12]. For all these materials, TTS diagrams of cold worked materials are shifted to the left (shorter time) and below that of the as received (0%) material. The nose temperature corresponding to maximum rate of sensitization is also shifted to lower temperature. Desensitization is faster at high levels of cold work especially at high ageing temperature. As the cold work level increases, critical linear cooling rate (Table 2) above which there is no risk of sensitization also increases because sensitization kinetics are faster. From these results it was found that for 316 LN SS containing 860 ppm nitrogen, critical cooling rate is so low (0.43 to 0.97 K/h) and hence there is no risk of sensitization in HAZ during welding [11].

All the above results reveal that the effect of cold work on sensitization kinetics is to enhance the rate of sensitization. Deformation of austenitic stainless steels result in lot of changes in the defect structure. The mill annealed sample has a low dislocation density which increases sharply on cold working [26, 33, 35]. The dislocation density at the grain boundary is higher than that of matrix. The presence of such a defective structure containing dislocations, stacking faults etc. is known to enhance the overall diffusion of alloying elements and result in faster sensitization. At higher levels of cold work, the nose temperature and upper temperature boundary decreases and these could be explained by the well-known Hart dislocation pipe diffusion equation [36].

$$D_{tot} = D_{o,l}\, e^{-Q_a/RT} + nA\, D_{o,p}\, e^{-Q_{a,p}/RT} \tag{4}$$

where D_{tot} is the total diffusivity, $D_{o,l}$ is diffusion coefficient of lattice diffusion, $D_{o,p}$ is that of pipe diffusion, Q_a is the activation barrier for lattice diffusion, $Q_{a,p}$ that for pipe diffusion, 'n' the dislocation density, 'A' area of the dislocation pipe. When temperature is high, partial recovery from the effect of cold work sets in. Therefore the effect of cold work on sensitization behaviour at high temperature is less pronounced. The observation of faster desensitization kinetics at high levels of cold work can be attributed to the fact that austenitic stainless steels have low stacking fault energy; High levels of cold work result in large dislocation pile ups on slip planes. Due to this slip planes become additional favourable sites for carbide precipitation within the grain. This leads to short diffusion path for carbon. Once the carbon activity is reduced, chromium activity near the intragranular carbide particle increase due to quicker homogenisation and the material no longer shows marked depletion of chromium at the grain boundaries and faster desensitization results.

To understand the influence of morphological features of grain boundary carbides on the enhanced rate of sensitization with deformation, Transmission Electron Microscopic studies were carried out for nuclear grade 316 SS [34]. Materials aged at nose temperature (corresponding to onset of sensitization) were characterised in terms of D (average diameter), inter carbide spacing λ, shape factor S (length/width). These parameters were found to decrease with deformation reaching a saturation at 15% cold work. A minimum intercarbide spacing λ_{crit} was found to characterise the onset of sensitization at each ageing temperature and % cold work. The λ_{crit} was larger for elliptical carbides than spherical ones, thus making the presence of elliptical carbides more detrimental than spherical carbides.

Reynalda Beltran et al studied the simultaneous effects of cold work and grain size on carbide precipitation and sensitization kinetics in 304 SS [37]. They have reported that as the degree of cold work increased, systematic increase in carbide precipitation and rate of sensitization/desensitization were observed. In the case of very small grains size, strain effects were found to be far less prominent. Straining at liquid nitrogen temperature produced submicron two phase α'/γ subgrain size which demonstrated an almost instantaneous sensitization/desensitization behaviour.

HEATING AND COOLING RATES

Heating and cooling rates will have pronounced effect on the sensitization behaviour. Rapid cooling through the sensitization temperature range in austenitic stainless steel, normally does not result in chromium carbide precipitation provided the cross section of the work piece is sufficiently small ensuring uniformly rapid cooling rate throughout the alloy. At faster cooling rates, the time spent in the sensitization region is insufficient for chromium carbide to precipitate and hence degree of sensitization would be far less.

PRE-EXISTING CARBIDE NUCLEI

It has been observed that at temperatures well below the established isothermal temperature range of sensitization (573 K-748 K), austenitic stainless steels could become sensitized after very long duration. This is known as low temperature sensitization (LTS). LTS may occur in materials which have been exposed to time-temperature combinations sufficient for nucleation of carbides but insufficient for their appreciable growth. If this material is now held at a low temperature for a long period, it would lead to the growth of these precipitates resulting in sensitization. The pre-requisite for LTS is the presence of chromium carbide along grain boundaries. Thermal exposures during welding or hot working may nucleate grain boundary carbides necessary for LTS [38, 39]. This phenomenon is schematically illustrated in Fig. 6. The high temperature heat treatment path A does not lead to sensitization because the time of exposure is less than that required to cause sensitization at that temperature. Path B is well below the temperature required for sensitization and even prolonged exposure to this temperature does not cause sensitization. Combination of path A and B shown as path C causes severe sensitization. However it may take years before this effect becomes significant. It has been reported that carbides of optimum size and distribution are the essential pre requisites for LTS and prior cold work enhances susceptibility of austenitic stainless steel to LTS [12].

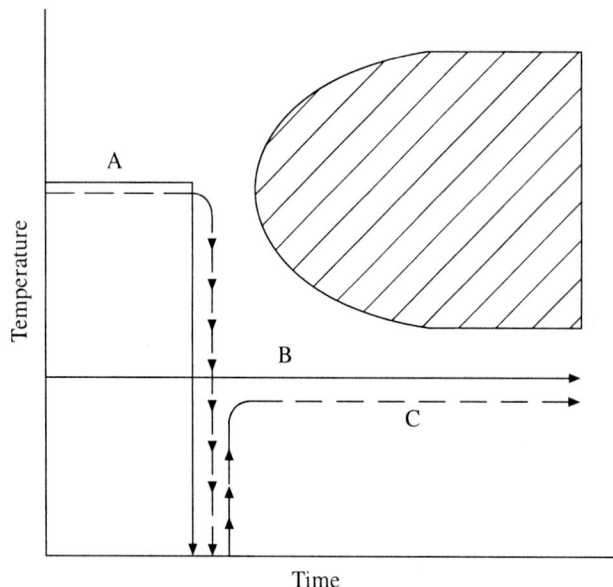

Fig. 6.   Schematic representation of low temperature sensitization (LTS). Paths A and B do not cause sensitization. However, path C causes severe sensitization. The hatched region is the normal isothermal time-temperature-sensitization (TTS) zone.

At temperatures sufficiently below the nose temperature of the TTS diagram, log t vs $1/T$ plots (t being the time required for sensitization at temperature T) follow a linear relationship. From the slope, the activation energy can be calculated and it was estimated to be of the same order as that of volume diffusion of chromium in austenitic stainless steels [12]. It can be inferred that the rate controlling step for sensitization at low temperature is the lattice diffusion of chromium. The validity of extrapolating these linear plots to lower temperatures has been verified by the author. It was found that sensitization kinetics is much slower than what is expected from log t vs $1/T$ plots. Similar observations are reported by Fullman et al for 304 SS [40]. They attribute this slower sensitization kinetics to the reduction in effective area of dislocations. An alternate explanation offered for this was that as the temperature becomes lower, a shift from diffusion limited kinetics to interface limited kinetics may occur.

THEORIES OF IGC

Several theories and models are available in literature to explain how carbides, nitrides and other phases are responsible for IGC.

CHROMIUM DEPLETION THEORY

A qualitative model of this theory was originally proposed by Bain et al [41]. According to this theory, chromium content is reduced in regions adjacent to the precipitating chromium carbide. When chromium level falls below that required for passivation the material becomes susceptible to IGC. This theory is very useful and is valid in most instances. Based on this theory two quantitative models were developed independently. Strawstrom and Hillert developed a model based on chromium diffusion control to calculate time to sensitization/desensitization [42]. They calculated the chromium content at carbide interface by considering the changing equilibrium during the precipitation of $M_{23}C_6$ in austenite in Fe-Cr-Ni alloys. They emphasised the kinetic features of carbide precipitation and obtained good agreement between theory and experiments. Tedmon et al emphasised the thermodynamic aspects of precipitation process and their model describes local chromium-carbon-carbide equilibrium at grain boundary [43]. Recent analytical and quantitative refinement to this theory for IGC in sensitized stainless steels have demonstrated the widespread applicability and validity of this model.

STRESS THEORY

According to this theory, local stresses arise in those areas where secondary phases begin to precipitate and grow [7]. Due to this considerable energy differentials are produced in these zones. This leads to imperfect passivation which results in poor corrosion resistance.

MICROCELL THEORY

This theory is based on the dissolution of grain boundary due to the formation of local cells, the precipitate acting either as anode or cathode [44]. Stickler and Vinckier proposed that IGC is an electrochemical reaction between the noble carbide particles acting as cathode and the matrix acting as anodes and proceeds rapidly along grain boundary where there is a continuous path provided by carbide particles. This theory could not explain desensitization as well as resistance to IGC during the onset of precipitation when chromium level is above 12% near chromium carbides. However, it can adequately explain IGC in sensitized material.

SEGREGATION THEORY

According to this theory, IGC takes place due to the presence of continuous grain boundary path of either second phase or soluble segregate resulting from solute-vacancy interaction. This model can explain the IGC arising due to the segregation of impurity elements. The observation that IGC takes place even in annealed steel in highly oxidising solution can be explained by this model. Although this model holds good for IGC in non sensitized austenitic stainless steel, attempts to extend the model to carbide-sensitized stainless steel have been inconclusive.

It can be concluded that chromium depletion theory is able to explain most of the cases of IGC resulting from chromium carbide precipitation. Various theories do not contradict each other but rather supplement each other in explaining IGC problems.

SENSITIZATION EVALUATION TESTS FOR STAINLESS STEELS

For most of the applications of austenitic stainless steels it is required to assess whether a fabricated component is sensitized and has become susceptible to IGC. ASTM has standardised the test procedure and the specifications are detailed in ASTM A262 (practice A-F) and G108 [45]. Muraleedharan has reviewed these tests and has compared the various test procedures for both conventional and electrochemical tests [46]. These standard tests are commonly used as qualification/acceptance tests during purchase/fabrication stages. However non-inclusion of acceptance limits in these standards leaves the interpretation of the results open to the users. The salient features of these procedures are discussed briefly.

ASTM A262 Practice A test

This test consists of electrolytically etching a polished specimen in 10 wt% oxalic acid solution at room temperature at a current density of 1 A/cm^2 for 1.5 minutes. The etched structure is then examined at 200 ×. In this test, chromium carbide is dissolved preferentially and the microstructure gives an idea of chromium depletion which is responsible for IGC. The different microstructures which can be obtained are presented in Fig. 7 (a-f). If there is no carbide precipitation 'step' structure (a) is obtained, because of the differences in the rate of etching of variously oriented grains. Dual structure (b) is obtained, if chromium carbide precipitation is discontinuous. Ditch structure (c) is obtained if grain boundaries are completely surrounded by chromium carbide. Even if one grain is completely surrounded by ditch, it is characterised as 'ditch' structure. Step and dual structure are acceptable but if the structure is 'ditch' the material may or may not be sensitized and hence it has to be further tested by any one of the ASTM tests (B to F). This test is only a qualitative test but is very useful as a screening test. This cannot detect sigma phase in molybdenum bearing alloys. Since titanium and niobium carbides do not dissolve appreciably in this test, this can be used to detect chromium carbide precipitation even in stabilised stainless steels. ASTM further recommends a heat treatment at 950 K for 1 h and water quenching for low carbon SS, for weld simulation before carrying out this test. This test also characterises the microstructure with inclusions as end grains. End grain pitting (II) (transverse section) (Fig. 7g) is not considered to be acceptable because although these steels do not contain chromium carbide precipitates at grain boundaries, active inclusions in the form of stringers undergo IGC in oxidising environments such as nitric acid. For instance, (Fe, Mn) sulphide and oxide inclusions stringers lead to catastrophic IGC in HNO$_3$ medium but not in other environment.

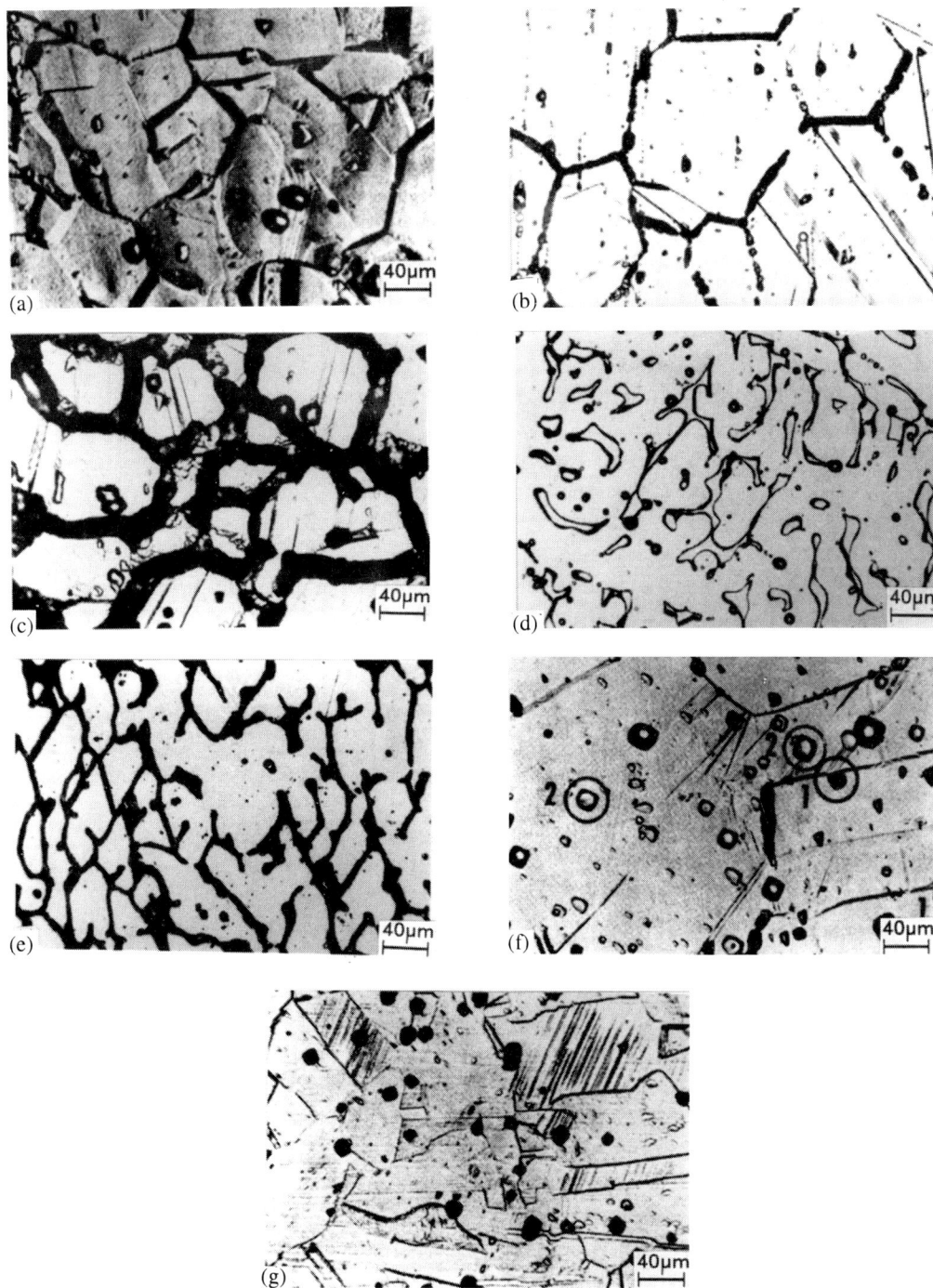

Fig. 7. Classification of etch structure after oxalic acid etching (ASTM A262 practice A): (a) Step structure; (b) Dual structure; (c) Ditch structure; (d) Isolated ferrite pools; (e) Inter-dendritic ditches; (f) End grain pitting I; (g) End grain pitting II [45].

ASTM A262 Practice B test

In this test a sample of surface area 5–20 cm^2 is exposed for a period of 120 h to boiling solution of 50% H_2SO_4 + 2.5% $Fe_2(SO_4)_3$. Corrosion rate is calculated from weight loss measurements. ASTM practice does not specify the criterion to judge the susceptibility of the material to IGC. Normally accepted limit for 304 SS is 48 mpy. Streicher has reported [47] that if the ratio of the weight losses of sensitized to annealed material is greater than 1.5 to 2.0, the material is considered as susceptible. This test is applicable to austenitic stainless steels and it detects the IGC associated with chromium carbide precipitation and chromium depletion. Sigma phase in 321 and 347 SS are attacked whereas that in Mo bearing 316 SS is not attacked.

ASTM A262 Practice C test

In this test, a sample of 20–30 cm^2 area is exposed to 65 wt% HNO_3 for five 48 h period. After every 48 h, the solution is changed and the sample is weighed. The corrosion rate for each period and the average for the five periods is determined. ASTM does not state the acceptance criteria. Experience has shown that corrosion rate < 18 mpy for 304 SS and < 24 mpy for 304L does not lead to IGC. The material is not acceptable, if the corrosion rate is increasing rapidly for the successive periods. Besides chromium depleted zone, carbides and sigma phase in molybdenum bearing alloys are attacked in this test. These alloys can give high corrosion rates even when they are immune to IGC in other tests, which reveal sensitization caused by chromium depleted zones. Submicroscopic sigma may also form in stabilized grades of 321 and 347 and show susceptibility to IGA in HNO_3 test. This test has to be followed only when the alloy is intended to be used for nitric acid service.

ASTM standard A262 Practice D test

In this test samples are tested in 10% HNO_3-3% HF solution at 343K for two 2 h periods (fresh solution is used for each period). If the ratio of the weight loss of the sensitized to annealed material is greater than 1.5, the sample is considered to be susceptible to IGC. This detects only chromium depletion from carbide precipitation and not submicroscopic sigma.

ASTM standard A262 Practice E test

In this test, austenitic stainless steel specimen is embedded in metallic copper chips and then exposed to boiling 16% H_2SO_4 + 10% $CuSO_4$ for 24 h. After the test, the specimen is bent through 180° over a mandrel of diameter equal to the thickness of the specimen. The bent specimen is examined under low magnification. If cracks are seen, the material is considered to be sensitized. Although this is not a quantitative test, ASTM gives acceptance criterion for this test. Electrical resistivity and tensile properties are changed considerably by the IGC. These can be used for quantifying the degree of sensitization (DOS). Muraleedharan et al have suggested a modified version to determine the DOS quantitatively [48]. Flat tensile specimens can be exposed to the test solution and can be pulled to fracture at a strain rate of 6.6×10^{-4} s^{-1} and DOS can be correlated to % loss in strength as follows:

$$DOS = \% \text{ loss in strength} = [1 - \sigma_{UTS.exp} / \sigma_{UTS.unexp}] \times 100$$

ASTM A262 Practice F test

This test is useful for Mo bearing SS for which practice B and D have been used so far. Since practice B shows corrosion rates due to the presence of molybdenum associated phases in SS, IGC arising exclusively due to chromium depletion can be obtained from this test. This test may also be used to evaluate resistance of extra low carbon grades to sensitization and IGC caused by welding or heat treatment. It involves exposing the specimen to boiling Cu-CuSO$_4$-50% H$_2$SO$_4$ for 120h and measuring the weight loss. Similar to the other tests it does not indicate the rejection criteria.

From the above details, it is clear that the ASTM standard practices have three draw backs: (i) they are only qualitative, (ii) destructive and (iii) time-consuming (except practice A). Hence several electrochemical techniques were developed to determine the susceptibility of a material to IGC which are fast, non-destructive and quantitative.

ELECTROCHEMICAL METHODS FOR DETECTING SENSITIZATION

For the electrochemical test methods, different electrolytes such as H$_2$SO$_4$/KSCN; HClO$_4$/NaCl are being used. The following electrochemical methods have been adopted by various investigators:

 (i) Anodic polarisation under slow scan rate to determine the critical current density and passivation current density.
 (ii) Constant potential etching at the secondary anodic peak and measuring the charge value.
 (iii) Electrochemical potentiokinetic reactivation technique (EPR). This has been standardised and has been incorporated in the ASTM standards as G108.

Most of the investigators have followed test conditions similar to the one adopted by Clark et al [49]. In this method the surface to be tested is immersed in a deaerated solution of 0.5M H$_2$SO$_4$ + 0.01 M KSCN at ambient temperature. The specimen is then passivated by making it anodic. After holding it in this range, the potential is reversed at a fast, constant rate to bring it back to the corrosion potential. If Cr content exceeds 12–13 (wt%) protective chromium oxide film is formed on the alloy and it is this film which protects the alloy from corrosion when this is in passive state. With Cr content less than this value, oxide film becomes less protective. Hence during the reactivation, the chromium depleted area is more readily dissolved than the undepleted surface. As a result, sensitized SS generates a larger corrosion current than solution-annealed stainless steel.

Depending upon the potential scanning mode, there are two types of EPR experiments, namely single loop EPR and double loop EPR method. In single loop EPR method, the reactivation of the sensitized region is reflected as an anodic peak in the plot of potential vs current. The area under the reactivation peak is directly proportional to the electric charge (Q) which is a measure of the DOS. Clarke et al have suggested that the total charge derived from the area under the reactivation peak should be adjusted for the grain size of the material while comparing alloys with different grain size. He proposed that the normalised reactivation charge P_a be obtained from the relation

$$P_a = Q/\text{GBA} \tag{6}$$

where Q is reactivation charge and GBA is total grain boundary area. GBA is given by the equation

$$\text{GBA} = A_s\,[5.09544 \times 10^{-3} \exp(0.34696n)] \tag{7}$$

where A_s is the area of the specimen and n is ASTM grain size number, measured at 100 ×. Here it is assumed that the width of the attack of sensitized material is always $2(5 \times 10^{-5})$ cm, attack is distributed uniformly over the entire GBA and grains are spherical.

In double loop EPR method, the reactivation scan from the passive potential is preceded by an anodic scan from open circuit potential. Instead of using area under the peak to determine DOS, the ratio I_r/I_a is used, where I_r and I_a are the peak currents during reverse and forward scans. Since I_r/I_a is not sensitive to surface finish, even 120 grit finish is enough for the experiment.

Kain and De has reviewed the recent developments in EPR testing [50]. In order to correlate the DOS with the measured EPR parameters much more accurately the influence of test temperature has to be considered because reactivation charge density is highly dependent on the test temperature. Reactivation charge at 30°C (q_{30}) can be determined using the following equation:

$$q_{30} = q_T \exp [26.55(30 - T)/(273 + T)] \qquad (8)$$

where q_T is reactivation charge at temperature T (°C).

It should be noted that while normalising the reactivation charge with respect to grain boundary area, it was assumed that all the grain boundaries contribute to the reactivation charge. However this is not always true. For example as mentioned earlier, chromium depletion is not continuous in steels that show 'step' and 'dual' structure. Secondly, width of the attacked zone increases with increase in test temperature and is considerably greater than Cr depleted zone. Because of these reasons, width term was dropped and it has been proposed that reactivation charge should be normalised with respect to GB length. Finally to avoid the assumption that grains are circular in shape, equations are available for GBA relating length of the grain boundary per unit specimen area and mean linear intercept. It has been shown that in addition to length and width of the chromium depleted zone, the extent of chromium depletion (i.e., depth) also is important in correlating reactivation charge to DOS. Breummer has shown a volume depletion parameter correlates well with DOS [22]. Similar to practice A, EPR test results are strongly dependent on grain boundary coverage and length of grain boundary depleted of Cr when DOS is low. These saturate at higher degree of sensitization and discriminating power is lost. Under such conditions, length, width and extent of chromium depletion has to be taken into account to calculate normalised reactivation charge density to correlate DOS with IGC or IGSCC behaviour. Our studies have shown that the threshold EPR charge value above which a material can be considered as sensitized is dependent upon ageing temperature and prior thermomechanical history [48, 51].

COMBATING SENSITIZATION

IGC constitutes a major problem for SS but there are number of measures and IGC resistant materials to control this corrosion problem. As the mechanism of sensitization phenomenon is well-understood, several methods have been evolved in the recent past to control and minimise sensitization. By employing high temperature solutionising and rapidly cooling through the sensitization range, sensitization can be avoided. However such a bulk treatment is not always feasible because large thermal stresses may be produced in the components. In such cases, other ways must be used to prevent sensitization. By utilising low carbon varieties of SS, sensitization can be delayed or weld decay can be avoided. But reduction in carbon content leads to lower mechanical strength. This can be compensated by addition of nitrogen to the low carbon variety to produce LN stainless steels. Nitrogen also delays the onset of sensitization and avoids sensitization in the HAZ. However, prolonged thermal ageing of nitrogen added SS also results in sensitization. By alloying with elements which have greater tendency than chromium to form carbides (Ti and Nb in Type 321 and 347) stable carbides are formed and sensitization can be avoided. However welding and subsequent stress relieving treatment make them more prone to KLA. Rapid induction heating has also been attempted by which acceptable recovery of the properties of the sensitized material can be achieved.

However, if the components are found to be sensitized during final stages of fabrication or commissioning or during service, they cannot be used in hostile environments without solution treatment. In such a case, in-situ method is required to selectively eliminate sensitized structure without affecting the bulk properties. Laser treatment is one among the methods using heat treatment to eliminate sensitization. By the proper choice of laser parameters like beam power, size and traverse speed, desensitization can be effected in components which are sensitized during fabrication. With the advent of fiber optics that can transmit high energy beams to locations remote from the laser, in-situ laser melting has become much easier even in high radiation fields and even components with complicated geometry can be desensitized.

Laser surface melting was used to desensitize the surface of austenitic stainless steels and improve their IGC resistance [52, 53, 54]. Kamachi Mudali et al. have carried out detailed studies on 304 and 316 SS which were laser surface melted in (i) solution annealed + sensitized, (ii) cold worked + sensitized condition using a Nd-YAG laser of 300W power and 9 ms pulse width at a traverse speed of 2.5 mm s^{-1}. This resulted in a dendritic cellular structure with HAZ free from sensitization. The improvement in IGC resistance is attributed to the dissolution of $M_{23}C_6$ carbides and homogenisation of Cr depleted zones [52, 53, 55]. Similar results have been obtained for 304 SS which was surface melted using multibeam CO_2 laser [56]. The optical micrograph of (i) 10% cold worked and sensitized, (ii) 10% cold worked, sensitized and laser melted 316 SS specimens are shown in Fig. 8 ((a) and (b)) [53]. The desensitization effect was seen in the melted zone and with cellular dendritic structure and melt affected zone was without the ditch structure. It has also been established that laser surface melting of sensitized 316 LN SS increased critical pitting potential (E_{pp}) significantly compared to as-sensitized material due to the elimination of sensitized heterogenous microstructure and vulnerable pitting sites [54] (Fig. 9). Jeng et al have reported the surface melting of type 347 SS using 2 kW continuous wave CO_2 laser to improve the resistance of materials against propagation of IGC particularly the end grain attack in HNO_3 medium [57]. This technique was also applied on type 304L SS plates and it was reported to have resulted in the elimination of sensitized material in the surface layers, thereby preventing the initiation of IGC or stopping corrosion which had already started. In the untreated material the directional nature of end grain attack was related to the dissolution of NbC and sulphide inclusions in 304 SS besides dissolution of enriched phosphorous and chromium depleted

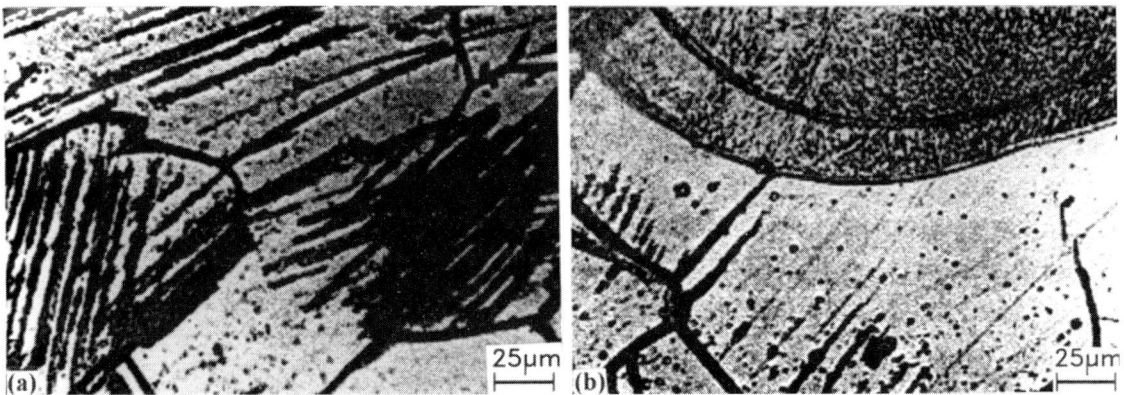

Fig. 8   Optical micrographs of (a) as-cold worked (10%) and sensitized (923 K/25 h) (b) cross section of 10% cold worked, sensitized (923 K/ 25 h) and laser-melted 316 SS specimens.

regions near the grain boundaries when exposed to HNO_3. By laser melting surface pits causing localised attack were sealed and also the susceptible sites were eliminated.

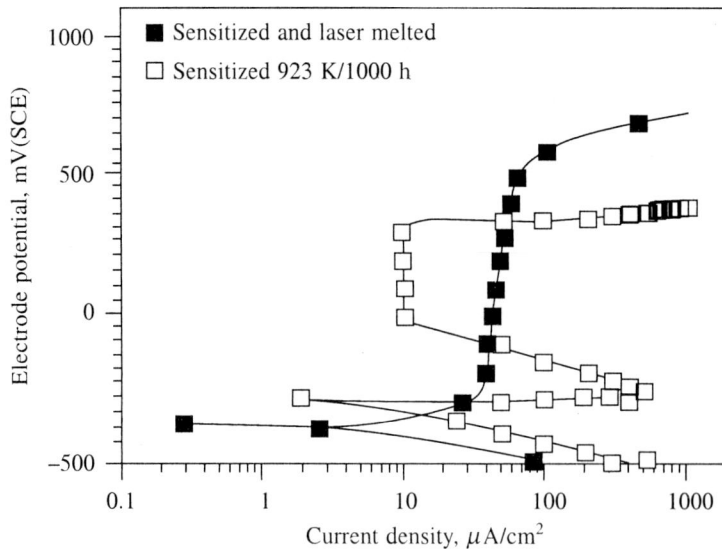

Fig. 9 Potentiodynamic anodic polarization curves of as-sensitized and sensitized and laser-melted (923 K/1000 h) 316 LN SS in acidic chloride medium [54].

Dayal has published a comprehensive review of the laser surface modification carried out on materials such as carbon steel, low alloy steel, and SS and the important developments in this new area of studies during the past one and half decades [58]. Improvement in IGC is due to the dissolution of chromium-rich carbides and redistribution of such carbides into finer particles and redistribution of alloying elements. Since melting during laser processing occurs in a very short time and only in the surface region, the bulk of the material remains cool, thus serving as an infinite heat sink. Large temperature gradient exist across the boundary between melted surface layer and the underlying solid substrate. This produces rapid self-quenching and resolidification with quench rates as large as 10^{11} Ks^{-1} and accompanying resolidification velocity in the range of 20 ms^{-1}. Hence resensitization is not possible. Since laser melting results in melted layer which has compressive residual stress, resistance to IGSCC is also improved.

REFERENCES

1. Durgam G. Chakrapani, Hand Book of Case Histories of Failure Analysis, ed. Kholefa A.esakul, ASM International, Materials Park, OH, Dec. 1992 pp. 164–170.
2. Harry, E. Ebert, Ibid, pp. 278–83.
3. Charles, J. Novak, Handbook of Stainless Steels, ed. Peckner and Bernstein (McGraw-Hill), 1977, p. 1.
4. R.K. Dayal and J.B. Gnanamoorthy, Corrosion, 36 (1980) 104–105.
5. Erich Folkhard, "Welding Metallurgy of SS, Springer-Verlag/Wien, 1988.
6. C.L. Briant, R.A. Mulford and E.L. Hall, Corrosion, 38 (1982) 468–477.
7. R.A. Lula, A.J. Lena and G.G. Kiefer, Trans. Am. Soc. Met., 46 (1954) 197.
8. J.S. Armijo, Corrosion, 24 (1968) 24–30.

9. A. Joshi and D.J. Stein, Corrosion, 28 (1972) 321–330.
10. T.A. Mozhi, W.A.T. Clark, K. Nishimoto, W.B. John and D.D. McDonald, Corrosion, 41 (1985) 555–559.
11. N. Parvathavarthini, R.K. Dayal and J.B. Gnanamoorthy, J. of Nucl. Matls., 208 (1994) 251–258.
12. N. Parvathavarthini, R.K. Dayal, S.K. Seshadri and J.B. Gnanamoorthy, J. of Nucl. Matls., 168 (1989) 83.
13. U. Kamachi Mudali, R.K. Dayal, J.B. Gnanamoorthy and P. Rodriguez, Metallurgy and Materials Trans., 27A (1996) 2881–2887.
14. H.S. Betrabet, K. Nishimoto, B.E. Wilde and W.A.T. Clark, Corrosion, 43 (1987) 77.
15. J.J. Eckenrod and C.W. Kouach, Effect of Nitrogen on the Sensitization, Corrosion and Mechanical Properties of 18Cr-8Ni SS, ASTM-STP 679, eds. C.R. Brinkman and H.W. Garvin, (ASTM, Philadelphia, 1979) p. 17.
16. A. Kendal, J.E. Truman and K.B. Lomax, Int. Conf. on High Nitrogen Steel, HNS-88, Lille, 1988, eds. J. Foct and A. Hendry (The Institue of Metals, York, 1989), p. 403.
17. P. Gumpel and T. Ladwein, Proc. Int. Conf. on High Nitrogen Steel, HNS-88, Lille, 1988, eds. J. Foct and A. Hendry (The Institute of Metals, York, 1989) p. 272.
18. J.E. Truman, Proc. Int. Conf. on High Nitrogen Steel, HNS-88, Lille, 1988, eds. J. Foct and A. Hendry (The Institute of Metals, York, 1989) p. 225.
19. R.F.A. Jargelius, Proc. on Stainless Steel 87 (Institute of metals, York, 1987) p. 266.
20. R.S. Dutta, P.K. De and H.S. Gadiyar, Corrosion Sci., 34 (1995) 51.
21. P. Marshall, Austenitic Stainless Steels: Microstructure and mechanical properties, Elsevier Applied Science, London, 1984, p. 23.
22. S.M. Breummer, L.A. Charlott and B.W. Arey, Corrosion, 44 (1988) 328.
23. V. Kain, R.C. Prasad, P.K. De and H.S. Gadiyar, ASTM J. of Testing and Evaluation, 23 (1995) 50.
24. A.J. Sedricks, Corrosion of Stainless Steels, John Wiley, New York, 1979.
25. C.L. Briant, Corrosion, 38 (1982) 596–597.
26. V. Cihal and J. Kubelker, Pract. Metallography, 12 (1975) 148.
27. C.S. Tedmon, D.A. Vermilyea and D.E. Broecker, Corrosion, 27 (1971) 104–106.
28. H.D. Solomon, Corrosion, 36 (1980) 356–361.
29. H.D. Solomon and D.C. Lord, Corrosion, 36 (1980) 395–399.
30. C.L. Briant and A.M. Ritter, Met. Trans., 12A (1981) 910–913.
31. M.J. Povich and D.E. Broecker, Mat. Perform, 18 (1979) 41–48.
32. W.L. Clark and G.M. Gordon, Corrosion, 29 (1973) 1–12.
33. S. Pednekar and S. Smialowska, Corrosion, 36 (1980) 565–577.
34. S.K. Mannan, R.K. Dayal, M. Vijayalakshmi and N. Parvathavarthini, J. Nucl. Matls., 126 (1984) 1–8.
35. G. Rondelli, B. Mazza, T. Pastore and B. Vincentini, Mater. Sci. Forum, 8 (1986) 593.
36. A.H. Advani, L.E. Murr, D.G. Atteridge and R. Chelakara, Metall. Trans., 22A (1991) 2917.
37. R. Beltran, J.G. Maldonado, L.E. Murr and W.E. Fisher, Acta Mater., 45 (1997) 4351.
38. M.J. Povich and P. Rao, Corrosion, 34 (1978) 269–275.
39. M.J. Povich, Corrosion, 34 (1978) 60–65.
40. R.L. Fullman, Electric Power Research Inst. Report No. WS-79–17 (1980).
41. E.C. Bain, R.H. Aborn and J.B. Rutherford, Trans. Am. Soc. Steel Treating, 21 (1933) 481.
42. C. Strawstorm and M. Hillert, J. Iron Steel Inst., 207 (1969) 77–85.
43. C.S. Tedmon Jr., D.A. Vermilyea and T.H. Rosolowski, J. Electrochem. Soc., 118 (1971) 192–202.
44. R. Stickler and A. Vinckier, Corros. Sci., 3 (1963) 1–8.
45. Annual Book of ASTM Standards, ASTM, Philadelphia, PA, 1990, Vol. 03.02
46. P. Muraleedharan, Corrosion and Maintenance, Jan-Mar (1984), 47–57.
47. M.A. Streicher, Corrosion, 20 (1964) 57 (t)–72(t)
48. P. Muraleedharan, J.B. Gnanamoorthy and K. Prasad Rao, Corrosion, 45 (1989) 142–149.
49. W.L. Clarke, R.L. Cowan and W.L. Walker, Intergranular corrosion of stainless alloys, ASTM STP 656, ed. R.F. Steigerwald, (1978), 99–132.
50. V. Kain, P.K. De, Proc. Symp. on Localised Corrosion and Environment Cracking, January 22–24, 1997, Kalpakkam, paper no I3.
51. N. Parvathavarthini, R.K. Dayal, S.K. Seshadri and J.B. Gnanamoorthy, Br. Corr.J., 26 (1991)67–76.

52. U. Kamachi Mudali, R.K. Dayal and G.L. Goswami, Surface Engg., 11 (1995) 331.
53. U. Kamachi Mudali, R.K. Dayal and G.L. Goswami, Anti-corrosion Methods Mater., 45 (1998) 181.
54. U. Kamachi Mudali, M.G. Pujar and R.K. Dayal, J. Mater. Engg. Perf., 7 (1998) 214.
55. R.K. Dayal, Surface Engg., 13 (1997) 299.
56. S.V. Deshmukh, C. Rajagopalan, R.V. Subbarao, R.K. Dayal and J.B. Gnanamoorthy, Proc. Material Laser Symp., IRDE, Dehradun, Feb., (1995) P. 312.
57. J.Y. Jeng, B.E. Quayle, P.J. Modem, W.M. Steen and B.D. Bastow, Corros. Sci., 35 (1993) 1289.
58. R.K. Dayal, Trans. Ind. Inst. Metals, 50 (1997) 1.

6. Metallurgical Influences on Stress Corrosion Cracking

P. Muraleedharan[1]

Abstract This paper will present the results of several investigations on SCC of austenitic stainless steels. It is a review of SCC work accomplished in our laboratory—with a focus on the relationship between metallurgical microstructure and SCC. Cold work was deleterious for SCC resistance whereas long term ageing resulting in uniform carbide precipitation in an austenite matrix, was found to improve the resistance. Both sensitization and grain boundary segregation made stainless steels susceptible to intergranular stress corrosion cracking. A transition in SCC mode from transgranular to intergranular was observed in both 304 and 316 stainless steels, when tested in magnesium chloride solution boiling at 155°C. This transition was influenced by the metallurgy of the steel; it was facilitated by cold work and inhibited by carbide precipitation. A generalisation of this transition phenomenon vis-a-vis the metallurgical condition of the steel leads to a dissolution-controlled mechanism of SCC of stainless steels in chloride media.

Key Words Stress corrosion cracking (SCC), transgranular SCC, intergranular SCC, stainless steel, cold work, sensitization, segregation, precipitation, transition in crack morphology.

STRESS CORROSION CRACKING

Introduction

Stress corrosion cracking (SCC) is a term used to describe failure of engineering materials that takes place by environmentally induced crack propagation. This occurs mainly in metals and alloys that are protected against uniform corrosion by the formation of passive films. In such materials, localized breakdown of the passivity by mechanical means (stresses present in the material) can lead to accelerated attack in a very narrow region, the rest of the area still remaining protected. A general definition of the phenomenon is that SCC is the fracture of a material by the simultaneous action of a tensile stress and a corrosive environment. It is a synergistic process in the sense that the time-to-fracture, the decrease in load-bearing capacity, and other effects manifested in the phenomenon, are different from similar effects by stress or corrosion acting alone. The phenomenon is of great industrial significance since stresses are invariably induced during the fabrication of plant components and structures and it is not always possible to remove the stresses or to make reliable measurements of it in the engineering structures.

The phenomenon of general corrosion (uniform corrosion), by which a material either dissolves

---

[1]Scientific Officer, Corrosion Science and Technology Division, Indira Gandhi Centre for Atomic Research, Kalpakkam-603 102, India.

away more or less uniformly or the surface layers are converted into scales of corrosion products (e.g., rusting of iron), is fairly well known. Figure 1 is a schematic of the development of uniform corrosion and localized corrosion [1]. Uniform corrosion data for many industrial materials are readily available in hand books. Therefore, allowances for loss in material thickness, can be given based on average corrosion rates, at the design stage itself. However, SCC is a localized form of corrosion and, hence, the extreme value approach must be taken for the analysis. As shown in Fig. 1, the service life of a material can be considered finished when the fastest growing localized corrosion attack has caused the first perforation. Therefore, it is not the average crack growth rate or crack initiation time, but the maximum crack growth rate or the minimum time for crack initiation that must be considered in the life estimation of components. In the absence of a crack, it is safe to assume that a component has infinite life from the SCC considerations.

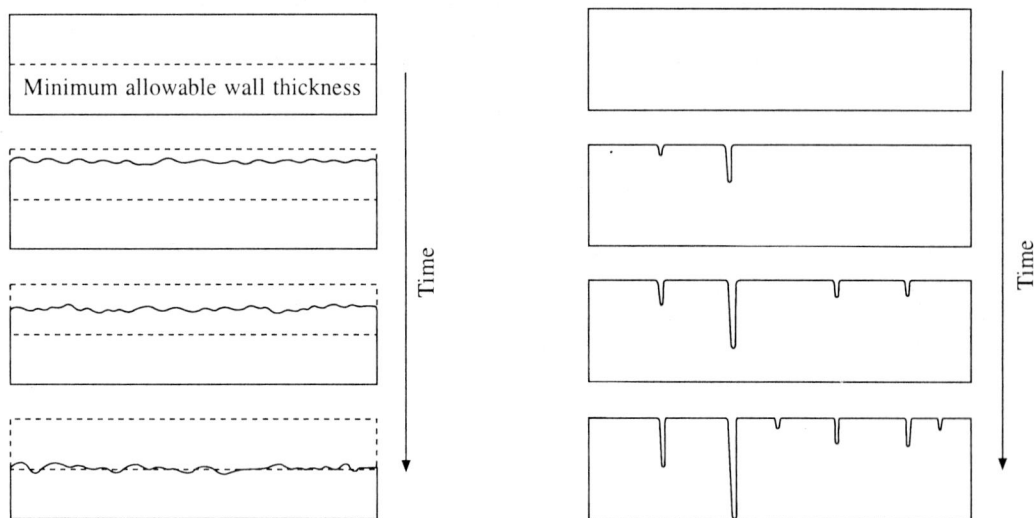

Fig. 1.   Schematic of the development of general corrosion and localized corrosion [1].

Stress corrosion cracking can take place in highly corrosion resistant alloys in seemingly innocuous environments. Highly branched and tight cracks in the direction perpendicular to tensile stress axis are developed during SCC and, in many cases, it is difficult to detect these cracks in the initial stages of development, by non-destructive techniques. An idea about the highly localized nature of SCC can be had from the aspect ratio of a growing crack. If an aspect ratio of 1 between penetration and lateral corrosion corresponds to general corrosion and a ratio above 1 but less than 10 for pitting corrosion, an aspect ratio as high as 1000 has been observed for growing stress corrosion cracks [2]. Another important characteristic of SCC is that the material other than that in the cracked region will be as sound as the original material. In other words, there is no accumulation of SCC damage in the material unlike in the cases of creep and fatigue. Even in ductile materials, SCC is macroscopically brittle in the engineering sense that a crack can initiate and propagate to failure of a component without significant dimensional changes. All these features of SCC phenomenon points to the fact that failure by SCC may take place unnoticed if such an eventuality is not anticipated.

Until recently it was thought that pure metals were immune to SCC. However there are exceptions to this rule, e.g. iron in anhydrous ammonia [3] and copper in sodium nitrite solution [4, 5] undergo transgranular SCC.

Parameters Affecting SCC

For a stress corrosion crack to nucleate and propagate a suitable combination of tensile stress, environment , alloy chemistry and microstructure is essential.

Stress

For fracture by SCC, a tensile stress (nominally static) is generally considered essential. However, static stress does not exclude slow monotonic straining or low-amplitude cycling (ripple loading) which accelerates SCC in metallic systems, for example, by promoting oxide film rupture at the crack tip [6]. Stresses arise in practice from applied loads or from residual stress due to welding, forming or heat treatment. Stresses below the macroscopic yield stress is sufficient to cause SCC. But, SCC propagation rates are influenced by the magnitude of the stress. Introduction of compressive stresses by controlled shot peening of component surfaces is reported to prevent the initiation of SCC in engineering materials [7,8]. It is also recognized that it is the strain rate at the crack tip rather than stress per se, that is important in SCC. For a variety of systems, SCC occurs only if the strain rates are within an initial range limited by the upper and lower bound critical strain rates as shown in Fig. 2 [9].

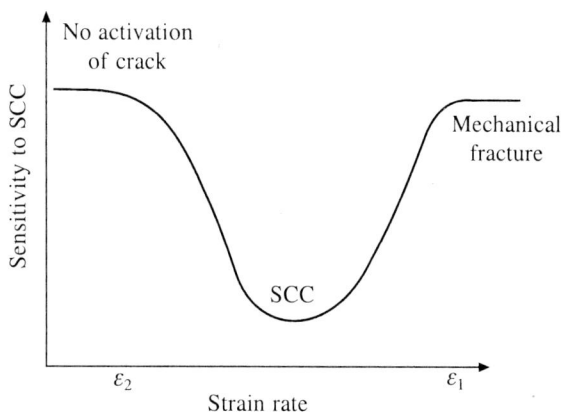

Fig. 2.   Effect of strain rate on stress corrosion cracking.

The magnitude of stress required for SCC depends on the material microstructure and the environment. For example, sensitized austenitic stainless steels undergo SCC in high temperature water if the stresses are of the order of yield strength of the material. However, in hot chloride environments the stresses required are much lower and cracking has been observed in the laboratory at stress values of the order of 20% yield stress.

Environment

Environments that cause SCC are usually aqueous solutions although there are many practical instances of SCC in nonaqueous systems such as Ti and Zr alloys in methanol-halide solutions, Zr alloys in iodine vapour etc. It is generally observed that some specific chemical species in the environment is required for SCC. Thus, the SCC of copper alloys, traditionally referred to as season cracking, is virtually always due to the presence of ammonia in the environment, and chloride ions cause cracking in stainless steel and Al alloys. Hot concentrated hydroxides have been identified to be the causative factor in the caustic cracking of boilers and other steam generating equipments. Also, a chemical species that causes SCC in one alloy may not cause SCC in another alloy. Changing the temperature, the degree of aeration and/or the concentration of ionic species may change an innocuous environment into one that causes SCC failure.

The solubility of the reaction product in the environment appears to be an important factor in the crack propagation by SCC. However, for the crack to propagate in the forward direction general corrosion has to be stifled along the crack faces. This is accomplished by the formation of some surface films which may be a thin passive film or a layer of precipitated corrosion products. If the stability of the film is high, the alloy becomes resistant to cracking. Such a condition of the alloy/ environment system, generally referred to as border-line passivity, is essential for the cracking to take place. Such border-line passivity is possible only in certain regimes of electrochemical potential as depicted in Fig. 3 [10].

Alloy Chemistry and Microstructure

As discussed above, environmental criteria for SCC are necessary but not sufficient. Stress corrosion cracking will not occur without a susceptible metallurgical condition. The only exception is the transgranular cracking process in pure metals like iron and copper [6]. Alloy composition including concentration of impurity and trace elements influence the SCC properties of metals and alloys.

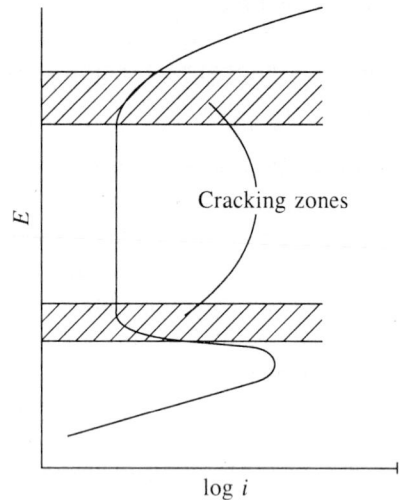

Fig. 3. Various electrochemical potential regimes that cause SCC.

Metallurgical conditions, which include strength levels, second phases present in the matrix and the grain boundaries, composition of phases, grain boundary segregation; all of these factors affect SCC. In fact, it is possible to make a susceptible alloy relatively immune or highly resistant to SCC or vice versa, by altering the alloy chemistry and/or the microstructure. A classic example is the susceptibility of austenitic stainless steels to IGSCC on sensitization even in innocuous environment like high purity water at high temperatures depending on the oxygen content. Austenitic stainless steels, when devoid of trace impurities, such as P, S and As, become highly resistant to chloride-induced SCC. Carbon or nitrogen segregation and/or precipitations at grain boundaries influence caustic intergranular SCC of carbon steel, as it interferes with passive film formation and affects plasticity. The effect of P-segregation in carbon steel is to introduce SCC in a new more oxidizing range of potentials [6].

The Phenomenon of SCC

The overall stress corrosion process may be divided into two stages: the initiation process, which in many cases is preceded by an incubation period, and the propagation process. It is widely accepted by researchers that the most important change taking place prior to the initiation process is the establishment of a local chemistry of the environment suitable for cracking. From a practical view point, this stage is very important especially for active - passive materials such as austenitic stainless steels because the bulk environment which the material usually encounters is usually harmless; and it is the locally produced environment that causes cracking. For such systems, the time taken for crack initiation forms a major part of the total time - to - fracture. However, crack initiation time is reduced in the accelerated tests conducted in the laboratory, as the environments used for these tests are very severe.

During the initiation process, development of an occluded cell with its attendant acidification and concentration of anionic species takes place. The development of this occluded cell is strongly correlated with pitting or crevice corrosion in stainless steels and other active-passive materials. In fact, many SCC failures in service have been reported to initiate from pits in austenitic stainless steels. Pits can form at inclusions that intersect the free surface or by breakdown of the protective film. However, it does not mean that all the cracks initiate this way. Rather, film rupture by slip is an alternative mechanical means for initiation of SCC from smooth surfaces. Stress corrosion cracking can also initiate at pre-existing surface discontinuities such as grooves, laps or burrs resulting from fabrication processes. This is because these surface flaws can be points of stress concentration. Crack initiation from pre-existing surface features are common in high strength steels whereas in low strength materials, like austenitic stainless steels, establishment of a suitable local chemistry rather than stress concentration is the deciding factor.

There are many models that may account for most of the known cases of SCC in metals; slip dissolution [10–12], hydrogen embrittlement [13–15] and film-induced cleavage [16–19] being the most popular ones.

SCC OF AUSTENITIC STAINLESS STEELS

Austenitic stainless steels (ASS) have excellent resistance to general corrosion. However, these steels are highly susceptible to localized corrosion such as pitting, crevice corrosion and SCC. Stress corrosion cracking of these steels, in chloride media, is the most prevalent, probably because of the prevalence of chloride ions everywhere. Other than the chloride present in water, that is often used as a heat transfer medium in industries, chloride can also get deposited on the external surfaces of stainless steel (SS) components either from atmosphere (at coastal sites) or as a result of chloride leaching from thermal insulation. Other sources of chlorides are industrial chemicals, dyes and lubricants used for various purposes. Temperature is one of the important influencing parameters in the chloride SCC. Austenitic stainless steels in the sensitized condition are known to crack by IGSCC in coastal atmosphere at ambient temperatures. It is generally believed that a temperature of 60°C or above is required for SCC of non-sensitized stainless steels. However, this is a controversial point and many researchers are of the opinion that there is nothing sacred about 60°C. It is reported that stress corrosion crack growth rate in chloride media increases radically above 80°C; but below 80°C the threshold temperature observed depends on the patience of the observer. Our own experience has shown that SCC failure can occur even at ambient temperatures provided there are surface impurities like embedded iron particles on steel (SS) components [20–22]. Rusting of these iron particles provides sites for accumulation of chloride from the atmosphere. Figure 4 shows an optical micrograph illustrating the initiation of stress corrosion cracks from a corrosion deposit formed on the surface of a stainless steel by rusting of the embedded iron particles. A typical branched transgranular stress corrosion crack is also shown in the same figure.

Another question often asked is whether there is any threshold concentration of chloride below which there would be no SCC of ASS. This is a difficult question to answer. It is because, as we have seen in the preceding discussion, it is the local environment and not the bulk environment that is responsible for the SCC of ASS in majority of service failures. Accumulation of chloride can take place inside crevices or under corrosion deposits on stainless steel surfaces. During service, there can be evaporative concentration or a concentration in the surface film on a heat rejecting surface. If none of the concentrating mechanisms are operating it can be assumed that stainless steel can withstand normal water containing less than 1000 ppm chloride under flowing conditions. In many cases, chloride get deposited on stainless surfaces as a result of leaching from the thermal insulation. The amount of chloride that can be allowed in the insulation depends on the silicate content.

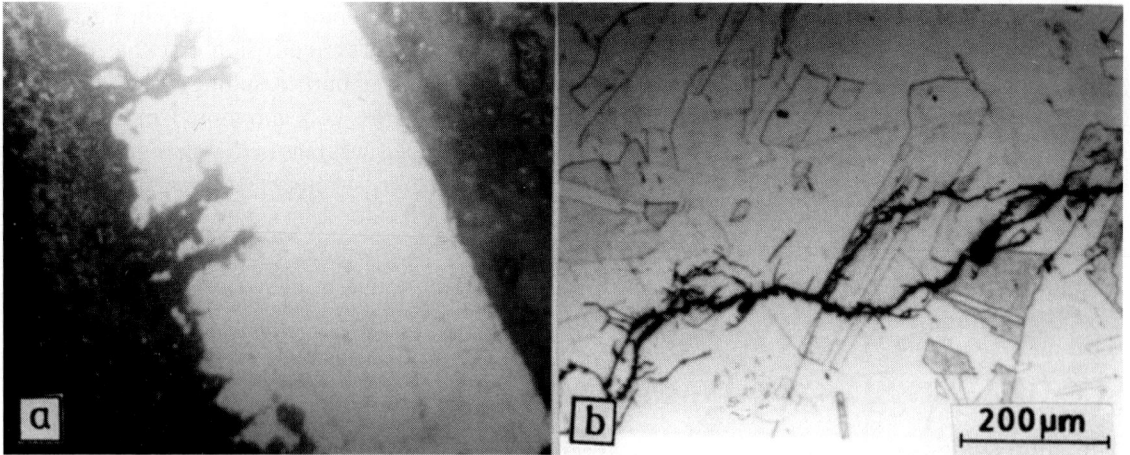

Fig. 4. Optical micrographs of (a) SCC crack initiating from a corrosion deposit and (b) typical branched transgranular SCC in chloride environment.

Metallurgical Influences on SCC

Many studies correlating metallurgical variables and SCC have been reported in literature. Such studies not only help in evaluating the SCC resistance of a SS under different metallurgical conditions, but, by way of generalization of the SCC behaviour of the material under different microstructural conditions, also enable an understanding of the mechanisms of SCC. Austenitic stainless steels of the 300-series are amenable only to very little variations in microstructure. As a result, studies on the effect of microstructure are limited for these steels. In addition to the normally used solution annealed structure, the most widely studied microstructures in ASS are those resulting from deformation at room temperature and from sensitization, often following cold work. This is, perhaps, due to the fact that these microstructures are produced during the various stages of fabrication of SS components. We have carried out extensive studies [23–31] on the influence of metallurgical variables on SCC of austenitic stainless steels. The metallurgical variables studied are:

- Alloy chemistry
- Sensitization
- Grain boundary segregation
- Cold work
- Precipitation

Most of these studies were carried out using commercial alloys. The only exception is the study of P-segregation that was conducted in a model alloy. The chemistry of the various commercial ASS used in our study are given in Table 1.

Alloy Composition

Alloy chemistry and microstructure have very significant influence on the SCC characteristics of SS. Austenitic stainless steels are highly susceptible to chloride - SCC whereas ferritic stainless steels are

Table 1.   Chemical compositions of the stainless steels used (wt. %)

Element	Steel A 304 SS	Steel B 316 SS	Steel C 304 SS	Steel D 316 SS	Steel E 304 SS
Cr	18.2	16.3	18.2	15.6	18.6
Ni	9.9	10.9	8.5	12.5	9.2
Mn	0.95	1.13	1.16	1.57	0.82
Mo	–	2.04	–	2.27	–
Si	0.38	0.31	0.35	0.64	0.79
C	0.049	0.053	0.070	0.050	0.079
N	0.030	0.044	0.058	0.030	0.039
P	0.036	0.036	0.045	0.026	0.040
S	0.019	0.012	0.008	0.011	0.009
Fe	Bal.	Bal.	Bal.	Bal.	Bal.

highly resistant or even immune. Even in ASS, it is possible to increase the SCC resistance by changing the chemical composition. The classic example is the increase in the chloride - SCC resistance of ASS on increasing the nickel content above 8% [32]. Similarly, addition of molybdenum and nitrogen also improves the SCC resistance, possibly through the beneficial effect of these elements on pitting resistance; because as mentioned earlier, in many practical situations pits are formed as precursors to SCC. Service performance of 300-series SS would suggest that molybdenum containing type 316 SS is more resistant to chloride SCC than 304 SS [33]. Sulphur and phosphorus contents of the steel affect very adversely the resistance of ASS to SCC. Stress corrosion studies conducted on four commercial ASS of types 304 and 316 in standard boiling magnesium chloride solution as per ASTM G36 have shown that maximum susceptibility is exhibited by the steel containing the highest phosphorus content as shown in Fig. 5 [26]. In a review on SCC by Newman and Procter [34], it is indicated that minimisation of phosphorus content is a key part of the recent Japanese strategy to develop SS resistant to chloride SCC.

Sensitization

Austenitic stainless steels of the 300-series undergo sensitization during thermal exposure in the temperature range of 500-850°C. When sensitized, these steels become susceptible to intergranular corrosion (IGC) and intergranular stress corrosion cracking (IGSCC) in a variety of environments such as high purity water, polythionic acid and aqueous solutions containing chloride or fluoride ions.

In our study [29], solution annealed specimens of a type 304 SS (steel E in Table 1) were sensitized by isothermal heat treatments at 500, 550, 600 ,650 and 700°C for different time durations. The sensitized specimens were tested for their IGSCC susceptibility using the SSRT method in 20% sodium chloride solution boiling at 105°C at a nominal strain rate of 2.4×10^{-5}. The degree of sensitization (DOS) in the specimens were measured by the electrochemical potentiokinetic reactivation (EPR) test [35].

A summary of the SSRT results for various sensitized specimens are presented in Table 2. The IGSCC susceptibility index (I_{scc}) values are also plotted against log of the heat treatment time for various temperatures in Figure 6 which shows that the susceptibility to IGSCC increases with heat treatment time, saturates and decreases thereafter. Figure 6 also shows that for sensitizing times of 8h and 100h, the lower the sensitizing temperature, the higher is the IGSCC susceptibility. The number

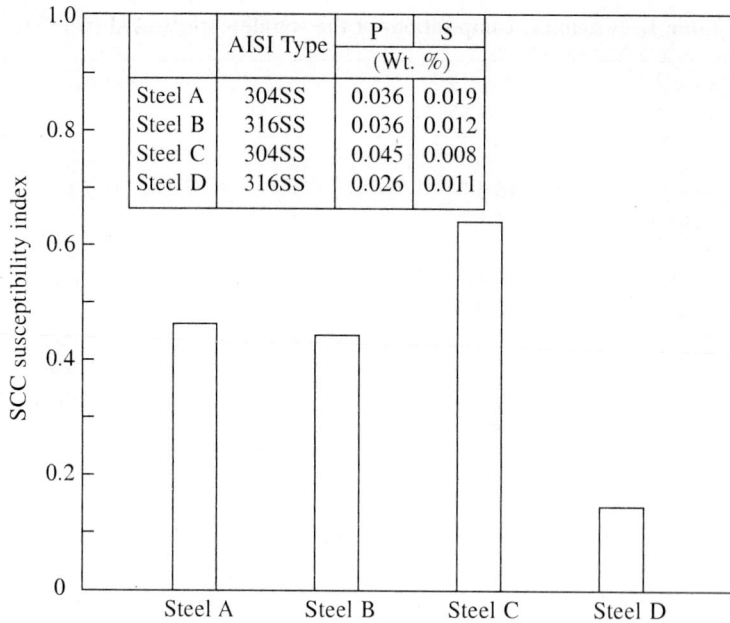

Fig. 5. SCC susceptibility of various commercial stainless steels.

of cracks on the gauge length of the specimen is also indicative of the IGSCC propensity. For specimens with high SCC susceptibility, the number of cracks was also high (Table 2). Reduction-in-area is another parameter that reflects the SCC susceptibility. Table 2 shows that as the susceptibility of the sensitized steel to IGSCC increases there is a corresponding decrease in RA.

Table 2. Summary of the SSRT results for various sensitized specimens

Heat Treatment	Strain to Failure %	Reduction in Area %	Failure Mode	No. of Cracks
500°C/ 240h	50	–	Ductile	–
1000h	16	9	IGSCC	100
550°C/ 8h	48	–	Ductile	–
20h	50	–	Ductile	–
100h	14.2	8.3	IGSCC	100
600°C/ 1h	50	–	Ductile	–
8h	20.2	13.3	IGSCC	50–100
100h	20.4	10.8	IGSCC	100
650°C/ 1h	37	31	IGSCC	5–10
8h	25.5	21	IGSCC	50–100
100h	22.6	11.5	IGSCC	100
700°C/ 1h	42	41	IGSCC	5
2h	36	27	IGSCC	5–10
8h	27.6	21	IGSCC	10–15
100h	32.2	29	IGSCC	10–40

Scanning electron microscopic examination of the fracture surfaces revealed that in every SSRT test that indicated SCC susceptibility, the corresponding fracture surface exhibited IGSCC. The area of IGSCC on the fracture surface was also found to increase with increasing susceptibility. EPR- etch structures and the corresponding fractographs for the specimens sensitized at 550°C are given in Fig. 7.

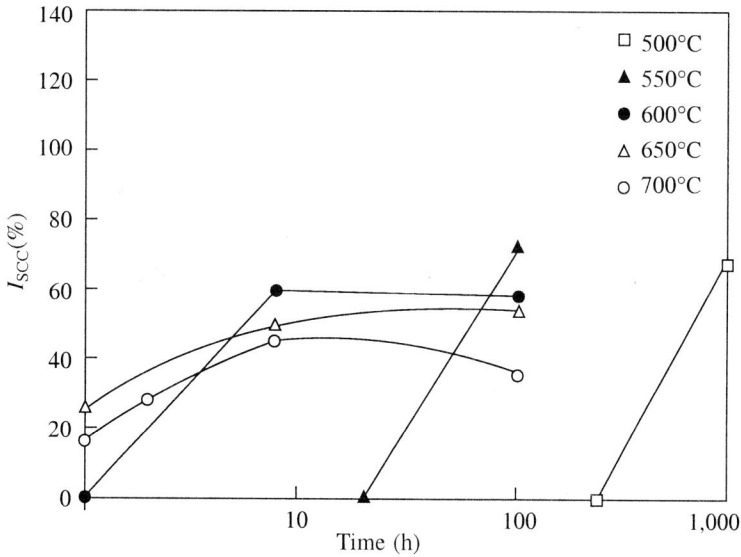

Fig. 6. Plot of IGSCC susceptibility index versus duration of heat treatment for various sensitizing temperatures.

Another observation was that the IGSCC susceptibility increased with decrease in the sensitizing temperature for longer heat treatment times, as shown in Fig. 8. Thus, for 100-h heat treatment, the highest SCC susceptibility was observed when the sensitizing temperature was 550°C, followed by 600, 650 and 700°C. This indicates that the level of Cr-depletion in the grain boundary zones has a significant influence on the IGSCC susceptibility of sensitized stainless steel. The chromium content near the grain boundary decreases with decrease in the sensitizing temperature. A decrease in the Cr-content of the depleted zone at the grain boundaries causes a decrease in the stability of the passive film and increases the chemical activity of the region. According to the anodic dissolution mechanism of SCC the above condition promotes cracking susceptibility [10–12].

The values of reduction in area (Table 2) also suggest a mechanism of anodic dissolution of Cr-depleted grain boundary zones under the influence of stress. Here, it is assumed that if the extent of Cr-depletion at the grain boundaries is very high, IGSCC initiation is possible at very low values of plastic strain owing to the increased chemical activity of the depleted zones. However, large amounts of grain boundary deformation will be required to initiate cracking in lightly sensitized boundaries. It is also true that once a crack grows substantially (say, above 10 to 20% of the specimen cross section), the strain concentrates at the crack tip and further extension is "used up" more in the crack extension than in overall reduction in area. Therefore, the RA of a severely sensitized specimen will be less and that of a lightly sensitized specimen is expected to be high. Indeed, this has been shown to be true from the values of RA in Table 2. These results support the view that IGSCC is initiated in sensitized stainless steels when there is a certain amount of grain boundary activity produced by the combined

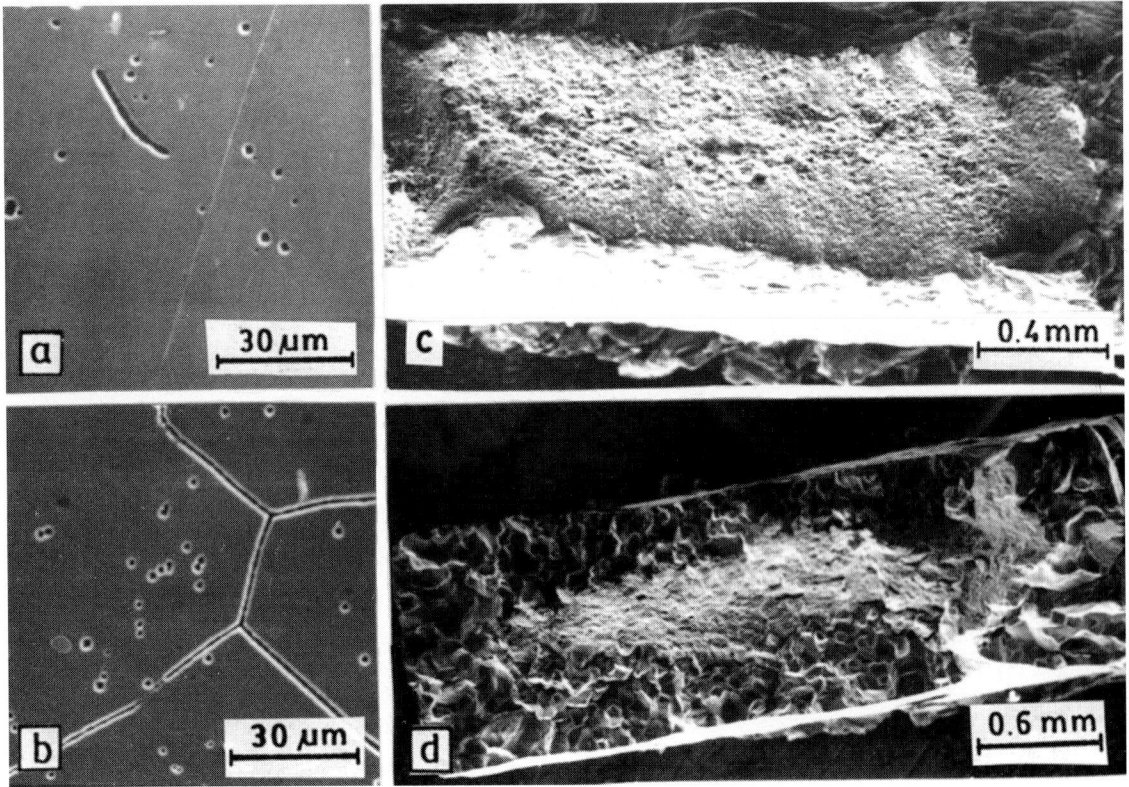

Fig. 7.   EPR etch structures and the corresponding fractographs of specimens sensitized at 550°C.

effects of Cr-depletion and grain boundary deformation. This suggests that the predominant mechanism of IGSCC of sensitized stainless steels is one of active dissolution of Cr-depleted grain boundaries.

Since the EPR test for measuring DOS as well as the SSRT method for evaluating IGSCC susceptibility give quantitative results, it is possible to compare the results from these two tests. Figure 9 shows the plots of I_{scc} versus EPR charge for various sensitizing heat treatments. In general, IGSCC susceptibility and EPR charge values are linearly related, for low degrees of sensitization. However, for higher DOS (as indicated by EPR), there is no corresponding increase in IGSCC susceptibility, and in some cases (e.g. 700°C /100h), the IGSCC susceptibility decreases for a higher EPR value. Also, for a given EPR value, the IGSCC susceptibility increases with decrease in the sensitizing temperature. This suggests that the EPR value is not a unique indicator of the propensity to IGSCC of sensitized ASS in the environment studied. For example, for a measured EPR charge value of 0.2C/cm^2 in specimens heat treated at different sensitizing temperatures, the specimen sensitized at 700°C showed the least IGSCC susceptibility; the susceptibility was found to increase with decreasing sensitizing temperature.

Segregation

Segregation of impurity elements such as phosphorus and sulphur at the grain boundaries can also facilitate intergranular attack even in the absence of Cr-depletion [36]. Enrichment of these elements at grain boundaries is possible even in solution annealed steel [37], provided the bulk concentration

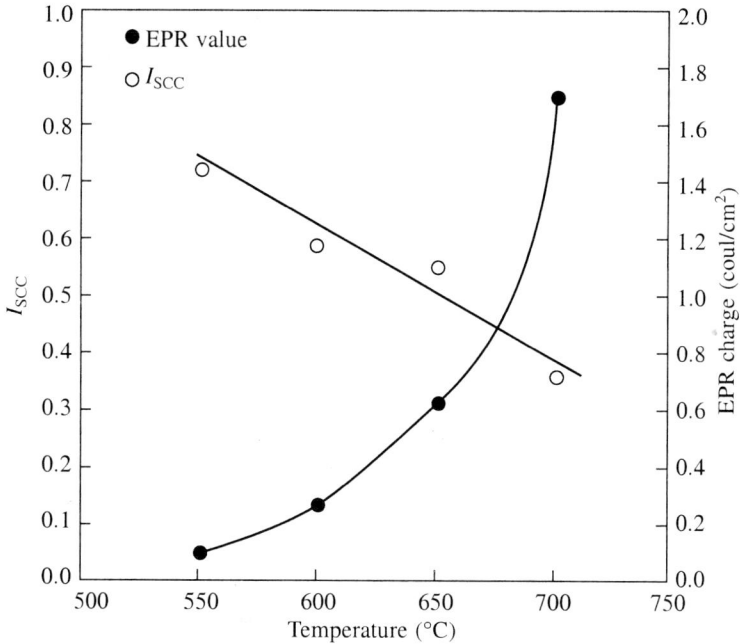

Fig. 8.   Plot of IGSCC susceptibility index and EPR values of specimens subjected to sensitizing heat treatment for 100 h.

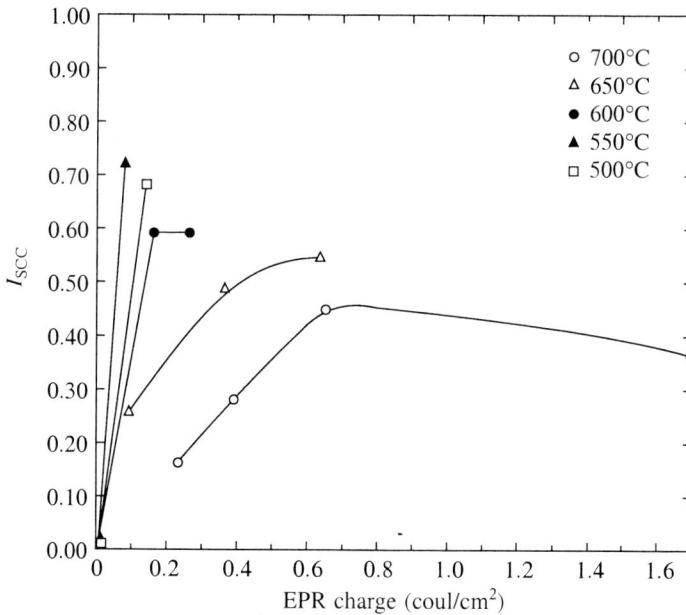

Fig. 9.   Plot of IGSCC susceptibility index against EPR charge values for various sensitized specimens.

is high. However, it is thermal ageing, especially in the low temperature regime (400–600°C), that facilitates segregation at the grain boundaries. Besides, neutron irradiation enhance impurity segregation to the grain boundaries [38]. Therefore, segregation effects are very pronounced in nuclear reactor core components. Since Cr carbide precipitation and segregation take place in the same temperature regime, it is often difficult to study the influence of segregation alone on the corrosion behaviour of SS.

The susceptibility of a P-doped type 304L SS to IGSCC in different aged conditions was studied in hot water environments [30, 31]. The chemical composition of the alloy was (wt %): 70.0 Fe, 17.8 Cr, 10.9 Ni, 1.1 Mn, 0.112 P, 0.009 S and < 0.005 C. The material was subjected to various heat treatments viz. 600°C/140 h, 550°C/1000 h and 500°C/1000 h. Two corrosion tests were used in this study for detecting segregation : boiling $HNO_3 + Cr^{+6}$ test and an electrochemical etch test at a transpassive potential [39, 40]. Besides, a copper-copper sulphate -sulphuric acid test was used for detecting Cr-depletion. The SCC tests were conducted using a screw-driven machine (Instron Universal Testing Machine) in two test environments viz., 0.01 M NaCl solution at 250°C and oxygenated 0.01 M sodium sulphate solution at 150°C, using a nominal strain rate of 7×10^{-7} s^{-1}.

Scanning electron microscopic examination of the specimens after exposure to copper-copper sulphate -sulphuric acid solution indicated intergranular attack only in the case of the 550°C/1000h heat treatment. This is indicative of Cr-depletion in the material. Since there was no carbides present at the grain boundaries, the Cr-depletion might have been caused by the formation of Cr-P phase at the boundaries, as suggested by the analytical electron microscopy results. Line scan across grain boundaries in Scanning Transmission Electron Microscopy (STEM) showed enrichment of phosphorus and Cr at the grain boundaries in some places, in the case of 550°C/1000 h aged specimen. Area scan at the grain boundaries also showed enrichment of both P and Cr together at many grain boundary sites and corresponding depletion of Fe and Ni at these sites (Figure 10).

Enrichment of Cr and P together at some points may be because of the formation of some Cr-P compounds. Formation of Cr-P compounds has been reported in thermally aged type 304 stainless steel with high P content [41]. Formation of Cr-rich phosphides can deplete the nearby region of Cr.

The summary of constant extension rate test results in oxygenated 0.01 M Na_2SO_4 solution at 150°C are given in Table 3. In the tests conducted at 250°C, there was no significant difference in the parameters such as % reduction in area, % elongation, UTS etc. for various heat treatments. Also, the mode of fracture was TGSCC for all the heat treatment conditions. However, in the tests conducted in oxygenated environment at 150°C, % elongation, % reduction-in-area and UTS showed significant reduction for the 550°C/1000 h heat treatment over the other heat treatments. The mode of fracture was also IGSCC. However, all the other heat treatments showed only TGSCC in the oxygenated environment at 150°C.

Scanning electron microscopic observations of the fracture surfaces has shown typical TGSCC for all the specimens except for the 550°C /1000 h specimen tested in oxygenated solution. The SEM fractograph showing IGSCC in the 550°C/1000 h specimen after the test in oxygenated solution at 150°C, is given in Figure 11(a). An SEM picture of the specimen surface, from an interrupted test of a specimen under the same heat treatment, showing intergranular crack initiation, is given in Fig. 11(b).

One possible reason for the absence of IGSCC in the solution annealed, 500°C/1000 h and 600°C/ 140 h specimens can be the insufficient grain boundary segregation in these cases. It is also possible that segregation alone cannot cause IGSCC in the hot water environments studied and a synergistic effect of P-segregation and Cr-depletion is important. The only instance of IGSCC in this study was

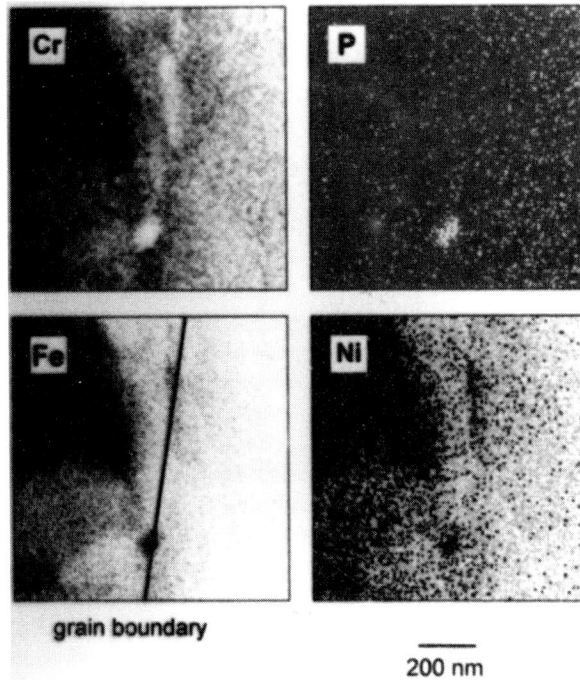

Fig. 10. Elemental distribution maps in STEM analysis of a 550°C/1000 h specimen. Bright areas show enrichment and dark areas depletion.

Table 3. Results of slow strain rate tests in oxygenated 0.01M Na_2SO_4 solution at 150°C

Heat treatment	Reduction in area (%)	Elongation (%)	UTS (MPa)	SCC morphology
Soln. annealed	28.9	50.0	418	Transgranular
600°C/140 h	28.9	50.0	423	Transgranular
550°C/1000 h	21.9	36.0	413	Intergranular
500°C/1000 h	24.3	47.0	435	Transgranular

in the case of 550°C/1000 h specimen in which case the Cu-CuSO$_4$ -H$_2$SO$_4$ test showed Cr-depletion. This view is shared by many researchers. Briant [42] and Briant and Andresen [43] have shown that phosphorus segregation enhances corrosion in the Huey test. But phosphorus segregation appears to have little effect on the IGSCC of SS in high temperature water. Many researchers [36, 37, 42, 43] have shown that ASS containing phosphorus corrode much more rapidly in the Huey test (boiling 65% HNO$_3$ test) than those steels that do not contain phosphorus. However it is not clear from the literature whether segregation of P or other impurity elements alone can cause IGSCC of austenitic stainless steels. Tice et al. [44] have shown that phosphorus enrichment alone cannot significantly affect the dissolution or passivation behaviour in either near neutral or low pH solutions, suggesting that segregation of impurity elements alone cannot explain cracking observed in operating LWR plants. According to Kuroda et al., [45] intergranular attack in the Huey test correlated with segregation of P at the grain boundaries. But IGSCC hardly occurred in such cases. However, in the cases where P segregation is taking place in synergy with Cr-depletion, increased susceptibility to IGSCC was observed.

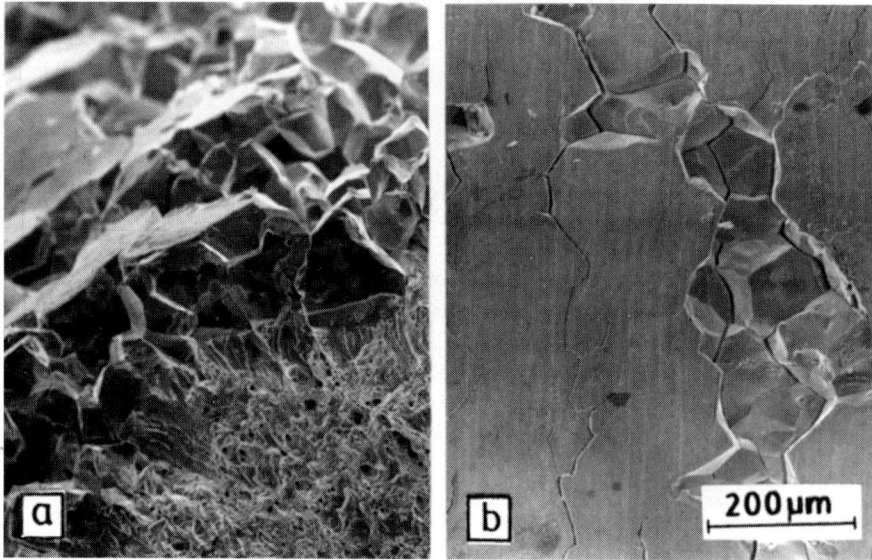

Fig. 11.   **SEM micrograph of IGSCC in 550°C/1000 h specimen, tested in oxygenated solution at 150°C: (a) fracture surface and (b)cracks on the specimen surface.**

Cold Work

It is well known that cold work has an influence on the SCC behaviour of ASS. The subject has been discussed in the extensive reviews on SCC of these materials by a number of authors; the reviews by Latanision and Staehle [46] and Hanninen [47] being notable among them. Unfortunately, a direct comparison of the results from different investigations is not possible because of the variations in alloy composition, methods of deformation, test procedures, environment etc. that have been used in different studies. In most of the cases, where a constant load method is employed, comparison is made on the basis of a given applied stress. But it is clear that since the yield strength of the material increases with deformation, a given applied stress in the SCC test would mean decreasing fractions of the yield strength for increasing levels of cold work. From an engineering point of view, comparison of results at stresses which are chosen fractions of yield strength is more indicative of material performance since yield strength is a primary design parameter. Tests at a chosen fraction of yield strength have an advantage from the mechanistic aspect also since the strain rate at the crack tip rather than the stress is the controlling factor in stress corrosion crack propagation.

The effect of prestraining on the SCC susceptibility of types 304 and 316 stainless (steel A and steel B in Table 1) has been studied using the constant load method at a constant initial stress as well as at a stress equivalent to a constant percentage of yield strength of the material [23, 26]. The cold working of the specimens were done by deforming the specimens at room temperature in uniaxial tension in an Instron tensile testing machine at a nominal strain rate of 5.2×10^{-4} s^{-1} at nominal strain levels of 2.3, 6.9, 11.6, 16, 26 and 56 pct. Tests for stress corrosion susceptibility were done in boiling MgCl$_2$ solution at 155°C using the constant load method at two initial stresses of 112MPa and 40% of the yield strength of the material.

Figure 12 illustrates the time-to-fracture (t_f) versus percentage prestrain curve for type 316 stainless steel. At an applied stress of 112 MPa, t_f decreased with increasing cold work, reached a minimum around 15 pct cold work, and then increased for higher deformation levels. This decrease in t_f may be due to the increase in the probability of crack nucleation because of the availability of a large number of defect sites. In fact, Cigada, et al., [48] have found that crack initiation time decreases with increasing degree of deformation. But, the t_f reported in the present study includes both initiation time as well as propagation time. As the strength of the material increases with deformation, the crack tip strain rate will be lower in tests employing a constant initial stress for cold worked materials. Thus, the crack propagation rate decreases with increasing degree of pre-training. The observed minimum in Fig. 12 can therefore be attributed to the greater influence of initiation time than propagation time at low levels of deformation [23, 27]. Such minima in t_f vs cold work plots have also been observed in earlier studies employing a constant initial stress. But in type 304 stainless steel no significant change in t_f was observed with increasing cold work at 112 MPa as shown in Fig.13 [23, 25]. Although a little amount of martensite was detected metallographically in 304 specimens prestrained to 56%, it could hardly explain the difference in the SCC behaviour of the steel tested at 112 MPa. However, when the applied stress was 40% of the yield strength of the material, t_f decreased with increasing prestrain for both types 304 and 316 stainless steels, up to a certain degree of deformation, and there was no significant change thereafter.

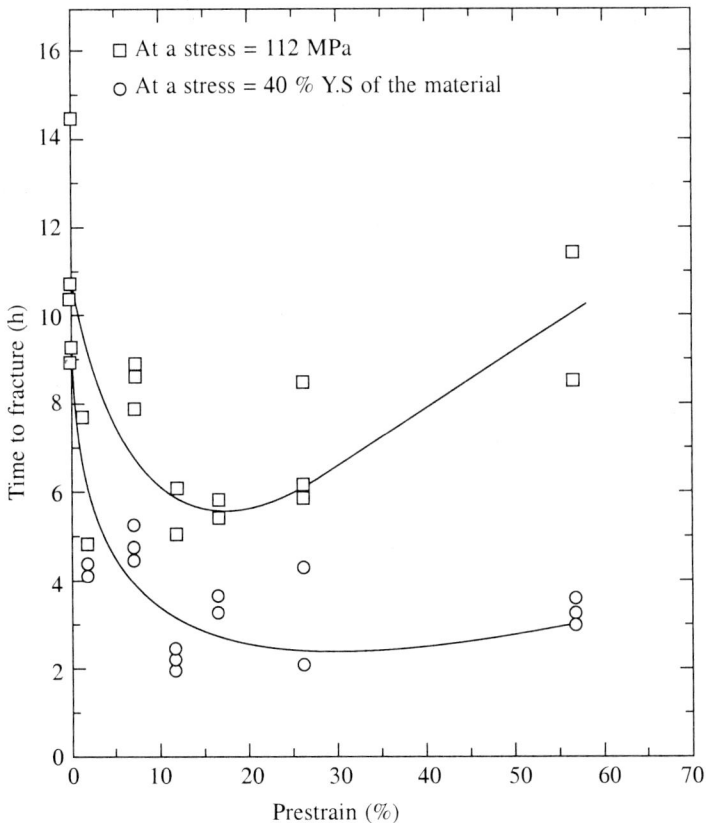

Fig.12.   Effect of cold work on time-to-fracture for type 316 stainless steel.

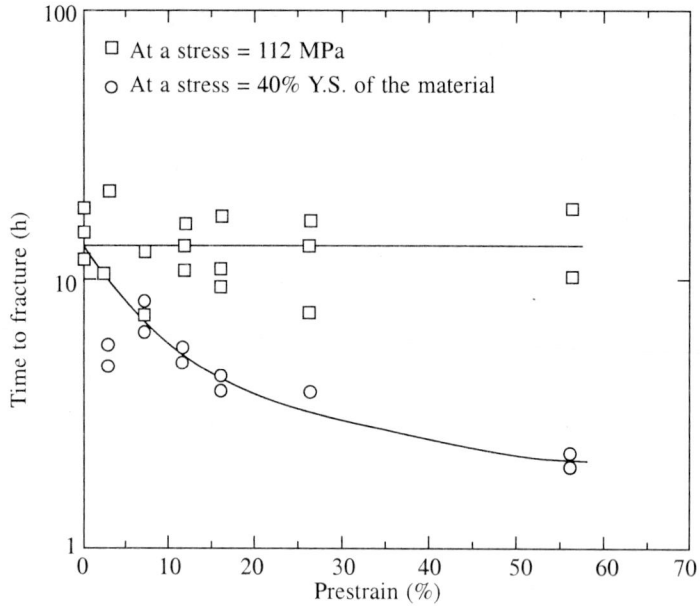

Fig. 13. Effect of cold work on time-to-fracture for 304 stainless steel.

Prestraining influences the SCC morphology also, in both 304 and 316 SS. The crack path in 316 SS was transgranular initially, changed over to intergranular mode as the crack proceeded and finally ended in ductile mode of fracture (Fig. 14) [23]. Such transitions in cracking mode from transgranular

Fig. 14. SEM showing three different modes of fracture in 316 stainless steel with 6.9% cold work, tested at 112 MPa.

to intergranular was observed in all the type 316 SS specimens irrespective of the extent of prestraining and of the stress applied. But in type 304 SS, the transition was observed reproducibly only above 16 pct prestraining i.e. only in the cases of specimens prestrained to 26 pct and 56 pct [23]. The fraction of the area of the fracture surface, occupied by intergranular mode of fracture, varied slightly from test to test at the same applied stress level for a particular level of deformation. But, on an average, the ratio of intergranular to transgranular SCC areas was found to increase with the increase in the extent of deformation in both 304 and 316 stainless steels. Increasing the applied stress was also found to facilitate such a transition. Table 4 gives the fracture morphology of specimens tested under different conditions.

Table 4. SCC morphology of stainless steels with various cold work levels

% Cold Work	Initial Applied Stress	Transition in Fracture Morphology	
		Steel A	Steel B
0	a. 112 MPa	T–D	T–I–D
	b. —		
2.3	a. 112 MPa	T–D	T–I–D
	b. 40% Y.S.	T–D	T–I–D
6.9	a. 112 MPa	T–D	T–I–D
	b. 40% Y.S.	T–D	T–I–D
11.6	a. 112 MPa	T–D	T–I–D
	b. 40% Y.S.	T–D	T–I–D
16.0	a. 112 MPa	T–D	T–I–D
	b. 40% Y.S.	T–D	T–I–D
26.0	a. 112 MPa	T–I–D	T–I–D
	b. 40% Y.S.	T–I–D	T–I–D
56.0	a. 112 MPa	T–I–D	T–I–D
	b. 40% Y.S.	T–I–D	T–I–D

T = Transgranular SCC, I = Intergranular SCC and D = Ductile dimple fracture

Prestraining did not affect the intergranular fracture morphology; whereas the transgranular SCC morphology was significantly affected as illustrated by Fig. 15. In solution annealed and mildly cold worked specimens, typical fan-shaped features which are characteristic of SCC, has been observed. Although there was an increase in the number of 'fans' with cold work, at very high degrees of deformation these fan patterns became less and less clear. The relation between the direction of crack propagation and the direction of fan features was studied by observing them from the periphery, where the cracks were initiated, to the interior of the specimen cross section. It was found that the direction of 'fans' (i.e., the direction in which the fans are diverging) coincided with the macroscopic direction of crack propagation. Scully et al.[49] have reported that the direction of fan shaped patterns in the fracture surfaces of stressed U-bend specimens, tested in boiling $MgCl_2$ solution is opposite to that of macroscopic crack propagation. But, as mentioned earlier, the fracture surface in our tests exhibited fans radiating outward in the direction of crack propagation. Similar directional relationship has been reported in constant load tests by Mukai et al. [50].

Precipitation

Ageing of ASS, especially of type 304 SS, at elevated temperatures for long durations of time produces

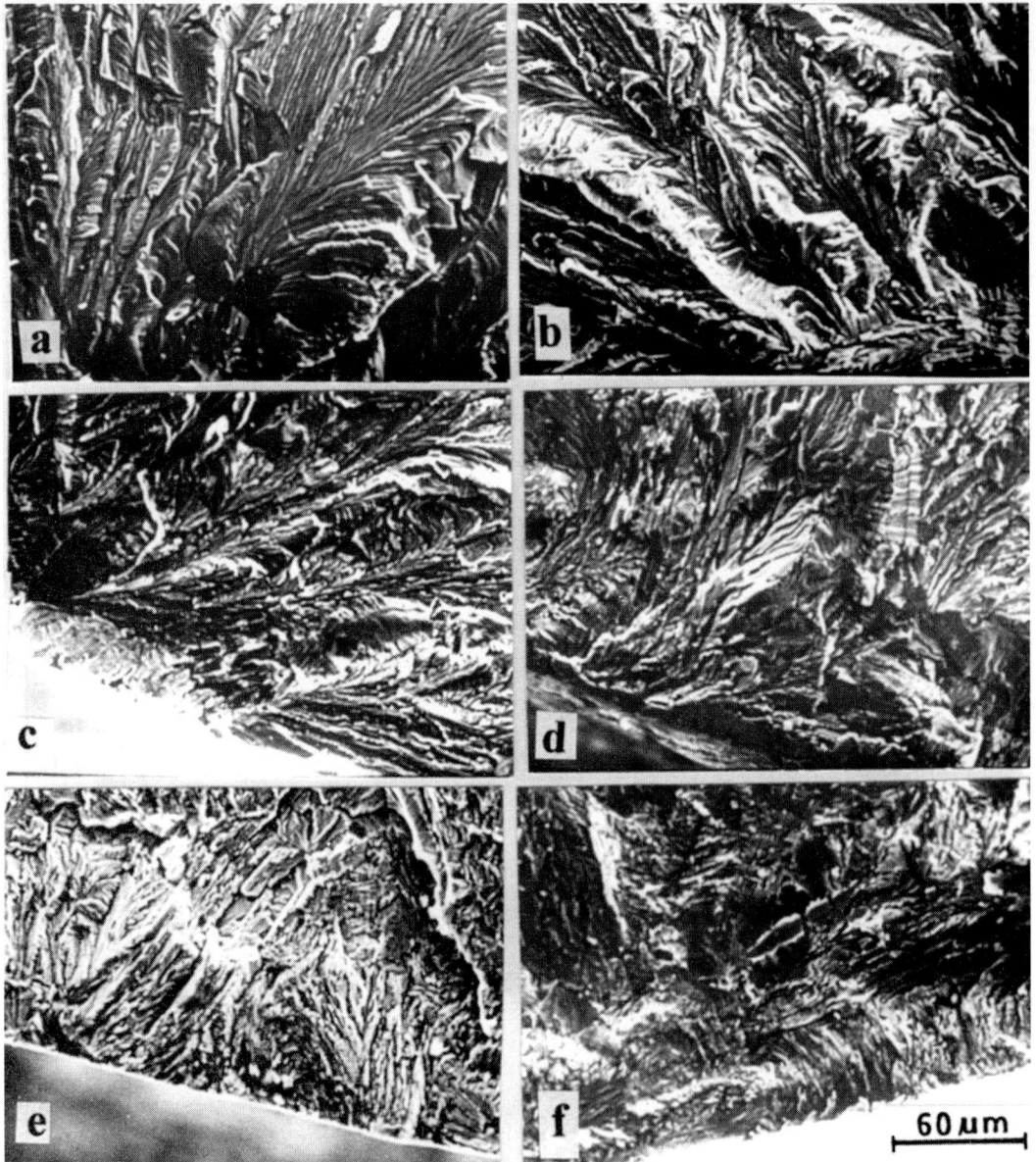

Fig.15.   SEM of transgranular fracture of 316 stainless steel specimens, prestrained to different levels and tested at 112 MPa.

a microstructure consisting of carbide precipitates at the grain boundary as well as in the matrix. The precipitates may affect significantly the SCC behaviour of these steels because the carbides can act as barriers to dislocation motion at low stress levels employed in SCC studies. The SCC properties of a type 304 SS (steel C in Table 1) aged at 700°C, for time duration ranging from 100 to 1000 h was studied in boiling $MgCl_2$ solution at free corrosion potential using constant load and slow strain rate methods [26, 29].

Figure 16a shows the variation of t_f (time-to-fracture) with ageing time, in boiling MgCl$_2$ solution at 155°C at an applied stress level of 112 MPa. Similar t_f versus ageing time plots for the tests performed in MgCl$_2$ solution at 125°C and at an applied stress of 150 MPa are given in Fig. 16b. It was observed that in the tests conducted in MgCl$_2$ solution at 155°C, t_f increased with ageing time up to a duration of 500 h. There was considerable scatter in the data. But, in general, the steel in the aged condition was more resistant to SCC than in the solution annealed condition. In the MgCl$_2$ solution boiling at 125°C also, t_f increased with ageing time up to 500 h and decreased for 1000 h aged steel, showing a maximum after ageing for 500 h. However, the effect of ageing on t_f values was not as marked as that exhibited in tests performed at a nominal stress of 112 MPa in MgCl$_2$ solution boiling at 155°C.

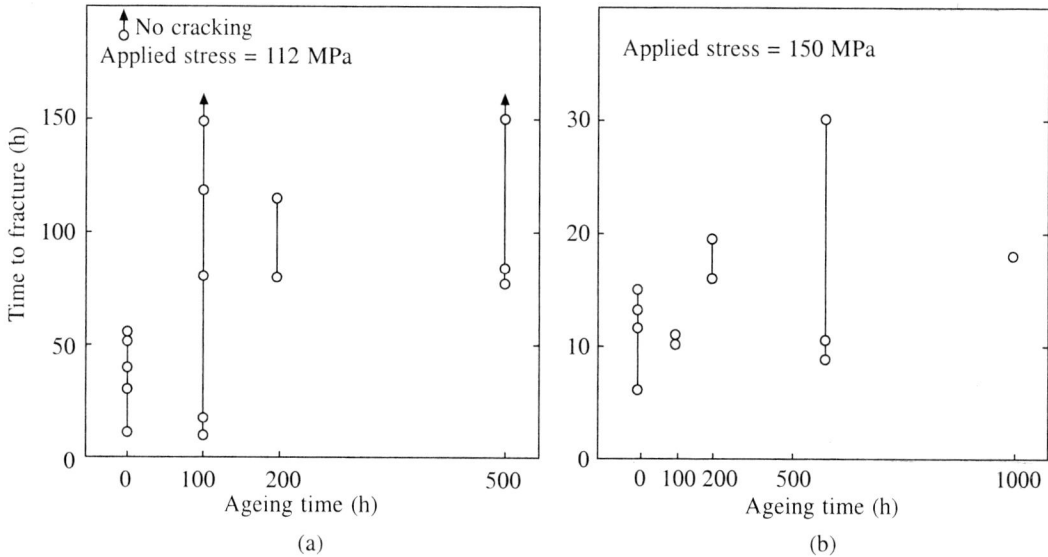

Fig. 16. **Variation of time-to-fracture with ageing time in magnesium chloride solution boiling at (a) 155°C and (b) 125°C.**

Figure 17 shows the plot of SCC susceptibility index I_{SCC} versus ageing time. The I_{SCC} values were found to decrease with ageing time up to 500 h indicating a decrease in the SCC susceptibility and then increased for higher ageing times. This behaviour was similar to that exhibited in constant load tests.

Ageing of the steel at 700°C improve the SCC resistance significantly. Ageing of a type 304 SS produces two important changes in the alloy from SCC considerations. First, carbide precipitation takes place initially at the grain boundaries, and on long term ageing, throughout the matrix. Secondly, precipitation of chromium-rich carbides produces attendant Cr-depleted regions. The Cr-depleted regions undergo preferential anodic dissolution and this is the reason for the accelerated intergranular attack in sensitized ASS. In boiling MgCl$_2$ solution, Cr-depleted regions do not significantly influence the SCC behaviour, because austenitic stainless steels sensitized to intergranular corrosion are found to crack by transgranular SCC in hot chloride solutions [47]. Therefore, the influence of ageing on SCC in MgCl$_2$ solution is due to the production of a microstructure with intergranular as well as intragranular carbides. No other secondary phases have been reported for a type 304 SS for ageing times up to 1000 h at 700°C [51]. The intragranular carbides can act as barriers to moving dislocations, thereby increasing the SCC resistance. In other words, the pinning of dislocations by the precipitates reduces the number of dislocations moving out at the crack tip, producing fine slip and the material becomes more resistant to SCC as the slip steps are fewer. Since the velocity of dislocations is reduced by the presence of carbide precipitates, the slip-step generation rate is also decreased. The improvement in SCC resistance of Inconel alloy X-750, after heat treatment for 96 h at 760°C, has been attributed to homogenisation of slip in the matrix by the precipitated gamma prime [52]. In the slip dissolution mechanism of SCC, slip-step height and slip step-generation rate are important controlling factors.

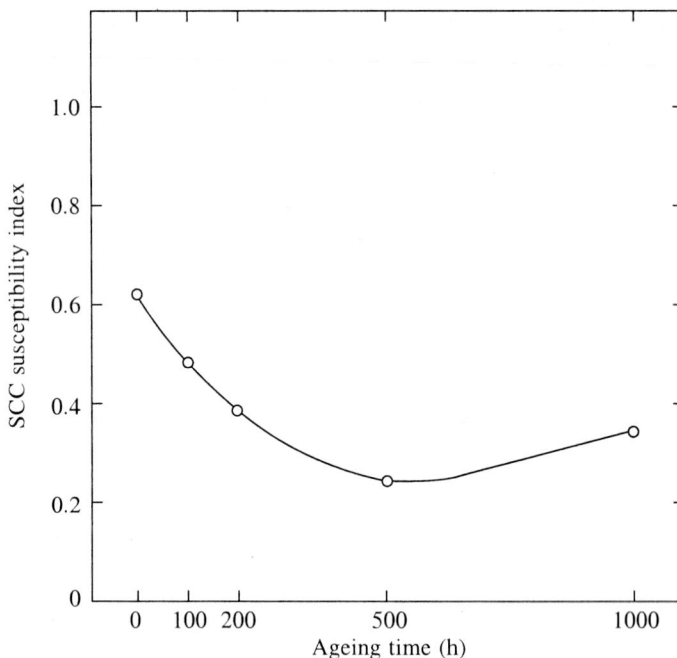

Fig. 17.   SCC susceptibility index versus ageing time for 304 SS.

The trend in the variation of SCC resistance with ageing time (Figs. 16(a) and (b)) is similar to typical mechanical property variations of age-hardening materials. The hardness or strength increases with ageing time, reaches a maximum, and then decreases for higher ageing times. This has been explained based on the increase in the number density of precipitates in the early periods of ageing and a coarsening of the precipitate in the later periods of ageing. It seems that the carbide precipitates influence the deformation behaviour of SS through dislocation-precipitate interactions and it is the predominant factor that affects the SCC properties of aged steels in $MgCl_2$ solution.

Scanning electron microscopic studies of the fracture surfaces revealed initiation of SCC in the transgranular mode in solution annealed and aged steel under all the test conditions. A transition in SCC morphology from transgranular to intergranular mode was observed in the solution annealed steel under all the test conditions. Aged specimens did not indicate any transition in SCC morphology in $MgCl_2$ solution boiling at 155°C in both SSRT and constant load tests. However, when the test was carried out in $MgCl_2$ solution boiling at 125°C, transition in cracking mode was observed in specimens aged for 100 and 200 h at an applied stress of 150 MPa and the transition was observed in all the specimens when the applied stress was raised to 200 MPa. Nevertheless, the area of intergranular fracture was less for aged specimens, compared to the solution annealed material, under similar conditions of testing. The various modes of fracture observed in the solution annealed and aged specimens under different test conditions employed in this study are given in Table 6.

Table 6.   Various fracture modes observed in solution annealed and aged specimens

Ageing time	Constant load tests in MgCl₂ at 155°C			Constant load tests in MgCl₂ at 125°C		Slow strain rate tests in MgCl₂ at 155°C
	112 MPa	*150 MPa*	*200MPa*	*150 MPa*	*200 MPa*	
0 h	T-I-D	T-I-D	T-I-D	T-I-D	T-I-D	T-I-D
100 h	T-I-D	T-D	–	T-I-D	T-I-D	T-D
200 h	T-D	–	–	T-I-D	T-I-D	T-D
500 h	T-D	–	–	T-D	T-I-D	T-D
1000 h	T-D	–	–	–	T-I-D	T-D

The predominantly intergranular SCC region of solution annealed steel exhibited features of TGSCC in many places. Also, grain facets revealed different degrees of slip line attack when examined at magnifications higher than 600 ×. Specimens aged for 500 h exhibited very fine fractographic features especially near the regions of transgranular-to-intergranular transition and the grain facets had a 'fibrous' appearance (Fig. 18). When the grain facets on the stress corrosion fracture surface of 1000 h-aged specimens were examined at high magnification, carbide particles and/or fine dimples were seen on the facets. Higher magnification pictures of grain facets in the IGSCC region of solution annealed and aged steels are given in Fig. 19.

Fig. 18.   Morphology of transgranular to intergranular transition zone with fibrous appearance.

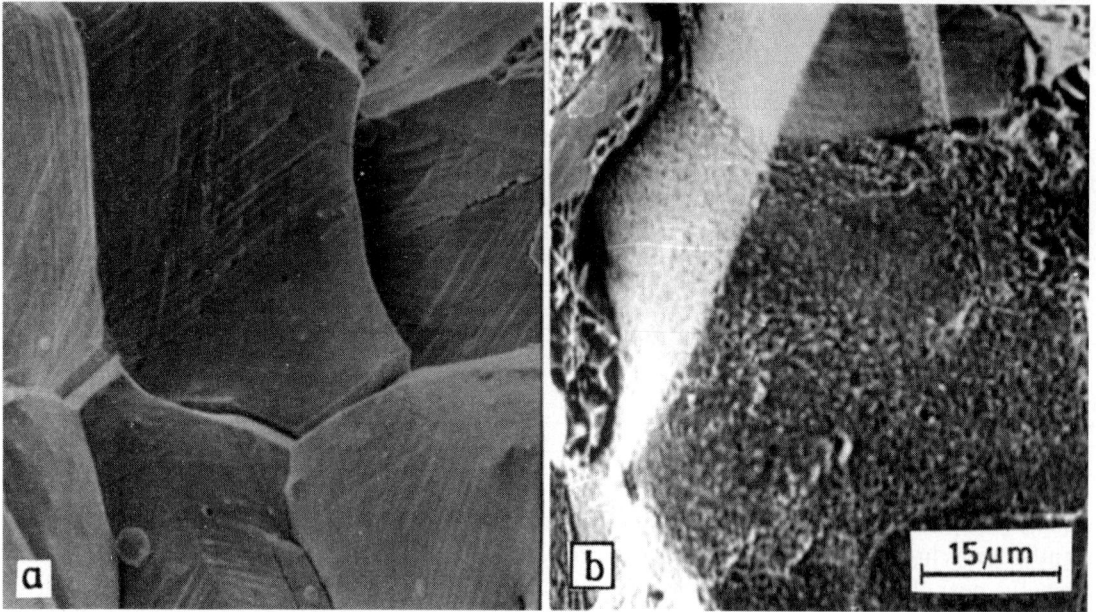

Fig. 19. Magnified picture of intergranular fracture facets in: (a) solution annealed and (b) aged steel.

Transgranular SCC surfaces generally exhibited typical cleavage-like morphology with fan-shaped features consisting of facets and steps. Ageing reduced the size of the facets and steps making the cleavage appearance finer (Fig. 18). This, again, can be attributed to the formation of large number of dislocation pile-ups as the number of matrix precipitates increases. When the pile-ups are not large enough to allow the crack to grow to a longer distance in aged steels, cracks have to be re-nucleated at other sites for continued propagation. Similarly, fine fractographic features were also observed in heavily deformed stainless steels with a tangled dislocation structure [23].

TRANSITION IN CRACK MORPHOLOGY: A KEY TO THE UNDERSTANDING OF SCC MECHANISM

Austenitic stainless steels exhibit a transition in SCC morphology when tested in boiling magnesium chloride solution. A schematic of the transition process is shown in Fig.20. The crack is invariably initiated in the transgranular mode and changes over to intergranular mode during the propagation and final fracture is by mechanical overload (dimple fracture). Our studies [23, 26, 29] have shown that metallurgical variables influence this transition. Types 304 and 316 ASS exhibit the transition in the solution annealed condition, both Mo and P contents facilitating the transition [23, 26]. However, when a 304 SS that exhibited the transition phenomenon in the solution annealed condition, was aged at 700°C to produce carbides at the grain boundaries as well as in the matrix, no transition was observed [29]. Prestraining facilitates the transition in SCC morphology in types 304 and 316 SS [23].

This phenomenon of transition in cracking mode is important from the mechanistic viewpoint since both TGSCC and IGSCC take place under the same set of experimental conditions and hence forms part of the same mechano-chemical continuum. Therefore, the transition phenomenon may provide a key to the understanding of the SCC mechanism. The same view is shared by Parkins [53] who suggests that, where different crack paths are followed in the same piece of material under nominally identical testing conditions, the mechanism of growth along the different paths may not be essentially different and the crack path may be simply a reflection of the distribution of microplastic deformation. However, the general trend in the literature is of treating TGSCC and IGSCC as isolated processes. Many researchers are of the opinion that the mechanisms of TGSCC and IGSCC are fundamentally different [13]. With respect to IGSCC, there is a general agreement among researchers that the crack propagates by preferential anodic dissolution at the crack tip[10–12]. In contrast, the most favoured mechanism for TGSCC is environmentally induced cleavage [13–15]. This is evident from the fact that the mechanistic models for SCC that have been proposed in the recent years [16-19] deal exclusively with the environmental interaction resulting in TGSCC. In this context, the author feels that a closer look at the afore-mentioned transition in cracking morphology will be rewarding.

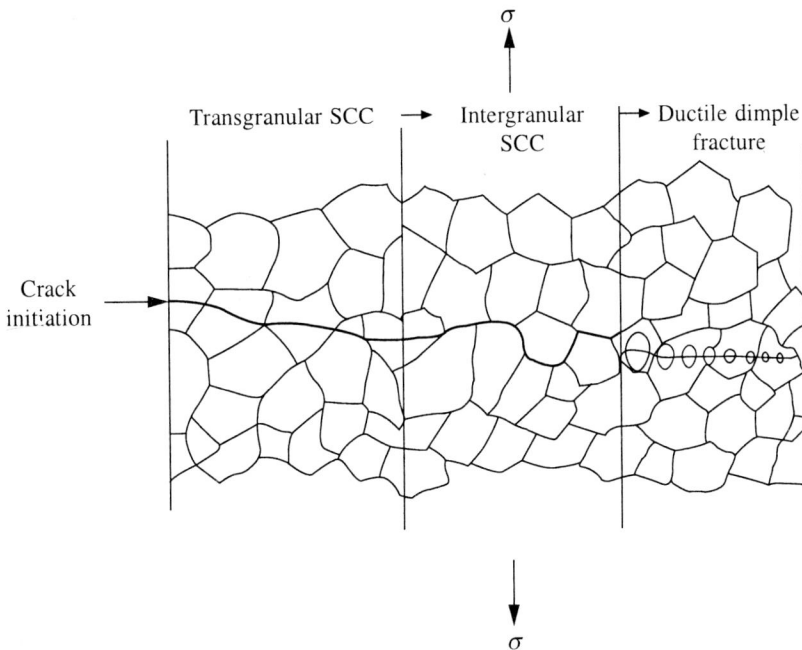

Fig. 20. A schematic of transition in SCC morphology.

Majority of the publications [54–63] discuss the phenomenon of transition in SCC morphology in ASS from an environmental viewpoint employing the slip dissolution model for SCC. According to this, when the slip step generation rate (a function of applied stress/strain rate) is greater than the dissolution rate of the slip steps, dislocation pile-ups are created at the grain boundaries leading to a stress concentration there and intergranular dissolution is preferred. However, our work [23, 26, 29] has shown that metallurgical variables also influence the above phenomenon. The role of metallurgical factors on the transition can also be explained based on slip-dissolution mechanism. It is reported that high voltage electron microscopic study of stress corrosion crack tips of 304 SS in MgCl$_2$ solution has shown preferential attack on slip planes containing moving dislocations to form a crack [64]. When a crack is initiated on a smooth specimen the initial propagation, at low stress levels, will be in the transgranular mode, by the dissolution of the dislocation pile ups along slip planes. However, at higher stress levels near 0.2% offset yield strength of the material, intergranular cracking originating from triple points of grain boundaries has been reported in type 304 SS investigated in boiling $MgCl_2$ solution at 155°C [62]. But as the crack propagates, the stress intensity at the crack tip increases in a constant load test and the crack tip plastic zone becomes heavily deformed producing dislocation tangles. The absence of slip planes with sufficient dislocation pile-ups makes the inherently more active grain boundaries, the preferred path for crack propagation. Indeed, our studies have shown that prior deformation which produces a tangled dislocation structure facilitates intergranular transition [23]. The high stresses at the boundaries, together with the segregated impurities such as P and S, also make the boundary a susceptible path.

The observation of a larger fraction of IGSCC on the fracture surface of steel C containing 0.045% P compared to the same type of steel (steel A in Table 1) with a lower P content (0.036% P) in our study shows the importance of grain boundary chemistry in this transition phenomenon. The role of impurity elements at the grain boundaries has been recognized by Kowaka et al. [63] and Manfredi et al. [54]. Kowaka has reported an increase in IGSCC propensity with increase in phosphorus content of the steel. Enrichment of phosphorus at the grain boundaries is possible by the non-equilibrium segregation as per the solute-vacancy interaction mechanism proposed by Aust et al.[65]. It has also been reported that phosphorus increases the thermal equilibrium concentration of vacancies [66], facilitating segregation by the above mechanism. This segregation of impurities to the grain boundary will have a pronounced effect on the electrochemical behaviour of the grain boundary.

The predominant role of grain boundary enrichment of P in the transition process is supported by the observed inhibition of intergranular transition in the aged steel when tested in boiling $MgCl_2$ solution at 155°C. It is known that $M_{23}C_6$ type of carbides precipitated at the grain boundaries during the ageing has a high solubility for phosphorus. Thus, grain boundary carbides can effectively sequester the phosphorus segregated at the boundaries [47,67] making intergranular propagation no more favourable. It is also possible that since carbide precipitates at the grain boundaries reduce the free energy of the boundaries, the balance in the anodicities of the grain boundaries and the slip steps tilts in favour of the slip steps producing transgranular SCC in aged steel. There is ample data in the literature to illustrate that in austenitic stainless steel/$MgCl_2$ system, TGSCC and IGSCC are highly competing processes and slight changes in environmental and/or metallurgical parameters can tilt the cracking mode in either way. For example, SCC tests [54, 60] conducted under potential control have shown that slight changes in applied potential to the extent of 40mV in the anodic or cathodic direction can change the crack morphology to one of predominantly TGSCC or IGSCC. This is also evident from the observed slip line attack (Figure 19) in the predominantly intergranular region of the fracture surface in the our study, in which case the local variations in grain boundary characteristics and/or microplastic deformation favour corrosion attack at the slip steps.

From a microstructural viewpoint, the presence of carbides in the matrix homogenises slip with in the marix, thus reducing the number of dislocations at the intersections of the slip band with the grain boundaries. Since the local stress concentration at slip band intersections with grain boundaries is proportional to the number of dislocations, the grain boundary stresses and the driving force for producing crack growth along the boundaries will be reduced. This argument is also supported by the observation of the transition in SCC morphology in the aged steel when the tests were conducted in magnesium chloride solution at 125°C and at higher applied stresses viz., 150 MPa for 100 h and 200 h aged specimens and 200 MPa for 500 h and 1000 h aged specimens. When the applied stress is increased, the dislocations get sufficient energy to surmount the intragranular barriers and the grain boundaries will again become the regions of stress concentration. Moreover, the transition is also assisted by the decreased 'corrosivity' of the environment, by making it more selective in attack.

Extensive studies in the past [68–71] have shown that the $M_{23}C_6$ precipitates nucleate as a thin lath on a {111} plane in one of the grains at the boundary. One surface of this lath will be a low energy interface and the other a high energy interface. The growth of carbides generally takes place in such a way that the grain boundary will be moving into the grain with the high energy interface. During the SCC propagation, since anodic dissolution is preferred along high energy boundaries, the carbide particles will be left behind on one of the grain facets. Dissolution of carbides is not possible at the electrochemical potential encountered in the study. Higher magnification SEM pictures from the IGSCC region of the aged (700°C/1000h) showed carbide particles on the facets of the grains (Figure 19). These figures resemble the intergranular fracture surfaces of sensitized 304 SS produced by hydrogen embrittlement [72] in which case the crack propagation was known to take place along the carbide/matrix interface.

The transition in SCC morphology from transgranular to intergranular has been discussed based on the slip dissolution model for SCC. However it is possible that the extent of dissolution taking place during TGSCC is low when compared to that during IGSCC. This may be because of the lower reactivity of the crack faces in TGSCC than that in IGSCC. It is also possible that the contribution from dissolution is only partial in the TGSCC process as proposed in some of the recent models for TGSCC [17, 19].

REFERENCES

1. Kowaka, M., Introduction to Life Prediction of Industrial Plant Materials, Ed. M. Kowaka, A llerton Press Inc., New York, 1994, p. 19.
2. Jones, R.H. and Ritter, R.E., Materials Hand Book, Vol. 13 (Corrosion), ASM International, Ohio, 1987, p. 145.
3. Newman, R.C., Zheng, W. and Procter, R.P.M., Corros. Sci., Vol. 33, 1992, p. 1003.
4. Sieradzki, K., Sabatini, R.L. and Newman, R.C., Metall.Trans. A, Vol. 15A, 1984, p. 1941.
5. Pednekar, S.P., Agrawal, A.K., Chaung, H.E. and Staehle, R.W., J.Electrochem.Soc, Vol. 126, 1979, p. 701
6. Newman, R.C., in Corrosion Mechanisms in Theory and Practice P. Marcus, J. Oudar eds: Marcel-Dekker Inc., New York,1995, p. 311.
7. Daly, J.J., Chem. Engg, Vol. 83, 1976, p. 113.
8. Staehle, R.W., Materials Sci. Engg., Vol. 25, 1976, p. 207.
9. Parkins, R.N., Mazza, F., Royuela, J.J. and Scully, J.C., Brit. Corros. J, Vol. 7, 1972, p. 154.
10. Staehle, R.W., The Theory of Stress Corrosion in Alloys, ed. J.C. Scully, NATO Scientific Affairs Division, Brussels 1971, p. 223.
11. Champion, F.A., Sympo. on Internal Stresses in Metals and Alloys Institute of Metals London,1948, p. 468.
12. Logan, H.L., J. Res. Natl. Bur. Stand., Vol. 4, 1952, p. 99.
13. Pugh, E.N., Corrosion, Vol. 41, 1985, p. 517.
14. Pugh, E.N., Atomistics of Fracture, eds: R.M. Latanision and J.R. Pickens, Plenum Press, New York 1983, p. 997.
15. Bursle, A.J. and Pugh, E.N., Environment Sensitive Fracture of Engineering Materials AIME, Pennsylvania, 1979, p. 18.
16. Sieradzki, K. and Newman R.C., Phil. Mag. A, Vol. 51, 1985, p. 95.
17. Flanagan, W.F., Bastias, P. and Lichter, B.D., Acta Metall. Mater., Vol. 39, 1991, p. 695.
18. Jane, S., Marek, M., Hochman, R.F. and Meletis, E.I., Metall.Trans. A ,Vol. 22A, 1991, p. 1453.
19. Magnin T., Chieragatti, R. and Oltra, R., Acta Metall. Mater., Vol. 38, 1990, p. 1313.

20. Muraleedharan, P., Khatak, H.S. and Gnanamoorthy, J.B., Hand Book of Case Histories in Failure Analysis, CHFA-2, ASM International, Ohio, 1993, p. 427.
21. Muraleedharan P. and Gnanamoorthy J.B., Handbook of Case Histories in Failure Analysis, CHFA-2, ASM International, Ohio, 1993, p. 225.
22. Gnanamoorthy, J.B., Materials Performance, Vol. 29, 1990, p. 63.
23. Muraleedharan, P., Khatak, H.S., Gnanamoorthy, J.B. and Rodriguez, P., Metall.Trans. A, Vol. 16A, 1985, p. 285.
24. Khatak, H.S., Muraleedharan, P., Gnanamoorthy, J.B., Rodriguez, P. and Padmanabhan, K.A., J. Nucl. Mater., Vol. 168, 1989, p. 157.
25. Muraleedharan, P., Khatak, H.S, Gnanamoorthy, J.B. and Rodriguez, P., in Atlas of Stress Corrosion and Corrosion Fatigue Curves, Ed., A.J. McEvily Jr., ASM International, 1990, p. 173.
26. Muraleedharan, P., Ph.D. Thesis, Madras University, 1993.
27. Muraleedharan, P., Gnanamoorthy, J.B. and Prasad Rao, K., Corrosion, Vol. 45, 1989, p. 142.
28. Muraleedharan, P., Gnanamoorthy, J.B. and Rodriguez, P., Corrosion, Vol. 52, 1996, p. 790.
29. Muraleedharan, P., Gnanamoorthy, J.B and Rodriguez, P., Corr. Sci., Vol. 38, 1996, p. 1187.
30. Muraleedharan, P., Schneider, F. and Mummert, K., J. Nucl. Mater., Vol. 270, 1999, p. 342.
31. Muraleedharan, P., Schneider, F., Khatak, H.S., Mummert, K., preprints of the Eighth Congress of the National Corrosion Council, Kochi , 1998.
32. Spiedel, M.O., Metall.Trans. A, Vol. 12A,1981, p. 779.
33. McIntyre, D., Corrosion of Metals Under Thermal Insulation ASTM STP 800 eds: W.I. Pollock and J.I. Barnhardt, ASTM, Philadelphia ,1985, p. 27.
34. Newman, R.C. and Procter, R.M., Br. Corros. J., Vol. 25, 1990, p. 259.
35. ASTM Standard G. 10–92 "Test Method for Electrochemical Reactivation (EPR) for Detecting Sensitization of AISI Type 304 and 304L Stainless Steels" ASTM Book of Standards, Vol. 3.02, Philadelphia, PA, ASTM, 1992, p. 499.
36. Armijo, J.S., Corrosion, Vol. 8, 1968, p. 24.
37. Armijo, J.S., Corros. Sci., Vol. 7, 1967, p. 143.
38. Jacobs, A.J., Corrosion, Vol. 46, 1990, p. 30.
39. Shoji, T., Yamaki, K., Ballinger, R.G. and Hwang I.S., Proc. Fifth Internatl. Symp.on Environmental Degradation of Materials in Nuclear Power Systems -Water Reactors, Monterey, California, 1991, p. 827.
40. Watanabe, Y., Ballinger, R.G., Harling, O.K. and Kohse, G.E., Corrosion ,Vol. 51, 1995, p. 651.
41. Abe, S. and Kaneko, M., proc. Internatl. Symp. on Plant Ageing and Life Predictions of Corrodible Structures, Sapporo, Japan, 1995, p. 559.
42. Briant, C.L., Met.Trans. A, Vol. 18A, 1987, p. 691.
43. Briant, C.L. and Andresen, P.L., Met. Trans. A, Vol. 19A, 1988, p. 495.
44. Tice, D.R., Hurst P. and Platts, N., Eurocorr. 96, IX OR 39–1.
45. Kuroda, T., Yeon, Y.M. and Enjo, T., Yosetzu Gakkai Ronbunshu, Vol. 8, 1990, p. 91 (Japanese).
46. Latanision, R.M and Staehle, R.W., Proc.Conf. on Fundamental Aspects of Stress Corrosion Cracking, eds:R.W. Staehle, A.J. Forty and D. Van Rooyen, NACE, Houston, TX , 1969, p. 214.
47. Hanninen, H.E., International Metals Reviews, Vol. 24, 1979, p. 85.
48. Cigada,. A., Mazza, B., Pedeferri, P., Salvago, G., Sinigaglia, D. and Zanini, G., Corro. Sci., Vol. 22, 1982, p. 559.
49. Scully, J.C., The Theory of Stress Corrosion Cracking in Alloys, ed: J.C. Scully, NATO, Brussels, 1971, p. 127.
50. Mukai, Y., Watanabe, M. and Murata, M., Fractography in Failure Analysis, ASTM STP 645, ASTM, Philadelphia, PA, 1978, p. 164.
51. Boeuf, A., Cacinffo, R.G.M., Coppola, R., Crico, S., Melone, S., Puliti, P., Rebonato, R. and Rustichelli, F., J. Nucl. Materials, Vol. 126, 1984, p. 276.
52. Floreen , S. and Nelson, J.L., Metall. Trans. A, Vol. 14A, 1983, p. 133.
53. Parkins, R.N., Parkins Symposium on Fundamental Aspects of Stress Corrosion Cracking, TMS Publication, Pennsylvania, 1992, p. 3.

54. Manfredi, C., Maier, I.A. and Galvele, J.R., Corros. Sci., Vol. 27, 1987, p. 887.

55. Okada, H., Hosoi, Y. and Abe, S., Corrosion, Vol. 27, 1971, p. 441.

56. Takano, M., Corrosion, Vol. 30, 1974, p. 441.

57. Nakayama, T. and Takano, M., Corrosion, Vol. 42, 1986, p. 10.

58. Stalder, F. and Duquette, D.J., Corrosion, Vol. 33, 1977, p. 67.

59. Kuwano, S., J. Soc. Mater. Sci. Japan, Vol. 27, 1978, p. 539

60. Kessler, K.J. and Kaesche, H., Werk. und Korros., Vol. 35. 1984, p. 171.

61. Russel, A.J. and Tromans, D., Metall. Trans. A, Vol. 10A, 1979, p. 1229.

62. Yagasaki, T., Kimura, Y. and Kunio, T., Advances in Fracture Research (ICF-6, India) ed: S.R. Valluri, D.M.R. Taplin, P. Rama Rao, J.F. Knott and R. Dubey, Pergamon, Vol. 4, 1986, p. 2411.

63. Kowaka, M., Localised Corrosion, ed: F. Hine, K.Komai and K. Yamanaka, Elsevier, New York, 1988, p. 149.

64. Nakayama, T. and Takano, M., Corrosion, Vol. 37, 1981, p. 226.

65. Aust, K.T., Hanneman, R.E., Niessen, P. and Westbrook, J.H., Acta Metall., Vol. 16, 1968, p. 291.

66. Garner, F.A. and Kumar, A.S., ASTM STP 955, ed: F.A. Garner, N.H. Packan and A.S. Kumar, American Society for Testing and Materials, Philadelphia, 1987, p. 289.

67. Aust, K.T., Armijo, J.S., Koch, E.F. and Westbrook J.H., Trans. ASM, Vol. 61, 1968, p. 270.

68. Weiss, B. and Stickler, R., Metall. Trans. A, Vol. 3A, 1972 p. 851.

69. Southwick, P.D. and Honeycombe, R.W.K., Metal Sci., Vol. 16, 1982, p. 475.

70. Beckitt, F.R. and Clark, B.R., Acta Metall., Vol. 15, 1967, p. 113.

71. Hall, E.L. and Briant, C.L., Metall.Trans. A, Vol. 15A, 1984, p. 793.

72. Hannula, S.M., Hanninen, H. and Tahtinen, S., Metall. Trans. A, Vol. 15A, 1984, p. 2205.

7. Stress Corrosion Cracking of Austenitic Stainless Steel Weldments

Hasan Shaikh[1]

Abstract At the outset, this paper reviews the available literature on stress corrosion cracking (SCC) behaviour of weldments of austenitic stainless steel. This section deals with the SCC resistance of weld metal as compared to that of the base metal of similar composition, effects of δ-ferrite, heat input, defects, welding variables and residual stresses on SCC resistance of weld metal. SCC behaviour of heat affected zone (HAZ) is also highlighted. Results of our work on effects of high temperature aging, cold work prior to welding, and applied potential on the SCC behaviour of type 316 austenitic stainless steel weldments are discussed. Matrix hardening caused by extensive sigma phase precipitation at high temperatures led to a decrease in SCC resistance. Application of potentials anodic to critical cracking potential led to increase in SCC susceptibility; while cathodic polarisation caused no failure or led to general corrosion. SCC susceptibility increased with increase in cold work level of the parent material in welded austenitic stainless steel.

Key Words Stress corrosion cracking, δ-ferrite, sensitisation, cold work, residual stresses, heat input, thermal aging, polarisation.

INTRODUCTION

Austenitic stainless steels are the most commonly used materials in the chemical, petrochemical, power and nuclear industries. The use of these iron-based alloys in industrial service requires the fabrication of components by welding. Austenitic stainless steels are generally regarded as readily weldable materials with considerable tolerance for variations in the welding conditions. Amongst the large number of available welding processes, shielded metal arc welding (SMAW) and gas tungsten arc welding (GTAW or TIG) processes are most commonly used for welding of these steels. TIG welding is most often preferred over SMAW process because of the lower heat input associated with it. Lower heat input results in faster cooling rates which helps in overcoming the problem of sensitisation, thus leading to a remedy to the problems of intergranular corrosion (IGC) and intergranular stress corrosion cracking (IGSCC).

Generally, a welded joint is required to perform either equal to or better than the base metal it joins. However, in practice, this objective is never achieved since the welding process itself introduces

[1]Scientific Officer, Corrosion Science and Technology Division, Metallurgy and Materials Group, Indira Gandhi Centre for Atomic Research, Kalpakkam-603 102, India.

features, which degrade corrosion properties of the welded joints as compared to the wrought base metal. During fusion welding processes, the molten metal is shielded by a gas or by slag to prevent contamination of the weld metal. Despite this protection, weld metal contamination occurs in the form of slag inclusions, tungsten inclusions etc. The deposited weld metal normally cools at a rate determined by the welding parameters. The fast cooling rates associated with the weld metal causes the formation of a dendritic structure besides straining of the weld metal. Also, several metallurgical transformations can take place in the weld metal during this cooling. Apart from weld contamination and metallurgical changes, improper welding procedures can leave behind a host of defects, such as porosities, undercuts, microfissures, in the weld metal. All these detrimental features do not augur well from the mechanical and corrosion properties point of view. In fact, a majority of the corrosion failures in components could be directly or indirectly related to the corrosion of the weld metal or the heat affected zone(HAZ).

The schematic illustration in Fig. 1 shows that a heterogeneous weldment comprises four regions: a composite region, an unmixed zone, a partially melted region and a true heat affected zone (HAZ).

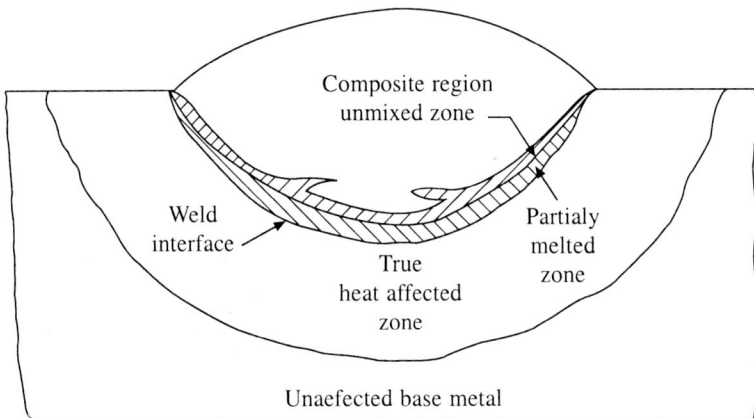

Fig. 1.   A schematic illustrating the regions of a heterogeneous weld.

In the composite region, also known as mixed or fusion zone, hydro-dynamic mixing of the molten filler and base metals results in a relatively uniform chemical composition. Next to the mixed zone is the unmixed zone in which a small amount of the base metal melts and resolidifies without undergoing filler metal dilution. In the partially melted region, the base metal melts partially in some localised regions and when subjected to the faster cooling rate yields a microstructure, which is different from the unmelted regions. In the true HAZ, thermally-induced solid-state changes occur in the base metal adjacent to the fusion line.

The welding of austenitic stainless steel is not bereft of problems. In fact, during welding of this class of steels, the two most commonly encountered problems are hot cracking of the weld metal and sensitisation of the HAZ. The problem of hot cracking can be overcome by retaining some amount of high temperature δ-ferrite to room temperature. δ-ferrite can be retained by appropriate choice of the filler metal composition, which could be done in consultation with the 70% iron isopleth (Fig. 2) [1]. This diagram determines the position of the alloy composition with respect to the liquidus minimum. Alloys with compositions located on the Ni-rich side solidify as primary austenite while those located on the Cr-rich side solidify as primary ferrite. In both the solidification modes, the δ-ferrite phase is enriched in ferritisers like Cr, Mo, Si, while the austenite is enriched in austenitisers, such as Ni, Mn,

C, N etc.[2]. The extent of partitioning of alloying elements would also depend on heat input [3]. Apart from partitioning of alloying elements, impurites such as S and P segregate to the δ-ferrite /austenite (DF/A) interface. The extent of segregation is governed by the solidification mode and heat input [2, 3]. Kujanpaa et al. have reported that the stainless steel weld metal solidifies as primary δ-ferrite when the Cr eq./Ni eq. Ratio exceeds 1.95 and primary austenitic solidification mode occurs when this ratio is less than 1.45.

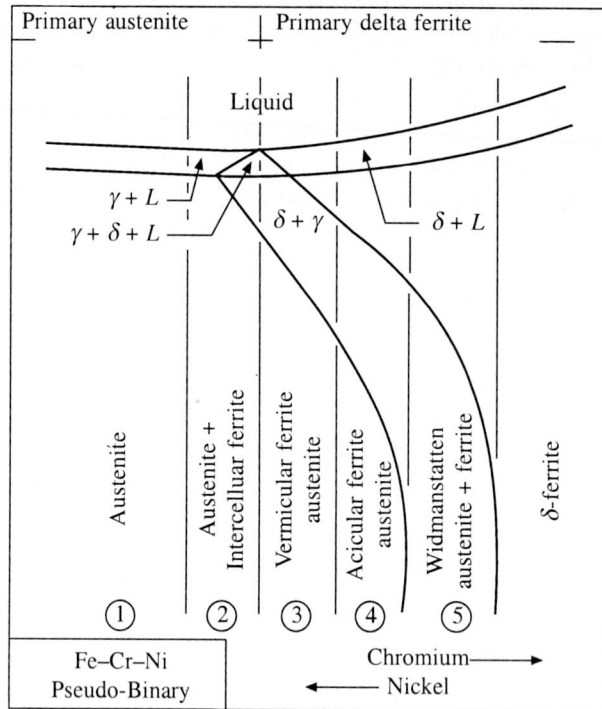

Fig. 2.   A schematic of the 70% iron showing the effect of composition on the austenite and the ferrite morphology in austenitic stainless steel weld metal.

STRESS CORROSION CRACKING OF WELDMENTS OF AUSTENITIC STAINLESS STEEL

Stress corrosion cracking (SCC) is the premature degradation of a material under the conjoint action of a tensile stress and a corrosive medium, neither of which when acting alone would have caused a failure. SCC has many typical characteristics, e.g. it occurs only in (1) specific medium, (2) alloys showing active-passive behaviour etc. Normally two types of SCC are encountered in service. They are transgranular SCC (TGSCC) and intergranular SCC (IGSCC). Their occurrence would largely depend on a number of environmental factors, such as concentration of corrosive species, temperature, pH, oxygen level etc., apart from stress level present in the component and microstructural features such as sensitisation [4–7].

Welded joints are commonly regarded as particularly prone to SCC because the welding operation introduces a residual tensile stress field in the weld area unless effective post weld stress relief, typically by heat treatment, is carried out. In the immediate weld area, peak stresses will be of yield magnitude and, hence, frequently exceed any threshold stress for SCC for a given metal-environment combination. This is particularly true for austenitic stainless steels in many environments and hence the significance of residual stresses must be fully appreciated. Inevitably, the rapid thermal cycle experienced in the HAZ promotes some metallurgical changes. In austenitic stainless steel, a degree of grain coarsening may be experienced, but is usually minor, although some large grains may be observed very adjacent to the fusion line. The welding cycle also introduces formation of secondary phases, of which precipitation of $Cr_{23}C_6$ is of major concern from the corrosion and SCC point of view.

The SCC behaviour of base metal of austenitic stainless steel in the solution annealed, cold worked and heat treated conditions have been investigated in detail by slow strain rate testing (SSRT) and constant load testing techniques [8–11]. However, studies on weld metal and welded joints have been few and far between. The SCC behaviour of the weld metal is expected to be different from that of the base metal because the weld metal is effectively a casting; has high temperature δ-ferrite retained in it; residual stresses are present in it; and, metallurgical defects like vacancies and dislocations are much higher in it than in the base metal.

Stress Corrosion Cracking Behaviour of Weld Metal vis-à-vis Base Metal

The difference in SCC behaviour of base and weld metals has been a subject of much disagreement in literature. Investigations by Flowers et al. indicated that cast duplex stainless steels were more resistant to SCC than single-phase austenitic alloys because the globular ferrite present in the cast steel provided a "keying" action, which inhibited SCC propagation [12]. Baeslack III et al. reported that of the various regions shown in Fig. 1, the duplex unmixed zone is most susceptible to SCC due to the complementary and simultaneous occurrence of SCC of austenite and dissolution of δ-ferrite [13]. Sherman et al. reported that in magnesium chloride solution boiling at 427 K, weld metal of type 304 stainless steels possessed better SCC resistance than the base metal of similar composition when tested by SSRT technique [14]. They and Gooch [15] attributed the better SCC resistance of the weld metal to the cathodic protection offered to the austenite by the corroding DF/A interface. However, Stalder et al. reported that the weld metals of types 304 and 304 L stainless steel had a SCC resistance similar to that of the base metal when tested by SSRT in 45% magnesium chloride solution [16]. Tests on the same materials in magnesium chloride boiling at 408 K showed a much higher SCC resistance for the base metal. Baeslack III et al. also reported very little difference, in maximum engineering stress and % reduction in area, between types 304 stainless steel base and weld metals that were tested at various strain rates, using the SSRT technique, in magnesium chloride boiling at 427 K [17]. The same authors reported better SCC resistance for types 304 stainless steels base metal vis-à-vis its weld metal of the same composition when tested in de-aerated 1N HCl at room temperature [18]. Raja et al. reported that the base metal of type 316 stainless steel had better SCC resistance than its autogeneous weld in a 5 N H_2SO_4 + 0.5 N NaCl solution at room temperature [19]. In essence, the differences in SCC behaviour between the base and weld metals depend on the chemical composition, environment and testing techniques.

Effect of δ-Ferrite on the Stress Corrosion Cracking Resistance of Weld Metal

The presence of δ-ferrite can appreciably alter both the SCC resistance and the crack morphology of the weld metal. The SCC resistance of the weld metal depends on the δ-ferrite content, its distribution and the solidification mode. Weld metal of AISI type 304 stainless steel, which solidifies in the primary ferritic solidification mode, has SCC resistance similar to that of the base metal [20]. Duplex weld metal that solidifies as primary austenite also behaves similarly [20]. Fully austenitic weld metal has the most degraded SCC property and fails by IGSCC due to extensive segregation of S and P at the grain boundaries [20]. In NaCl and HCl solutions, both at ambient and high temperatures, the weld metal was reported to fail by stress-assisted dissolution of δ-ferrite (SAD) and SCC of austenite [19, 21]. In boiling 45% magnesium chloride solution, the failure occurred due to cracking of the δ-ferrite /austenite (DF/A) interface and SCC of austenite [16–18]. Baeslack III et al. reported cracking of δ-ferrite when it is oriented in a direction perpendicular to the direction of crack propagation [18]. Raja et al. reported that in 5N H_2SO_4 + 0.5 M NaCl solution at room temperature, weld metal failure occurred by SAD of δ-ferrite and SCC of austenite at open circuit potentials (OCP) [19]. The amount, morphology and continuity of the δ-ferrite network influences the SCC resistance of the weld metal. Baeslack III et al. reported that a continuous network of δ-ferrite was most harmful for SCC of weld metal as it provided a continuous path for crack propagation [18]. The effects of amount of δ-ferrite on SCC of weld metal are confusing. Krishnan et al. observed a degradation of SCC resistance as the ferrite content increased from 3 FN to 8 FN [22]. Franco et al. found that increasing the δ-ferrite content increased the SCC resistance in boiling 45% magnesium chloride solution [23]. They associated this increase more to morphology and distribution of the ferrite phase than to its contents. They stated that with increasing ferrite content the δ-ferrite network became continuous thus causing decerease in SCC resistance. However, Baeslack III et al. [20] and Sherman et al. [14] reported that the resistance to SCC of autogenous weld of type 304 stainless steel in boiling 45% magnesium chloride solution increased with increasing δ-ferrite content. Vishwanathan et al. reported better SCC resistance for weld metals of types 304 and 316 stainless steels which contained lower amounts of ferrite [24].

Effect of Strain Rate on SCC Resistance of Weld Metal

Strain rate plays a significant role in determining the SCC resistance of the weld metal, vis-à-vis base metal, and also on the fracture morphology of the weld metal. Baeslack III et al. reported very little difference in the maximum engineering stress and % reduction in area between type 304 stainless steel base and weld metals, which were tested in boiling 45% magnesium chloride soultion at various strain rates using the SSRT technique [17]. The same authors reported better SCC property for the base metal of type 304 stainless steel, with respect to its weld metal of the same composition, in de-aerated 1N HCl at room temperature [18]. Fig. 3 shows that the difference in SCC resistance was larger at lower strain rates. Baeslack III et al. have reported strain rate dependence of SCC fracture morphology of weld metal tested in 45% magnesium chloride at 427 K [17]. At low strain rates (1.5*10^{-5}/s), SCC initiated and propagated by TGSCC of austenite.

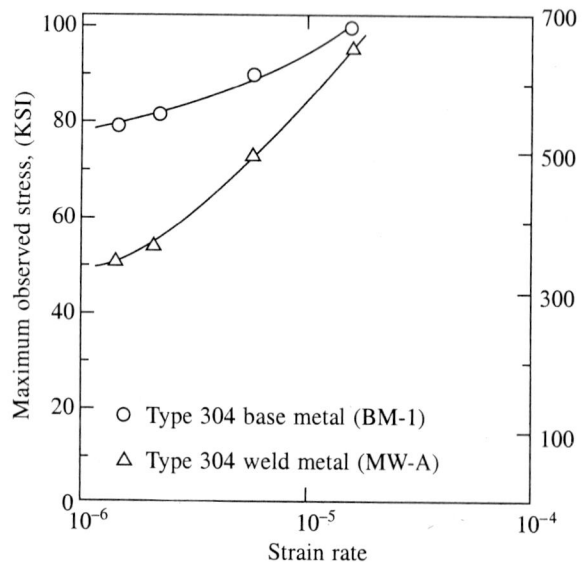

Fig. 3.   Maximum observed stress vs. initial strain rate for type 304 stainless steel base and weld metals.

No evidence of ferrite was available on the fracture surface. At high strain rates ($7.5*10^{-5}$/s), SCC initiated and propagated entirely by cracking of DF/A interfaces. The cracking of the DF/A interface at high strain rates was attributed to the alteration of interfacial defect structure, which caused not only the dissolution of the DF/A interface but also the fracture of the near continuous ferrite network by hydrogen-induced mechanisms. A mixed mode of cracking was observed at intermediate strain rates.

Effect of welding defects on SCC behaviour of weld metal

Different types of defects on the weld surface, such as adhesion and edge defects, slag inclusions, macro and micro-fissures, are points of initiation of corrosion attack in the weld metal. The initiation mechanism of the corrosion attack is that of crevice corrosion [25]. Also, these defects act as regions of stress concentration, which aid in faster initiation of SCC. Non-metallic inclusions, such as sulphides, are undesirable from SCC point of view. Influence of non-metallic inclusions in reducing the SCC life is directly related to the easy initiation of pitting corrosion at inclusion sites [26]. In acid solutions, and hence in occluded cells, sulphide inclusions dissolve to form H_2S, which has an accelerating effect on corrosion of steel [27]. H_2S is reported to accelerate both the anodic and cathodic processes [26]. In many service applications, SCC initiated from pits, which act as stress raisers [28, 29]. In these cases, the induction times for SCC were longer than those for pitting [28]. Apparently, the non-metallic inclusions do not participate directly in crack nucleation, but their presence is undesirable, as they give rise to pitting. However, Clarke and Gordon reported a strong effect of secondary phases and inclusions on crack nucleation [30]. They reported that cracking nucleated from the crevice corrosion attack around the included particle. Non-metallic inclusions also play a role in hydrogen embrittlement (HE) of stainless steel welds. Surface inclusions facilitate entry of hydrogen into the weld and thus induce cracking [26]. Bulk inclusions may act as trap sites for hydrogen and thus assist in nucleation and development of internal crevices and cracks [26]. The shape of the sulphide inclusion influences the crack initiation time. Elongated sulphide particles may cause six-fold increase hydrogen entry into the metal as compared to spherical inclusions [31]. The non-metallic inclusion also cause stress concentration in the material. The extent of stress concentration would depend on the shape of the inclusion [26]. Sharp edged inclusions act as more effective notches than spherical or elliptically shaped inclusions. Apart from the shape, the ratio of thermal expansion coefficients and modulus of elasticity of the inclusion to that of the matrix also contribute to a lesser extent in determining the magnitude of stress concentration [26].

Effect of high temperature aging on the SCC resistance of weld metal

The understanding of the SCC behaviour of weld metal aged at high temperatures requires an understanding of the precipitation of secondary phases at those temperatures. In our studies, on bead-on-plate weld deposits of AISI type 316 L stainless steel filler wire (composition in Table 1), the average ferrite content of the as-deposited weld metal was 7 FN. The ferrite content before and after aging at 873 K was converted to fraction of δ-ferrite transformed as a function of aging time.

Table 1. **Chemical compositions (in weight %) of AISI type 316 L weld deposit**

Element	C	Cr	Ni	Mo	Mn	Si	S	P
Type 316 L Weld deposit	0.022	18.7	10.8	2.1	–	0.38	0.002	0.02

Figure 4 shows a decrease in the transformation rate of δ-ferrite with increasing aging time. Fig. 5 shows a continuous increase in the amount of precipitates, present in the weld metal, with increasing aging time. The various phases present in the electrochemical residue (Table 2) were identified in accordance with the Standard ASTM X-ray diffraction (XRD) practices. It is seen that δ-ferrite was the dominant phase at 2 hours of aging and carbides/carbonitrides (C/CN) phase dominated at 20 hours of aging. At 200 and 2000 hours, σ phase was the dominant phase; the amount of σ phase being greater in the latter case. Another phase, whose index values corresponded to the R-phase, was found in increasing quantities up to 200 hours of aging. On aging for 2000 hours, this phase was not detected.

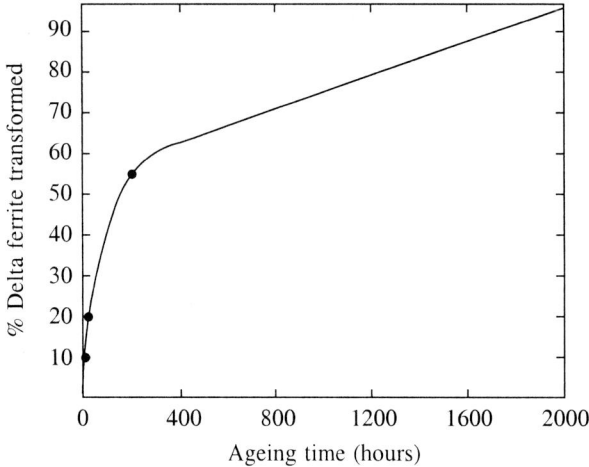

Fig. 4.   **Transformation kinetics of δ-ferrite at 873 K.**

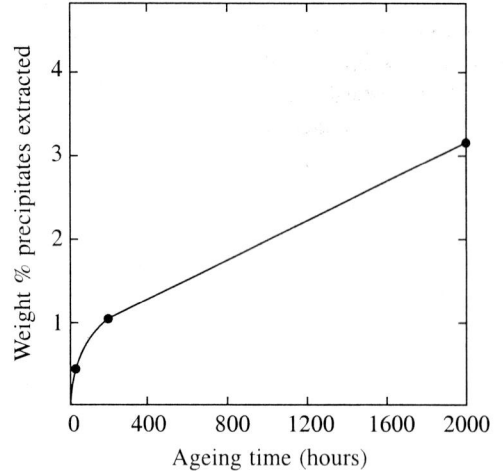

Fig. 5.   **Dependence of amount of precipitates extracted on aging time at 873 K.**

Based on the dominance of C/CN in the electrochemical residues of weld metal aged up to 20 hours, it was inferred that up to this aging time, the ferrite transformed by the reaction

$$\delta\text{-ferrite} \rightarrow \text{C/CN} + \text{R-phase}$$

The preponderance of σ phase on aging at 200 and 200 hours indicated that at higher aging times, the following transformation reaction occurred

Table 2.   **Results of X-ray diffraction studies showing the abundance of various phases in the weld metal aged at 873 K for various durations of time**

Aging Time	Phases identified (with relative abundance)
2	$\delta > $ C
20	C > R
200	$\sigma > $ C > R
2000	$\sigma > $ C

$\delta = \delta$-ferrite; $\sigma = \sigma$ phase; R = R-phase; C = carbide phase

$$\delta\text{-ferrite} \rightarrow \sigma + C/CN$$

with C/CN precipitating during early stages of aging.

The results of tensile tests in liquid paraffin at 427 K are presented in Table 3. It is seen that yield strength (YS) decreased up to 200 hours of aging beyond which a very slight increase was observed. Ultimate tensile strength (UTS) and work hardening exponent, 'n', showed very little change on aging up to 200 hours. Significant increase in UTS and 'n' was observed on aging to 2000 hours. Ductility (% total elongation) increased by about 30% (vis-à-vis as-deposited weld metal) on aging up to 200 hours, and decreased thereafter. The decrease in YS and increase in ductility up to 200 hours of aging indicated softening of the austenite matrix. Increase in UTS and decrease in ductility at 2000 hours of aging indicated hardening of the austenite matrix. So from the tensile results, it is clear that two processes, namely, matrix softening and matrix hardening occur simultaneously in the weld metal during aging. The process, which dominates, governs the tensile property of the weld metal. Matrix softening resulted from: (a) dissolution of δ-ferrite network and (b) depletion of solid solution strengtheners due to secondary phase precipitation. Matrix hardening was caused by copious precipitation of non-deformable σ phase.

Table 3. Tensile data of type 316 L weld metal, aged at 873 K, after testing in liquid paraffin at 427 K

Aging Time (h)	YS (MPa)	UTS (MPa)	Ductility (%)	Work Hardening Exponent, n
0	399	487	27.13	0.15612
2	335	478	27.56	0.19524
20	325	498	27.9	0.20774
200	315	503	31.6	0.23632
2000	323	544	28.83	0.40153

SCC resistance of the weld metal was evaluated in boiling 45% $MgCl_2$ solution using the SSRT technique. The SSRT technique results in liquid paraffin and in boiling 45% $MgCl_2$, were compared to evaluate the SCC resistance based on the ratios of values of UTS and % total elongation in the two tests, alongwith SCC susceptibility index, I, which is defined as the ratio of values of UTS* % elongation in the two tests [32]. Fig. 6 shows a deterioration in SCC resistance on aging to 2 hours with little change thereafter on further aging.

The observations in Fig. 6 were verified by calculating the average crack propagation rates (CPR) using the formulae proposed by Desestret and Oltra and by Hishida et al. [33, 34]. The formula proposed by Desestret and Oltra bases the calculation of CPR on the ratios of values of fracture stresses in $MgCl_2$ and in liquid paraffin. Hishida et al. suggested the use of the value of 'n' to calculate the average CPR. Since flat tensile specimens were used in this study and since SCC initiated at all four corners, the calculation of the CPR was based on the initial area of the specimen and not on the initial radius of the specimen as suggested [33, 34]. The time-to-failure (Table 4), used to calculate the CPR was taken as the time spent by the weld metal between YS and UTS in $MgCl_2$. YS was used as the initial criteria because little difference was observed in the values of YS in the the tests carried out both in liquid paraffin and in $MgCl_2$. Beyond UTS, the material failed by pure ductile failure. Fig. 7 shows an improvement in the SCC resistance up to 200 hours of aging (decreasing CPR) and a deterioration thereafter. So, it is clear that the results are contrary in trend to those observed by using

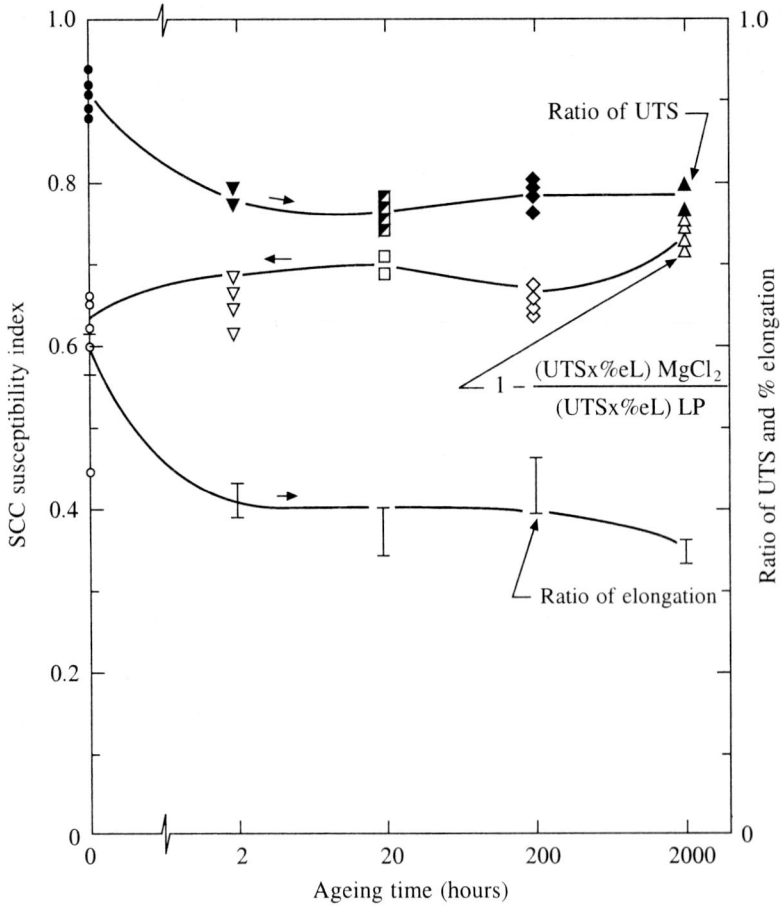

Fig. 6. Variation in SCC susceptibility as a function of aging time at an aging temperature of 873 K.

assessment parameters in Fig. 6. This indicates that the evaluation parameters are of paramount importance in assessing the results of the SSRT experiments. A careful choice of the evaluation parameters is necessary when assessing materials with differing tensile properties, using the SSRT

Table 4. Average residence time in the region between YS and UTS in boiling 45% magnesium chloride for weld metals heat treated at 873 K for various times

Aging time/(h)	Average residence time (min)
0	6.0
2	6.9
20	10.4
200	15.0
2000	12.0

tests as in the present case. The conventionally used parameters such as ratios of UTS and % elongation in MgCl$_2$ and liquid paraffin alongwith the cracking index could give error-prone conclusions in such cases. Kim and Wilde [35] and Khatak et al. [8] expressed skepticism on the use of the SSRT technique for comparing the susceptibility of alloys with widely differing microstructures, ductility and strength levels. Khatak et al. suggested the use of the average CPR calculated by the Desestret and Oltra formula [36].

The contradictions in the assessment of the SSRT data was resolved by testing for SCC using the constant load technique at 40% YS. These tests indicate that the time-to-failure per unit area of the test specimen increases with increasing aging time up to 200 hours of aging and decreases thereafter on aging to 2000 hours (Fig. 8). Since the constant load tests were carried out at a certain fraction of the YS, the results obtained were considered reliable [36]. The trend observed in Fig. 7 is in agreement with the results obtained in constant load tests.

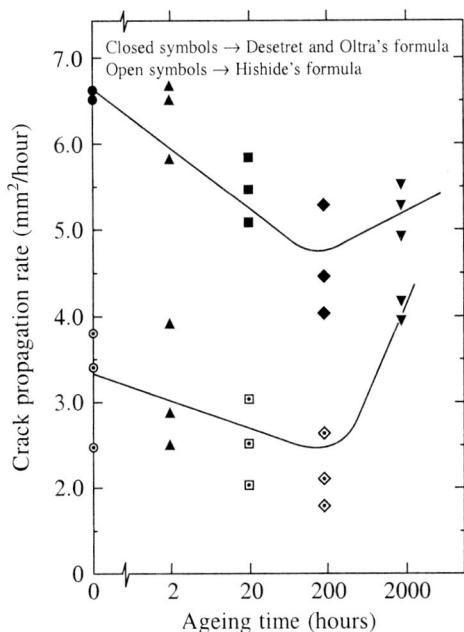

Fig. 7. Dependence of crack propagation rates on aging time at 873 K.

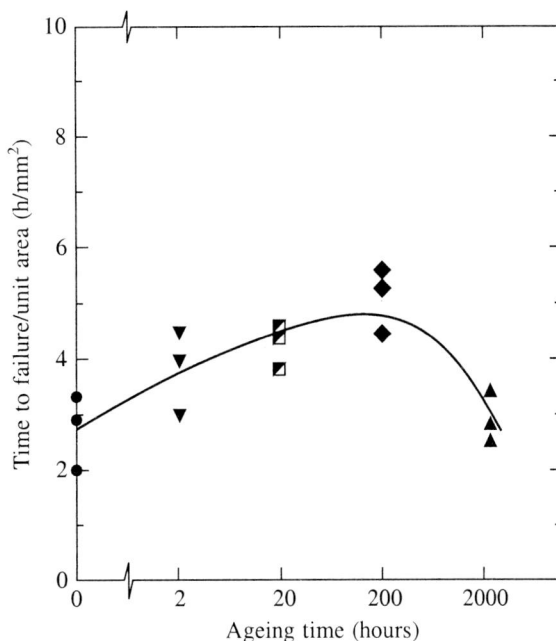

Fig. 8. Dependence of time-to-failure, in constant load tests, on aging time at 873 K.

The increase in SCC resistance of the weld metal with increasing aging time, on aging between 0 and 200 hours, is related to the decrease in YS with increasing aging time. The increase in SCC resistance could be explained as follows : depending on the applied load, there exists a concentration of stress at the tip of an advancing crack. The stress distribution ahead of the crack tip causes plastic deformation in the region around the crack tip, where the YS of the material is exceeded. The plastic deformation causes crack blunting and leads to relaxation of stresses in the deformed zone. As a result an impediment in crack growth takes place. The size of the plastic zone ahead of a growing crack increases with decreasing YS. The larger the plastic zone, the greater the crack tip blunting and higher the resistance to crack growth. Thus, matrix softening, which caused a decrease in YS on aging up to

200 hours, led to increased SCC resistance. Increasing Cr depletion in the matrix due to secondary phase precipitation was expected to decrease the SCC resistance because Cr depleted regions provide a pre-existing active path for crack propagation. But this was not the case because of the use of hot concentrated chloride solution for testing. Hot concentrated chloride solutions make a sensitised stainless steel fail by TGSCC indicating that they do not respond to Cr depletion in the steel [37]. The nil effect of Cr depletion on the SCC resistance was a result of the environment, which, due to its high corrosivity, was not able to differentiate the pre-existing active paths from the strain-generated ones, thus attacking both equally. Cr depletion had an indirect influence in determining the SCC resistance in this case. Increasing Cr depletion softened the austenite matrix, thus increasing the SCC resistance. The small quantity of σ did not have any effect on the SCC resistance of the weld metal aged up to 200 hours.

The decrease in the SCC resistance on aging to 2000 hours was a direct consequence of the increase in the value of 'n', as a result of extensive σ phase precipitation. This decrease in the SCC resistance could have been due to decreased stress relaxation as a result of increased particle-dislocation or dislocation-dislocation interaction. Hence, faster crack propagation took place due to reduced blunting of the crack tip. Also, the large amounts of σ phase would have contributed to faster crack propagation by making more amounts of austenite-σ interfaces available for cracking.

Fractographic examination indicated that failure had occurred by a combination of TGSCC of austenite and interphase-interface cracking in all the heat treated conditions (Fig. 9). This transition to interface cracking was also reported by Stalder et al. and Sherman et al. [14, 16]. Sherman et al. reported that crack propagated preferentially along interfaces which were oriented near normal to the tensile stress. Similar, but less severe interface attack was observed in unstressed specimens. Hence, they attributed the SCC growth to the SAD of the DF/A interfaces. Stalder and Duquette [16] attributed the preferential attack of the interface to the non-equilibrium segregation of solute elements in the weld metal. In the present study, the interfaces involved were those of δ-ferrite and σ with austenite.

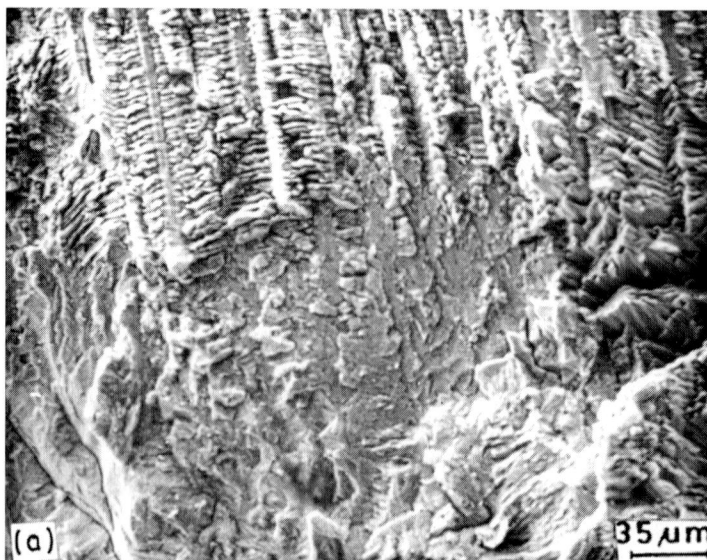

Fig. 9. A typical high magnification fractograph of SCC tested weld metal showing a combination of TGSCC of austenite and interface cracking.

The transition to interface cracking can be explained based on reasons that explain transition to intergranular cracking of austenite [38, 39, 40].

STRESS CORROSION CRACKING OF THE HEAT AFFECTED ZONE OF WELDMENTS OF AUSTENITIC STAINLESS STEEL

The rapid thermal cycles experienced in the HAZ of a fusion weld promote some metallurgical changes that would significantly affect the corrosion resistance of the weld joint. In austenitic stainless steel, a degree of grain coarsening may be experienced very close to the fusion line. While sensitivity of austenitic stainless steel to chloride SCC is increased with increasing grain size, the extent of HAZ grain coarsening is not significant enough to impair the SCC resistance [15]. The more often encountered phenomena, which degrade the corrosion resistance of an austenitic stainless steel weld joint, is the sensitisation of the HAZ. Sensitisation occurs when the austenitic stainless steel is exposed to a temperature range of 723 to 1123 K. During this high temperature exposure, depletion of Cr to less than 12% occurs in the region around the grain boundary, due to the precipitation of a continuous network of $M_{23}C_6$ carbides. Sensitisation makes the steel susceptible to intergranular attack.

The probability of the HAZ being sensitised during welding would depend on the time it spends in the sensitisation temperature range. The residence time spent by the HAZ in this temperature range depends on the heat input, which in turn depends on the heating and cooling rates. Figure 10 illustrates the maximum time spent by four different regions of the HAZ in the sensitisation temperature range. The region, which experiences a peak temperature of 1123 K, is sensitised maximum. Chromium carbides do not form in the regions experiencing temperatures beyond 1123 K and if any carbides are formed during heating between 723 and 1123 K, they dissolve beyond 1123 K. Below 723 K, carbide precipitation is too sluggish to cause any concern. Figure 11 shows that for a butt weld configuration for thicker sections, increasing heat input increases the time spent by the HAZ in the sensitisation

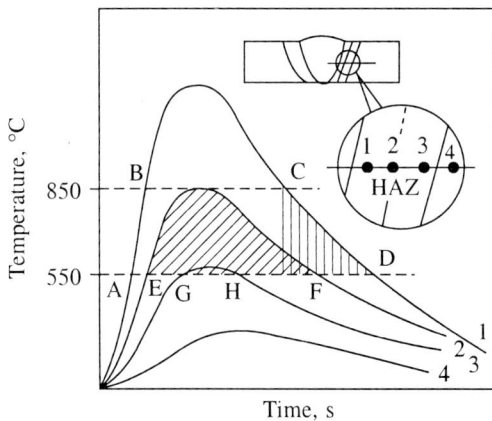

Fig. 10.   A schematic diagram indicating the effect of thermal cycling in inducing sensitisation in different regions of HAZ exposed to temperatures (1) above 1123 K, (2) 1123 K, (3) between 823 and 1123 K and (4) below 823 K.

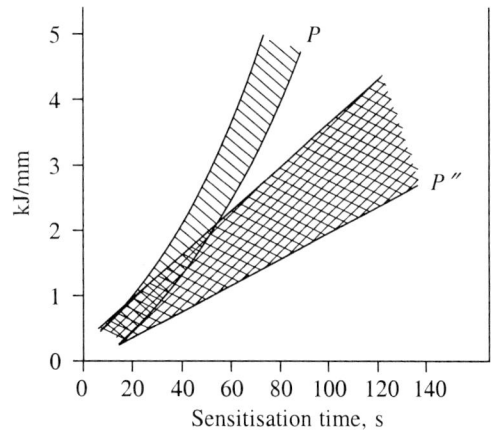

Fig. 11.   Relationship between heat input and maximum time spent by HAZ between 823 K and 1123 K for welds made with 373 K (*P′*) and room temperature (*P*) interpass temperatures.

temperature range. Also, the use of higher interpass temperatures increases the probability of sensitisation of HAZ [41].

The susceptibility of the stainless steel to sensitisation can be known from the time-temperature-sensitisation (TTS) diagram [42, 43]. The sensitisation behaviour on continuous cooling, as in welding, can be predicted by evolving a continuous cooling sensitisation (CCS) diagrams by superimposing different cooling curves on the TTS diagram [44]. The kinetics of sensitisation of HAZ is influenced by prior deformation [42, 43], grain size [45], alloying elements that alter the activities of C and Cr [46], and changes in corrosive environment [47].

During welding, the austenitic stainless steel may not undergo sensitisation but a few carbide nuclei may be present along the grain boundaries. During service, at low temperatures (temperatures lesser than the sensitisation temperature range), such as in boiling water reactors (BWRs), the carbide nuclei grow with time and encompass the whole grain boundary and cause Cr depletion. This process is called low temperature sensitisation (LTS) and is of prime concern for IGSCC of nuclear reactor components. Though the problem of sensitisation can be reduced by stabilising the C with Ti and/or Nb, as in types 321 and 347 stainless steels, the stabilised steels face the problem of knife line attack in the base metal very adjacent to the fusion line.

Intergranular Stress Corrosion Cracking of Heat Affected Zone
A sensitised HAZ could fail by TGSCC or IGSCC. The most commonly encountered environments that cause failure in austenitic stainless steels are those, which contain chlorides. It is very difficult to specify chloride levels below which SCC will not occur since other environmental factors such as oxygen content and pH of environment play a role in the failure. Failures could occur even in environments containing only a few ppm of chlorides, especially if the chlorides get concentrated in crevices or under rust scales. A feature of chloride SCC is that threshold stress level for cracking may be extremely low. The threshold stress depends on the environment. In fact, the question arises whether a threshold stress for SCC really exists. This is because the elastic limit of austenitic stainless steels is low and the residual stresses developed during welding are so high that the threshold value is crossed.

IGSCC of HAZ is found in components, which have been sensitised to some degree by the welding cycle [48]. This problem is normally encountered during two types of services viz. exposure to polythionic acid in refinery industry and to high temperature water in BWRs [49, 50]. IGSCC of sensitised HAZ of austenitic stainless steel piping is a major problem in the safe and economical operation of BWRs. It occurs mainly in small diameter pipes where the stresses are expected to be very high. The cracks are mostly circumferential due to axial stresses. Whether or not IGSCC occurs depends on the conjoint action of three factors viz. tensile stress, sensitised metallurgical condition and, environmental conditions close to active-passive transition on the material surface [15]. During welding, the first two conditions for IGSCC to occur are met. The threshold stress for IGSCC to occur in a sensitised HAZ of austenitic stainless steel, under BWR condition, is close to 0.2% proof stress depending on the degree of sensitisation [15]. Hence, stress relieving of the weld joint is necessary after the welding operation.

The remedial measures to overcome IGSCC in BWRs can be classified into three groups. They are: (1) stress-related remedies, (2) material-related remedies and (3) environmental-related remedies.

Stress-related remedies These methods aim to impart compressive residual stresses at the inner surface of the pipe as well as partially through the wall. Inducing compressive residual stresses on the surface results in improved resistance to IGSCC. Some of the techniques [51] include:

Last pass heat sink welding A TIG welding arc is used as the heat source to heat the outer surface while simultaneously melting the filler metal. During the process, the inner surface is flushed with water, thus cooling it. A temperature difference is thus established between the outer and inner surfaces. The resulting thermal stresses produce localised plasticity inducing compressive stresses on the inner surface of the pipe.

Induction heat stress improvement In this process, the weld area in the pipe is inductively heated from outside. The pipe is simultaneously cooled from inside with water. Just as in the above process, compressive stresses are induced on the inner surface of the pipe.

Mechanical stress improvement In this process, the pipe is radially compressed a slight amount on one side of the weld by means of hydraulic jaws to produce a permanent deformation. The deformation involved is less than 2%. The resulting curvature reduces the tensile stresses, produced by welding, on the root side of the weld area and produces compressive stresses in both the axial and radial directions.

Material-related remedies These methods aim to eliminate the material flaws which cause IGSCC.

Solution annealing treatment: Helps in dissolving the grain boundary carbide network and evens out the material composition.

Corrosion resistant cladding The weld joint is deposited with weld overlays, which have a duplex microstructure. In the BWR environment, the weld overlay may crack but complete resistance to IGSCC is ensured for the pipe.

Alternate pipe material: Choose better materials for the application, such as types 304 LN stainless steel, which is resistant to sensitisation and hence to IGC.

Environmental-related remedies The method aims to remove the IGSCC causative species from the environment. Some of the methods are: start-up de-aeration, hydrogen water chemistry, control of impurities (chlorides, sulphates etc.).

Effect of Applied Potential

The potential-time behaviour of MMA and TIG weldments of AISI type 316 stainless steel (chemical composition in Table 5), during SCC tests in boiling 5 M NaCl + 0.15 M Na_2SO_4 + 2.5 ml/l HCl (b.p. = 381 K) solution using the constant load technique at an initial stress level of 200 MPa, is shown in Fig. 12. HCl was added to lower the pH of the solution to about 1.3, to aid corrosion. Sodium sulphate was added to facilitate easy repassivation. All potentials were measured against the reference saturated calomel electrode (SCE). The OCP was –428 mV and –423 mV for TIG and MMA weldments respectively, at the start of the test. Thereafter, there was an initial shift in the negative direction till a maximum negative value of –430 mV for MMA weldment and –434 mV for TIG weldment was reached. During this period, a copious evolution of hydrogen was observed on the sample surface suggesting active dissolution. A non-adherent blackish-green film was formed during this period. Subsequently, the OCP moved in the positive direction till a maximum, transient value of –315 mV for MMA weldment and –340 mV for TIG weldment was reached. This change of potential in the positive direction corresponded to the formation of a protective adherent film [52]. The system then stablilised

Table 5.   Chemical compositions (in weight %) of AISI type 316 base metals, MMA flux coated electrode and TIG filler wire

Element	C	Cr	Ni	Mo	Mn	Si	S	P
Type 316 stainless steel base metal	0.065	16.28	10.92	2.04	1.38	0.5	0.01	0.02
TIG filler wire	0.04	19.1	12.7	2.4	1.3	0.38	0.01	0.02
MMA flux coated electrode	0.06	17.6	11.05	2.25	0.45	0.97	0.05	0.03

to more negative values at which SCC occurred. The stable potential in the potential-time curves was defined as the critical cracking potential (CCP). The CCP was determined to be –336 mV for MMA weldment and –369 mV for the TIG weldment. The more negative CCP of the TIG weldments was thought to be an effect of difficult repassivation kinetics of the Cr depleted regions in the HAZ. This was confirmed by carrying out SCC tests on furnace-sensitised base metal samples that showed a CCP of –375 mV, which was closer to that of the TIG weldments.

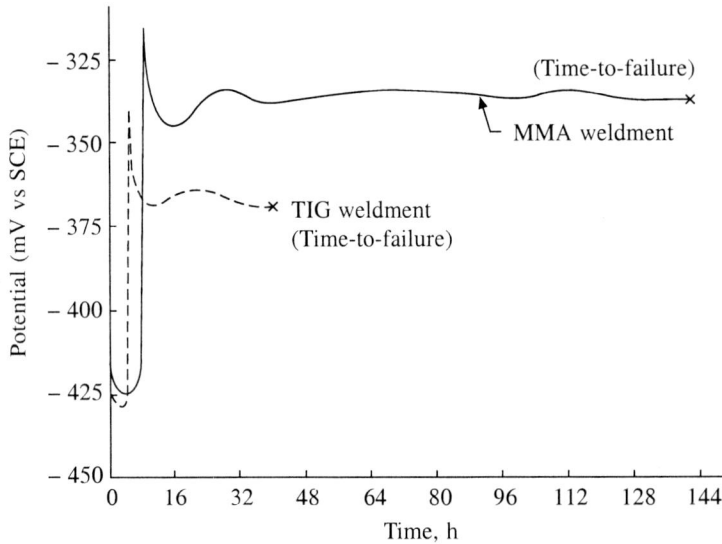

Fig. 12.   Variation of open circuit potential (vs. SCE) with time for stressed samples of TIG and MMA weldments in boiling 5 M NaCl + 0.15 M Na$_2$SO$_4$ + 2.5 ml/l HCl.

Anodic polarisation led to a decrease in SCC time-to-failure, the decrease increasing with increasing anodic polarisation, as seen in Fig. 13. The figure also shows that slight cathodic polarisation (about 5 mV) prevented SCC. High anodic polarisation led to heavy pitting while extensive cathodic polarisation led to general corrosion. This suggested the existence of a narrow potential range (–369 to –361 for TIG and –336 to –327 for MMA weldments) in which SCC occurred. During anodic polarisation, a drop in the value of current, which is an indication of film formation [7], from 8 mA to 0.1 mA was observed.

The results of impressed potential tests suggested that active path corrosion mechanism is operative during SCC of the weldment [53], and that corrosion rate is the controlling process which determines

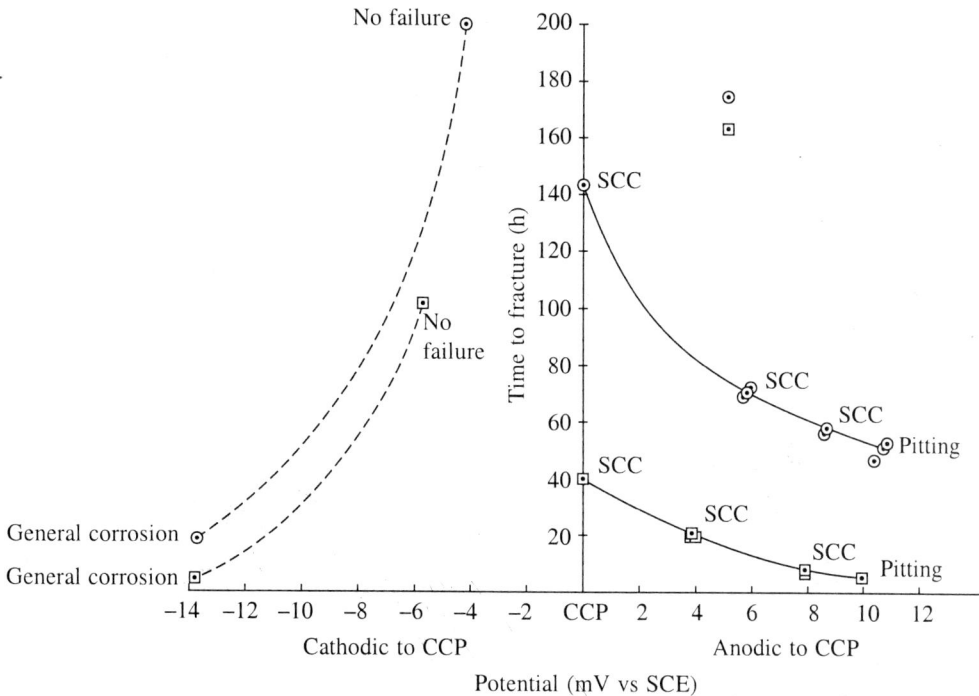

Fig. 13. **Effect of impressed potential on time-to-failure for TIG and MMA weldments in boiling 5 M NaCl + 0.15 M Na$_2$SO$_4$ + 2.5 ml/l HCl.**

the SCC behaviour. The corrosion rate is increased on anodic polarisation and the corrosion process is inhibited during cathodic polarisation. The increase in corrosion rate on anodic polarisation is due to rapid reduction of the pH, which is caused by increased ingress of chloride ions to the crack-tip as compared to the OCP condition [52, 53]. The effect of applied anodic potential on SCC can be understood by its influence on both crack initiation and propagation. On anodic polarisation, crack initiation time is reduced due to decrease in pit initiation time. Increased dissolution of slip steps generated on the surface of the pits, due to enhanced corrosion rate of the environment, enhanced crack propagation rate. Raja et al. reported similar effect of applied potentials on SCC of autogenous weld metal of type 316 stainless steel in a 5 N H$_2$SO$_4$ + 0.5 N NaCl solution at room temperature [19]. They reported that under anodic applied potentials, the time-to-failure reduced; while under cathodic applied potentials, the time-to-failure increased vis-à-vis OCP condition. At very high cathodic potentials, no failure occurred. They reported that the potential range in which SCC occurred was 50 mV for base metal and 220 mV for weld metal.

Optical microscopic examination of the as-received weldments indicated a coarser δ-ferrite and a greater inter-dendritic spacing in the TIG weld metal than in MMA weld metal. However, in both the cases the δ-ferrite was vermicular in shape and discontinuous, thus indicating a Type A microstructure [54]. The δ-ferrite was located on the austenite grain boundary and not in the austenite matrix, thus suggesting a primary austenitic solidification mode [55]. Oxalic acid etch test of the HAZ indicated deeply etched grain boundaries in the TIG weldment and a lightly etched structure in the MMA weldment. This indicated a sensitised HAZ for TIG weldments.

Optical and electron-optical microscopic examination of the SCC-failed samples indicated that SCC initiated through pits in both the weldments under the OCP and IP conditions. In the TIG weldment, cracking initiated and propagated in the sensitised HAZ by transgranular mode in the OCP tests. However, in some cases, mixed mode of transgranular and intergranular cracking was observed (Fig. 14). At all applied potentials, crack initiation and propagation occurred in the HAZ wholly by intergranular cracking (Fig. 15). In MMA weldments, failure occurred in the weld metal under both the OCP and IP conditions. Under OCP conditions, failure occurred due to SAD of δ-ferrite (Fig. 16) and TGSCC of austenite. Under applied anodic potentials, failure occurred due to SAD of δ-ferrite and IGSCC of austenite.

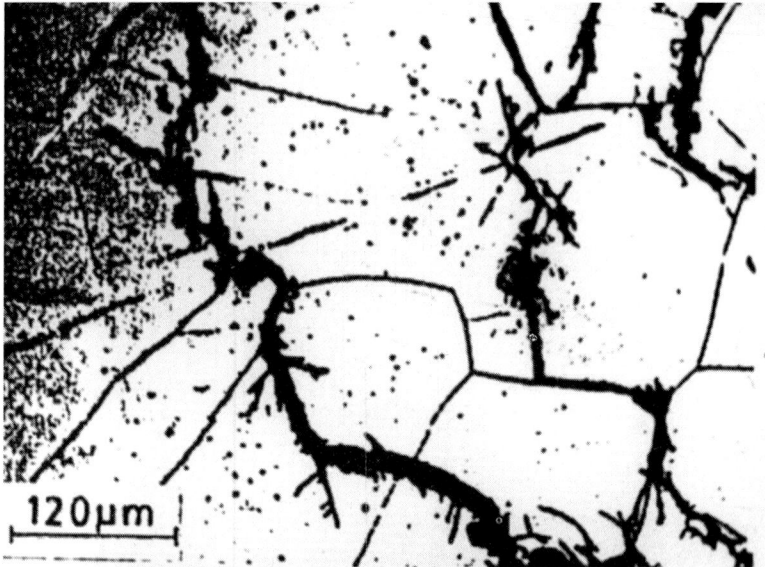

Fig. 14.   Optical micrograph showing mixed mode of cracking in the HAZ of TIG weldment in 5 M NaCl + 0.15 M Na$_2$SO$_4$ + 2.5 ml/l HCl at 200 MPa.

The transgranular cracking observed in the weldments was typical of the 'fan-shaped' patterns, which are normally observed in hot concentrated chloride environments [38, 56, 57]. In MMA weldment, failure by SAD of δ-ferrite was caused by its higher electrochemical activity due to sulphur and phosphorous segregation. The higher inclusion content would have initiated pits in the weld metal and caused cracking. Cr depletion at the grain boundaries in the HAZ of TIG weldments caused the initiation of pits. Intergranular SCC initiation and propagation from pits on anodic polarisation of both types of weldments could be attributed to the higher oxidising potential of the system which resulted in faster repassivation of generated slip steps. This causes dislocation pile-up at grain boundaries which in conjunction with the latter's special characterestics (such as chemical heterogeneities and presence of defect structure at the grain boundaries) resulted in intergranular cracking.

From the above results, it was clear that in case of an unsensitised weldment, weld metal is the weakest link; while in case of a sensitised weldment, the sensitised HAZ is the weakest link from SCC point of view. This would suggest that the different regions in a weldment in the order of decreasing SCC resistance are: base metal > weld metal > sensitised HAZ. However, Gooch had suggested that

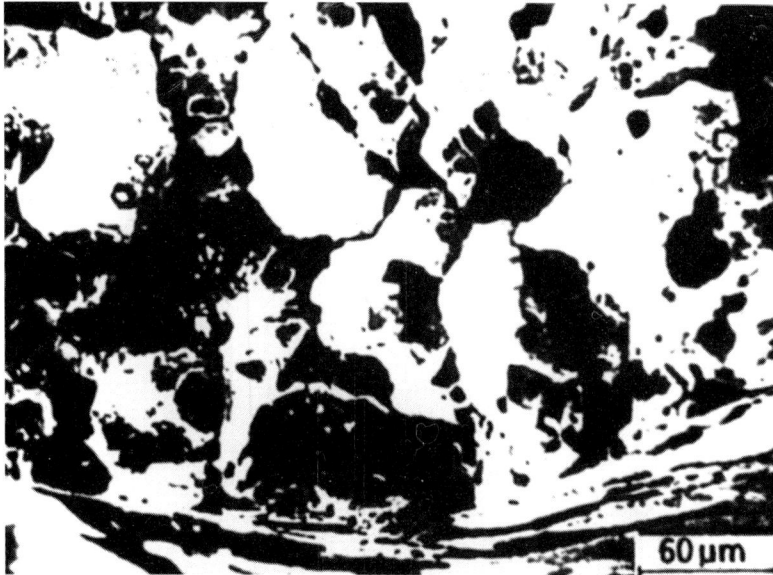

Fig. 15. SEM micrograph of TIG weldment tested at an impressed potential of –365 mV in 5M NaCl + 0.15M Na_2SO_4 + 2.5 ml/l HCl at 200 MPa.

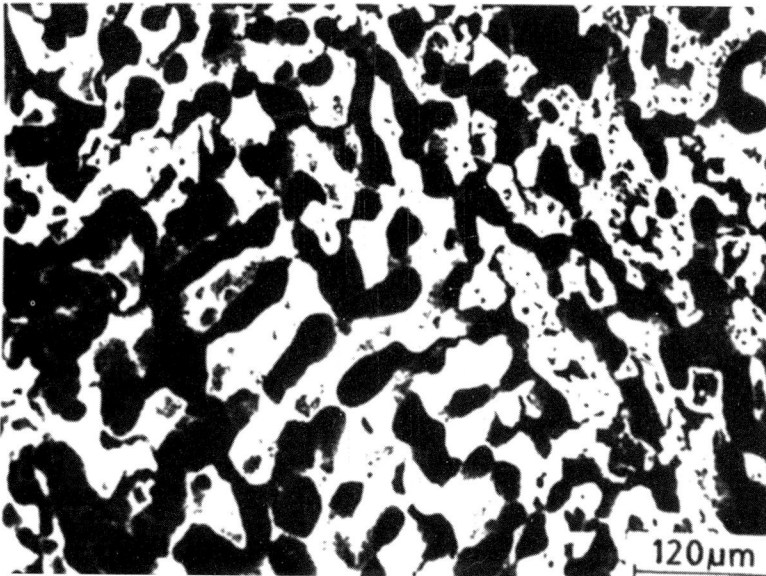

Fig. 16. SEM fractograph of MMA weldment at OCP tested at 200 MPa in 5 M NaCl + 0.15 M Na_2SO_4 + 2.5 ml/l HCl.

the weld metal had better SCC resistance than the base metal because the δ-ferrite dissolves and cathodically protects the austenite in the weld metal thus delaying the failure [15]. However, our work along with those of Raja et al. [19] recently, and Stalder et al. [19] and Baeslack III et al. [18] earlier, showed that the homogeneity of microstructure and chemical composition in the base metal gives it a higher SCC resistance vis-à-vis weld metal. Raja et al. reported a more noble CCP and a narrower range of potentials (about 85 mV) in which SCC occurred for sensitised base metal vis-à-vis the weld metal [19]. This suggested a better SCC resistance for a sensitised HAZ than the weld metal. The effect of environmental composition may have played a role in this case.

Effect of Cold Work Prior to Welding

SCC tests were carried out by the constant load technique in boiling 45% magnesium chloride solution (b.p. = 398 K) at an initial stress level of 245 MPa, and in boiling 5 M NaCl + 0.15 M Na_2SO_4 + 2.5 ml/l HCl (b.p. = 381 K) solution at an initial stress level of 390 MPa on round tension specimens, machined from weldments (composition in Table 6) of AISI type 316 austenitic stainless steel base metal which were cold rolled to 5, 10 and 15% prior to welding. Some of the weldments were heat treated at 1023 K for 10 minutes to sensitise the material.

Table 6.   Chemical compositions (in weight %) of AISI type 316 stainless steel base metal, MMA flux coated electrode

Element	C	Cr	Ni	Mo	Mn	Si	S	P
Type 316 stainless steel base metal	0.05	15.56	12.59	2.27	1.2	0.38	0.011	0.026
MMA flux Coated Electrode	0.065	18.19	11.12	1.9	1.5	0.5	–	–

The results of the SCC tests in magnesium chloride are shown in Fig. 17. It is seen that there is no change in time-to-failure with an increase in the cold work level of the base metal. A decrease in time-to-fracture with increasing cold work has been reported earlier [8, 36]. This has been attributed to the increase in crack initiation sites because of the presence of a large number of defects in cold worked materials. In the case of weldments tested in magnesium chloride solution, failure occurred in weld metal. Cold work in the base metal is not expected to influence the chemistry or structure of the weld metal. Therefore, no difference in time-to-fracture was observed with an increase in level of cold work of base metal. The lower time-to- failures observed in case of heat treated weldments, was because of precipitation of secondary phases in the weld metal. Cracking occurred by a mixture of TGSCC of austenite and cracking of DF/A interfaces. Magnesium chloride solution of high concentrations is reported to produce TGSCC even in sensitised materials [37].

SCC behaviour in boiling 5 M NaCl + 0.15 M Na_2SO_4 + 2.5 ml/l HCl solution is shown in Fig. 18. It is seen that time-to-failure decreases with increase in the cold work of the base metal. Failure always occurred in the HAZ near the weld metal. No cracks were seen in the weld metal. Cracking occurred in the base metal both in the sensitised and non-sensitised conditions. The lower time-to-failure with increasing degree of cold work could be due to lower time required for crack initiation or higher crack propagation rates or both. The cracks were found to initiate from pits. The lower resistance of cold worked material to pitting and the increased susceptibility of the Cr depleted regions in a sensitised structure to pitting would have led to lower crack initiation times [58]. Moreover, a sensitised structure provides a pre-existing active path for crack propagation, which is faster due to the higher dissolution rates of chromium-depleted regions. Increase in intergranular cracking with increasing cold work was attributed to the increase in degree of sensitisation with increasing cold work.

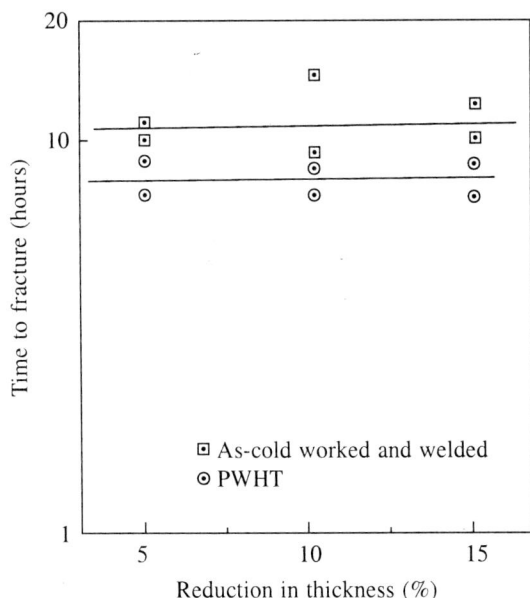

Fig. 17. Effect of the degree of cold rolling (before welding) on time-to-fracture in boiling 45% magnesium chloride solution at 245 MPa for the as-deposited and heat treated weldments.

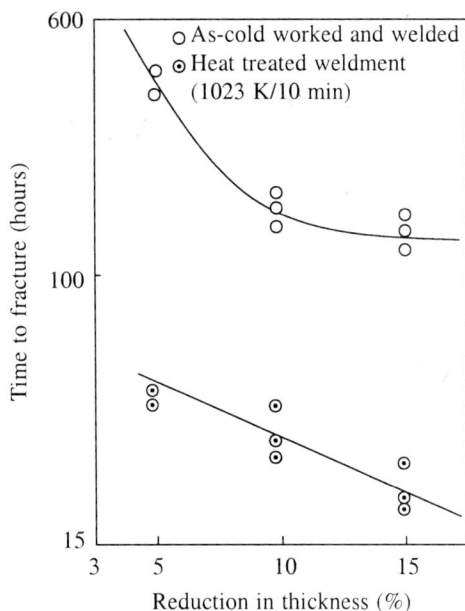

Fig. 18. Effect of degree of cold rolling (before welding) on time-to-failure in boiling 5 M NaCl + 0.15 M Na_2SO_4 + 2.5 ml/l HCl at 390 MPa in as-deposited and heat treated weldments.

Typical TGSCC of austenite was observed in all non-heat treated weldments (Fig. 19) except for a few grains, which showed IGSCC in weldments of 15% cold work base metal. The heat-treated weldments showed a mixed mode of failure with intergranular cracking increasing with increasing cold work (Fig. 20). The transition from transgranular to intergranular cracking in the base metal and weld metals have been explained based on the competitive processes such as increase in crack-tip strain rate, dissolution rate, repassivation kinetics and grain boundary characterestics such as segregatiom, sensitisation etc. [52, 56, 57].

EFFECT OF WELDING VARIABLES AND POSTWELD HEAT TREATMENT ON SCC OF WELD JOINTS OF AUSTENITIC STAINLESS STEELS

Very little work has been done on the influence of the welding process on the SCC susceptibility of welds of austenitic stainless steel. Sensitivity to IGC is increased with increasing heat input and it would seem the risk of IGSCC is also promoted at higher heat inputs [15]. To a lesser extent, a similar adverse effect of high heat input has been found also for TGSCC [15]. However, Franco et al. reported an increase in SCC resistance of weld metal of AISI type 304 stainless steel with increasing heat input in boiling 45% magnesium chloride solution [23].

Post-weld heat treatment (PWHT) can be of substantial benefit in avoiding IGSCC. In this regard, full solution treatment at 1323 K may not be necessary, and, indeed, a stabilising anneal at 1143 to 1223 K is frequently applied to welded components in petrochemical industry to guard against polythionic acid attack. However, the situation with respect to chloride-induced SCC is less clear. It can be argued

Fig. 19. SEM fractograph of weldments of 5% cold worked base metal tested in boiling 5 M NaCl + 0.15 M Na$_2$SO$_4$ + 2.5 ml/l HCl solution at 390 MPa showing transgranular failure with slight evidence of intergranular failure.

Fig. 20. SEM micrograph of sensitised weldment with 15% cold work in base metal showing IGSCC in HAZ when tested in 5 M NaCl + 0.15 M Na$_2$SO$_4$ + 2.5 ml/l HCl solution.

that cracking occurs where environment conditions are adverse, and, because of low threshold stress for this form of failure, service loading would be sufficient to induce failure even away from loaded joints, particularly if local stresses are caused by other factors such as cold deformation. At the same time, service experience shows that when loaded components suffer chloride SCC, cracking preferentially occurs in weld areas, strongly implying weld residual stresses to be a significant factor. On this basis, PWHT is advisable, at least for critical applications. The high coefficient of thermal expansion and the extremely low limits of elasticity for austenitic stainless steels would mean that residual stresses would readily arise during cooling from peak PWHT temperature, unless the operation is carefully controlled. Heating to over 1203 K is necessary to achieve maximum stress relief, and these high temperatures can pose critical problems such as control of distortion, scaling etc. Slow heating and cooling is preferred, subject to avoidance of sensitisation during cooling part of the cycle. Heat treatment at intermediate temperature, say 823 to 1023 K, represents an easier operation but only about 60 % residual stress will be relieved. However, the problem of sensitisation and formation of intermetallic brittle phases such a sigma phase rule out this heat treatment cycle. Low temperature stress relief at 673 K has been advocated to avoid SCC even though less than 40% stress relief is achieved [59]. In principle, the entire component should be heat treated, but in practice, it is more commonly possible to apply only local area heat treatment, when thermal gradients around the heated area are to be minimised.

Due to the above considerations, it may not be possible to guarantee that PWHT can avoid chloride SCC under the said service condition [60, 61]. Gooch proposed elimination of chloride ions, minimise operating temperature and appropriate design to eliminate crevices as possible methods to prevent SCC of austenitic stainless steel weld joints in preference to PWHT. This is not to say that PWHT should be overlooked. PWHT is definitely beneficial in countering chloride SCC and should be resorted to when other properties, like creep and fatigue, render them essential.

CONCLUSIONS

1. The paper reviews many aspects of SCC behaviour of weld metal and weldments of austenitic stainless steel. The SCC behaviour of weld metal vis-à-vis base metal; effects of δ-ferrite, residual stresses and welding defects have been discussed. Effect of sensitisation of HAZ on SCC of weldment and remedial measures to overcome IGSCC of HAZ have been highlighted.
2. Results of work on SCC of weld metal and weldments, in the author's laboratory, are discussed in detail. Studies on effect of applied potentials on SCC of weldments of AISI type 316 austenitic stainless steel in boiling 5 M NaCl + 0.15 M Na_2SO_4 + 2.5 ml/l HCl solution showed that:

   (a) SCC occurred in narrow range of potentials of nearly 9 mV anodic to CCP.
   (b) Time-to-failure decreased with increasing anodic polarisation. Slight cathodic polarisation inhibited SCC. Pits initiated the failures. At OCP, both TGSCC and IGSCC of austenite were observed in the HAZ of TIG weldment. MMA weldment failed in the weld metal by δ-ferrite dissolution and TGSCC of austenite. Anodic potentials initiated cracking by IGSCC of austenite.

   SCC tests on weldments of cold rolled type 316 austenitic stainless steel in boiling magnesium chloride solution (b.p. = 398 K) and boiling 5 M NaCl + 0.15 M Na_2SO_4 + 2.5 ml/l HCl solution (b.p. = 381 K) showed that:

(a) In boiling $MgCl_2$ solution, failure occurred in the weld metal due to which no difference in time-to-failure was observed in weldments with different degrees of cold work. Sensitised weldments showed slightly lower time-to-fracture.

(b) In boiling 5 M NaCl + 0.15 M Na_2SO_4 + 2.5 ml/l HCl solution, cracking always occurred in the base metal. Time-to-fracture decreased with increasing cold work. A large difference in time-to-fracture was observed between the non-sensitised and sensitised weldments.

SCC tests in boiling 45% $MgCl_2$ solution using the SSRT and constant load techniques on aged weld metal of AISI type 316 L stainless steel, showed that:

(a) Matrix softening improved the SCC resistance up to 200 hours of aging; while matrix hardening decreased the SCC resistance on aging up to 2000 hours.

(b) The weld metal failed by a combination of TGSCC of austenite and interface cracking in both the CERT and constant load tests.

ACKNOWLEDGEMENTS

The author wishes to thank Dr. H.S. Khatak, Head, Corrosion Science and Technology Division, Indira Gandhi Centre for Atomic Research, Kalpakkam, for his constant encouragement and immense contribution during the experimental work.

REFERENCES

1. Lippold, J.C. and Savage, W.F., *Welding Journal*, 59 (1980), 362-s.
2. Brooks, J.A. and Thompson, A.W., *International Metals Review*, 36 (1991) 16.
3. Gill, T.P.S., Mudali, U.K., Seetharaman, V. and Gnanamoorthy, J.B., *Corrosion*, 44 (1988) 511
4. Nakayama, T. and Takano, M., *Corrosion*, 41 (1985) 592.
5. Nakayama, T. and Takano, M., *Corrosion*, 37 (1981) 226.
6. Takano, M., Teramoto, K., and Nakayama, T., *Corrosion Science*, 21 (1981) 459.
7. Voccaro, F.P., Hehemann, R.F. and Troiano, A.R., *Corrosion*, 36 (1980) 530.
8. Khatak, H.S., Muraleedharan, P., Gnanamoorthy, J.B., Rodriguez, P. and Padmanabhan, K.A., Journal *of Nuclear Materials*, 168 (1989) 157.
9. Khatak, H.S., Gnanamoorthy, J.B. and Rodriguez, P., *Metallurgical and Materials Transactions A*, 27A (1996) 1313.
10. Vinoy, T.V., Shaikh, H., Khatak, H.S., Sivabharasi, N. and Gnanamoorthy, J.B., *Journal of Nuclear Materials*, 238 (1996) 278.
11. Russels, A.J. and Tromans, D., *Metallurgical Transactions A*, 10A (1979) 1229.
12. Flowers, J.W., Beck, F.H. and Fontana, M., *Corrosion*, 19 (1963) 186-t.
13. Baeslack III, W.A., Lippold, J.C. and Savage,W.F., *Welding Journal*, 58 (1979) 168-s.
14. Sherman, D.H., Duquette, D.J. and Savage, W.F., *Corrosion*, 31 (1975) 376.
15. Gooch, T.G., *Welding in the World*, 22 (1984) 64.
16. Stalder, F. and Duquette, D.J., *Corrosion*, 33 (1977) 67.
17. Baeslack III, W.A., Savage, W.F. and Duquette, D.J., *Metallurgical Transactions A*, 10A (1979) 1429.
18. Baeslack III, W.A., Duquette, D.J. and Savage, W.F., *Corrosion*, 35 (1979) 45.
19. Raja, K.S. and Rao, K.P., *Corrosion*, 48 (1992) 634.
20. Baeslack III,W.A., Savage, W.F. and Duquette, D.J., *Welding Journal*, 58 (1979) 83-s.
21. Fang, Z., Wu, Y. and Zhu, R., *Corrosion*, 50 (1994) 171.
22. Krishnan, K.N. and Rao, K.P., *Corrosion*, 46 (1990) 734.

23. Franco, C.V., Barbosa, R.P., Martinelli, A.E. and Buschinelli, A.J.A., *Werkstoffe und Korrosion*, 49 (1998) 496.
24. Vishwanathan, R., Nurninen, J.I. and Aspden, R.G.; *Welding Journal*, 58 (1979) 118.
25. Rogne, T., Drugli, J.M. and Valen, S., *Corrosion*, 48 (1992) 864.
26. Szklarska-Sialowska, S. and Lunarska, E., *Werkstoffe und Korrosion*, 32 (1981) 478.
27. Holtan, H. and Sigurdsson, H., *Werkstoffe und Korrosion*, 28 (1977) 475.
28. Szklarska-Smialowska, S. and Gust, J., *Corrosion Science*, 19 (1979) 753.
29. Shamakian, R.L., Troiano, A.R. and Hehemann, R.F., *Corrosion*, 36 (1980) 279.
30. Clarke, W.L. and Gordon, G.M., *Corrosion*, 29 (1973) 1.
31. Craig, B.D., *Corrosion*, 34 (1978) 282.
32. Hishida, M. and Nakada, H., *Corrosion*, 33 (1977) 332.
33. Desestret, A. and Oltra, R., *Corrosion Science*, 20 (1980) 799.
34. Hishida, M., Begly, J.A., McCright, R.D. and Staehle, R.W., *Stress Corrosion Cracking—The Slow Strain Rate Technique*, Eds. Ugiansky, G. M. and Payers, J. H., ASTM STP 665, (1979) 47.
35. Kim, C.D. and Wilde, B.E., *Stress Corrosion Cracking—The Slow Strain Rate Technique*, Eds. Ugiansky, G.M. and Payers, J.H., ASTM STP 665, (1979) 97.
36. Khatak, H.S., Muraleedharan, P., Gnanamoorthy, J.B., Rodriguez, P. and Padmanabhan, K.A., *Proceedings of the 'Sixth International Conference on Fracture (ICF-6)'*, Poster Session papers, New Delhi, (1984) 153.
37. Kowoka, M. and Kudo, T., *Transactions of the Japan Institute of Metals*, 16 (1975) 36.
38. Muraleedharan, P., Khatak, H.S., Gnanamoorthy, J.B. and Rodriguez, P., *Metallurgical Transactions A*, 16A (1985) 285.
39. Takano, M., *Corrosion*, 30 (1974) 441.
40. Cigada, A., Mazza, B., Pedeferi, P., Salvago, G., Sinigaglia, G. and Zanini, G., *Corrosion Science*, 22 (1982) 559.
41. Gill, T.P.S, *Proceedings of the 'Corrosion Management Course'*, Indian Institute of Metals, Kalpakkam, October, 1995, Paper No. L-6.
42. Mannan, S.K., Dayal, R.K., Vijayalakshmi, M. and Parvathavarthini, N., *Journal of Nuclear Materials*, 126 (1984) 1.
43. Parvathavarthini, N., Dayal, R.K., Gnanamoorthy, J.B. and Seshadri, S.K., *Journal of Nuclear Materials*, 168 (1989) 83.
44. Dayal, R.K. and Gnanamoorthy, J.B., *Corrosion*, 36 (1980) 104.
45. Dayal, R.K., *Proceedings of the 'Corrosion Management Course'*, Indian Institute of Metals, Kalpakkam, October, 1995, Paper No. L-3.
46. Briant, C.L., Muldorf, R.A. and Hall, E.L., *Corrosion*, 38 (1982) 468.
47. Mahla, E.M. and Nilsen, N.A., *Transactions of ASM*, 43 (1950) 290.
48. Cragnolino, C. and McDonald, D.D., *Corrosion*, 38 (1982) 406.
49. Samans, C.H., *Corrosion*, 20 (1964) 256.
50. Heller, J.J. and Prescott, G.R., *Materials Protection*, 4 (1965) 14.
51. Schmidt, J., Pellkofer, D. and Weiss, E., *Nuclear Engineering and Design*, 174 (1997) 301.
52. Staehle, R.W., *Corrosion*, 26 (1970) 451.
53. Wilde, B.E., *Corrosion*, 27 (1971) 326.
54. Takalo, T., Suutala, N. and Moisio,T., *Metallurgical Transactions A*, 10A (1979) 1173.
55. Kujanpaa, V.P., Suutala, N.J., Takala, T. and Moisio, T.J.I., *Metal Construction*, 12 (1980) 282.
56. Scully, J.C., *The Theory of Stress Corrosion Cracking in Alloys*, ed. J.C.Scully, NATO, Brussels (1971) 127.
57. Mukai, Y., Watanabe, M. and Murata, M., *Fractography in Failure Analysis*, ASTM STP 645, ASTM, Philadelphia-PA (1978) 164.
58. Tedmon, C.S., Vermilea Jr., D.A. and Broecker, D.E., *Corrosion*, 27 (1971) 104.
59. Cole, C.L. and Jones, J.D., *Proceedings of Conference on "Stainless Steel"*, ISI Publication 117, (1969) 71.
60. Edeleanu, C., *Corrosion Technology*, 4 (1957) 49.
61. Gooch, T.G., *Proceedings of Conference "The Influence of Welding and Welds on Corrosion Behaviour of Constructions"*, IIW Annual Assembly, Tel Aviv, (1975) 1.52.

8. Applications of Fracture Mechanics in Stress Corrosion Cracking and Introduction to Life Prediction Approaches

H.S. Khatak[1]

Abstract Number of loading techniques such as constant strain, constant load, slow strain rate are being used for stress corrosion testing of materials. These techniques have been useful in comparision of stress corrosion susceptibility of various alloy/environment combinations and also to evaluate the influence of metallurgical factors for a particular alloy. Fracture mechanics approach in combination with above loading techniques has been successfully utilized in getting quantitative data with respect to threshold values and crack growth rates in addition to grading the alloys with respect to cracking susceptibility. This data could be used in predictive models. In this paper, data available up to date on austenitic steels is presented and analyzed. Work carried out at authors laboratory on AISI 304LN and AISI 316LN stainless steels and their welds has been highlighted. It has been found that cold work and sensitization decrease the threshold values and adversely influence the crack growth rates. For welds of AISI 316SS, it is seen that threshold values are almost half and crack growth rates are an order of magnitude higher as compared to solution annealed condition. Based on the data generated on crack growth rates, analysis of acoustic emission records during crack growth, fractographical features and activation energy measurements attempt has been made to understand the mechanism of stress corrosion cracking of austenitic steels in chloride medium. A brief introduction to life prediction approaches such as slip dissolution model, models based on crack growth rates and statistical approach has been given.
Key Words Stress corrosion, fracture mechanics, austenitic stainless steels, crack growth, life prediction.

In conventional tests at constant load and at constant strain, time to crack initiation and time to fracture give large scatter. Difficulty arises in determining the threshold stress as it involves testing for long durations. Also, since cold working increases the yield strength, it becomes evident that conclusions may vary even on the basis of applied stress because at the same applied stress time to fracture will be higher for materials with high yield strength because the load represents a smaller fraction of yield strength. Also, the applicability of slow strain rate tests for materials with different yield strengths has been questioned [1]. In view of the above, the fracture mechanics approach which involves measurement of crack growth rates, in combination with other techniques, is useful method for assessing the SCC susceptibility. The test environment should be one in which initial passivation conditions (thickness, composition and structure of the passive film) are close to those encountered in service. Low pH [2] and high concentration of salts [3] have been reported to be present in the cracks. Therefore, data

[1]Head, Corrosion Science and Technology Division, Indira Gandhi Centre for Atomic Research, Kalpakkam-603 702, India.

reported in the literature with the use of fracture mechanics approach, generated using such environments is justified. In this paper brief introduction to fracture mechanics, method of crack length measurement and up todate information published has been discussed. Work carried out at author's laboratory, particularly on LN varieties of stainless steels has been highlighted.

INTRODUCTION TO FRACTURE MECHANICS

Linear Elastic Fracture Mechanics

Structural failure, under loading conditions well below the yield stress of the structural material, can often be attributed to cracks or defects in structures. Such failures show that the conventional stress analysis of structure alone is not sufficient to guarantee the structural integrity under operational conditions. Structural studies, which consider crack extension behaviour as a function of applied loads is called fracture mechanics. In particular, in the absence of large plastically yielded regions surrounding cracks or flaws, such a study is referred to as linear elastic fracture mechanics (LEFM). Considerable research effort has been put in the study of LEFM, which can now be used to solve many engineering problems. The fracture mechanics analysis should be able to answer the following questions:

(a) What is the residual strength as a function of crack size?
(b) What size of the crack can be tolerated at the expected service loads i.e. what is the critical crack size?
(c) How long does it take for a crack to grow from a certain initial size to the critical size?
(d) What size of pre-existing flaw can be permitted at the moment the structure starts its service life?
(e) How often should the structured be inspected for cracks?

Detailed discussions on the principles of fracture mechanics are available elsewhere [4, 5] .The crack in a solid can be stressed in three different modes as shown in Fig. 1.

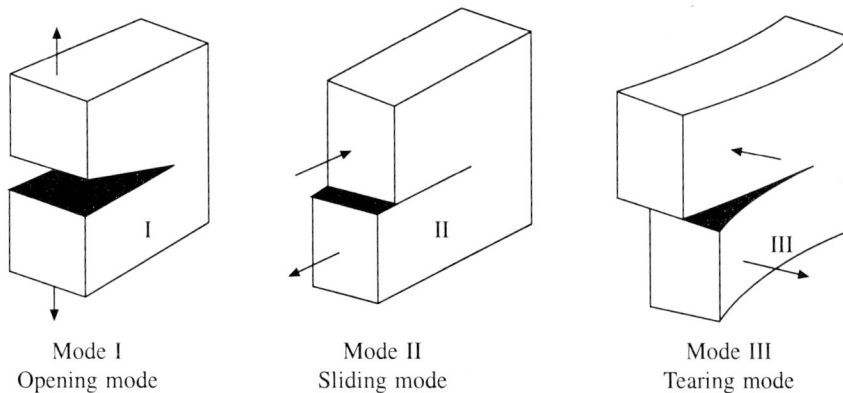

Mode I   Mode II   Mode III
Opening mode   Sliding mode   Tearing mode

Fig. 1.   The modes of loading.

Figure 2 shows the stress distribution in an infinite plate with a crack length 2a in mode-I. The stresses in the different directions can be represented

$$\sigma_x = \sigma \sqrt{a/2r} \cos \theta/2 \; [1 - \sin \theta/2 \cdot \sin 3\theta/2] \tag{1}$$

$$\sigma_y = \sqrt{a/2r} \cos \theta/2 \; [1 + \sin \theta/2 \cdot \sin 3\theta/2] \tag{2}$$

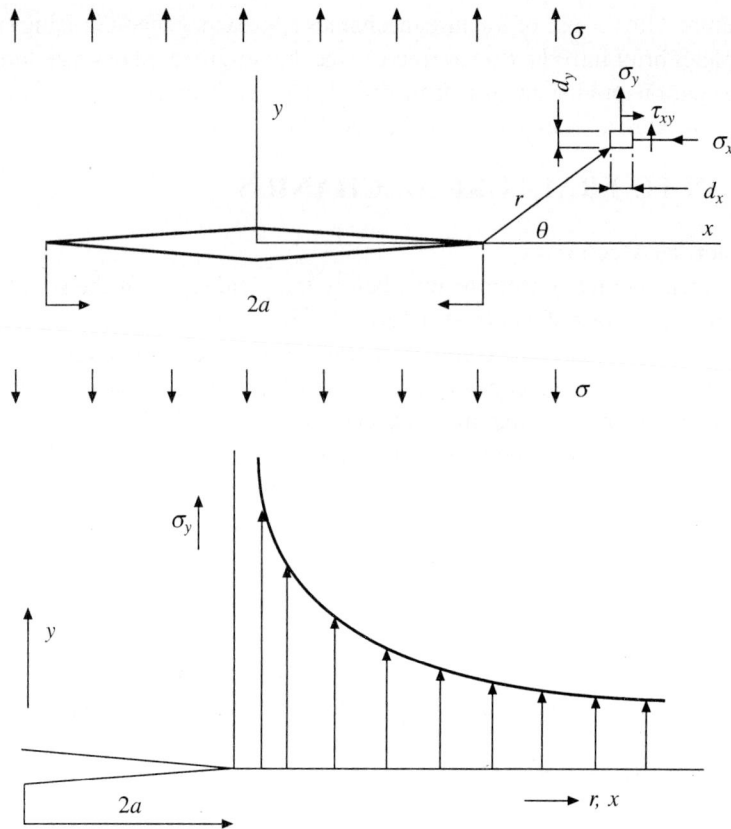

Fig. 2.   Elastic stress σ_y at the crack tip.

$$\tau_{xy} = \sigma \sqrt{\alpha/2r} \sin \theta/2 \cos \theta/2 \cos 3\,\theta/2 \tag{3}$$

$$\sigma_Z = 0 \text{ (plain stress)}$$

$$\sigma_Z = v\,(\sigma_x + \sigma_Y) \text{ (plain strain)}$$

The equations for mode I can be written in the form

$$\sigma_{Iij} = K_I/\sqrt{2\pi r} \cdot f_{ij}\,(\theta) \text{ with } K = \sigma \sqrt{\pi a} \tag{4}$$

In reality the situation is slightly different. The expression for K_I for finite size plate is

$$K_I = \sigma \sqrt{\pi a}\, f(a/W) \tag{5}$$

where W is the plate width and the function $f(a/W)$ approaches unity for small values of a/W. The factor K_I is known as "stress intensity factor" where the subscript stands for mode I. Thus, for linear elastic conditions, K_I constitutes a single parameter to represent the crack driving force for crack advance. For monotonic loading of stationary cracks, this approach has been applied to characterise the onset of brittle fracture, where for plain strain conditions $K_I = K_{IC}$, the fracture toughness [6], and to estimate the onset of crack instability in plain stress conditions through the use of K_I resistance curves [7]. Furthermore, for the subscritical crack growth, K_I has been used to correlate rates of crack growth for environmentally assisted fracture; corrosion cracking [8], hydrogen embrittlement [9], and in fatigue through an expression of the form $da/dN = CK^m$. However, there are certain limitations. In elastic analyses, crack tip stress tends to be infinite as r tends to zero. In reality this cannot occur. The plastic deformation taking place at the crack tip keeps the stress finite. Plastic zone size r_p can be written as

$$\gamma_p = 1/2\,\pi(K_I/\sigma_y)^2 \text{ (plane stress)} \tag{6}$$

$$\gamma_p = 1/6\,\pi\,(K_I/\sigma_y)^2 \text{ (plane strain)} \tag{7}$$

The plastic zone can be considered as merely a small perturbation in the linear elastic field. This situation, known as small scale yielding, is met when plastic zone is of the order of 15 times smaller than the crack length [10]. Additionally, for characterising the onset to brittle fracture at $K_I = K_{IC}$, the requirement of plain strain must also be met such that the plastic zone size is 15 times smaller than the thickness [11]. These limitations form the basis of the requirements of ASTME 399 [6].

$$a, B, b \geq 2.5(K_{IC}/\sigma_Y)^2 \tag{8}$$

where a is the crack length, B the thickness and b the remaining ligament.

K_{IC} is a measure of the crack resistance of a material. Therefore, it is called "plane strain fracture toughness". If deformation in the thickness direction can take place freely (plane stress situation) the critical stress intensity factor depends on plate thickness.

Calibration expressions for different geometries (K) are listed in the handbook of stress intensity factors [12]. Expressions for widely used geometries in mode I are shown in Figs. 3 to 5.

K can be expressed in terms of G, the elastic energy release rate as follows:

$$K^2/E = G \text{ (plane stress)} \tag{9}$$

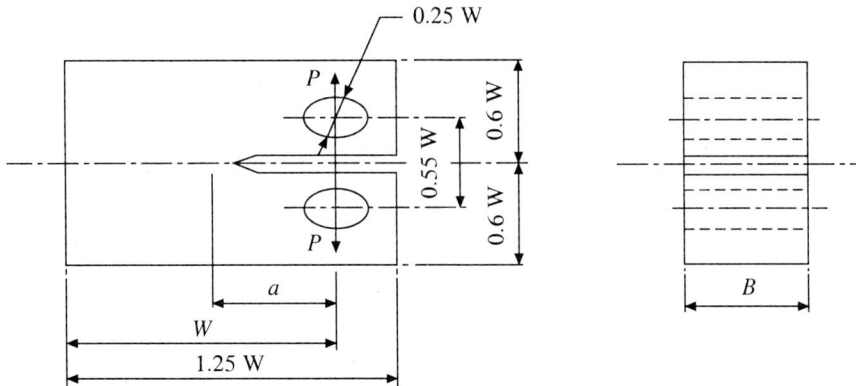

$$K_I = \frac{P}{BW^{1/2}}f(\alpha), \; \alpha = \frac{a}{W}$$

$$f_I(\alpha) = \frac{(2 + \alpha)(0.886 + 4.64\alpha - 13.32\alpha^2 + 14.72\alpha^3 - 5.6\alpha^4)}{(1 - \alpha)^{3/2}}$$

Fig. 3. Compact tension specimen.

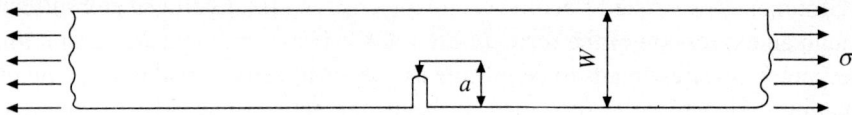

$$K_I = \sigma\sqrt{\Pi\alpha} \cdot F_I(\alpha) \cdot \alpha = \frac{a}{W}$$

$$F_I(\alpha) = 1.12 - 0.231\alpha + 10.55\alpha^2 - 21.72\alpha^3 + 30.39\alpha^4$$

Fig. 4. Single edge cracked plate tension specimen.

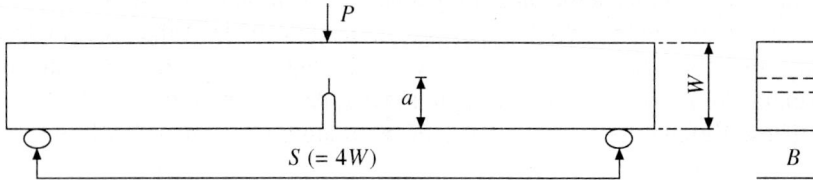

$$K_I = \frac{3SP}{2BW^2}\sqrt{\Pi a} \cdot F_I(\alpha) \cdot \alpha = \frac{a}{W}$$

$$F_I(\alpha) = \frac{1.99 - \alpha(1-\alpha)(2.15 - 3.93\alpha + 2.7\alpha^2)}{(1 + 2\alpha)(1 + \alpha)^{3/2}}$$

Fig. 5. Single edge cracked three point bending.

$$G = K_I^2(1 - v^2)/E \quad \text{(for plane strain)} \tag{10}$$

The expression for G per unit plate thickness for infinite plate with crack length $2a$ is

$$G = \pi\sigma^2 a/E \tag{11}$$

and for compact tension specimens it can be taken to be

$$G = 1.73\, p^2(2W + a)/E(W - a)^3 \tag{12}$$

where W is the width, a the crack length and P is the load.

Elastic Plastic Fracture Mechanics

The restrictions of small scale yielding places a severe limitation on the application of linear elastic fracture mechanics. This restriction excludes lower strength materials. However, several approaches have been suggested over the years to extend linear elastic fracture mechanics to situations where plastic zones are larger as in plain stress condition [13]. K_I-field solution in general cannot be utilised for large scale yielding conditions. The attempts to extend the linear elastic fracture mechanics to elastic plastic regime has led to the most important development of the J-Integral method of analysis. It can be viewed as a direct extension of the method of LEFM into elastic plastic regime. This approach is reviewed by Paris [14] and Ritchie [11]. J-integral first introduced by Rice [15] and developed further by Begley et al. [16] can be defined for any closed contour around a crack tip as path independent line integral for two dimensional problem.

$$J = f_r\, \text{wdy} - f_{ti}\, \delta ui/\delta x\, ds$$

where t is traction vector perpendicular to r and w is energy density.

By closing the contour to fall within the region dominated by K_I-field for small scale yielding, J can be directly related to strain energy release rate G and hence to stress intensity K_I for linear elastic behaviour as

$$J = G = K_I^2/E \quad \text{plane stress} \tag{13}$$

$$J = G = K_I^2(1 - v^2)/E \quad \text{plane strain} \tag{14}$$

Rice [12] has further showed that the J-integral can be interpreted as the difference in potential energy between two identically loaded bodies having infinitesimally differing crack lengths da

$$J = 1/B \cdot d \, (PE)/da \tag{15}$$

where PE is the potential energy, B the thickness and a the crack length. The energy interpretation of J is shown in Fig. 6. J can be evaluated as a function of load point displacement by measuring the

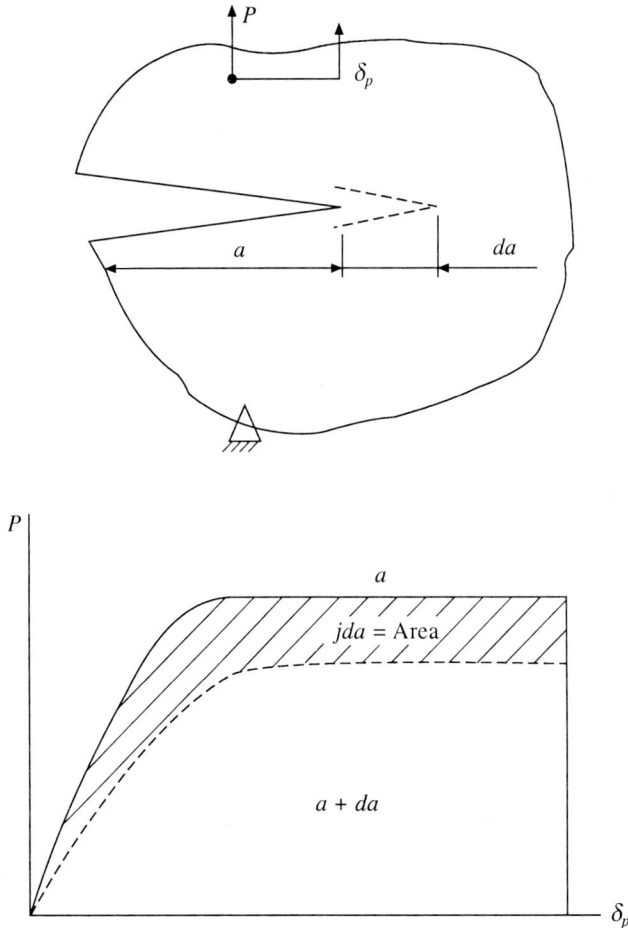

Fig. 6   A body with a crack subjected to a load.

difference in energy (at constant displacement) between specimens differing only in crack length. The critical value of J is taken at the point of initial crack extension and is labelled J_{IC}. A standard procedure for determination of J_{IC}, which can be used as toughness value at the initiation of crack growth for metallic materials has been developed by ASTM [14]. The expression used for calculating J is given as

$$J = A/Bb \cdot f(a/W) \qquad (16)$$

where A is the area under load point displacement record in energy units. Abramson et al. [10] calculated J using the expression

$$dJ = 2/b(1 + 0.26b/W)Pda - 1/b(1 + 0.76b/W) \; Gda$$

Similar to the conditions for K_I-dominance (small scale yielding) and valid K_{IC} measurement in linear elastic analysis, certain size requirement must be met for the J analysis to be relevant. In case of compact tension (CT) specimen $B \geq 25 \; J_{IC}/\sigma_y$ [17]. By analogy with stable mechanical crack growth rate, J_{ISC} is the value of J at which crack growth starts, and J_{SS} (steady state) is the value where crack growth can continue without further increase in J. A dimensionless parameter T, in terms of slope of initial J-a, has been proposed as a measure of resistance to crack growth [18]. This T is known as tear modulus.

$$T = (E/\sigma_y^2)d_j/da \qquad (17)$$

There are a number of other parameters such as C^*, σ_{net} and effective displacement rate $(dl/dt)_{eff}$ which have been used for characterisation of crack growth behaviour. C^* and σ_{net} have been used initially in creep crack growth studies [19, 20]. All these parameters have been used in stress corrosion crack growth studies [21]. The expressions for compact tension specimens are:

$$C^* = 2P/B(W - a). \; (dl/dt) \qquad (18)$$

$$(dt/dt)_{eff} = dl/dt \; (2W + a) \; (W - a)/2 \; (W^2 + aW + a^2) \qquad (19)$$

Effective extension rate is presumed to be equal to extension rate at crack tip.

For net section stress, the following expressions are given

$$(\sigma_n)_e = 2P(2W + a)/B(W - a)^2 \qquad (20)$$

$$(\sigma_n)_p = P[W + a + 2(W^2 + a^2)]/B(W - a)^2 \qquad (21)$$

$(\sigma_n)_e$ and $(\sigma_n)_p$ are based on the assumption that the material is either a perfectly elastic body or a perfectly plastic body respectively.

CRACK LENGTH MEASUREMENT

Accurate measurement of crack length is of utmost importance in fracture analysis. A detailed review of the different methods used is given in the book, "The measurement of crack length and shape during fracture and fatigue" [22]. Some guidelines to the selection of the techniques described by Richards [23] are given below:

Optical technique
Optical methods largely rely on the use of microscopes or telescopes. Crack length measurement is

often aided by markings etched or scribed on the specimen surface. The method can be applied to almost all test piece geometries. It is relatively inexpensive and it does not require calibration. However, surface measurements only are provided and underestimates of crack length would result because crack curvature is not accounted for.

Compliance methods

The basis of these methods is to measure the crack opening displacement (COD) between points along the loading line or, in the case of bend and wedge opening specimens, at the front face. The displacements are mostly measured by clip gauges or linear voltage displacement gauges. For high temperature testing, extension arms may be used to transfer the displacements to regions of lower temperature where conventional gauges can be used. The specimen does not have to be visually accessible and can provide an 'average' crack length where crack front curvature occurs. The method is easily incorporated in automatic crack length measurement systems. Disadvantages of this method are that separate calibration tests may be necessary in some cases, and this technique is only applicable to specimens where time dependent, time independent and reversed plasticity effects are small.

Compliance methods—Back face strain (BFS) measurement

Strains are measured on the back-face of CT or WOL specimens by strain gauges or possibly clip gauges or transducers in large test pieces. The method does not require the specimen to be visually accessible and can provide an 'average' crack length value. The method is probably the most sensitive one available provided the specimen behaves in a linear elastic manner and under favourable conditions crack length increases of the order of 10 microns have been resolved. However, the technique should be applied only to specimens where time dependent, time independent and reversed plasticity effects are small.

Compliance methods—Crack tip strain measurement

The strain close to but behind the crack tip is measured using surface mounted strain gauges. The method is suitable only for bend specimens. The initiation of crack growth is detected even in the presence of large scale plasticity. Problems occur in thick specimens where events on the surface of the specimen do not reflect crack growth in the interior.

Electrical methods—Strain gauge filaments

Electrically conducting wires are attached to the specimen in such a way that they are broken by an advancing crack thereby producing stepwise changes in resistance. The method can be used for all specimen geometries. Only surface measurements are made and the technique is not well suited to high temperature and aggressive environments.

Electrical methods—D.C. potential difference (resistance)

A constant D.C. current is passed through a specimen in such a way that a change in crack length alters the potential difference of suitably placed contact points usually in the vicinity of the crack tip. The method is suitable for all test piece geometries. 'Average' crack length values can be obtained. The system is highly stable and well suited for automatic control and long term, high temperature testing. Small relaxations from linear elastic behaviour are easily accommodated. However, there is some uncertainty in stress corrosion and corrosion fatigue studies over its use due to the possible interference with electro-chemical conditions adjacent to the crack tip. For decreasing K/constant COD tests, the crack faces may shot electrically thus producing under-estimates of crack length.

Electrical methods—A.C. potential difference (resistance)

A constant A.C. current is passed through a specimen in such a way that a change in crack length alters the potential difference of contact points suitably placed in the vicinity of the crack tip. The method is suitable for all test piece geometries.

The calibration for different specimen geometries is simple since there is a linear relationship between output and crack length and there is no specimen size dependence. Some relaxation from linear elastic behaviour is easily accommodated.

Limitations of the technique are that connecting wires have to be carefully placed and must not be moved during tests because of lead interaction effects. Although considerable improvements have recently been made, there is a requirement for high stability in the electronics and long term stability may be difficult to achieve. Bridging of crack surfaces by corrosion products may produce erroneous crack length readings. Electrical insulation of specimens is required.

Eddy currents

An eddy current probe positioned adjacent to a cracked surface produces an electrical signal indicating crack growth. One system involves a servo-system or stepping motor that moves the probe in such a way that a nil eddy current signal is maintained. It has been used for centre cracked sheets but should be adaptable to other geometries. The technique is expensive and enables only surface measurements.

Ultrasonic methods

Ultrasonic methods involve the transmission and reception of a pulsed ultrasonic beam intersected by a crack. For the amplitude calibration procedures the probe is fixed and the growing crack alters the relative amplitudes of the reflected and transmitted signals. Either signal can be calibrated against crack size. The probe displacement method is similar to the amplitude calibration procedure except that a motor drives the probe in step with the growing crack maintaining constant strength of ultrasonic signal. In another variation, the time of flight of the ultrasonic pulse from the transmitter to the receiver via the crack tip is measured. Simple trigonometric functions are used to calculate crack length.

Embedded cracks and crack profiles can be obtained using ultrasonic methods. Incorporation in automatic systems is relatively easy and non-metals can also be studied. Relaxation from linear elastic behaviour is readily accommodated and accuracies of + 0.2 mm can be achieved. The probe displacement variation does not require calibration and the time-of-flight variation is simply calibrated and very versatile.

Ultrasonic methods are not well suited to small, thin specimens. The methods are expensive and have not yet been developed for high temperature or aqueous corrosion studies. The amplitude calibration variation is sensitive only over small amounts of crack growth and careful calibration is necessary.

APPLICATION OF FRACTURE MECHANICS TO STRESS CORROSION CRACKING OF AUSTENITIC STAINLESS STEELS

Most of crack growth data for stress corrosion cracking has been generated using mode I. The crack growth rates vs K_I or J_I show a rapid increase (region-I) and then reach a stress-independent region (region-II) for all the metallurgical conditions reported in literature. This type of behaviour has been reported in many investigations [24–26]. Region-I is understandable in terms of the stress increasing the crack growth rate due to the accelerated corrosion process at the crack tip. Kinetics of the corrosion processes such as passivation/repassivation and dissolution rate are influenced by slip step generation rate at the crack tip. In region-II, the influence of mechanical factors reaches a plateau and the crack growth is controlled by environmental factors or it may also be because of lower effective K_I due to crack branching. Depending on the material/environment combinations, a large variation is possible in the plateau velocities.

Normal and Stabilised Grade Steels

Most of the corrosion studies on austenitic stainless steels involving the use of fracture mechanics have been carried out under the following conditions:

1. In high temperature water simulating the conditions of boiling water reactors (BWRs) and pressurised water reactors (PWRs).
2. In low temperature concentrated aqueous solutions of sodium chloride and magnesium chloride.

Andresen [27] measured the stress corrosion crack growth rates on 316 SS at 561 K in simulated water reactor environments, under fatigue and constant load conditions. At constant K_I of 27.5 MPa.m$^{1/2}$ the growth rate was 4.2×10^{-11}ms^{-1} which was about a factor of 8 lower than that for the sensitized 304 SS [28]. The measured crack growth rates were in good agreement with the prediction made by the same author [28] using slip dissolution model discussed later. Using fracture mechanics approach, Jewett et al. [29] measured the crack propagation rates of sensitized 304 SS in BWR environment at 561 K. The measured crack growth rate at 17.3 MPa.m$^{1/2}$ was 7×10^{-11}ms^{-1}. They reported that addition of hydrogen completely eliminated the crack initiation and propagation. In similar studies, Park et al. [30] measured the crack propagation rates in water at 502 K containing 8 ppm of dissolved oxygen. The crack propagation rates in 304 SS at 33 MPa.m$^{1/2}$ increased from 1.2×10^{-10}ms^{-1} to 2.2×10^{-10}ms^{-1} as the degree of sensitization increased from value of 1.4 C/cm^2 to 1.8 C/cm^2. Kawakubo et al.[31] showed that low frequency cyclic loading and slow strain rate tests (SSRT) accelerated the crack growth rates compared to static loading. These facts indicated that the SCC growth might be controlled by plastic deformation at crack tip, rather than by stress itself. From this point of view, the authors carried out SCC growth rate during SSRT, using centre-notched [32] and compact tension [31] thin plate specimens. They carried out the SSRT tests on sensitized AISI 304 stainless steel in oxygenated water at 523 K using extension rates between 5 nm/s and 167 nm/s. The crack growth data was analysed using different fracture mechanics parameters. They found that SCC growth rates are better expressed by dJ/dt parameter which relates to the crack tip deformation rate. The growth rates of intergranular and transgranular stress corrosion cracks in stabilized austenitic stainless steels exposed to 288°C water have been measured, using fracture mechanics techniques [33]. The parameters studied are stress intensity, alloy composition, sensitization and the increase of hardness and yield strength due to cold work. The fact that cold work of austenitic stainless steels can result in transgranular stress corrosion cracking in BWR water has been observed in several boiling water reactors, including the examples of Ringhals and Oskarshamn [34]. The cold work causes the increase in hardness. It has been suggested that the effect of hardness on the crack growth rates can be used to predict crack growth rates and residual lifetimes of BWR pipes and other non-sensitized components. This shed some light on irradiation assisted stress corrosion cracking because neutron irradiation results in increasing hardness which may be the reason for the increased susceptibility to stress corrosion crack growth.

Speidel [35, 36] applied the fracture mechanics approach for stress corrosion crack growth studies in austenitic stainless steels in aqueous solutions using DCB type of samples on Fe-Mn-Cr alloy in water with addition of different halides. Later he extended [37] these studies to Fe-Cr-Ni alloys in 22% boiling sodium chloride solution at 378K. The investigation included a wide range of chemical compositions, microstructures and yield strengths. It is reported that magnesium chloride is more aggressive compared to sodium chloride . This is attributed to high temperature and high chloride contents of magnesium chloride. Sensitization resulted in increase of crack propagation rates and decrease of K_{ISCC}. Molybdenum addition has a strong beneficial influence on SCC resistance because its addition not only prevents localised corrosion but also prevents the growth of preexisting cracks under SCC conditions. Other important conclusion from these studies was that the beneficial effect of nickel addition found in magnesium chloride solution (copson curve) was also observed in sodium chloride. Nickel addition has more influence on K_{ISCC} than on plateau crack growth rate. Russell et al. [24] carried out detailed investigations on 25% cold worked 316 SS in magnesium chloride solution. They found that in region-II, velocities were independent of K_I and were indifferent to small variations in potential. The activation energies (63 to 67 kJ/mol) were considered high for diffusion control in liquid phase. Lefakis et al. [25] had earlier measured the crack propagation rates in thin (0.45 mm) specimens in boiling magnesium chloride solution and found the K_{ISCC} corresponding to a crack growth rate of $4 \times 10^{-9} ms^{-1}$ as 11 MPa.m$^{1/2}$. Baladon et al [26] have determined K_{ISCC} and measured crack propagation rates of austenitic and austeno-ferritic stainless steels in boiling MgCl$_2$ solution. They obtained two plateau regions in K_I vs crack propagation rates curves. Such observations had been made by earlier workers [38] under similar conditions of testing. Based on the measurement of anodic dissolution rates (0.065–0.085 mm/h for 304 SS and 0.065 to 0.1 mm/h for 316 SS), Balladon et al. [26] emphasized the role of anodic dissolution in crack propagation mechanism.

Some studies have been carried out on austenitic stainless steels under modes I, II and III in MgCl$_2$ and under hydrogen charged conditions [39–41]. In these studies, only initiation part has been investigated mainly to understand the mechanism. It is generally agreed that tensile stress is a necessary condition for stress corrosion cracking. But in one case, it is reported that SCC crack can grow under compressive loading [42]. However, incubation periods were found to be ten to hundred times longer than those under tensile stress.

Nitrogen Containing Steels

The resistance to localized corrosion of austenitic steels can be increased, without sacrificing the strength levels, by lowering the carbon content and increasing the nitrogen content in them [43]. The improved resistance to localized corrosion attacks has made nitrogen-added austenitic stainless steels a candidate construction material in a wide variety of industries. In spite of large amount of data published on SCC of austenitic alloys, very little systematic work has been carried out on SCC crack growth behaviour of austenitic SS with high nitrogen contents by using the fracture mechanics approach. Most research on the SCC of austenitic stainless steels using a fracture mechanics approach has been carried out in magnesium chloride solution . However, magnesium chloride solution is reported to be insensitive to the effects of chromium depletion or impurity segregation in austenitic stainless steels [44]. On the other hand, a sodium chloride environment is found to be sensitive to the effects of chromium depletion or impurity segregation [45]. Moreover, sodium chloride environment is most often encountered in nature. Although the SCC behaviour of base metal of austenitic SS has been thoroughly investigated, studies on weld metal and welded joints have been few and far between. The differences in SCC behaviour between the base and weld metals depend on the chemical composition, environment and testing techniques [46, 47]. In NaCl and HCl solutions, base metal showed better SCC resistance for base metal vis-à-vis the weld metal [47, 48]. Delta-ferrite alters both the SCC resistance and the crack morphology of the weld metal. A systematic studies have been carried out in authors laboratory to generate crack growth data on nitrogen containing austenitic stainless steels, in various metallurgical conditions, and on weld metal has been generated using constant load, constant strain and slow strain rate (SSRT) testing techniques in acidified sodium chloride solution [49, 50]. Data generated on AISI 316 under similar conditions has also been included [51]. Crack growth analysis was carried out using stress intensity factor K_I and J-integral J_I described respectively in ASTM standard E-399 [6] and E-813 [17].

The chemical compositions of all the stainless steels of the present study are presented in Table 1.

Table 1. Chemical composition in weight percent of nitrogen-added AISI type 304 stainless steel, nitrogen-added AISI type 316 stainless steel and its weld metal and AISI type 316 stainless steel

Weight % of Element	C	Cr	Ni	Mo	Mn	Si	N	S+P
Nitrogen-added type 304 LN Stainless Steel	0.04	18.3	9.2	–	1.6	0.37	0.086	0.026
AISI type 316 Stainless Steel	0.054	16.5	11.4	2.3	1.7	0.64	–	0.031
AISI type 316 LN Base Metal	0.027	17.4	11.2	1.8	1.6	0.65	0.11	0.039
AISI type 316 LN Weld Metal	0.061	19.7	10.7	1.8	2.0	0.7	0.14	0.034

The results of the tests are discussed in terms of three parameters which are used to define the crack initiation and growth behaviour of a material viz. K_{ISCC}, J_{ISCC} and PCGR. The values of these parameters are listed in Table 2. In the present study, the applicability of the fracture mechanics approach has been assessed by calculating the maximum valid value of K_I (K_Q), based on linear elastic fracture mechanics, and J_I (J_Q), based on elastic-plastic fracture mechanics.

Table 2. SCC growth data for the three stainless steels in different metallurgical condition in 5 M NaCl + 0.15M Na_2SO_4 + 2.5 ml/l HCl

Parameter	SA NA type 304 stainless steel	Sensitised NA type 304 stainless steel	SA type 316 stainless steel	Sensitised type 316 stainless steel	Type 316 LN Base Metal	Type 316 LN Weld Metal
KISCC (MPa.m$^{0.5}$)	17.0	11.0	13.0	10.5	22.38	5.79
J_{ISCC}(kPa.m)	0.9	0.5	1.0	0.6	2.601	0.233
da/dt(m/s)	1.3E-8	2.3E-8	4.0E-9	1.0E-8	1.75E-9	2.8E-8
K_Q (mpa.m$^{0.5}$)	18.0	240.0	11.2	13.28	18.0	24.0
J_Q (kPa.m)	17.0	230.0	74.8	84.0	112.0	122.0

SA: Solution annealed; NA: Nitrogen added.

Crack Growth Behaviour of AISI Type 304LN Stainless Steel and AISI Type 316 Stainless Steel

Sensitization decreases K_{ISCC} and J_{ISCC} and increases the PCGR in both the types of austenitic SS, as shown in Table 2 and Figs. 7 and 8, respectively. This can be explained on the basis that in Region I the crack growth rate is controlled by (i) the rate of slip step generation and (ii) the rate of corrosion/repassivation at the crack-tip. Sensitisation results in a weaker passive film in the regions along the grain boundary vis-a-vis solution annealed material. Hence, a much higher slip step generation rate is required to cause breakdown of passive film on solution annealed material which, in turn, would mean a requirement of higher threshold stress parameters. The higher PCGR for sensitized material could be explained on the basis that the environment at the crack-tip has a low pH and high salt concentrations.

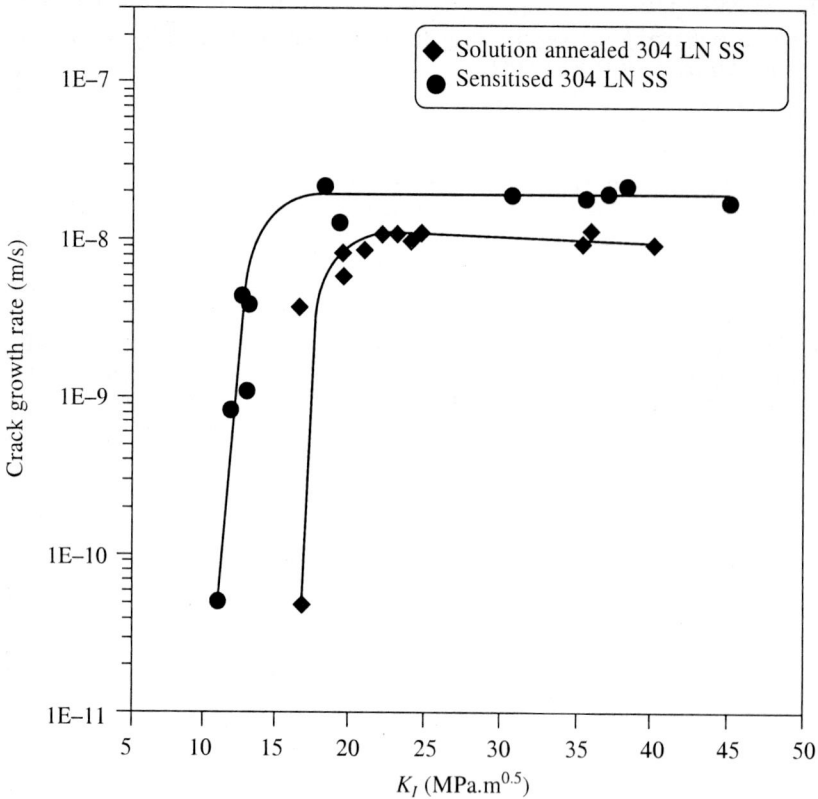

Fig. 7. K_I vs da/dt curve for annealed and sensitized nitrogen-added type 304 LN stainless steel.

Since the dissolution rate of the Cr-depleted regions is much faster, the required environmental crack-tip conditions are achieved more rapidly in a sensitized material than in a solution annealed material. This leads to higher PCGR. Also, in sensitized material, the presence of pre-existing active paths coupled with lesser stress relaxation, due to lesser branching, leads to higher PCGR. The values of K_I were valid only in Region-I while the values of J_I were valid over the whole range of values for both the nitrogen added type 304 LN and type 316 SS. Crack initiation and propagation occurred by transgranular SCC (TGSCC) in both the sensitized and solution annealed conditions, in both types of SS. In the sensitized condition, both the steels showed a transition to intergranular SCC (IGSCC).

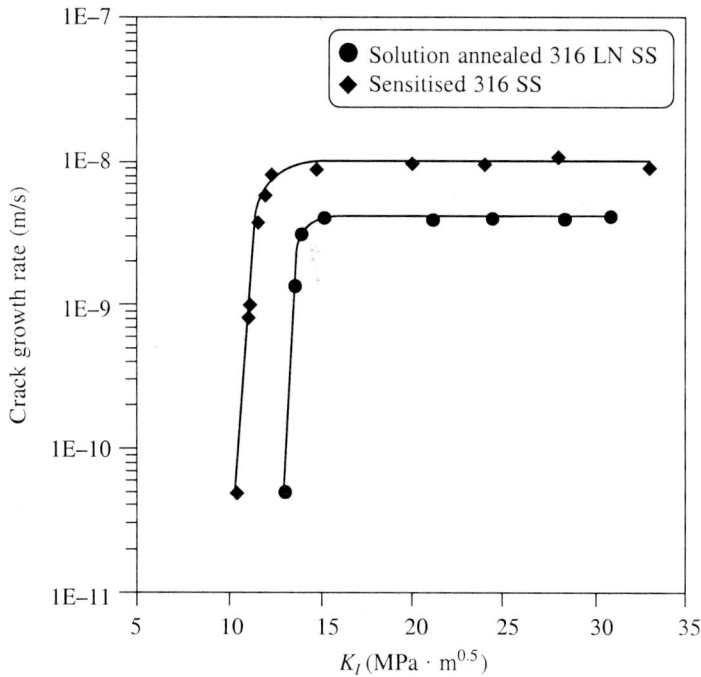

Fig. 8. Dependence of *da/dt* on K_I for solution annealed and sensitized type 316 stainless steel.

Crack Growth Behaviour of Nitrogen Added AISI Type 316 LN Stainless Steel and Its Weld Metal

The plot of average crack growth rates (*da/dt*) versus K_I and J_I for base and weld metals is shown in Fig. 9. The weld metal shows lower K_{ISCC} and J_{ISCC} than the base metal. This could be explained based on microstructural and microchemical heterogeneities in the weld metal caused by the presence of delta-ferrite which leads to partitioning of major alloying elements—ferritizers, such as Cr and Mo, in delta-ferrite and austenitisers, like Ni, C, N, in austenite. Also, segregation of S and P occurs at the delta-ferrite/austenite interface. The weld metal has a high concentration of microscopic defects, such as dislocations and vacancies and macroscopic defects, such as inclusions. All these result in formation of uneven and weakly adherent passive film on the weld metal surface as compared to base metal which in annealed condition has an even distribution of the alloying elements in the matrix.. The amount of slip steps required to be generated would be lesser to break the uneven and weakly adherent passive layer present on the weld metal, which could be attained at lower values of K_{ISCC} and J_{ISCC} and easier crack initiation would result in it vis-a-vis the base metal.

The higher PCGR for the weld metal vis-à-vis the base metal can be explained based on the higher strength and lower ductility of the weld metal. Stainless steels with higher yield strength are known to possess poorer SCC resistanc [44, 45]. The higher PCGR of the weld metal, vis-à-vis base metal, could be explained as follows:depending on the applied load, there exists a stress concentration at the tip of an advancing crack. The stress distribution ahead of the crack causes plastic deformation in the region around of the crack-tip, where the yield stress of the material is exceeded. This plastic deformation causes crack blunting and leads to relaxation of stresses in the deformed region. As a result, an impediment in crack growth takes place. The larger the plastic zone, greater the crack-tip blunting and

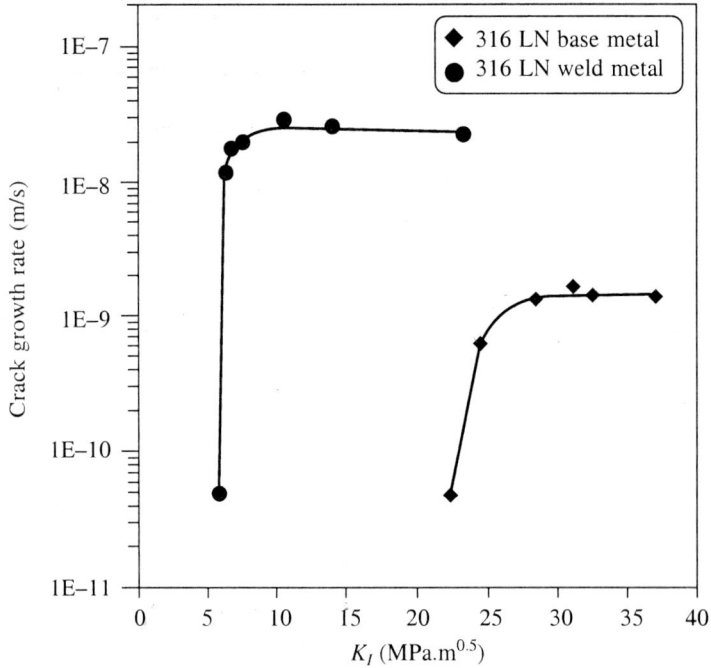

Fig. 9. K_I **vs** *da/dt* **curve for nitrogen-added type 316 L stainless steel base and weld metals.**

higher the resistance to crack growth. The size of the plastic zone ahead of the crack-tip decreases with increasing yield strength. Thus, weld metal, which has a higher yield strength than the base metal, experiences smaller plastic zone ahead of the crack-tip, resulting in lesser blunting of the crack and faster crack propagation. This leads to increased PCGR of the weld metal vis-à-vis base metal. The K_I and J_I values were valid over the whole range for weld metal. For base metal, the K_I values held no validity while J_I values were valid over the whole range of test. SCC initiation and propagation in both the base and weld metals was transgranular. No change in cracking mode was observed.

Comparison of the Crack Growth Data of Nitrogen Added Types 304 and 316 L Stainless Steel and Type 316 Stainless Steel

Figure 10 represents a comparison of the effects of K_I on the crack growth rates of nitrogen added AISI types 316 LN and 304 LN SS, and type 316 SS. It is seen that nitrogen added type 316 SS has higher K_{ISCC} and J_{ISCC} and lower PCGR than type 316 SS. This is because of lower carbon and higher nitrogen contents in the former. Lowering the carbon content increases the SCC resistance of the austenitic SS in NaCl and high temperature water environments [45]. Usually, SCC of austenitic SS in boiling acidified concentrated NaCl solution is reported to initiate through pits, which act as precursors to SCC. Thus, increase in resistance to SCC initiation can be explained based on improved pitting resistance. All the mechanisms proposed to explain improved pitting resistance of nitrogen added austenitic SS account for improved passive film stability. These theories include formation of ammonium ions, or nitrate/nitrite ions, blocking effect of nitrogen on the steel surface, nitrogen enrichment on the surface, and existence of nitrogen as Cr-N at metal surface or as a complex of ammonia or NO or as an ammonium salt. The chemistry of the environment at the crack-tip and that of pits are similar and this could explain the role of improved passive film stability in decreasing the crack growth rates of nitrogen added type 316 SS vis-à-vis type 316 SS. Nitrogen added types 304 SS show lower values of K_{ISCC} and J_{ISCC}, and higher PCGR than type 316 LN SS. This could be clearly attributed to the role of molybdenum in improving SCC resistance through its influence on improving pitting resistance by improving passive film stability [43].

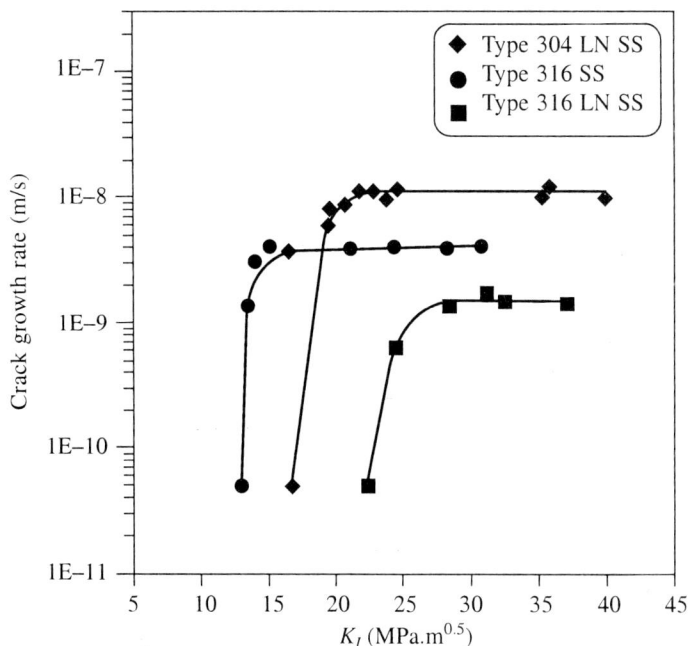

Fig. 10. **Comparison of variation of *da/dt* with K_I for the three stainless steels.**

Comparison of crack growth data of types 316 and nitrogen added type 304 LN SS gives an insight to the effects of Mo and N on the SCC behaviour of 18-10 austenitic SS. Type 316 SS shows a lower PCGR and lower K_{ISCC} and J_{ISCC} than nitrogen added type 304 SS. This suggests that nitrogen imparts a better resistance to SCC initiation, as compared to Mo, in 316 SS due to a tougher and more adherent passive film that it forms. However, Mo imparts better resistance to crack growth for an 316 SS. This could be explained based on the effect of nitrogen in lowering the stacking fault energy which, in turn, would promote planar slip. Planar slip is known to accelerate SCC crack growth. However, the presence of both these elements synergistically improves the SCC resistance of austenitic SS as evidenced by better SCC properties for nitrogen-added type 316 L stainless steel as compared to the other two steels.

Influence of Cold Work and Sensitization after Cold Work for 304 LN
The results on 10% and 20% cold worked and solution annealed materials are presented in Fig. 11. The threshold values are found to be significantly influenced by cold work and are somewhat lower than that of sensitized and solution annealed materials. The plateau velocities, however, are lower. Another difference is in the slope of the initial part of the curve. There is a slow increase in the crack growth rates with increasing K_I, unlike the sharp increase for solution annealed and sensitized material.

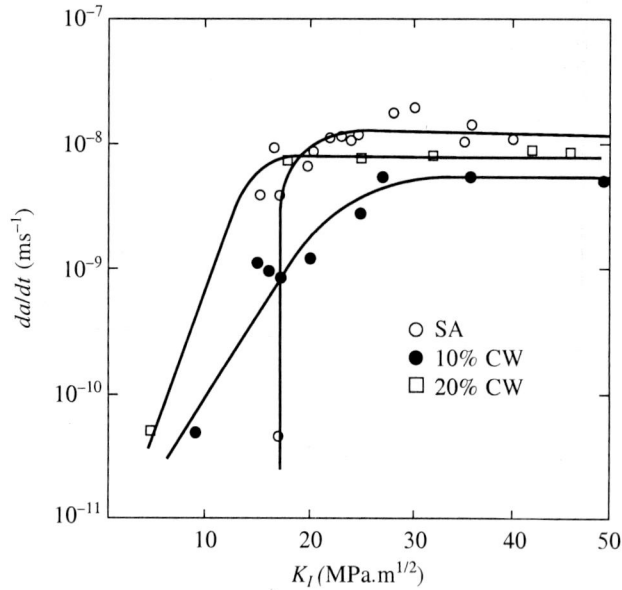

Fig. 11. Influence of cold work on crack growth rate of 304 LN SS as a function of stress intensity factor in 5 M NaCl + 0.15 M Na$_2$SO$_4$ + 3ml/l HCl at 381 K.

The results of the sensitized material with and without prior cold work are presented in Fig. 12. Threshold values are higher than the cold worked materials, indicating that the influence of cold work is overshadowed by the effects of sensitization. The slope of the initial part of the curves is similar to

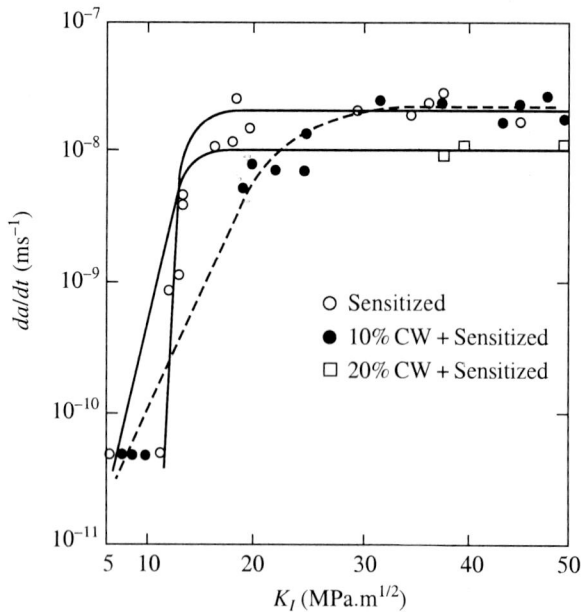

Fig.12 Influence of cold work + sensitization on crack growth rate of 304 LN SS as a function of stress intensity factor in 5 M NaCl + 0.15 M Na$_2$SO$_4$ + 3ml /l HCl at 381 K.

that of the cold worked material. The maximum crack growth rates in 20% cold worked + sensitized material are the same as for the solution annealed material. It is seen that the threshold values are reduced by cold working to values below sensitized material. The plateau crack growth rates are lower than for solution annealed material. However, 20% cold worked samples exhibit higher crack growth rates than those with 10% cold work.

Depending on the temperature and extent of deformation, cold working can affect the material in the following ways:

1. Cold work may cause a phase change from austenite to martensite.
2. Cold work induces residual stresses and increased defect density.
3. The dissolution rate may be increased due to increased strain energy.

Briant and Ritter [52, 53] have observed transformation to martensite from austenite at low temperatures. However, Muraleedharan et al. [54] and Seetharaman et al. [55] have not observed any appreciable transformation to martensite on deformation of 304 and 316 stainless steels at room temperature. In the present work, magnetic measurements and metallographic examination did not indicate any martensite in the steel used. Cold rolling introduces higher strains at the surface than at mid thickness resulting in residual stresses. Machining reduces the residual stresses due to removal of highly strained surface material. Potentiostatic measurements [56] have shown that corrosion rates increased with an increase in deformation. Therefore, it is concluded that an increase in dissolution rate and production of lattice defects are responsible for decreased threshold values on cold working. The lattice defects provide nucleation sites for crack initiation and therefore, cracking is possible at lower stress levels.

Within the crack tip plastic zone, the microstructure of the cold worked material plays an important role and it might not be significantly different from that of the solution annealed material at high stress levels. However, the stronger material might restrict liquid mass transport to the crack tip due to less load line displacement resulting in lower corrosion rate and hence lower crack propagation rates. Cold deformed AISI 316 stainless steel showed increased time to failure in $MgCl_2$ solution [54]. However, the higher plateau velocities for 20% cold work condition cannot be explained based on this argument. Also, no suitable explanation can be offered for the decreased slope of the initial part of the curve for the cold worked material.

The plateau crack growth rates for the sensitized material with and without 10 % cold work were found to be the same. However, the threshold values for 10% cold worked material were slightly lower. The material with 20% cold work showed a lower threshold and also lower plateau growth rates. Cold working is known to promote sensitization. Studies carried out in our laboratory on cold worked and sensitized material using flat and round samples showed that SCC susceptibility is related to enhanced degree of sensitization due to cold work [57, 58]. It is the extent of sensitization, whether caused by long ageing time or promoted by cold work, which is responsible for decreased resistance to SCC. As seen in the present work, 10% cold worked samples after sensitization showed the same crack growth rates as sensitized samples without prior cold work. Stress relieving during sensitization treatment is of minor consequence. Defect structure produced by cold work plays a role at low stress levels as reflected in lower threshold values. The slight differences in threshold values and crack growth rates observed in samples with higher cold work levels (20% cold work) are related to a decreased degree of sensitization due to carbide precipitation within the grains also. On comparing the threshold values of sensitized and cold worked materials, it is seen that cold working is more harmful than sensitization. Sensitization of cold worked material showed some improvement in SCC resistance with respect to threshold values. This is due to stress relieving and decrease in defect density during sensitization treatment.

On the Mechanism of Stress Corrosion Cracking

Further investigation were carried out to understand the mechanism of cracking. The activation energies measured in the temperature range of 363 to 381 K, using an Arrhenius rate equation, for solution annealed, sensitized and 10% cold worked materials were found to be in the range of 50 to 65 kJ/mol. Fractographic observations at high magnifications for solution annealed material showed well developed fan pattern. Crystallographic pits, secondary cracking and serrations on river lines were also seen. A number of parallel lines and cracks perpendicular to river lines are also seen (Fig. 13). The distance between the parallel lines varied between 1 and 7 microns in different micrographs. Such striations have been reported to be associated with crack arrest [59, 60]. The rate of emission of acoustic signals, measured in the plateau region, was found to be constant throughout the crack growth period. A typical record of the background noise level and acoustic signal for a solution annealed material is shown in Fig. 14. Time period per event and crack growth per event computed from acoustic emission and crack growth data have been presented in Table 3. The crack growth per event varies from less than a micron for a solution annealed material to 15 micron for 10% cold worked material.

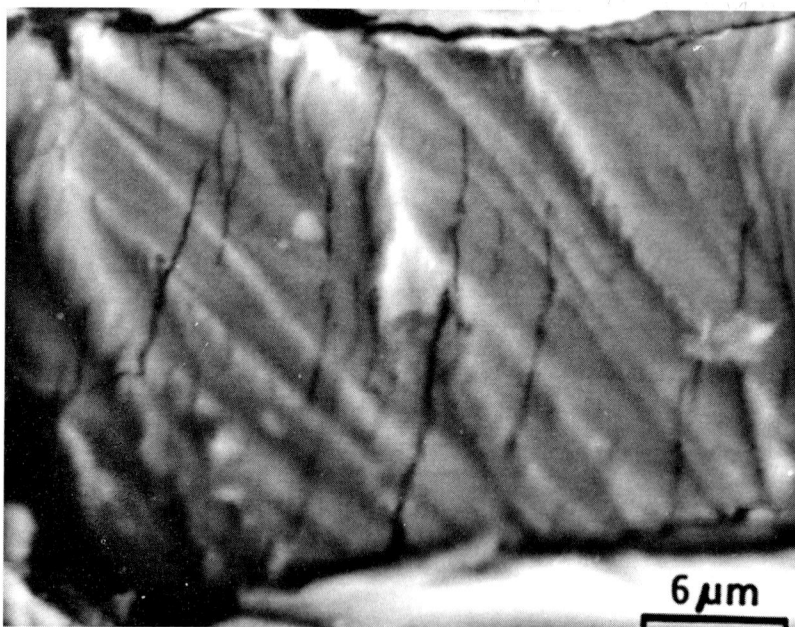

Fig. 13.   SEM fractographs of solution annealed 304 LN SS showing crack arrest marks.

The acoustic data indicated that the crack growth is discontinuous. Evidences of discontinuous crack growth have been widely cited in literature [59, 60, 61, 62]. In the plateau region crack growth per event is found to vary from less than a micron to 15 microns. These observations and crack arrest marks on the fracture surface (1 to 7 microns) suggest that cracking occurs by discontinuous jumps of the order of a few microns. The difference between fratographic features and acoustic signals may be due to (a) some crack arrest marks are dissolved and/or (b) some signals being missed due to threshold of 80 dB used in the present investigation. The critical question is, "What is the source of the acoustic events?". Gerberich et al. [63] attributed it to fracture of ligaments behind the advancing intergranular crack. These transgranular ligaments were observed on the fracture surfaces. However, no evidence of such ligaments was found in the present investigation. If deformation processes such as dislocation multiplication and twinning are considered to be the source, then the event rate would have increased with increasing K_I. Formation of river lines has been explained to be due to fracture of ligaments between two neighboring planes. Such a source is not likely to give detectable signals above 80 dB. The source, therefore, is probably related to some sort of microcleavage taking place in between the crack arrests. Identification of cleavage and parameters controlling it has been the main aim of many investigations [64-66].

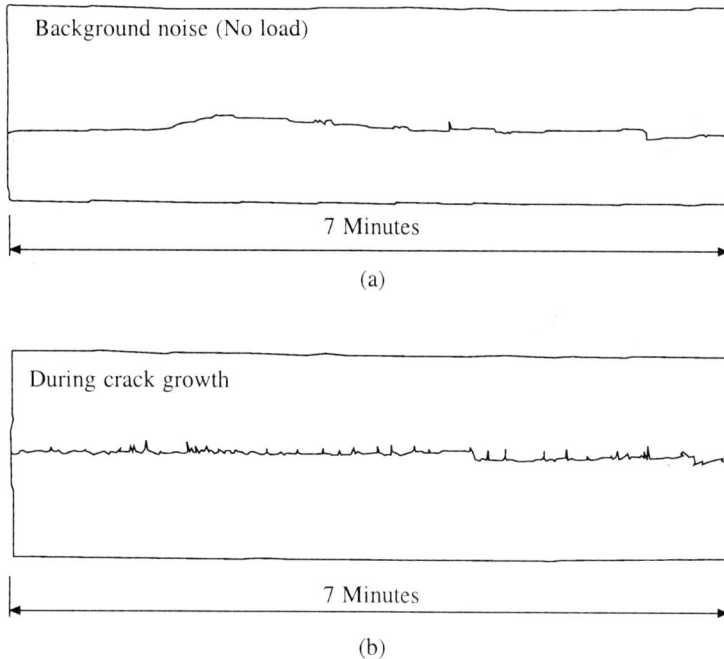

Background noise (No load)

7 Minutes

(a)

During crack growth

7 Minutes

(b)

Fig. 14. Typical record of acoustic events for 304 LN SS in solution annealed condition during crack growth

It is generally agreed that stress corrosion cracking involves the generation of a bare surface by plastic flow and is sensitive to the rate at which the fresh surface is generated. Stress corrosion is regarded as a balance between dissolution rate and rate of creation of new surface. However, actual crack growth rates are generally faster than would be predicted by Faraday's law and also dissolution alone cannot explain fractographic features. A brittle component, as mentioned above, along with anodic dissolution has been proposed by many authors [64–66]. The role of hydrogen in the brittle component of the SCC in austenitic stainless steels has been widely discussed. In this case, both anodic dissolution and microcleavage are considered. The source of hydrogen is the corrosion reaction at the crack tip. Mechanisms such as reduction in cohesive strength and enhancement of dislocation mobility at very low stress levels have also been proposed [67]. The activation energy for crack growth measured in the present investigation in solution annealed, sensitized and 10% cold worked materials

Table 3. Acoustic Emission Data Measured During Crack Growth

S. No.	Material Condition	a-SCC (mm)	da/dt (ms⁻¹)	da/dN (um/N)	Time period per event (sec.)	K_I (MPa. m^{1/2})	J_I (KPa.m)
1.	SA	1.85	1.10×10^{-8}	0.55	50	33–46	9.7–70
2.	SA	1.65	1.60×10^{-8}	1.15	72	24–30	65–130
3.	SA	1.86	7.7×10^{-9}	6.35	1716	27–33	10–13
4.	SA + sensitized 923 K/20h	2.90	8.6×10^{-9}	2.30	375	15–25	1–6.5
5.	10% CW	0.66	7×10^{-10}	0.27	380	18–20	2.6–5
6.	10% CW	5.32	4.8×10^{-9}	15	2870	43–80	15–107
7.	10% CW	3.70	3.6×10^{-9}	11	3120	24–43	6.7–27
8.	10% CW + sensitized 923 K/20 h	6.4	6.8×10^{-9}	2	293	18–23	0.5–11
9.	As 8 above		1.7×10^{-8}	10	514	17–27	1.5–419

SA: Solution annealed, CW: Cold worked.

was found to be in the range of 50 to 65 kJ/mol. These values correspond to hydrogen diffusion in iron and austenitic steels [68]. Hydrogen diffusion distances per event time for plateau region of crack growth curves calculated using the equation $X = (Dt)^{1/2}$ given by Louthan et al. [68] are presented in Table 4 along with the acoustic data for the K_I-independent region. A diffusion coefficient of 1.76×10^{-10} cm²/sec for hydrogen diffusion in austenitic stainless steels measured by the same authors [68] was used. During the cracking event hydrogen can diffuse to a distance more than the measured crack growth rate. Also, it has been shown that hydrogen transport rates in association with dislocation motion can be several orders of magnitude greater than that associated with lattice diffusion [69, 70].

Table 4. Acoustic data in K_I independent region

S. No.	Material condition	K_I range (MPa.m^{1/2})	Time period/event (sec.)	a/N (μm/event)	Hydrogen diffusion distance (μm/event)
1.	Solution Annealed	33–46	50	0.55	13
2.	As 1 above	24–30	72	1.15	159
3.	Sensitized	15–25	375	2.30	360
4.	10% CW	43–80	2870	15.00	1000
5.	10% CW	24–43	3120	11.00	1040

It is also possible that martensite formed at the crack tip increases hydrogen diffusion due to a higher diffusion co-efficient and thus increases crack propagation. But cracking in stable alloys such as AISI type 310 stainless steel shows that martensite formation is not a necessary condition for stress corrosion [59] and crack growth in hydrogen gas [71]. In the present investigation, magnetic measurements did not indicate the presence of martensite in the material (even after cold working) and on fracture surfaces. In the absence of martensite formation, hydrogen effects can be explained based on a decohesion concept [72] or locally enhanced plasticity [69, 70, 71, 73]. The influence of temperature and sensitization can be explained by their influence on corrosion rate thus influencing the hydrogen availability. Fractographic features (transgranular) are due to microcleavage caused by hydrogen. Sensitized material has shown intergranular failure which is similar to the observations of Briant [74] in hydrogen gas. Role of hydrogen in crack propagation through reduction in stacking fault energy (SFE) at crack tip has been reported by Jani et al. [75]. Hydrogen is the only environmental species which is capable of reducing SFE of austenitic steels. Based on the discussion above, it is concluded that hydrogen produced in the corrosion reaction played a significant role in the cracking process.

Summary of Results on LN Variety of Steels

1. Sensitisation of solution annealed base metal lowered K_{ISCC} and J_{ISCC} by about 60 to 70 %, and increased the PCGR by 2 to 3 times for types 316 LN and nitrogen added type 304LN SS.
2. K_{ISCC} and J_{ISCC} were about four times higher and PCGR were nearly one order of magnitude lower for the base metal vis-à-vis the weld metal of AISI type 316 LN SS.
3. Comparison of K_{ISCC} and J_{ISCC} and PCGR of base metal of nitrogen added type 316 L SS with corresponding values for type 316 SS, suggested the strong influence of nitrogen on improving the SCC properties of the former.
4. Cold working of solution annealed material further reduced the threshold values, below that observed for sensitized material. Threshold values for the sensitized material with prior cold work were lower than those for the sensitized material without cold work.
5. From the acoustic emission data it was inferred that the crack growth was discontinuous.
6. Activation energies measured for solution annealed, sensitized and 10% cold worked materials were found to be in the range of 50 to 65 kJ/mol. This correspond to hydrogen diffusion in stainless steels.

PROPOSED MODELS FOR LIFE PREDICTION

The crack propagation model based on slip dissolution proposed by Ford is summarised in [76].

Slip-Dissolution Model

The basis of slip dissolution model of crack propagation is that total charge density at the crack tip after the passive film rupture can be related paradoxically to amount of metal transfered to dissolved species or oxide i.e., it may be related to crack advance. Crack propagation rate will depend on the repassivation rate and frequency of film rupture at strain crack tip. The later parameter will be determined by fracture strain of the film and strain rate at the crack tip. Therefore, by invoking Faraday's law, the crack growth rate V_t is related to oxidation charge density passed between film rupture events, Q_f and strain rate at crack tip ε_{ct}

$$V_t = M/ZF\rho * Q_f \varepsilon_{ct}$$

where M is the atomic weight and density of crack tip model, F is Faraday's constant and Z the number of electrons involved in the overall oxidation of an atom of model

Because charge density varies with time depending on metal/environmental conditions, the equation has been written in more general form

$$a = A\ (\varepsilon_{ct})^n \tag{22}$$

where a is the crack growth rate (cm/s);

A and n are interrelated parameters which depend on the crack tip material and water chemistry conditions.

$$A = 2.924 \times 10^{-3}$$

$$n = 0.7612$$

$$\varepsilon_{ct} = \text{crack tip strain rate}$$

given by $4.1 \times 10^{-14} K^4$, where A and n are constants depending on material and environment conditions at the crack tip. The parameter n has been related to bulk parameter such as corrosion potential, anionic conductivity, degree of grain boundary sensitization and crack tip strain rate to stress intensity and various loading conditions [77].

For 304 stainless steel (EPR value between 15 and 30 C/cm^2) in high temperature water at 280°C, the parameters evaluated are $n = 0.55$, $a = 7.8 \times 10^{-3} n^{8-6}$ and a good correlation has been found with observed crack growth rates. However, there are certainly many doubts/questions in formulations of parameters, slip dissolution model has quantitative validity for several ductile alloy/aqueous environment system. Recently, crack tip strain rate due to creep formation have been calculated using fine element method and were employed in predicting SCC crack growth in welded pipe using the above model [78].

Methods Based on Crack Propagation

Several researchers have attempted to combine the different processes by means of simple models based on macroscopic crack growth data. These include process superposition, process competition and process interaction models. A brief review of these models is presented. These models combine influence of corrosion fatigue and stress corrosion cracking.

Process Superposition Model

This model was proposed by Wei [79] to decouple the mechanical and environmental components of crack growth rate. Modeling was based on the proposition that the rate of crack growth in a deleterious environment $(da/dN)_e$ is composed of sum of three components :

$$(da/dN)_e = (da/dN)_r \, (1 - \phi) + (da/dN)_c \, \phi + (da/dN)_{scc}$$

$(da/dN)_r$ = rate of fatigue crack growth in inert environment (pure fatigue)
$(da/dN)_c$ = cyclic dependent contribution which requires synergistic interaction of fatigue and environment
    (pure corrosion fatigue)
ϕ = fraction area of crack undergoing pure corrosion fatigue

In the absence of SCC contribution, equation also can be rewritten as

$$(da/dN)_{cf} = [(da/dN)_e - (da/dN)_r]$$

$$(da/dN)_{cf} = [da/dN)_c - (da/dN)_r] \, \phi$$

For $\qquad\qquad\qquad \phi = 0, \ (da/dN)_e = (da/dN)_r \ldots$ pure fatigue

For $\qquad\qquad\qquad \phi = 1, \ (da/dN)_e = (da/dN)_{e,s} \ldots$ maximum saturation limit

$$= (da/dN)_c \ \ldots \text{pure corrosion fatigue}$$

In essence, ϕ represents the materials response to changes in environmental conditions. It is equivalent to fraction of area coverage in gaseous environment and q (charge transfer/cycle)/qs (saturation change) for aqueous environment.

Process Competition Model

Here it is assumed that stress corrosion and mechanical fatigue processes are not additive. The crack will propagate by the easiest mechanisms possible. The borderline between true corrosion fatigue and a stress corrosion contribution to fatigue can be expressed as (da/dt) $1/f = C[K_p(1 - R)]^n$, where f and R represents frequency and ratio, respectively, at which the process domination changes, K_p is the stress intensity at the onset of stress corrosion plateau and da/dt is the plateau crack growth rate. Also for calculation of environmental contribution to corrosion fatigue, Paris equation $da/dN = C(\Delta K)^n$ has been used.

$$(da/dN)_e = C(\Delta K_{eff} + d\Delta K)^n$$

where $d\Delta K$ term is an adjustable parameter, representative of corrosion fatigue contribution.

Process Interaction Model

In alloys which are highly susceptible to SCC, a kind of interaction model has been proposed. This model allows interaction between different processes due to which one process may be inhibited or enhanced by the action of the other. An equation which allowed these interaction was developed [80]

$$(da/dN)_e = C(\Delta K_{eff})^m + {}_o f^f A \gamma \Delta K_{eff} \alpha dt$$

where ΔK_{eff} is effective stress intensity factor which is adjusted to account for blunting and microbranching etc. The term accounts for the influence of load cycling on SCC rate. A, c, m and α are constants which are experimentally determined.

EXTREME VALUE STATISTICAL METHOD FOR LIFE PREDICTION

Extreme value statistical method consists of finding out cumulative distribution function from the frequency distribution. Cumulative distribution as a function of critical parameters (pit depth or crack length) in turn is plotted on the commercially available probability plotting papers to find out the type of distribution (normal, exponential, weilbul distribution etc.). Once the type of distribution is known, extreme value i.e. Maximum value of the parameter (pit depth or crack length) can be determined. This procedure is discussed in the book by coordinating editor Masamichi Kowaka [81]. Flow chart below explains it systematically (Fig. 15). The extreme value evaluated from limited measurements can be corrected for the large number of measurements. These extreme values are evaluated for different intervals of time from data collected from plant experiences or laboratory measurements and linearly extrapolated for life prediction.

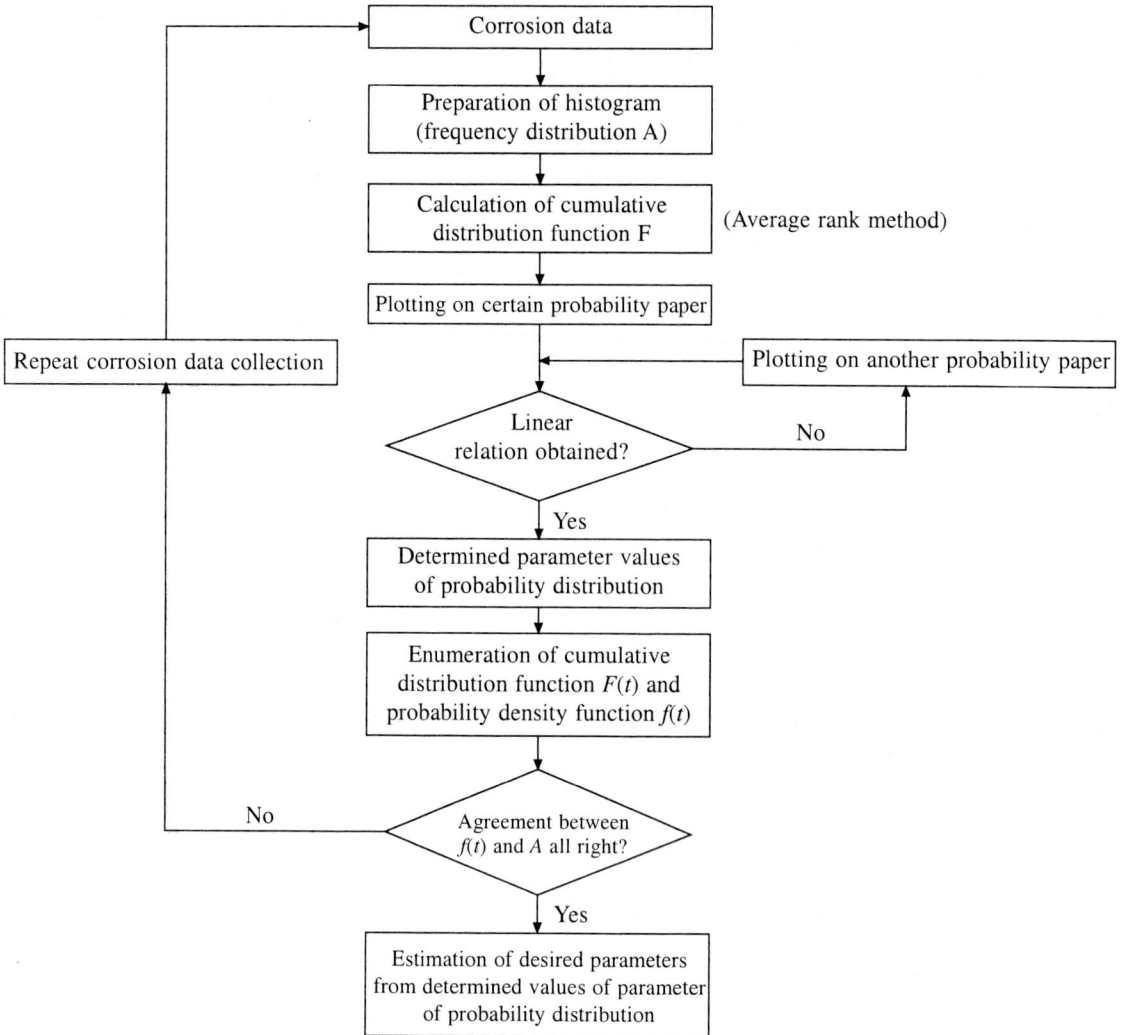

Fig. 15.   Flow chart for statistical processing of corrosion data.

REFERENCES

1. Khatak, H.S., Murleedharan, P., Gnanamoorthy, J.B., Rodriguez, P. and Padamanabhan, K., Evaluation of the stress corrosion resistance of cold rolled AISI type 316 stainless steel using constant load and slow strain rate tests. J. Nuclear Material, 1989, Vol. 168, pp. 157–161.
2. Baker, H.R., Bloom, M.C., Bolster, R.N. and Singleterzy, C.R. Corrosion 26 (1970), 470.
3. Turnbull, A. The solution composition and electrode potential in pits, crevices and cracks. Corrosion Science, 1983, Vol. 23, (8), pp. 833-870.
4. David Broik, Elementary Engineering Fracture Mechanics, Sijthoff and Noordhoff Alphen aan denRijn, The Netherlands 1978.
5. ASTM, STP 743, Fracture Mechanics, Editor Richards Roberts, 1980.
6. ASTM E 399-83, Test method for plane strain fracture toughness of metallic materials, American Society for Testing Material, 1986.

7. ASTM E 561-86, Practice for R-curve determination. American Society for Testing Material, 1986.
8. Speidel, M. O., Stress corrosion cracking of stainless steels in NaCl solutions.Metallurgical Transactions A, 12 A (1981) 779.
9. Jones, R.H., Analysis of hydrogen-induced subcritical intergranular crack growth of iron and nickel. Acta Met, 38,(1990) 1703.
10. Rhodes, D., Musuva, J.K., and Rodon, J.C., Significance of stress corrosion cracking in corrosion fatigue growth studies. Engg. Fract. Mechanics, Vol. 15, (1981) 407.
11. Ritchie, R.O., Trans. ASME, Journal of Engineering, Materials and Technology, 105 (1983) 1.
12. Murakami, Y., Ed., Stress Intesity Factors HandBook, Vol. 1, 1987.
13. Abramson, G., Evans, J.T. and Parkins, Investigation of stress corrosion crack growth in Mg alloys using J-integral estimations. Metallurgical Transaction A, (1985) 101.
14. Paris, P.C., in Flaw Growth and Fracture ASTM STP 631, 1977, page 1.
15. Rice, J.R., Journal of Applied Mechanics, 35 (1968), 379.
16. Begley , J.A. and Landes, I.D., In Fracture Toughness, ASTM STP 514, 1972, p. 1.
17. ASTM E 813-81, Test method for J_{ic}, a measure of fracture toughness. American Society for Testing Material, 1986.
18. Paris, P.C., Tada, Zahoor, A. and Ernst, H., In Elastic- Plastic Fracture, ASTM STP 668, 1979, p. 5.
19. Jones, P.L. and Tetelman, A.S., Characterisation of the elevated temperature static load crack extension behaviour of type 304. Engineering Fracture Mechanics 12, (1979) 79.
20. Curbishley, I., Lloyd, G.I., and Pilkington, R., Proc. Second International Conf. on 'Creep and Fracture of Engineering Materials', Swansea 1984, p. 913, part II, editors B.Wilschire and D.R.J. Owen.
21. Kawakubo, T. and Hishida, M., Transaction of ASME, J. Engg. Materials 107, (1985) 240.
22. Weevers, C.J., Ed. The Measurement of Crack Length and Shape during Fracture and Fatigue, EMAS Engineering Materials Advisory Services Ltd., United Kingdom, (1978).
23. Richards, C.E., Ibid. P. 461.
24. Russel, A.J. and Tromans, D., A fracture mechanical study of stress corrosion cracking of type 316 austenitic steel. Metallurgical Transaction A, 10A, (1979), 1229.
25. Lefakis, H. and Rostoker, W., Stress corrosion crack growth rates of brass and austenitic stainless steel at low stress intensity factors. Corrosion 33, (1977) 178.
26. Balladon, P., Freycenon, J. and Heritier, J., ASTM STP 743, 1980, p.167.
27. Anderson, P.L., Environmentally assisted growth rate response of non sensitized AISI 316 grade stainless steels in high temperature water. Corrosion 44, (1988) 450.
28. Anderson, P.L., and Ford, F.P., 'Predictive Capabilities in Environmentally Assisted Cracking, P. Rungta, ed. Nov. 17–22, 1985, DVP-Vol. 99, ASME New York, 1985.
29. Jewett, C.W., and Pickett, A.E., Trans. ASME., J. Engineering Materials and Technology, 108, (1988) 10.
30. Park, J.Y., Ruther,W.E., Kassner, T.F. and Schack, W.J., Trans. ASME, J. Engineering Materials and Technology, 108, (1986) 20.
31. Kawakubo, T. and Hishida, M., Trans. ASME, J. Engineering Materials and Technology, 107, (1985) 240.
32. Kawakubo, T. and Hishida, M., Crack initiation and growth analysis by direct optical observation during SSRT in high temperature water, Corrosion, 40, (1984) 120.
33. Speidel, M. O., Environmental degradation assessment and life prediction of nuclear piping made of stabilized austenitic stainless steels, Proc. Inter. Symp. On Plant aging and life predictions of corrodible Structures, May 15-18, 1995, Sapporo, Japan.
34. Tahtinen, S., Hanninen, H. and Trolle, M., Stress corrosion cracking of cold worked austenitic stainless steel pipes in BWR reactor water, Proc. of Sixth Intern. Symp. On Environmental Degradation of Materials in Nuclear Power Systems- Water Reactors Aug. 1993, San Diego, USA< TMS, pp. 265–275.
35. Speidel, M.O Stress corrosion cracking in Fe-Mn-Cr alloys Corrosion 32, (1976) 187.
36. Speidel, M.O. Corrosion 33, (1977) 199.
37. Speildel, M.O Stress corrosion cracking of stainless steels in NaCl solutions. Metallurgical Transaction A, 12, (1988) 779.
38. Robinson, M.J. and Scully, J.C., In Proc., 1973 Firminy Conference-Stress Corrosion Cracking and Hydrogen

Embrittlement of Iron Base Alloys, editors R.W. Staehle, J. Hochman, R.D. McCright and J.E. Slater, NACE, Houston, Tex., (1977), p. 1095.

39. Wu Yang Chu, He-Li Wang and Chi-Mei Hsiao, Mechanism of slow crack growth and stress corrosion cracking in austenitic stainless steel. Corrosion 40, (1984), 487.

40. Quiao, L.J., W.Y. Chu, Hsiao, C.H. and Lu, J.D., Stress corrosion cracking and H_2 induced cracking in austenitic stainless steel under mode II loading. Corrosion, 44, (1988), 51.

41. Quiao, L.J., H_2 induced cracking and stress corrosion cracking of austenitic stainless steel under mode III loading. Corrosion, 43, (1987), 479.

42. Wu Yang Chu, Jing Yao, and Chi-Mei Hsiao, Proceeding of International Congress on metallic Corrosion (ICMC-9) 1984, Vol. III, 550.

43. Dutta, R.S. De, P.K., and Gadyar, H.S., Corrosion Science, 34 (1993) 51

44. Shaikh, H., Khatak, H.S., Seshadri, S.K., Gnanamoorthy, J.B. and Rodriguez, P., Effect of ferrite transformation on the tensile and stress corrosion properties of type 316L stainless steel weld metal thermally aged at 873 K. Metall. Trans. A26 (1995) 1859.

45. Hanninan, H.E., International Metallurgical Reviews, 1979, Vol. 24, pp. 85–135.

46. Baeslack III, W.A., Dequette, D.A., and Savage, W.F., Welding journal, 58 (1979)168s.

47. Baeslack III, W.A., Dequette, D.A., and Savage, W.F., Corrosion 35 (1979) 45.

48. Shaikh, H., Khatak, H.S. and Gnanamoorthy, J.B., Stress corrosion cracking of weldments of AISI type 316 stainless steel. Werkst. Korros. 38 (1987) 183.

49. Khatak, H.S., Gnanamoorthy, J.B. and Rodriguez, P., Studies on the Influence of Metallurgical Variables on the Stress Corrosion Behaviour of AISI 304 Stainless Steel in Sodium Chloride Solution Using Fracture Mechanics Approach, Metallurgical and Materials Transaction A, Vol. 27A, May 1996, 1313-1325.

50. Shaikh, H., Schneider, F., Mummert, K., and Khatak, H.S., Stress corrosion crack growth behaviour of nitrogen added AISI type 316 stainless steel and its weld metal in hot chloride solution.

51. Vinoy, T.V., Shaikh, H., Khatak, H.S., Sivaibharasi, N. and Gnanamoorthy, J.B., Stress corrosion crack growth studies on AISI type 316 stainless steel in boiling acidified sodium chloride solution, J. Nucl. Mater., Vol. 238 (1996) 278–284.

52. Briant, C.L., Ritter, A.M. The effect of cold work on the sensitization of 304 stainless steel. Scripta Met., 1979, Vol. 13, pp. 177–181.

53. Briant, C.L. and Ritter, A.M. The effects of deformation induced martensite on the sensitization of austenitic stainless steels. Metallurgical Transactions A, Dec. 1980, Vol. 11A, pp. 2009–2017.

54. Muraleedharan, P., Khatak, H.S., Gnanamoorthy, J.B. and Rodriguez, P. Effect of cold work on stress corrosion cracking behaviour of types 304 and 316 stainles steels. Metallurgical Transactions A, Feb. 1985, Vol. 16A, pp. 285–289.

55. Seetharaman, V. and Krishnan, R.J. Material Science, 1981, Vol. 16, pp. 523–530.

56. Stefec, R. and Franz, F. A study of the pitting corrosion of cold-worked stainless steel. Corrosion Science, 1978, Vol. 18, pp. 161–167.

57. Khatak, H.S., Gnanamoorthy, J.B. Rodriguez, P. and Padmanabhan, K.A. Madras, India, Proc. 10th Int. Congress on Metallic Corrosion, 1988, Vol. III, paper 10.48, p. 2249.

58. Khatak, H.S., Gnanamoorthy, J.B., Padmanabhan, K.A. and Rodriguez, P. Evaluation of susceptibility to stress corrosion cracking of weldments of cold rolled AISI type 316 stainless steel. Trans. Indian Institute of Metals, Aug. 1991, Vol. 44, pp. 311–316.

59. Dickson, J.I., Groulx, D. and Li. Shiqiong, Evaluation of susceptibility to stress corrosion cracking of weldments of cold rolled AISI type 316 stainless steel. Material Science and Engineering, 1987, Vol. 94, pp. 155–173.

60. Meletis, E.I. and Hockman, R.F. The crystallographs of stress corrosion cracking in fcc crystals. Corrosion Science, 1984, Vol. 24 (10), pp. 843–862.

61. Kaufman, M.J. and Fink, J.L. Evidence for localized ductile fracture in the 'brittle' transgranular stress corrosion cracking of ductile f.c.c. alloys. Acta. Met. 1988, Vol. 36, No. 8, pp. 2213–2228.

62. Bursle, A.J. and Pugh, E.N. In Proceedings of conference on Mechanisms of Environment Sensitive Cracking of Materials, 1977, Metals Society London, pp. 471–481.

63. Gerberich, W.W., Jones, R.H. Friesel M.A. and Nozue, A. Acoustic emission monitoring of stress corrosion cracking. Material Science and Engineering, 1988, Vol. A103, pp. 185–191.

64. Jones, D.A. A unified mechanism of stress corrosion and corrosion fatigue cracking. Metallurgical Transaction A, 1985, Vol. 16A, pp. 1133–1149.

65. Magnin, T., Chieragatti, R. and OLtra, R. Mechanism of brittle fracture in a ductile 316 alloy during stress corrosion. Acta. Met., 1990, Vol. 38, No. 7, pp. 1313–1319.

66. Flanagan, W.F., Baslias, B. and Lichter, B.D. A theory of transgranular stress corrosion cracking. Acta. Met., 1991, Vol. 39, No. 4, pp. 695–705.

67. Beachem, C.D. A new model for hydrogen-assisted cracking (hydrogen embrittlement). Metallurgical Transaction A, Feb. 1972, Vol. 3A, pp. 437-451.

68. Louthan, M.R. Jr., and Devrick, R.G., Material Science, 1975, Vol. 15, p. 565.

69. Hirth, J.P., Effects of hydrogen on the properties of iron and steel. Metallurgical Transaction A, 1980, Vol. 11A, pp. 861–890.

70. Tien, J.K., Thomson, A.W., Bernstein, I.M. and Richard, R.J., Hydrogen transport by dislocations. Metallurgical Transaction A, 1976, Vol. 7A, pp. 821-829.

71. Huang , J.H. and Altstetter, C.J., Internal hydrogen-induced subcritical crack growth in austenitic stainless steels. Metallurgical Transaction A, 1991, Vol. 22A, pp. 2605-2618.

72. Oriani, R.A. and Josephic, P.A., Equilibrium aspects of hydrogen-induced cracking of steels, Acta. Met., 1974, Vol. 22, pp. 1065–1074.

73. Johnson, A.H. and Hirth, P.J., Internal H_2 supersaturation produced by dislocation transport. Metallurgical Transaction A, 1976, Vol. 7A, pp. 1543-1548.

74. Briant, C.L., Hydrogen assisted cracking of sensitized 304 stainless steel. Metallurgical Transaction A, 1978, Vol. 9A, pp. 731-733.

75. Jani,S., Merek, M., Hochman, R.F. and Meletes, E.I. A mechanistic study of transgranular stress corrosion cracking of type 304 stainless steel. Metallurgical Transaction A, 1991, Vol. 22A, pp. 1453–1462.

76. Ford, F.P. and P.L. Anderson, Electrochemical effects on environmentally assisted cracking, In: Proc. Parkins symposium on fundamental aspects of stress corrosion cracking, Ed. S.M. Bruemmer, E.I. Meletis, R.H. Jones, W.W. Geeberich, F.P. Ford and R.W. Staehle, TMS Publication, 1992, pp. 43–67.

77. Ford, F.P., Taylor D.F., Anderson P.L. and Ballinger R.G., Corrosion assisted cracking of stainless steels and low alloy steel, EPRI contract RP-2006-6, Report NP 5064 S, Feb.1987, Palo Alto.

78. Ishikawa, K., Hayashi, M. and kano, S., Life Prediction Method of Stress Corrosion Cracking Based on Crack tip strain Plate Analysis: In: Proceedings of International Symposium on Plant aging and life Prediction of Corrodable Structures, May 15–18, 1995, Sapporo, Japan.

79. Wei, R.P., "Environmentally Assisted fatigue Crack Growth" in Fatigue 87, Vol. III, and Editors: Ritchi, R.O. and Srarke, A., Jr., EMS, West Midland, UK.

80. Rhodes, D., Musava, J.K. and Randon, J.C., Significance of Stress Corrosion Cracking in Corrosion Fatigue Crack Growth Studies, Engineering Fracture Mechanics, 1981, Vol. 15, 3–4, pp. 407–419.

81. Kowaka, M., Introduction to life prediction of industrial plant material; Application of the extreme value statistical method for corrosion analysis, (ed.) Allerton Press Inc. 1994.

9. Microbiologically Influenced Corrosion

Rani P. George[1] and P. Muraleedharan[1]

Abstract Austenitic stainless steels are susceptible to microbiologically influenced corrosion (MIC) when it is used in contact with natural waters. This is due to the changes in the chemistry of the environment at the metal surface because of the settlement and activities of microorganisms. After introducing the mechanism of microbial mediation in the corrosion of metals and alloys, the paper summarizes the work carried out in our laboratory in the area of MIC of austenitic stainless steels. The thrust of our work was in understanding the changes in the electrochemical behaviour of a type 304 stainless steel in the presence of a natural biofilm as well as the influence of metallurgical characteristics on microbial adhesion and MIC .

Key Words MIC, stainless steel, AISI Type 304 SS, biofilm, passive film, localized corrosion, welding, sensitization.

INTRODUCTION

Austenitic stainless steels (SS) are widely used in industrial applications in which the material is in contact with natural waters. For example, AISI 300-series SS are often used as tubing material in cooling water systems of power plants that use fresh water for cooling. Stainless steels are susceptible to microbiologically influenced corrosion (MIC) under such conditions. This is because these systems provide ideal conditions for the growth of microorganisms: highly oxygenated water, good exposure to sunlight and air, with a temperature maintained in the range 27-60°C, and a pH between 6 and 9 [1, 2]. Microbiologically influenced corrosion has also been reported in stainless steels that are used in contact with high purity water, like in the case of reactor coolant systems, emergency systems, reactor auxiliary systems, feedwater heaters etc. Although some of these systems do not support microbial growth during operation because of the high temperature or the high purity water that is used, SS often seems to be vulnerable to MIC during pre-operational testing and start up phases of the plant, which often use untreated water [2].

Austenitic stainless steels are compatible with natural freshwater environments and the corrosion rate is negligible. However, microorganisms such as bacteria, algae and fungi present in these waters can settle on the material surfaces and generate environmental conditions conducive for accelerated corrosion. This phenomenon is known as microbiologically influenced corrosion. Initially, the

[1]Scientific Officers, Corrosion Science and Technology Division, Materials Characterisation Group, Indira Gandhi Centre for Atomic Research, Kalpakkam-603 102, India.

microorganisms adhere to metal surfaces randomly, grow, multiply and secrete fibrous polymeric material known as extracellular polymeric substances (EPS). The EPS helps the organisms to anchor themselves to the substratum as well as to make contact with the other organisms. A biological deposit on a material surface, consisting of attached microbial cells, their metabolic products, EPS etc. are collectively known as biofilm. These biofilms are not uniform either in space or in time and it is a very important factor that facilitates localized corrosion of SS.

The presence of biofilm on material surfaces have two serious consequences; biofouling and MIC [1–7]. If the growth of organisms and the biological deposit is having an undesirable effect such as reduction in the flow of water through heat exchangers, which in turn reduces the heat transfer properties and degrades the system, then it is called biofouling. In other words, biofouling is the physical obstruction that causes a reduction in flow, heat transfer or movement. The MIC is caused by the chemical changes in the environment at the metal/biofilm interface arising from the metabolic activities of the microorganisms inhabiting the biofilm. The presence of a biofilm on material surface can influence the corrosion behaviour since the value of a given parameter such as temperature, pressure, concentration of a solute and pH at the water/substrate interface under the biofilm may be different from that in the bulk environment. The non-uniform nature of biofilm thus helps in generating heterogeneity in the environment at the surface. Thus, biofilms are known to aid in the initiation of corrosion, change the mode of corrosion or cause changes in the corrosion rate [8, 9]. In the case of stainless steels the patchy nature of biofilms helps in generating differential aeration cell, with the metal surface under an actively respiring microbial film forming the anode and non-biofilmed surfaces exposed to the aerated bulk environment forming the cathode.

In the case of macrofouling, the macro organisms by themselves may not cause any corrosion. However, during the initial period of settlement of these macrofouling organisms, localized corrosion such as pitting and crevice corrosion can take place especially in active-passive materials such as stainless steels, aluminum and titanium alloys [10]. Moreover, macrofouling can create conditions that favour the growth of microorganisms like anaerobic bacteria beneath the fouling layers, causing corrosion of the materials [11].

MICROBIOLOGY RELEVANT TO MIC

Microorganisms like algae, bacteria, fungi etc. are ubiquitous and occur in soil, freshwater, seawater and air [12,13]. They use a variety of nutrient sources and survive, grow and reproduce under a wide range of temperature, pH, dissolved oxygen concentration, pressure and salinity [1]. However, a particular species of microorganism can thrive and multiply only in a particular range of environments (Table 1).

Two of the major class of nutrients which are essential for growth and survival of microorganisms are those that provide a source of carbon and nitrogen. Microorganisms can be broadly classified into autotrophs and heterotrophs depending on whether the organism can utilize atmospheric CO_2 for making cell materials or they require organic material for its source of cell carbon and energy. Autotrophs are those which use natural or artificial light and the catalytic action of chlorophyll pigments to reduce atmospheric carbon dioxide and water to sugar phosphate. Photosynthetic algae like blue-greens and green algae are examples of autotrophs. Chemolithotrophs oxidize inorganic compounds and couple these to the reduction of carbon dioxide in an analogous way to photosynthesis. Sulphur-oxidizing bacteria and iron-oxidizing bacteria are good examples of chemolithotrophs, utilizing

Table 1.   Tolerable range of environmental parameters for some microorganisms [21]

Genus or species	pH range (optimum)	Temperature range (°C)	Oxygen requirement
Desulfovibrio sp.	4–8	10–20	anaerobic
Desufotomaculum sp.	6–8	10–40	anaerobic
Desulfomonas sp.	–	10–40	anaerobic
Thiobacillus sp.	0.5–8	10–40	aerobic
Gallionella sp.	7–10	20–40	aerobic
Pseudomonas sp.	4–9	20–40	aerobic
Cladosporium resinae	3–7	10–45	–
Mixed flora (Crenothrix and Leptothrix sp.)	–	5–40	aerobic

the energy released from the oxidation of chemical compounds. The metabolic activities of these microorganisms are coupled with corrosion reactions. The metabolic products such as organic acids, inorganic acids and sulphides resulting from the bioactivity of organisms modify the environment in which they live [14]. A well-known iron/manganese bacterium, *Siderocapsa* sp. [1] can shift the chemical equilibrium between ferrous and ferric ions which will influence the corrosion rate.

Another interesting fact is that the microorganisms influencing corrosion can flourish at the corrosion site by associating with other microorganisms in a microbial community even when the bulk environment is not conducive to their growth and survival. For example, even in the presence of oxygen in the system, the biofilms of aerobic bacteria (require oxygen) can support anaerobic bacteria (do not require oxygen) at the metal/biofilm interface [15]. All these special properties of microorganisms make them very difficult to control.

The most important bacteria involved in the mediation of corrosion reactions by virtue of their metabolic activities are the methane producers, nitrate reducers, sulphur oxidizers (*Thiobacillus* sp.) as well as the most notorious sulphate reducers (*Desulfovibrio* sp.) and the iron oxidizers (*Gallionella* sp., *Sphaerotilus* sp.) [1,16]. Photosynthetic organisms like algae generally adhere to wetted surfaces such as cooling towers, mist eliminators, screens and distribution systems leading to biofouling [4]. Algae also play a significant role in MIC due to their capability to generate oxygen, corrosive organic acids and nutrients to support the life of other corrosion causing microorganisms. Some fungi metabolize wood components causing wood deterioration [17]. A species of fungi, *Cladosporium resinae* is reported to cause corrosion of aluminum tanks in aircraft by its metabolic activity [5].

MICROBES AND CORROSION MECHANISM

Corrosion by microbial action usually occurs by a number of mechanisms that are reasonably well understood as electrochemical corrosion phenomena. The microorganisms often contribute to corrosion without being solely responsible for it. Hence microbiologically influenced corrosion is not fundamentally different from any other known type of electrochemical corrosion. Biological activities can accelerate corrosion under anaerobic and aerobic conditions by modification of anodic and cathodic half reactions directly or indirectly [18].

Acceleration of Anodic Half Reactions

The anodic reaction involves oxidation, the process of removal of electrons from atoms or ions. In aerobic conditions, iron-oxidizing bacteria (IOB) can oxidize ferrous ions to ferric ions, which being insoluble precipitates as ferric oxide. This combines with organic matter to form a tubercle over the source of ferrous ions. The small area of the metal beneath the tubercle becomes the anodic site and sustains severe localized corrosion, even causing perforations of pipe wall. In anaerobic conditions sulphate-reducing bacteria (SRB) cause anodic stimulation by precipitation of ferrous ions produced at the anode with biological sulphur [9].

Acceleration of anodic half reactions is occurring by tubercle formation, under-deposit corrosion, acid production and breakdown of protective films and coatings by the metabolic activities of microorganisms. Bacteria such as *Pseudomonas* sp., *Aerobacter* sp., *Flavobacterium* sp., and fungi such as *Cladosporium* sp. and *Aspergillus* sp. adhere firmly and randomly to wet surfaces, grow, multiply and produce extracellular polymeric substances leading to the formation of dense fibrous deposits that are patchy in nature. The area under the deposit becomes depleted in oxygen, creating differential aeration cells. Metabolites of many microorganisms are invariably acids, which lower the pH of solutions where corrosion is occurring and so accelerate corrosion. Sulphuric acid is generated by sulphur-oxidizing, chemosynthetic bacterium, *Thiobacillus* sp. [10] and fungi secrete organic acids like acetic acid during their normal oxidative metabolism (*Cladosporium resinae*) [11]. Natural and artificial films may serve as sites for microbiological attack. Breakdown of protective iron-sulphide films (Mackinawite) produced by SRB is well known [19]. Corrosion due to destruction of protective ferric coatings by a *Pseudomonas* sp. that reduces iron from insoluble ferric to soluble ferrous state has been reported [20]. Artificially applied protective coatings on materials may also be susceptible to biodegradation by metabolic products of microbial activity (principally by H_2S and H_2SO_4) [21].

Acceleration of Cathodic Half Reactions

In the anaerobic environments, SRBs are able to take up molecular hydrogen while reducing sulphate for the synthesis of sulphur containing substances [22]. In aerobic systems, where oxygen reduction is the cathodic reaction, microbes can catalyse the oxygen reduction, and accelerate corrosion processes [23].

In a medium containing high ferrous ion concentration, sulphides produced by SRB may precipitate as iron sulphide and a galvanic couple can form between naked metal and FeS precipitate (acting as cathode) consequently accelerating the cathodic reaction [9].

Other Corrosion Pocesses

Hydrogen sulphide formed by SRB activity on the metal surface can retard the formation of molecular hydrogen. This facilitates the diffusion of atomic H produced by material/ environment interactions into the metal, causing hydrogen embrittlement and hydrogen induced cracking, especially in high strength steels [3]. Along with this effect of SRB, hydrodynamic loading of offshore structures with micro and macroalgae can result in enhanced fatigue crack growth rates [24]. Stress corrosion cracking of a stainless steel tank storing demineralized water, in which the cracks were initiated at pits beneath bacterial deposits, has been reported [25]. In another instance, ammonia produced by nitrate reducing bacteria has been identified as the cause for stress corrosion cracking of admiralty brass condenser tubes [26].

MIC OF STAINLESS STEELS

The basic cause of the corrosion resistance of stainless steels is the tightly adherent and protective film, composed of mainly chromium oxide, formed on the surface in oxidizing environments. This passive film acts as a barrier between the metal and its' environment, providing 'kinetic stability'. Stainless steels are passivated based on the presence of a critical amount of Cr in the alloy; with elements such as nickel, molybdenum and nitrogen improving this passivity of the alloy. However, if this passive film is damaged locally, accelerated corrosion takes place. The presence of slime-forming bacteria such as *Pseudomonas* and iron bacteria in localized deposits can consume oxygen diffusing into the deposit and thus make the passive film under the deposit unstable. This creates a differential aeration cell that can sustain crevice corrosion under the microbial film [7]. Metal-oxidizing and sulphate-reducing bacteria have been reported as causal agents for corrosion of these materials [3]. Many investigators have also reported that MIC is often associated with weld seams in fabricated components [2, 23, 27, 28]. Stainless steels containing molybdenum, which effectively resist crevice corrosion in seawater, have been found susceptible to MIC [29, 30]. In the following sections of this paper, a brief review of MIC work carried out in our laboratory is presented. The following questions are addressed in these studies.

1. Effect of natural biofilm on the electrochemical and corrosion behaviour of a type 304 SS
2. Influence of metallurgical condition of SS on MIC
3. Control of MIC

Electrochemical Studies

OCP ennoblement

It has often been reported that microbial films formed in natural waters shift the open circuit potential (OCP) of passive metals and alloys in the positive (or noble) direction [31–46]. This ennoblement in free corrosion potential has been observed both in freshwater and seawater. Another aspect of OCP ennoblement [31] is that the potential ennoblement is sustained for long periods of time only in the absence of corrosion initiation. In alloys with less resistance (304 SS), OCP first shifts in the noble direction, but as soon as pitting or crevice corrosion is initiated, the current supplied by the active corrosion polarizes the potential back to a more active value. The phenomenon of ennoblement of OCP of SS is important because it gives rise to

- an increase in the risk of localized corrosion initiation and propagation.
- an increase of the galvanic currents in the coupling of passive materials with ones having poor corrosion resistance.

Results from various researchers on OCP measurements have indicated that there is large amount of variation in the extent of ennoblement. The factors that promote or suppress this effect are also not clear. Dexter et al. [32] showed that 30 to 40 percent coverage of the metal surface by biofilm was required for the OCP to rise above +150 mV SCE level, while substantially complete coverage was required for maximum amount of ennoblement. According to Dexter and Zhang [33], biofilms grown at all salinity upto 3.2% were able to ennoble the OCP. However, the most noble potentials (about +500 mV SCE) and the largest amount of ennoblement (over 500 mV) were found in fresh water, and both decreased with increasing salinity. Temperatures in the range of 2 to 30°C and dissolved oxygen content in the bulk water from 0.5 to 10 ppm have small effects, which are both in the direction predicted by electrochemical thermodynamics. However, results of various investigators do not agree on the effect of sunlight as well as microbial ecology of biofilm, on the ennoblement. Dexter and

Zhang [33] have indicated loss of ennoblement in the presence of sunlight and enhanced ennoblement under dark conditions. Little et al. [34] and Maruthamuthu et al. [35] have reported enhanced ennoblement in the presence of light and loss of ennoblement in dark exposures. Our own studies [36] have shown that freshwater biofilm formed in the presence of sunlight, with high content of photosynthetic organisms (Fig. 1), cause large anodic shift in OCP, whereas biofilms formed under dark condition with high content of bacteria (Fig. 2), do not ennoble a type 304 SS. Type 304 stainless steel specimens exposed in raw reservoir water showed a positive shift of OCP for about 200 mV above the initial OCP, whereas, in sterile reservoir water, this shift was not observed (Fig. 3).

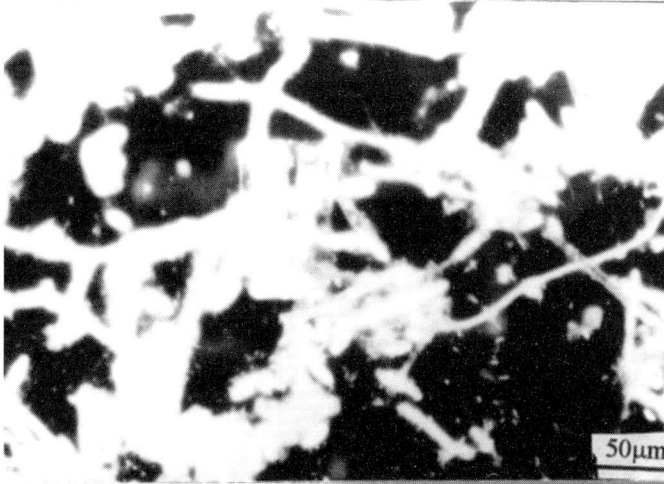

Fig. 1. Epifluorescence micrograph of algal- dominated biofilm on stainless steel, exposed in the reservoir water in the sun-lit condition for 120 h [36].

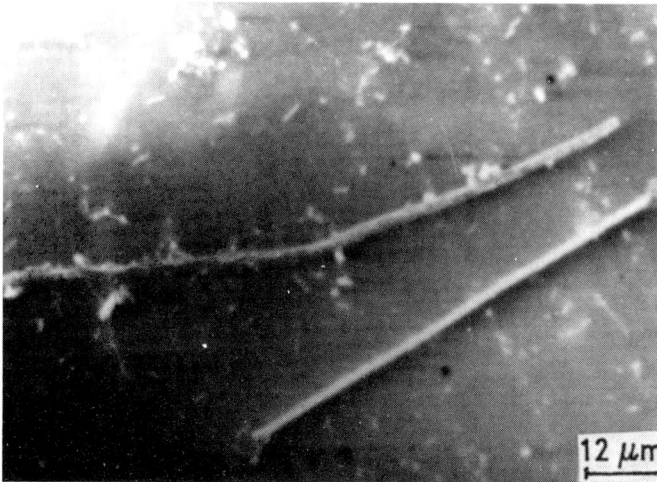

Fig. 2. SEM picture of the biofilm developed on perspex, exposed in the reservoir water in the dark conditions for 120 h [36].

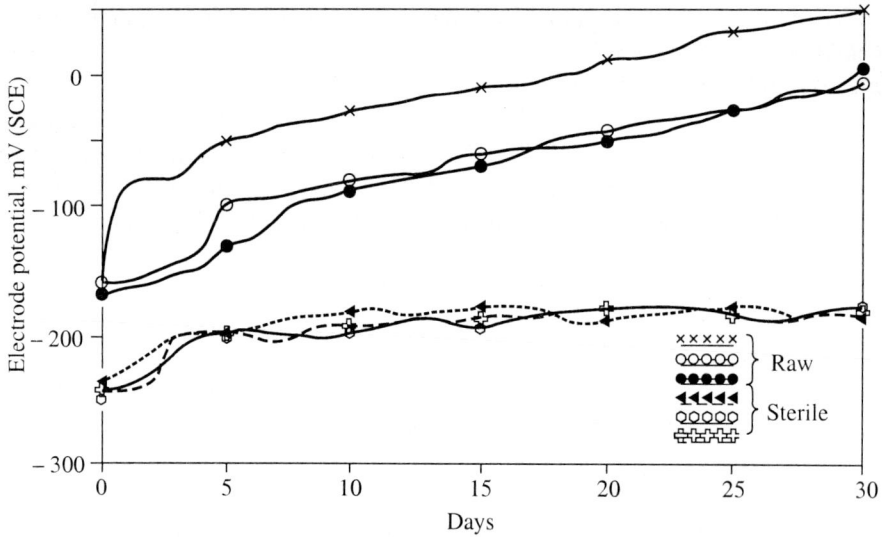

Fig. 3. Variation of open circuit potential (OCP) of 304 SS in reservoir water under raw and sterile conditions [36].

Theoretically ennoblement of OCP can be due to either thermodynamic or kinetic effects (Fig. 4) [37]. The thermodynamic effect can play a role when an increase of the partial pressure of oxygen (pO_2) will shift the reversible potential of the oxygen electrode (eqn.1) in the positive direction.

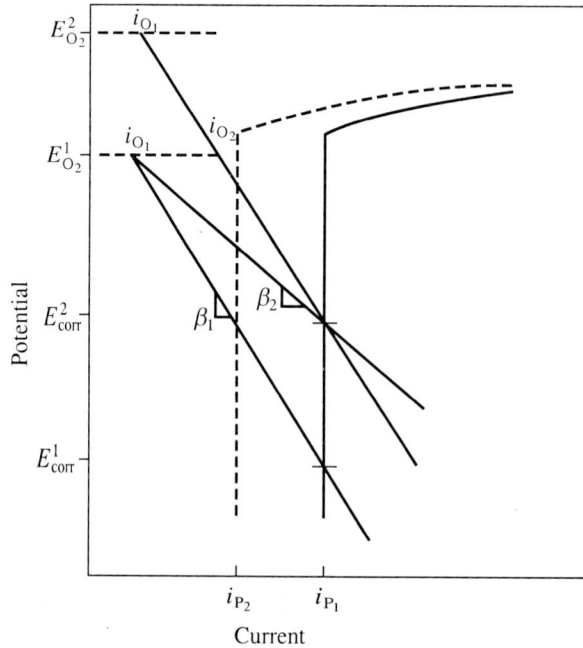

$$O_2 + 2\,H_2O + 4e^- \rightarrow 4OH^- \tag{1}$$

Fig. 4. Thermodynamic and kinetic influences on cathodic overvoltage [37].

When the reversible cathode potential for O_2 reduction changes from $E_{O_2}^1$ to $E_{O_2}^2$, then the OCP changes from E_{corr}^1 to E_{corr}^2 (Fig. 4). The kinetic effect of shifting the OCP in the noble direction is also shown in the same figure. This can happen when the exchange current density for oxygen reduction reaction changes from i_{O_1} to i_{O_2} or when the tafel slope of the cathodic curve changes from β_1 to β_2 or by the decrease of passive current from i_{p_1} to i_{p_2}. In all these cases the E_{corr}^1 is shifted to E_{corr}^2 showing an ennoblement of OCP.

Biofilms and associated metabolism of microorganisms can bring about many of the above changes leading to OCP ennoblement. Photosynthesis by algal dominated biofilms can increase the partial pressure of oxygen in the biofilms leading to increase in the reversible potential of the oxygen electrode. Studies on the effect of photosynthetic biofilms on OCP of stainless steel under periodic illumination by Dowling et al. [38] showed that positive shift in OCP, induced by periodic illumination of photosynthetic biofilms, are primarily the result of oxygen production during photosynthesis. The extent of potential change depended on the interval of illumination. They observed a drop in OCP value during the dark cycle. Illumination did not affect the OCP in sterile systems or in systems that contained only nonphotosynthetic eubacteria. As mentioned earlier, In our studies [36] also, only the algal-dominated natural biofilms and the pure cultured algal biofilms raised the potential of SS to a higher level.

Dexter et al. have proposed [39, 40] that the mechanism for ennoblement involve both a decrease in pH and an increase in hydrogen peroxide concentration at the metal surface under the biofilm. It is well known that the reversible oxygen potential, under acidic conditions (eqn. 2), shifts about 60 mV in the noble direction for each unit decrease in pH. Based on thermodynamics, the maximum calculated ennoblement for a reversible oxygen electrode would be 280 mV for changes in pH and O_2 from 8 (and 0.2 atm) to 3 (and 0.02 atm). These data show that the often observed ennoblements of 250 mV or less on common stainless steels in natural sea water can be explained by the pH mechanism alone for a pH of 3 at the metal surface. Dexter et al. have given experimental evidence to show that a pH less than 3 will be possible under a natural biofilm. A shift in the OCP to values in the range +300 to +500 mV SCE can be explained if reduction of H_2O_2 to water as the cathodic reaction is considered. Peroxide would contribute to the mechanism by virtue of its noble redox potential (eqn. 3).

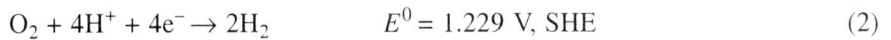

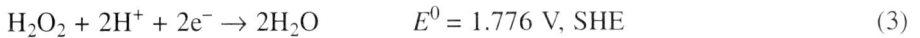

$$O_2 + 4H^+ + 4e^- \rightarrow 2H_2 \qquad\qquad E^0 = 1.229 \text{ V, SHE} \qquad\qquad (2)$$

$$H_2O_2 + 2H^+ + 2e^- \rightarrow 2H_2O \qquad\qquad E^0 = 1.776 \text{ V, SHE} \qquad\qquad (3)$$

The kinetic effect of shifting the OCP in the noble direction may result from an increase of the exchange current density (i_o) for the oxygen reduction reaction. According to Mollica and Trevis [41] and Scotto et al. [42], extracellular slimes produced by bacteria bind heavy metals like Fe, Co and Ni, which can serve as oxygen reduction catalysts and this can increase i_0 for the oxygen reduction reaction. Dexter and Gao [43] also suggested that a decrease in tafel slopes for the oxygen reduction or a decrease in the passive current density i_p in the presence of biofilm would result in the ennoblement. Maruthamuthu et al. [44] have also suggested that passivity improvement through the action of microbially produced inhibitors could be the major mechanism of ennoblement. According to them the "mixed inhibitor" involving dissolved nutrients like phosphate may strengthen the n-type semiconducting oxide film on the metal surfaces during the processes of ennoblement. This model is consistent with the observation by Eashwar et al. [45] that sunlight irradiation of sea water eliminates the ennoblement-causing capacity of biofilms due to photodecomposition of dissolved organics.

Dickinson et al. [46] have recently demonstrated that ennoblement of stainless steel coupons was caused by the deposition of manganese-rich material on the coupon surface by the bacterial species *Leptothrix discophora*. This study shows a chemical mechanism for ennoblement in which manganese dioxide acts as a galvanic cathode to elevate cathodic current and shift E_{corr} in the noble direction. The MnO_2-$MnOOH$ redox couple (E^0 = + 335 mV SCE at pH 8) was proposed as the reaction that fixes E_{corr} near +350 mV SCE.

Studies on the Susceptibility of SS to Localized Corrosion

Based on the measurement of OCP alone, it is not possible to determine the influence of biofilm on localized corrosion. The susceptibility of stainless steels to localized corrosion can be determined by potentiodynamic studies, and these techniques are most useful in systems in which the metal passivates by the formation of a passive film. A schematic diagram for anodic polarization of SS is shown in Fig. 5. In natural fresh water environments with a pH value near the neutral, the corrosion potential of SS is within the passive range. For a given corroding metal, the corrosion potential and corrosion current will be determined by the point at which cathodic polarization curve intersects the anodic curve as shown in Fig. 5. As for case 1 in Fig. 5, the corrosion potential falls in the stable passive region and this is the most desirable situation. But according to Dexter and Gao [43] the presence of microorganisms on the metal surface may change case 1 to case 2 condition where cathodic potential is noble enough to place the intersection in the transpassive region leading to localized corrosion. Potentiodynamic sweep techniques can provide more information on how microorganisms bring about such changes. The parameters of interest are OCP, E_b (breakdown potential) and the E_{repass} (repassivation potential). A typical cyclic polarization curve is shown in Fig. 6 which illustrates the key parameters and the hysterisis behaviour [47]. Pitting or crevice corrosion occurs only in alloys that display a well-defined hysteresis in the polarization curve [48]. A measure of the resistance to pitting, in a thermodynamic sense, has been suggested by Hoar et al. [49] to be the value of $E_{diff} = (E_{repass} - E_{corr})$. A positive value

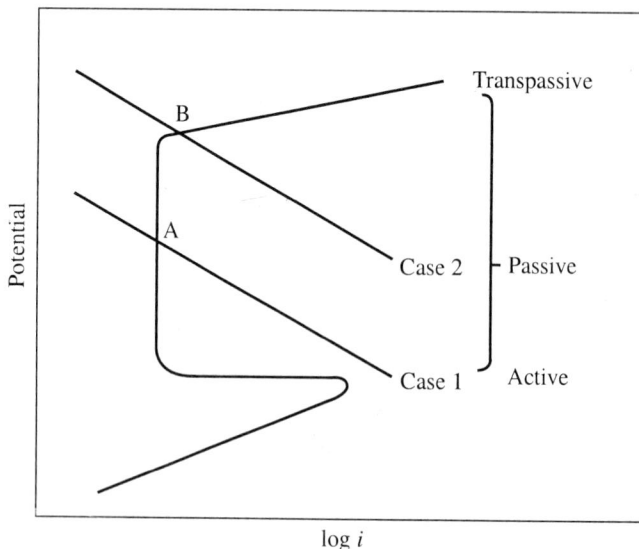

Fig. 5. Schematic diagram of the intersection of 2 cathodic curves with the anodic curve for a passivatable metal [43].

of E_{diff} is a desirable situation, since the initiated pits get repassivated easily. The breakdown potential of a material gives a measure of its resistance to pit initiation, the more noble the value of E_b the more resistant the material to pit initiation.

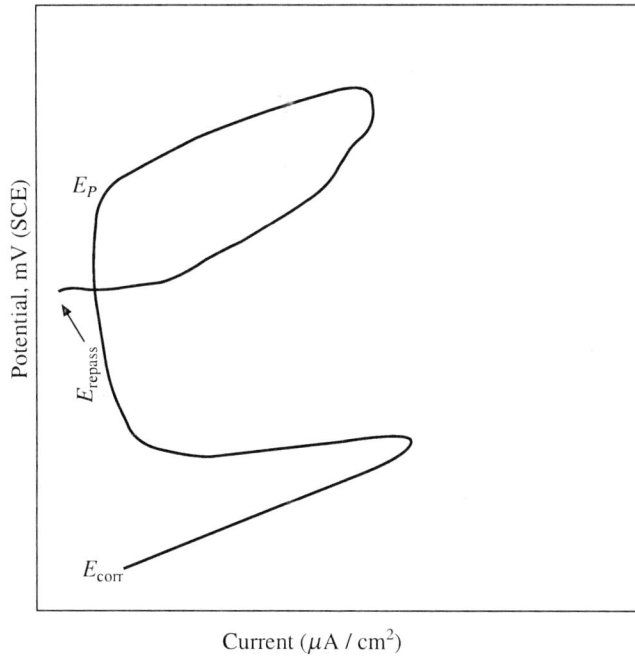

Fig. 6. Potentiodynamic polarization curve of a metal in a pitting environment [47].

Investigations were carried out in our laboratory to understand the role of microorganisms in the corrosion of stainless steel in fresh water using potentiodynamic polarization technique. In order to compare the aggressivity of three different types of fresh water viz., raw reservoir water, sterilized reservoir water and Palar water (underground water forming the source water of reservoir water), anodic polarization tests were performed on freshly polished specimens in the above three waters.

Polarization tests were also done on specimens with natural biofilms, developed by exposure in raw reservoir water for various durations ranging from 1 day to 123 days, both in natural dark/light condition ('sun-lit') and in continuous dark ('dark') condition. The experimental details are given elsewhere [36]. In the raw reservoir water with the highest bacterial and algal density, SS showed (Table 2) the most noble OCP value, lowest E_b and most active E_{repass}. The value of E_{diff} in this electrolyte was negative or in other words E_{repass} has gone below E_{corr} relating to an undesirable situation, where even in the absence of polarization, pit could be formed and propagated, with little tendency to repassivate, leading to complete penetration of the material [50]. The presence of a well defined hysterisis and an active E_{repass} also showed that the biofilm forming components in the reservoir water can cause severe localized corrosion of the material. In the sterile reservoir water and the Palar river water where the biofilm-forming components were very low or absent, polarization curves of SS showed noble E_b and positive E_{diff}. This suggested that the probability of pitting and crevice corrosion of SS in both these environments were very low.

Table 2.   Electrochemical polarization results [36]

Material/Environment	E_{corr} (OCP) Open circuit potential mV (SCE)	E_b Breakdown potential mV (SCE)	E_{repass} Repassivation potential mV (SCE)	E_{diff} "Resistance to pitting" mV
SS/Reservoir water	− 180	+ 600	− 220	− 40
SS/Sterile reservoir water	− 220	+ 800	+ 200	+ 420
SS/Palar water	− 200	+ 980	+ 250	+ 450

When the polarization curves of freshly polished SS in reservoir water were compared with those of biofilmed SS ('sun-lit' and 'dark' biofilms) significant changes were found in the electrochemical characteristics due to the presence of biofilms. The OCP and E_b of biofilmed specimens showed positive shifts compared to the specimens without biofilms (Table 3). This positive shift of OCP is significant because of its influence on localized corrosion initiation and propagation. In chloride bearing waters, probability of initiation of localized corrosion increase directly with chloride ion activity and OCP [51]. Thus, at a given chloride level, we can expect that the OCP value may move closer to the E_b or even cross E_b as the ennoblement takes place, leading to pitting. However, contrary to expectation, it was found that E_b also has becomes ennobled along with OCP (as shown in the polarization curve of 1-day biofilm in Fig. 7). Biofilms can affect electrochemical parameters at a metal surface both by their physical presence and by their metabolic activity. Physically, the film acts as a diffusion barrier, tending to concentrate chemical species produced at the metal-film interface and to retard diffusion of species from the bulk water towards the metal surface. Thus, increase in E_b can be due to the chloride ions by the biofilm. However, the most important observation was that, as the thickness and age of the 'sun-lit' biofilm (algal-dominated) increased with the duration of exposure in

Table 3.   Electrochemical polarization characteristics of AISI Type 304 SS specimens with 'dark' and 'sun-lit' biofilms [36]

Material/Environment	OCP Open circuit potential mV (SCE)	E_b Breakdown potential mV (SCE)	E_{repass} Repassivation potential mV (SCE)	i_p Passive current density $\mu A/cm^2$
Freshly polished SS/ Reservoir water\	− 180	+ 600	− 220	0.1
SS with 'sun-lit' biofilm/Reservoir water				
1 day biofilm	− 100	+ 1100	+ 50	0.2
34 day biofilm	− 75	+ 1050	+ 75	1.2
78 day biofilm	− 100	+ 1000	− 25	8.0
123 day biofilm	− 175	+ 900	0	20.0
SS with 'dark' biofilm/Reservoir water				
1 day biofilm	− 150	+ 950	+ 100	0.2
34 day biofilm	− 178	+ 700	+ 200	0.3
78 day biofilm	− 163	+ 700	+ 50	0.2
123 day biofilm	− 150	+ 1000	+ 50	0.1

the reservoir, the passive current showed an increasing trend. Finally, in the specimen with 123-day normal biofilm (Fig. 8 (a)), there was a steady increase in the applied current density starting from the OCP which indicated weakening of passivity and initiation of localized corrosion at the OCP itself. Scanning electron microscopic examination of the specimen surface of 123-day old biofilm, without polarization, after complete removal of the biofilm showed isolated regions of surface attack indicating crevice initiation (Fig. 8 (b)). Though initially the normal biofilm showed protective behaviour by its physical presence, the increased metabolic activities of the microorganisms in the growing biofilm was found to affect the integrity of the passive film on the SS specimen.

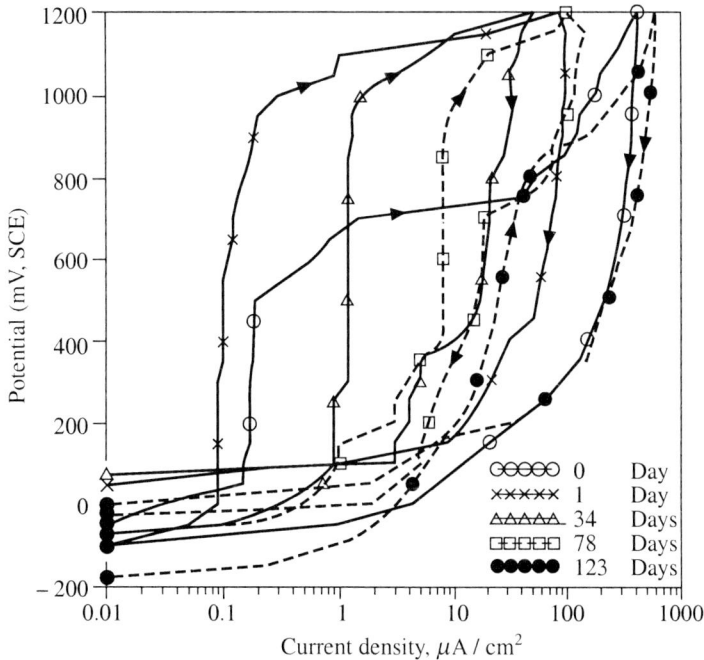

Fig. 7. Cyclic polarization curves for AISI Type 304 SS with biofilms, formed by exposure to reservoir water under sun-lit condition for various durations [36].

Potentiodynamic polarization studies of the specimens with dark biofilm helped to throw more light on the effect of biofilms on the passivity of stainless steel [52]. Though the specimens with 'dark' biofilms also showed ennoblement of E_b compared to polished specimen (Table 3), it was lesser compared to the 'sun-lit' biofilmed specimens. This can be attributed to the lesser thick biofilms in the dark condition (19.0–27.0 μm) compared to those in the normal condition (52.0–128.0 μm) [53]. The physical, biological and biochemical features of the biofilm formed under sun-lit and dark conditions are given in Table 4. Another significant difference was that, as the thickness and the age of the 'dark' biofilm (bacterial-dominated) increased with the duration of exposure in the reservoir, the applied current density did not show any significant increase. Scanning electron microscopic examination of the specimen surface with 123-day old 'dark' biofilm did not show any such corrosion initiation on the surface under the biofilm.

The increase in the passive current density (i_p) with the ageing of the biofilm and crevice corrosion initiation in type 304 SS in open reservoir water was observed only when the biofilm was developed

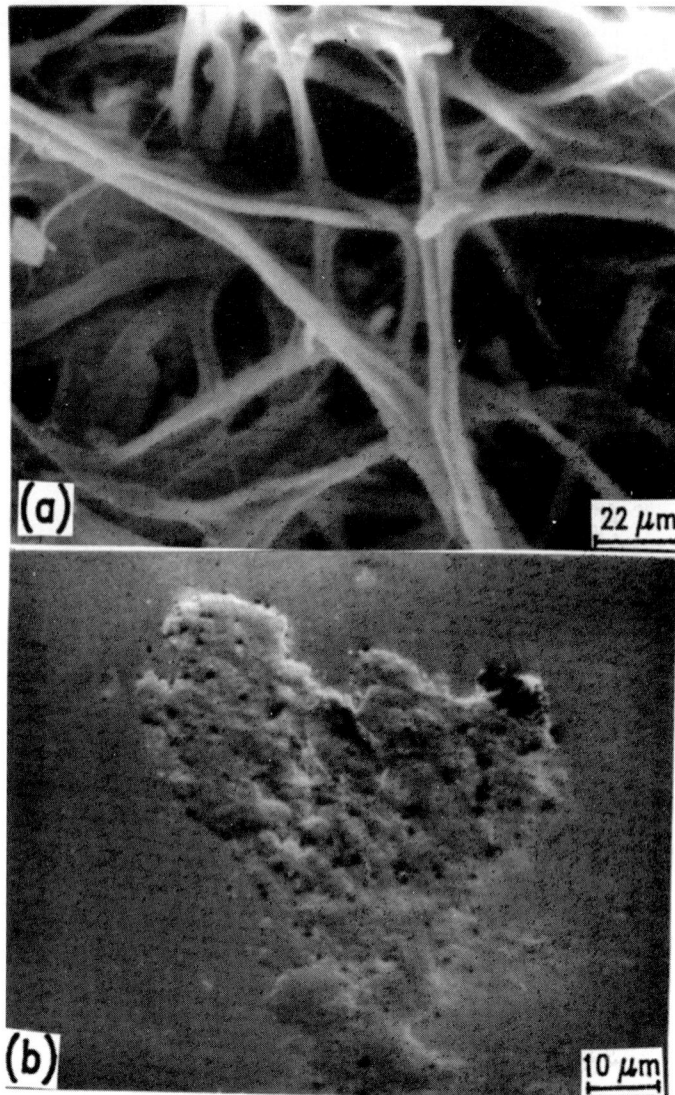

Fig. 8. SEM photographs of 304 SS surface after exposure in reservoir water under sun-lit condition for 123 days: (a) with biofilm and (b) after removal of biofilm [36].

under sun-lit conditions, containing photosynthetic algae, aerobic and anaerobic bacteria (Fig. 1). The biofilm developed under continuous dark conditions (Fig. 2), in which photosynthetic algae were absent, did not show such an increase in the i_p, nor any corrosion initiation. The role of algal photosynthesis in the release of oxygen is well known and this can cause an increase in pH and oxygen concentration in the biofilm which is found to contribute to the ennoblement of SS in this study. According to Dexter [51] OCP ennoblement reduces crevice initiation time and increases crevice propagation on alloys with low resistance. It is also reported that [54] ennoblement of a material can stimulate crevice corrosion under more active or anodic areas if the material is heterogeneous. In our study specimens with natural biofilms can be considered to consist of both algal-dominated regions that ennoble the

Table 4. **Physical, biological and biochemical features of the biofilm formed under sun-lit and dark conditions [36]**

Parameter		Biofilm		
		24 h	*72 h*	*120 h*
Biofilm thickness	Sun-lit	52.0	103.0	128.0
(μm)	Dark	19.0	30.0	27.0
Diatom count	Sun-lit	9	62	276
(Cells/cm^2)	Dark	NG	NG	5
SRB count	Sun-lit	0.7	1.6	2.3
(cfu/cm^2)	Dark	1.4	1.6	2.7
Total viable	Sun-lit	5×10^5	2×10^5	1×10^3
count	Dark	5×10^5	5×10^4	5×10^2
(cfu/cm^2)				
Particulate	Sun-lit	1.32	7.0	9.4
organic carbon	Dark	0..37	1.05	4 .35
(μg C/cm^2)				
Chlorophyll	Sun-lit	0.03	0.05	0.16
(μg/cm^2)	Dark	0.002	0.005	0.02

NG: No growth.

material by increasing the oxygen concentration and bacterial-dominated regions that form active surfaces by decreasing oxygen concentration and pH because of their metabolic activities. Therefore, in a specimen with natural fresh water biofilm, ennoblement due to algal species can initiate crevice corrosion at the smaller anodic areas occupied by bacteria. Thus the present studies suggests that, it is the activities of a consortium of microorganisms consisting of different types of bacteria and algae present in a natural biofilm, that influences the passivation behaviour of the stainless steel.

Influence of Metallurgical Variables on MIC

As mentioned earlier, MIC is not a new form of corrosion and it manifests as one of the known forms of corrosion. Since corrosion is influenced by microstructure, it is expected that microstructure plays a role in MIC also. This is evident from the analysis of many pitting failures from power plants, chemical process plants and pulp and paper mills which indicated that welded regions of SS components are preferentially attacked by microbes [55]. Pits in such cases often exhibit a special morphology having very small entrance and exit penetrations with very large subsurface cavities [56]. Pits may be associated with attack of the weld material, generally near the fusion line or in the heat affected zone. Sensitization has been recognized as a factor related to the level of pitting in the heat affected zone [3, 56]. The two-phase weld metal appears to be the most susceptible area, although the relative susceptibilities of the austenite and delta ferrite have not been clearly defined. As shown in the papers referred above, in some situations the austenite is preferentially attacked. In other situations, the delta ferrite is removed and the austenite is not attacked. Of seven cases with pitting in the weld metal, studied by Borenstein [57], two of these demonstrated preferential attack in the austenite, whereas five of them demonstrated preferential attack in the ferrite. The behaviour of weld metal also has been mentioned by other authors [58, 59].The variable behaviour in terms of the phase being attacked is attributed to the environmental conditions at the metal/biofilm interface. In oxidizing acidic environments, the interdentritic austenite corrodes, whereas in reducing environments, the δ ferrite shows signs of preferential attack [57, 60, 61].

Investigations carried out in our laboratory has shown that type 304 SS weldments exposed for an year in freshwater reservoir show preferential corrosion at the welds in a few cases [62]. The SEM examination of the corroded region has shown a honeycomb like structure (Fig. 9) suggesting preferential removal of austenite leaving behind the dentritic delta ferrite. Characterization of biofilm formed on weld and base metal regions of the specimen showed higher total bacterial density and chlorophyll content (indicating algal population) in the welded regions compared to the base metal. Total viable counts of bacterial species such as *Pseudomonas* sp., iron oxidizing bacteria were also higher in the welded region (Table 5). Such a biofilm can produce acidic oxidizing environment that promote preferential corrosion of austenite.

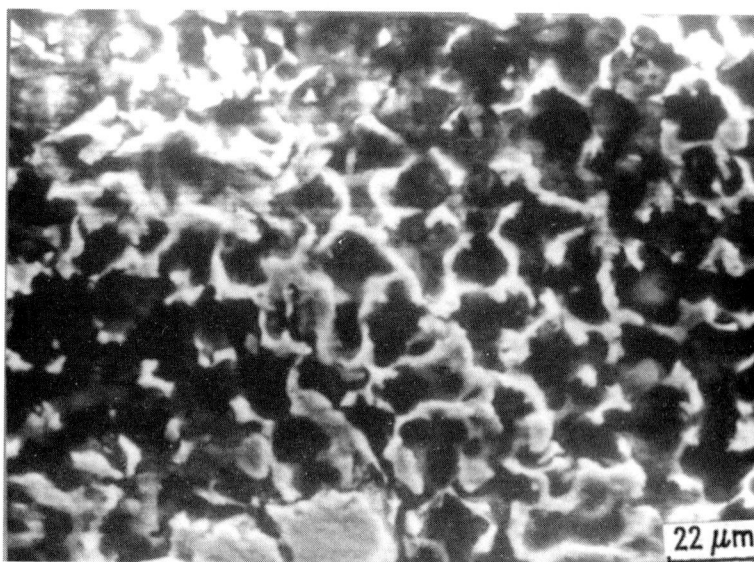

Fig. 9.   SEM photograph of the corroded region of a stainless steel weld exposed to reservoir water for one year showing honeycomb like structure [62].

The question often asked is that why welds undergo preferential attack during MIC. It is well known that surface and microstructural inhomogeneities of a weld makes it a high energy area and consequently it forms the anode of the corrosion cell even during abiotic corrosion. Also, the surface condition, altered by the welding process, seems to be influential, at least in the initial stages of the preferential attack [60].

Table 5   Results of the microbiological analysis of the AISI Type 304 SS welded frame [62]

Region	Total bacterial density (cfu/cm²)	Density of Pseudomonas sp. (cfu/cm²)	Density of iron oxidizing bacteria (cfu/cm²)
Corroded weld	3.0×10^6	2.4×10^6	1.4×10^5
Weld not corroded	2.0×10^5	8.0×10^5	3.8×10^4
Base metal	2.4×10^4	2.0×10^4	9.4×10^4

Buchanan et al. [63] present the results of electrochemical tests carried out in a laboratory using bacterial concentrations. Of the different types of superficial finish, the as welded state seemed to be most sensitive to the attack. But is there something more than that in the sense that whether the microstructural and microchemical heterogeneities of the weld facilitate preferential attachment and proliferation of microorganisms on the metal surface. Although attempts have been made to correlate microstructural features like residual stress, cold work and sensitization with MIC susceptibility, all these studies have looked at the problem from the corrosion angle only. Since adhesion of microbial cells onto a substratum is the first step in MIC, it is very important to examine whether micoorganisms are guided by any of the microstructural/microchemical features of the metal substratum in the initial attachment as well as growth and proliferation.

Influence of Surface Characteristics/Microstructure on Bacterial Adhesion

There are not many studies relating microstructure and MIC behaviour in wrought stainless steels. Stein [61] has pointed out that the sensitization of type 304 and 316 stainless steels does not affect their susceptibility to MIC. However, type 304 is susceptible to MIC when marked deformation lines are present in its microstructure. It has also been noted that high annealing temperatures applied to eliminate or reduce deformation lines may improve the resistance of austenitic stainless steels to MIC. However, tests which were carried out in the laboratory [64, 65] have shown that there is a clear relationship between the sensitization state of a type 304 SS and the pitting potential measured in a biogenic sulphide–containing electrolyte (SRB metabolites).

Since adhesion of microorganisms is the first step in biofilm formation that leads to MIC, it is essential to understand the influence of microstructural and surface properties on the attachment. Surface characteristics such as roughness [66-68], wettability [69, 70] and charge [71] are known to have an effect on microbial attachment. Since sensitization and surface oxidation are two important changes taking place during welding, the effect of these parameters on the adhesion of bacterial cells in a Type 304 SS was investigated in our laboratory by exposing coupons in a culture of *Pseudomonas* sp. in a 0.1% nutrient broth. This particular strain of bacteria used in the study was isolated from the natural biofilm formed on the 304 SS specimens exposed to fresh water in the open reservoir.

Epifluorescence microscopic pictures (Fig. 10) of the exposed coupons illustrates the differences in the bacterial attachment, proliferation and biofilm development on the various surfaces exposed for 7 days in nutrient broth. The bacterial density, size of attached cells and the density of aggregation and biofilm coverage was highest in the case of sensitized coupons followed by solution annealed and oxidized specimens. A difference in the colonization pattern was noticed with sensitized and solution annealed specimens showing bacterial cells embedded in exopolymers whereas cells were seen scattered on the oxidized surfaces. The smaller size of cells and lower bacterial density in the case of oxidized specimens suggests inhibition of bacterial growth on these specimens. Total viable counts of bacteria, estimated by the standard plate count technique, on 304 SS coupons with various microstructural/ surface characteristics and its variation with time of exposure were plotted in Fig. 11 . Total viable counts were the lowest on oxidized specimens and the highest counts were on sensitized specimens. The solution-annealed material had values in between. In addition, the rate of growth of bacteria on the sensitized surface was the maximum, followed by solution-annealed and oxidized.

The differences in the colonization behaviour of the three stainless steel surfaces studied suggest that attachment and proliferation of bacterial cells is influenced by the release of iron ions from the stainless steel surfaces, with the sensitized surface releasing the maximum and the oxidized the

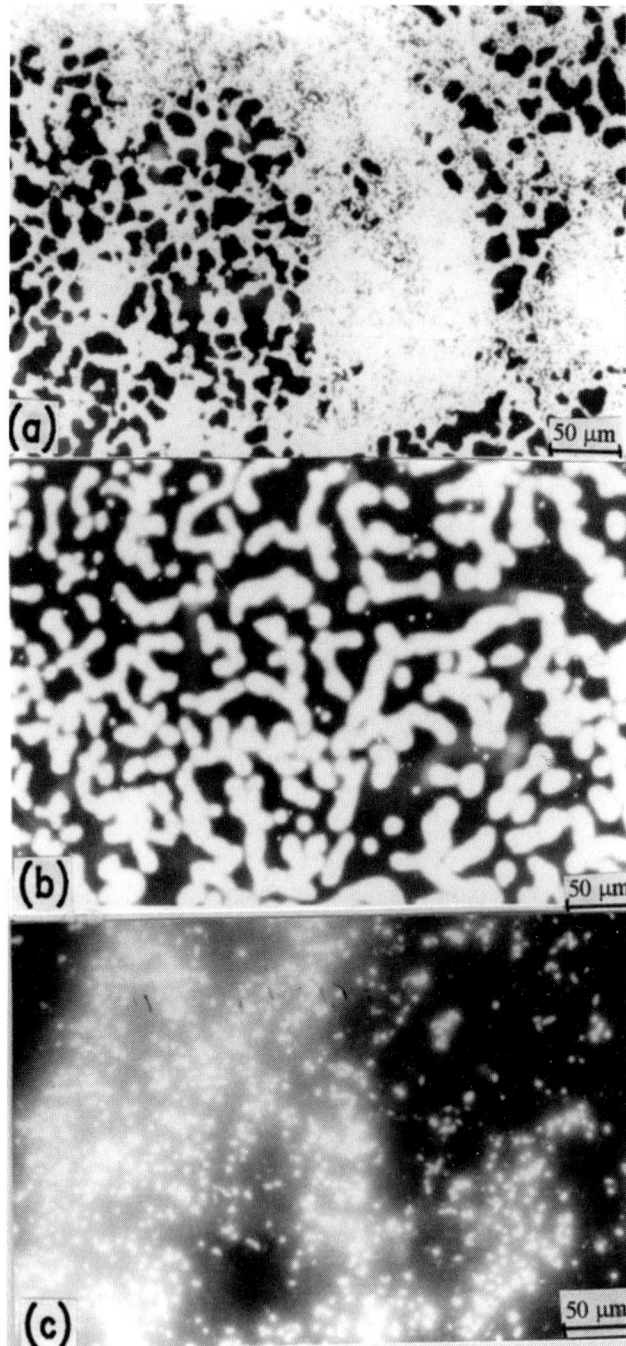

Fig. 10. Epifluorescence photographs showing adhesion, growth and biofilm formation of *Pseudomonas* sp. on AISI Type 304 SS specimens with different surfaces: (a) sensitized, (b) solution annealed and (c) oxidized.

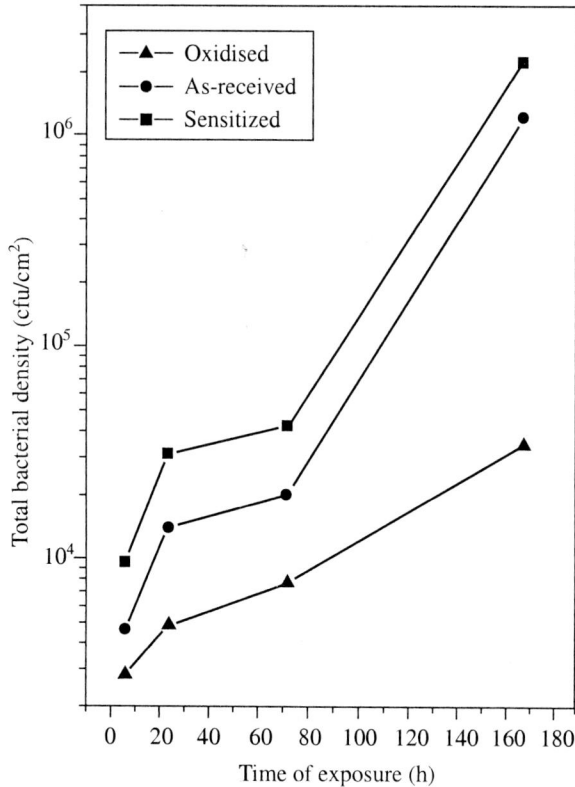

Fig. 11. **Total viable counts of *Pseudomonas* sp. on AISI Type 304 SS specimens with various microstructures and surface characteristics and its variation with time of exposure.**

minimum. Iron plays a fundamental role in microbial growth and metabolism. Studies on microbial physiology has shown that iron play a role in cytochrome structure and function, alternative pathways for respiratory electron transfer and pathogenicity [72]. The present results seem to explain the preferential attack of welds during MIC of fabricated components. It is possible that the microstructural and surface changes that are taking place during welding makes the weld more anodic, favouring preferential attachment and proliferation of microbes and the initiation of corrosion.

Control of MIC

The most effective means of control of biofouling and biocorrosion is to prevent the growth of the causative organisms, by a proper water treatment programme. Biocide treatments are often used as a countermeasure to kill or arrest the growth of these organisms [73]. Certain halogens and their compounds have been used as biocides for the last few decades. Chlorine has been a chemical of choice used due to its cost benefit and availability. However, there are several demerits of chlorine such as its corrosive nature, lack of penetrating power, delignification of wood and carcinogenic nature. Some of these problems can be avoided by using non-oxidizing biocides. However, the non-availability of proper biocide evaluation method is often a major drawback in many control programmes [74]. We have evaluated the influence of a non-oxidizing biocide (Legocide) on microbial colonization of stainless steel surfaces.

Biocide treatment is essential in water recirculating systems to minimize biofouling, which reduces heat transfer, increases pumping costs, causes MIC, degrades water treatment chemicals such as corrosion inhibitors and descalants, cause structural damage if microbial growth becomes heavy and can even have health implications [75]. Many workers have extensively reviewed [73] the data on the use of various biocides and found that most of them fail in field conditions. According to them, this is mainly because of improper evaluations often based on bacterial counts in water. Microorganisms present in biofilms are more resistant to biocide treatments due to various reasons. Many investigators believe that microorganisms present in biofilms have different physiological activities compared to planktonic microorganisms which may provide them some sort of biocide sensitivity [76].

In our study, planktonic and sessile bacterial counts were evaluated to test the efficacy of legocide in static and dynamic fresh water systems. 90% reduction of total viable counts (TVC) of bacteria in water at 20 ppm concentration of legocide was achieved after 5 hours. However, assays on SS specimens showed 90% reduction only at 40 ppm biocide concentration and after a contact time of 24 hours. These results clearly indicated that evaluations based on planktonic bacterial counts often did not give a realistic picture of the actual control achieved.

Another noteworthy observation was the effect of legocide concentrations on developed biofilms on SS specimens. Results showed less than 50% reduction in TVC on specimens. Microorganisms present in biofilms appears to be protected from the action of biocides, either by lack of penetration of the chemicals through the polysaccharide matrix and/or because of an altered cell sensitivity [77]. Studies with 80 ppm and 100 ppm biocide concentrations in the recirculating system showed that total control of TVC on the specimens and in water was achieved only at 100 ppm concentration compared to 40 ppm biocide concentration in the static studies. Time-series studies on the reduction of cell numbers at 100 ppm legocide concentration in the recirculating system showed that 50% reduction in bacterial density in the water was achieved after 30 minute contact time, whereas on specimens, 50% reduction was seen only after 4h contact time. Thus, as shown in Fig. 12, for the same biocide concentration the reduction in cell number in the biofilm would never be equal to that in the planktonic phase. According to Gaylarde et al. [78] this is because of the time taken by the biocide to reach half-maximum concentration in the biofilm by diffusion was 100 times that of cell half-life. Gaylarde used two biocides and showed that the phenolic biocide having a horizontal line of sessile cell death rate indicate that, the cells within the biofilm were killed extremely slowly. Such biocides are considered to have a high concentration exponent (the effect of concentration on the rate of cell death varies with the power of concentration) and their reduced concentration in the biofilm will affect their biocidal action. However, a heavy metal biocide was more penetrative and the plot of sessile cell death rate against time was more sloping compared to phenolic biocide. When the performance of the biocide (legocide) was compared with these two biocides, it was found that legocide showed a sloping sessile cell death rate like the heavy metal biocide and hence its performance is good in these waters.

According to Gaylarde et al. [78], this departure of sessile cell death rate line from linearity indicates that the diffusional delays are an important factor in determining biocide activity. But the effect of diffusion rate of the biocide on cell death is complex since factors determining the rate of increase of free biocide in the biofilm are also complex. Gaylarde and his co-workers have inferred that biocides are held in the biofilm by ionic forces as a complex. In some cases, hydrogen bonding and Van der Walls forces may be important. Many a times the biocides may form inert complexes. Hence, in the case of a biocide with a lot of difficulty in diffusing into the biofilms, even after dosing the waters with high concentration of the biocide, sufficient concentration of biocide within the

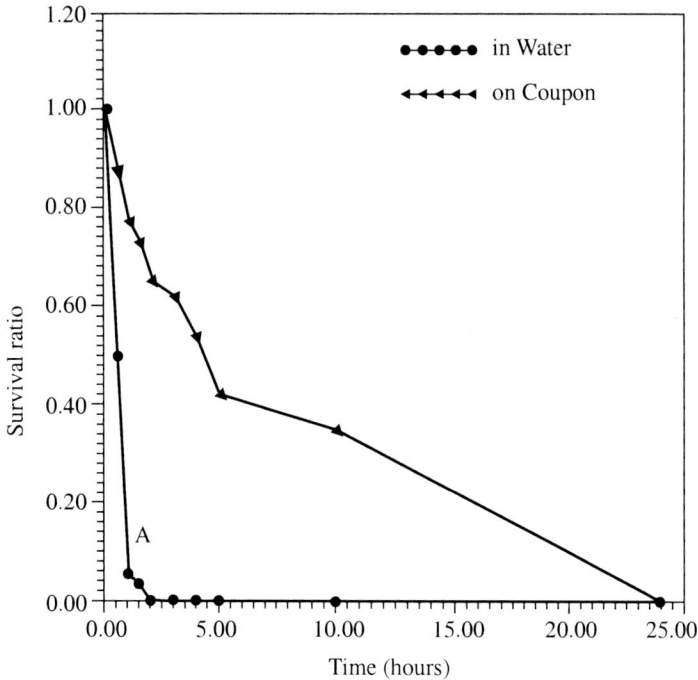

Fig. 12 **Effect of 100 ppm legocide on the survival ratio of bacteria in the biofilm and water.**

biofilm for sufficient time to kill the microbes may not be attained. Hence detectable biocidal death of cells in biofilm will occur only after a long period, even though the concentration of the biocide in the medium, outside the biofilm is many times higher than what is required to kill the microbes in the planktonic phase. Another interesting observation was the total control of slime forming gram-negative *Pseudomonas* sp. with this biocide. This has great relevance in the MIC studies, as slime formers are the predominant biofilm formers. The survivors of this biocide treatments were gram-positive microbes. The presence of susceptible gram-negative and tolerant gram-positive bacteria might explain the concavity of the plot of survival ratio with time.

CONCLUSIONS

Microbial colonization of stainless steel (SS) surfaces changes it's corrosion behaviour and makes the alloy susceptible to localized corrosion even in freshwater having low chloride content. Investigations that were carried out in our laboratory has shown that microbial film formed under sun-lit conditions is rich in photosynthetic organisms such as algae. A positive shift in open circuit potential (ennoblement) of a type 304 SS was observed in the presence of such a biofilm. Cyclic polarization studies on type 304 SS with biofilms of different age and thickness have shown that although a young biofilm gives certain protection against localized corrosion attack, crevice corrosion was found to initiate as the age and thickness of the biofilm increased. Exposure of 304 SS coupons with different microstructure in a nutrient broth of *Pseudomonas* species has shown higher attachment and growth of the bacterial species on sensitized specimens, compared to oxidized and solution annealed specimens. This is possibly one of the reasons for the preferential attack of the welded region of fabricated SS components during MIC. However, more studies are required to establish the influence of microstructure on microbial adhesion and MIC.

REFERENCES

1. Dexter, S.C., *Corrosion, Metals Handbook* (9th ed.), ASM International, Metals Park, Ohio, Vol. 13, 1987, p. 118.
2. Borenstein, S.W., *Microbiologically Influenced Corrosion Handbook*, Woodhead Publishing Limited, Cambridge, England, 1994.
3. Iverson, W.P., *Advances in Corrosion Science and Technology*, Vol. 2, Fontana, M.G. and Stalhle, R.W., (eds.), Plenum Press, New York, 1972, p. 1.
4. Miller, J.D.A., *Microbial deterioration*, Rose, A.H. (ed.), Academic press Inc., London, 1981.
5. Elphick, J.J., *Microbial aspects of metallurgy*, Miller, J.D.A., (ed.), Elsevier, New York, 1970, p. 157.
6. Characklis, W.G. and Marshall, K.C. (eds.), *Biofilms*, John Wiley and Sons Inc., New York, 1990.
7. Thierry, D. and Sand, W., *Microbiologically Influenced Corrosion, Corrosion Mechanisms in Theory and Practice*, Marcus, P. and Oudar, J. (eds.) Marcel Dekker Inc., New York, 1995, p. 458–499.
8. Charackilis, W.G. and Cooksey, K.C., *Advances in Applied Microbiology*, Vol. 29, 1983, p. 93.
9. Daumas, S., Massiani, Y. and Crousier, J., *Corrosion Science*, Vol. 28, No. 11, 1988, p. 1041.
10. Booth, G.J., *Journal of Applied Bacteriology*, Vol. 27, No. 1, 1964, p. 174.
11. Tiller, A.K., *Corrosion Processes*, Applied Science Publishers, 1982, p. 115.
12. Decker, R.F., *Metallurgical Transactions*, Vol. 17A, 1986, p. 20.
13. Pelczar, M.J. Jr., Reid, R.D. and Chan, E.C.S., *Microbiology*, TMH Edition, New Delhi, 1982, p. 3.
14. Ehlert, I., *Material U. Organismen*, Vol. 2, 1967, p. 297.
15. Hamilton, W.A., *Annual Review in Microbiology*, Vol. 39, 1985, p. 185.
16. Maale, O., *The Bacteria*, Vol. IV, Gunsalus, I.C. and Stainer, R.Y. (eds.), Academic Press Inc., London, 1981.
17. Hennington, B.O., *International Biodeterioration 7*, Houghton, D.R., Smith, R.N. and Eggins, H.O.W. (eds.), Elsevier, London, 1988, p. 703.
18. Costerton, J.W., Geesey, G.G. and Jones P.A., Corrosion/87, paper No. 54, NACE Publications, Houston, TX, 1987.
19. Schriffrin, D.J. and Desanchez, S.R., *Corrosion*, Vol. 31, 1985, p. 31.
20. Obuekwe, C.O., Westlake, D.W.S., Plambeck, J.A. and Cook, F.D., *Corrosion*, Vol. 37, 1981, p. 461.
21. Natarajan, K.A. and Ramesh, T., *Marine Biofouling and Power Plants*, BARC, Bombay, 1990, p. 122.
22. Von Wolzogen Kuhr, C.A.H. and Van der Vlugt, L.S., *Water Den. Haag*, Vol. 18, No. 16, 1934, p. 147.
23. Tatnall, R.A., *Material Performance*, Vol. 20, 1981 b, p. 41.
24. Terry, L.A. and Edyvean, R.G.J., *Algal Biofouling*, Evans, L.V. and Hoagland, K.D. (eds.), Elsevier, Amsterdam, 1986, p. 179.
25. Stoecker, J.G. and Pope, D.H., *Material Performance*, Vol. 25, No. 6, 1986, p. 51.
26. Tiller, A.K., *Microbial Corrosion I.*, Sequiera, C.A.C. and Tiller, A.K. (eds.), Elsevier Applied Science, New York, 1988, p. 6.
27. Garner, A., *Materials Performance*, Vol. 21, No. 8, 1982, p. 9.
28. Kobrin, G., *Biologically Induced Corrosion*, S.C. Dexter (ed.), NACE, Houston, 1986, p. 33.
29. Dundas, J.J. and Bond, A.B, *Materials Performance*, Vol. 24, No.10, 1985, p. 54.
30. Walsch, D., Willis, E., VanDiepen, T., and Sanders, J., *CORROSION 94*, NACE, 1994, paper 612.
31. Dexter, S.C. and Zhang, H.J., *Effects of biofilms, sunlight and salinity on corrosion potential and corrosion initiation of stainless steels*, EPRI NP-7275, Final Rep., Proj. 2939-4, Electric Power Research Institute, PaloAlto, CA, 1991.
32. Dexter, S.C., *Bulletin of Electrochemistry*, Vol. 12, 1996, p. 1.
33. Dexter, S.C. and Zhang, H.J., *Proceedings of* 11th *International Corrosion Congress*, Florence, Italy, Vol. 4, 1990, p. 333.

34. Little, B., Ray, R., Wagner, P., Lewandowski, Z., Lee, W.C., Charackilis, W.G. and Mansfeld, F., *Biofouling*, Vol. 3, 1991, p. 45.
35. Motoda, S., Suzuki, Y., Shinohara, T. and Tsujikawa, S., *Corrosion Science*, Vol. 33, No. 3, 1992, p. 445.
36. George, R.P. , Muraleedharan, P., Parvathavarthini, N., Khatak, H.S. and Rao, T.S., *Materials and Corrosion*, Vol. 51, No. 1-6, 2000, p. 1.
37. Mansfeld, F. and Little, B., *Corrosion Science*, Vol. 32, 1991, p. 247.
38. Dowling, N.J.E., Guezennec, J., Bullen, J., Little, B. and White, D.C., *Biofouling*, Vol. 5, 1992, p. 315.
39. Dexter, S.C., Chandrasekharan, P. , Zhang, H. and Wood, S., *Proceedings of 1992 Biocorrosion and Biofouling Workshop*, Videla, Lewandowski and Lutey (eds.), Buckman laboratories, Memphis, TN, 1993, p. 171.
40. Chandrasekharan, P. and Dexter, S.C., *CORROSION/93*, NACE, Houston, TX, Paper No. 493, 1993.
41. Mollica, H. and Trevis, A., *Proceedings of 4th International Corrosion Congress on Marine Corrosion and Fouling*, Antibes, France, 1976, p. 351.
42. Scotto, V., Dicintio, R. and Marcenaro, G., *Corrosion Science*, Vol. 25, 1985, p. 185.
43. Dexter, S.C. and Gao, G.Y., *Corrosion*, Vol. 40, 1988, p. 717.
44. Maruthamuthu, S., Rajagopal, G., Sathianarayanan, S., Eashwar, M. and Balakrishnan, K., *Biofouling*, Vol. 8, 1995, p. 223.
45. Eashwar, M., Maruthamuthu, S., Palanichamy, S. and Balakrishnan, K., *Biofouling*, Vol.8, 1995, p. 215.
46. Dickinson, W.H., Caccavo, S., Jr., Bo Olesen and Lewandowski, Z., *Applied and Environmental Microbiology*, Vol. 63, No. 7, 1997, p. 2502.
47. Baboian, R. and Haynes, G.S., *Electrochemical Corrosion Testing*, Mansfeld and bertocci (eds.), STP 727, ASTM, 1979, p. 274.
48. Morris, D.W., *Proceedings of Seminar on Nuclear Plant Layup and Service Water System Maintenance*, North Carolina, 1987, p. 4.
49. Hoar, T.P., Mears, D.C. and Rothwell, G.P., *Corrosion Science*, Vol. 5, 1965, p. 279.
50. Carew, J. and Abdullah, A., *Corrosion Science*, Vol. 34, No. 2, 1993, p. 217.
51. Dexter, S.C., *Biofouling*, Vol. 7, 1993, p. 97.
52. George, R.P., Muraleedharan, P., Parvathavarthini, N., Khatak, H.S. and Newman, R.C., *Microbial Corrosion*, European Federation of Corrosion series, EFC 29, Institute of Materials, UK, 2000, p. 116.
53. Rao, T.S., Rani P. George, Venugopal, V.P. and Nair, K.V.K., *Biofouling*, Vol. 11, No. 4, 1997, p. 265.
54. Schutz, R.W., *Materials Performance*, Vol. 31, 1992, p. 58.
55. Syrett, B.C. and Colt, R.L., *Materials Performance*, Vol. 22, 1983, p. 44.
56. Licina, G.J. and Cubicciotti , D., *Journal of Materials (JOM)*, 1989, p. 23.
57. Borenstein, S.W., *Materials Performance*, Vol. 27, 1988, p. 62.
58. Soracco, R.J., Pope, D.H., Eggers, J.M. and Effinger, T.N., *CORROSION/88*, NACE, Houston, TX, 1988, paper 83.
59. Licina , G.J, *Materials Performance*, Vol. 28, 1989, p. 55.
60. Dowling, N.J.E., Franklin, M., White, D.C., Lee, C.H. and Lundin, C., *CORROSION/89*, NACE, Houston, TX, 1989, paper 187.
61. Stein, A.A., *CORROSION/91*, NACE, Houston, TX, paper 107.
62. George, R.P., Sreekumari, K.R., Muraleedharan, P., Khatak, H.S., Gnanamoorthy, J.B. and Nair, K.V.K., *CORSIONON-97*, NACE (India), Mumbai, 1997, p. 924.
63. Buchanan, R.A., Zhang, X., Li, P., Stansbury, E.E., Dowling, N.J.E., Hall, T. and Lindberg, A., *Microbiologically Influenced Corrosion and Biodeterioration*, N.J. Dowling, M.W. Mittleman and J.C. Danko, (eds), The University of Tennessee, Knoxville, 1990, p. 3/99.
64. Mele, M.F.L.de, Moreno, D.A., Ibars, J.R. and Videla, H.A., *Corrosion*, Vol. 47, 1991, p. 24.
65. Videla, H.A., Mele, M.F. L.de, Moreno, D.A., Ibars, J.R. and Ranninger,C., *CORROSION/91*, NACE, Houston, TX, 1991, paper 104.
66. Hunt, A.P. and Parry, J.,D., *Biofouling*, Vol. 12, No. 4, 1998, p. 287.
67. Vanhaecke, F., Remon, J.P., Moors, M., Raes, F., De Rubber, D. and Van Peteghem, A., *Applied and Environmental Microbiology*, Vol. 56, 1990, p. 788.
68. Holah, J.T. and Thorpe, R.H., *Journal of Applied Bacteriology*, Vol. 69, 1990, p. 599.

69. Boulange-Petermann, L., *Biofouling,* Vol. 10, 1996, p. 275.

70. Busscher, H.J., Bellon-Fontaine, M.N., Mozes, N., Van der Mei, H.C., Sjollema, J., Cerf, O. and Rouxhet, P.G., *Biofouling*, Vol. 2, 1990, p. 55.

71. Fletcher, M. and Loeb, G.I., *Applied and Environmental Microbiology*, Vol. 37, 1979, p. 67.

72. Hughes, M.N. and Poole, R.K., *Metals and Microorganisms*, Chapman and Hall, London , 1989.

73. England, A.C., Frazer, D.W., Mallison, G.F., Mackel, D.C., Skaily, P. and Gorman, G.W., *Applied and Environmental Microbiology*, Vol. 43, 1982, p. 240.

74. Costerton, J.W., Geesey, G.G. and Jones, P.A., *Materials Performance*, 1988, p. 49.

75. Fliermans, C.B., Cherry, W.B., Orrison, L.H., Smith, S.J., Tison, D.L. and Pope, D.H., *Applied and Environmental Microbiology*, Vol. 41, 1981, p. 9.

76. Anwar, H., Strap, J.L. and Costerton, J.W., *Antimicrob. Chemth.*, Vol. 36, 1992, p. 1347.

77. Hoyle, B.D., Jass, J. and Costerton, J.W., *Antomicrob. Aq. Chemob.*, Vol. 26, 1990, p. 1.

78. Gaylarde, P.M. and Gaylarde, C.C., *International Biodeterioration and Biodegradation,* Vol. 29, 1992, p. 273.

10. Corrosion of Austenitic Stainless Steel in Liquid Sodium

S. Rajendran Pillai[1] and H.S. Khatak[2]

Abstract Several factors influence the corrosion of austenitic stainless steel by liquid sodium such as temperature, flow rate and the contents of various impurities in sodium. The paper analyses the various factors. A study has been carried out in Indian sodium to examine the different modes of corrosion. Stainless steel (AISI 316) samples were exposed to sodium in a dynamic loop at temperatures in the range of 583 to 823 K for 16000 h. The velocity of sodium was maintained at 5 m/s. These samples were retrieved and analyzed for the various modes of corrosion. The leaching and depletion of elements on the surface exposed to sodium was analyzed by Energy Dispersive X-ray Spectrometry (EDS). The preferential leaching of nickel followed by the formation of ferrite phase was noticed at 823 K. The nature of the corrosion product formed on the surface was examined by X-ray Photoelectron Spectroscopy (XPS). The compounds containing Fe-O and Cr-O bonds were noticed. The XPS spectra also revealed the segregation of carbon at some locations. The carbon profile was analyzed using Secondary Ion Mass Spectrometry (SIMS). The value of the effective diffusion coefficient of carbon was calculated and the carbon profile likely to be achieved on exposure to sodium up to 30 years (at a temperature of 823 K) was evaluated. The cumulative effect of three modes of corrosion (complete leaching, formation of degraded layer and formation of carburised austenite) has been analyzed.

Key Words Sodium corrosion, carburization, leaching, surface analysis, carbon profile, degraded layer.

INTRODUCTION

Sodium possesses several favourable heat transport and neutronic properties that makes it an appropriate choice as coolant in Fast Breeder Reactors (FBRs). In these reactors, austenitic stainless steels of different grades are used as structural material. This choice is based on the reasonably good compatibility of stainless steel with high temperature sodium [1, 2]. The exposure of stainless steel to high temperature sodium does not result in any chemical reaction. However, the presence of certain impurities (which can dissolve in sodium) causes detrimental effects on the mechanical behaviour of the structural materials. Moreover, liquid sodium can dissolve some of the constituents of stainless steel [3, 4], which are transported and deposited at different regions. Such mass transfer of metallic element is influenced by the contents of non-metallic impurities such as oxygen and carbon in sodium [5, 6] because leaching is preceded by the generation of ternary compounds of oxygen with sodium and the

[1]Scientific Officer, [2]Head, Corrosion Science and Technology Division, Materials Characterization Group, Indira Gandhi Centre for Atomic Research, Kalpakkam, India-603 102.

transition metal (one or more of the constituents of stainless steel). The most commonly encountered corrosion product in sodium is sodium chromite [7–10]. Ternary compounds of other transition metals are formed only if the oxygen content is very high [9, 10]. Such high concentration of oxygen is not usually encountered in sodium systems where the purity of sodium is controlled by continuously removing impurities through cold trapping. However a reliable understanding of the nature of impurities and their contribution to material behaviour is essential to gain confidence in running a system even under conditions of accidental ingress.

The consequences of corrosion by sodium are many folds. In addition to affecting the heat transport properties, the preferential leaching of elements causes changes in mechanical properties, [11–18]. Hence, control of the corrosion rate is vital for ensuring acceptable mechanical properties particularly of thin components, during the life of the reactor. A brief description of the consequences of exposure of stainless steel to liquid sodium are schematically shown in Fig. 1. The different processes involved in the corrosion by liquid sodium are:

(i) Complete loss of material by leaching and the consequent reduction in the wall thickness.
(ii) Preferential leaching of elements and generation of modified layer of inferior properties.
(iii) Formation of carburized or decarburized austenite layer, the depth of which depends on the duration of exposure, the temperature and the carbon potential of sodium.

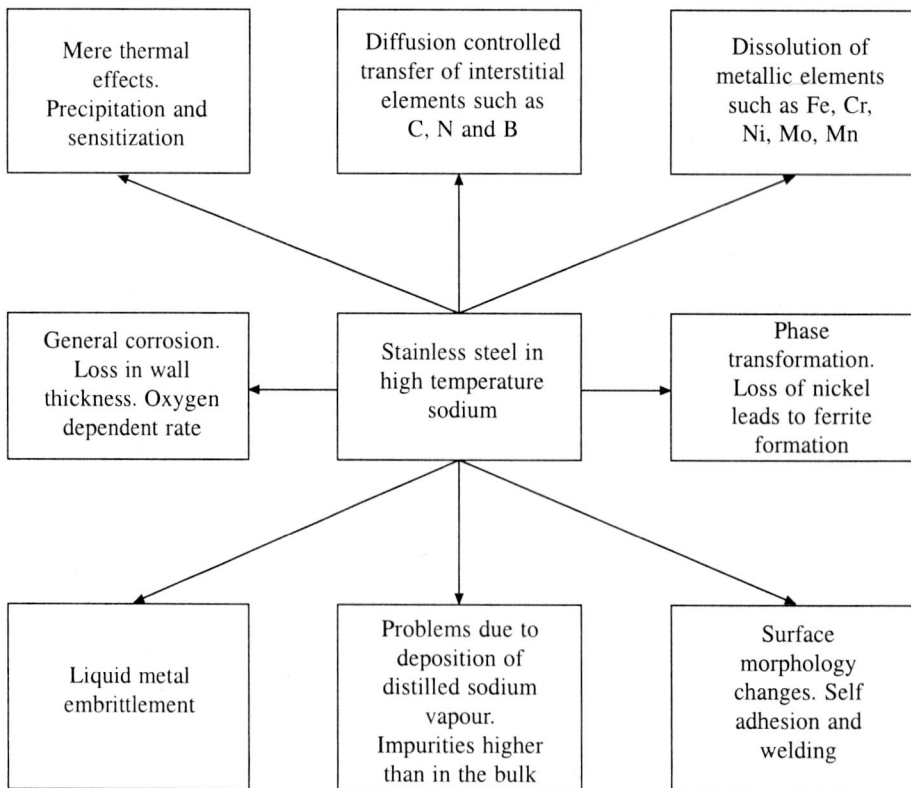

Fig. 1.   Consequence of exposure of stainless steel to high temperature sodium.

(iv) Precipitation of carbides in the matrix brought about by thermal effects coupled with the carbon transfer through sodium.

(v) Accelerated corrosion rate due to erosion by the high velocity sodium which causes preferential corrosion and internal corrosion.

FACTORS INFLUENCING CORROSION BY SODIUM

Even though the corrosion by sodium has been studied by several countries, there are very little attempt to quantify the result based on test parameters. Evidently, different countries have reported different rates of corrosion because of a strong dependence on the experimental parameters. Pure liquid sodium is nearly compatible with steels of both austenitic and ferritic grades, leaving apart slight dissolution of the elements at elevated temperatures. The marked difference in the corrosion properties arises from the impurity contents of sodium and process parameters of the test system. Such a situation rendered the generalized application of the corrosion equation very difficult. The different factors are described below.

SOLUBILITY OF METALS IN SODIUM

Iron, nickel, chromium, molybdenum and manganese are the main components of the structural material. Precise data on solubility of these elements are required to evolve a corrosion model and to explain the process that can undergo in a sodium-steel interface. Solubility of elements such as manganese is extremely important because of the existence of this element in the radioactive form in the primary circuit of a reactor. Iron and chromium can also exist in the activated form.

Several laboratories of the world have embarked on activities related to the measurement of solubility of metals in sodium. However, all the published data are not in good agreement mainly because of the strong dependence on various process parameters of the system, which influence the solubility. These are

(i) Presence of non-metallic contaminants such as oxygen, carbon and hydrogen which interact with metallic elements to form complex compounds.

(ii) Temperature.

(iii) Formation of fine suspension of the metal that affects the analytical determination of solubility.

The solubility-temperature relationship can be expressed mathematically by an Arrhenius-type equation.

$$S_0 = A \exp\left(- Q/RT\right) \qquad (1)$$

where S_0 is the solubility at temperature T (K), Q the activation energy for the solution process, R the gas constant and A the pre-exponential factor.

A comprehensive survey on the solubility of different elements in sodium measured by different laboratories has been reviewed by Claar [3]. A review of the solubility of iron, nickel, manganese and chromium has been carried out by Awasti and Borgstedt [4]. The table below lists the solubility expression of major elements encountered in sodium systems.

Solubility of metals is strongly dependent on the concentration of non-metals in sodium. This is more evident in the case of metals that form ternary oxygen compounds on reaction with sodium and oxygen. The most commonly encountered corrosion products in sodium system are $NaCrO_2$ and

Table 1.   Solubility of different constituents of stainless steel in sodium

Element	Ref.	Solubility expression	Typical values of solubility (PPM) at temperatures of (K)		
			673	773	873
Iron	[19]	log S (PPM) = 5.16 − 4310/T	0.06	0.38	1.7
Chromium	[20]	log S (PPM) = 9.35 − 9010/T	0.000092	0.00049	0.11
Manganese	[21]	log S (PPM) = 2.325 − 2017/T	0.2	0.52	1.03
Nickel	[22]	log S (PPM) = 1.47 − 918/T	1.3	1.9	2.06

Na_4FeO_3. The threshold concentration of oxygen in sodium that is required for the formation of $NaCrO_2$ is very small [23] and hence this compound is encountered even if the quality of sodium is stringently controlled through effective cold trapping. As such, the leaching and the dissolution of chromium invariably occurs through the generation of sodium chromite and a special influence of increase of concentration of oxygen on the leaching mechanism of chromium is not expected. On the other hand, the influence of concentration of oxygen on the solubility of iron in sodium is quite significant. At high concentration of oxygen, the formation of Na_4FeO_3 [24, 25] occurs and the leaching rate of iron increases significantly. Another constituent of stainless steel that forms a ternary compound is manganese [24] and therefore its solubility is also strongly dependent on the oxygen level in sodium.

The possible reactions that result in the generation of ternary compounds are given below.

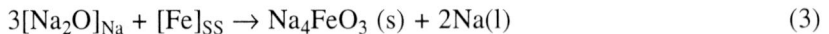

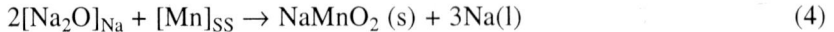

$$2[Na_2O]_{Na} + [Cr]_{SS} \rightarrow NaCrO_2 \text{ (s)} + 3Na(l) \tag{2}$$

$$3[Na_2O]_{Na} + [Fe]_{SS} \rightarrow Na_4FeO_3 \text{ (s)} + 2Na(l) \tag{3}$$

$$2[Na_2O]_{Na} + [Mn]_{SS} \rightarrow NaMnO_2 \text{ (s)} + 3Na(l) \tag{4}$$

The threshold concentration of oxygen in sodium for the formation of the different ternary compounds when liquid sodium comes in contact with type AISI 316 SS are given in Table 2 below.

Table 2. Threshold oxygen levels in Na(l)/AISI 316 SS system for the formation of different corrosion products

Compound formed	log $[O]_{thresh}$ = A+B/T	Threshold concentration of oxygen (PPM) at temperatures of (K)	
		773	973
$NaCrO_2$	12.3233 − 9949.8/T	0.576	8.126
Na_4FeO_3	4.973 − 1494.8/T	1095	2732
$NaMnO_2$	3.4541+286.0/T	(6669)	5598

Note: The value of threshold oxygen levels for $NaMnO_2$ at 773 K is indicated in parenthesis as this value is higher than the solubility limit of Na_2O in sodium. Hence, Na_2O will precipitate even before the formation of $NaMnO_2$.

Solubility of Non-Metals in Sodium
Among the constituents of steel, carbon is the only non-metallic element that dissolves and transports through sodium. Nitrogen does not dissolve in sodium. However, its transport is reported to occur

through sodium by the formation of a molecular adduct with sodium. Further studies are required to understand the actual mechanism. The dissolution of oxygen in sodium merits serious attention as this element can directly participate in the leaching of the metals. Different investigators have estimated the solubility of oxygen in sodium. Sreedharan and Gnanamoorthy [26] have carried out a critical examination of the various determinations and concluded that the data reported by Noden [27] is the most reliable. The expression relating solubility and temperature proposed by Noden is given by,

$$\log C_0^{\text{sat}} \text{ (PPM)} = 6.2571 - 2444.5/T \tag{5}$$

where C_0^{sat} is the solubility of oxygen in sodium.

Carbon, dissolved in sodium, also participates in the corrosion process. The transport of carbon to the different regions promotes the formation of carbides. Elements such as chromium readily form carbide and participate in the corrosion process. As in the case of oxygen, the solubility of carbon in sodium has also been determined by different investigators. Initial investigators reported wide variation in solubility. The accurate determination of the solubility of carbon in sodium has been rendered difficult because of the poor rate of dissolution of carbon coupled with its capacity to form insoluble suspension. Hence, in most of the determinations the solubility has been evaluated by an isopiestic method that involves simultaneous equilibration of a metal of known carbon activity concentration relationship. Thompson [28] carried out a critical analysis of the different solubility data and proposed an averaged expression based on the data reported in the literature which is given by

$$\log C_c^{\text{sat}} \text{ (PPM)} = 7.449 - 5858/T \tag{6}$$

where C_c^{sat} is the solubility of carbon in sodium.

The solubility of carbon is in the range of 1 PPM at 773 K to 7 PPM at 973 K.

Hydrogen is encountered in sodium system mainly through the reaction with moisture from the surrounding atmosphere. Hydrogen readily dissolves in sodium and is transported to different regions of the system. Also, reaction of water with sodium promotes the generation of both sodium hydride and hydroxide. Both these species can undergo decomposition at high temperature, thus releasing free hydrogen into the cover gas of the sodium system.

Vissers et al. [29] measured the solubility of hydrogen in sodium with the help of a diffusion-type hydrogen meter. The solubility values as determined in the temperature range 383 to 673 K is represented by the equation

$$\log S_{\text{H}} = 6.067 - 2880/T \tag{7}$$

where S_{H} is the solubility of hydrogen in PPM and T the temperature in K. The solubility varies from 1 PPM at 473 to 60 PPM at 673 K.

Thermophysical Condition of the Material
The materials used in sodium systems, in general, possess wide variation in the thermal history. Most of the materials are subjected to differing thermal and mechanical treatments to arrive at optimum properties. The content of nickel is reported to have a direct bearing on the corrosion of the material in sodium. Material with high content of nickel exhibited high rate of corrosion [30]. Further studies also revealed that, thermal history such as precipitation hardening also affects the rate of corrosion [31]. The carbon transfer behaviour of austenitic steel has been studied by Roy and Wozadlo [32] who

reported an influence of decarburization rate based on the nature of the carbide species formed in steel and sodium. The generation of a particular species is largely decided by the prior heat treatment history of the material. Surface condition of the material also has a strong influence on the rate of corrosion.

Exposure Temperature

When all other factors remain constant, the corrosion rate-temperature relation is expected to follow the Arrhenius expression (log R is directly proportional to the reciprocal of absolute temperature) [33]. Mass loss due to corrosion is not the true indication of corrosion in many materials. Certain alloys exhibit a sub-surface metal removal producing a porous region in the depleted zone. If a ferrous alloy contains more than 3% Mo, a sub-surface layer of Fe-Mo nodes are reported to be generated due to corrosion by sodium [34]. These modes of corrosion have a strong bearing on the temperature of investigation.

Down-Stream Effect

The corrosion rate is reported to decrease in the direction of lowering temperature. This effect is known as down-stream effect and is used always in reference to an isothermal section. This effect is related to the hydraulic length-to-diameter ratio in the isothermal zone, which can be expressed as L/D, where L is linear distance from the end of the maximum temperature zone and D the hydraulic diameter.

$$D = 4 \times \text{Cross sectional area of flow/Wetted perimeter}$$

The decrease of rate of corrosion with down-stream position is approximated by the equation,

$$\log R_c = \log R_0 - k \, (L/D) \tag{8}$$

where R_c is the corrosion rate at any value of L/D, R_0 the corrosion rate at $L/D = 0$ and k the linear down-stream coefficient.

In order to compare the corrosion data among different laboratories, the corrosion rate at zero down-stream effect R_0 must be established corresponding to a particular temperature.

Axial Rate of Heating of Sodium

Axial heating rate (increase of temperature divided by the axial heat input distance) has been reported to have an effect on the rate of corrosion [35]. High value of the axial heating rate is found to enhance the rate of corrosion.

It is known that the rate of corrosion across a sodium/steel interface is mainly dependent on the thermodynamic activity differential of the dissolving species. Therefore, the increase of corrosion rate with increase of axial heating rate is attributable to a decrease in the activity of the species in sodium at high temperatures. When the activity differential is high the corrosion rate shows a corresponding increase.

Mechanical Stress on Materials

The effect of sodium environment and mechanical stress on the material has been investigated in several laboratories from the point of examining the impact on mechanical properties due to corrosion [36-41]. As the wall of the specimen is thinned on account of corrosion, it is more likely to creep under the influence of stress.

A systematic study on the role of stress on corrosion rate in sodium has not been attempted so far. Stress is expected to increase the diffusion process, precipitation reaction and the dissolution kinetics. Such increase is attributed to lattice dilation introduced by stress. Additional investigations are required to clearly establish the influence of residual, compressive and cyclic stresses on the rate of corrosion.

Effect of Velocity of Sodium

It has been reported [42-44] that at temperatures above 923 K, the corrosion of steel increases with increase in the velocity of sodium. This increase continues only up to a critical velocity and at higher velocities corrosion becomes independent of velocity. A steady state rate of corrosion is achieved when the velocity of sodium is in the range of 3-4 m/s. Based on this observation, it has been proposed that in the velocity dependent range, diffusion of reactants (corrosion product or oxygen) across the laminar sub-layer is the rate determining step. At higher velocities the laminar sub-layer diminishes substantially so that surface reaction becomes the rate controlling step. The velocity effect must be considered in conjunction with other factors influencing corrosion such as specimen geometry and composition.

Temperature Differential in the Loop

Under isothermal condition the corrosion of structural material by sodium is limited by the solubility of the elements at this particular temperature. However, in a non-isothermal system, especially under dynamic sodium conditions, leaching of elements at one point followed by their deposition at another point which is at a lower temperature can proceed unabated. Existence of a relative corrosion source (region of high thermodynamic activity of elements when compared to its activity in sodium) and sinks (of lower activity of elements when compared to the activity in sodium) are essential prerequisite for corrosion to be aided by the temperature differential. Precipitation of impurities at lower temperature may also lead to plating of the elements on the walls of the sodium-exposed components. It is desirable that the area of the sink is much higher than the corroding region to result in the maximum rate of corrosion at the particular temperature.

Duration of Exposure

Experiences by different investigators have shown that accelerated corrosion occurs in the initial period of exposure of iron-base alloys to sodium. After the initial period a steady state rate of corrosion is achieved. Thus, any corrosion data extrapolated from initial rate data (taken in the first 1000 h) leads to erroneous predictions. Hence any corrosion test has to be conducted at least for 2000 h so that the result can be safely extrapolated.

Dissimilar Metal Contact

The presence of a different alloys in the system can mask the rate of corrosion. Masking is defined as the decrease in the corrosion rate of a specimen in the down-stream region due to saturation, or a higher than normal concentration of an element in sodium. Such a situation is possible if a specimen with a higher activity is present in the up-stream region.

Typical example is the corrosion by nickel-base alloys. If an alloy with high nickel content is situated in the up-stream region, then the corrosion of the alloys in the down-stream region occurs at a reduced rate. Thus, if several alloys with different contents of nickel are chosen for studies related to rate of corrosion, it is desirable to arrange them in the order of increasing nickel content in the direction of flow of sodium. The analysis of corrosion data should take into account the activities of nickel in these alloys.

INTERACTION OF STEEL WITH SODIUM

Several countries carried out experiments in flowing sodium in order to assess the extent of corrosion and mass transfer [45–51] and the consequent changes in mechanical properties.

However, inter-comparison of different results is rendered difficult on account of the wide variation of the quality of sodium used by the different countries. Indian sodium contains a higher concentration of carbon. Hence in the authors' laboratory we have embarked on a programme to examine the behaviour of stainless steel in sodium procured from Indian sources [50]. The sodium employed contained nearly 10 PPM oxygen, 27 PPM carbon and 2 PPM calcium. This purity level was achieved by slagging the commercially procured sodium bricks after melting.

The exposure of the stainless steel specimen was carried out in a loop constructed using AISI 316 stainless steel. A schematic flow diagram of the loop is shown in Fig. 2. The loop consisted predominantly of two sections, viz. purification section and experimental section. Part of the sodium in the loop was continuously by-passed through the purification section to control the content of impurities to an acceptable level. The purification section consisted of a cold trap, cold trap economizer and a plugging indicator to measure the total content of the impurities. The main part of the loop contained several test sections where samples could be exposed at different temperatures. Comprehensive details about the loop have been reported in the literature [50].

The exposure of the specimen was carried out in the different test sections. The test section comprised of a cylindrical chamber (Fig. 3) of diameter 200 mm and length 250 mm. The specimens of stainless steel (dimension $150 \times 240 \times 12$ mm) were contained in the holder. Sodium was allowed to pass through the surface of the specimen. A 5 mm wide and 1 mm deep slit was cut on the surface of the plate through which sodium flowed at a velocity of 5 m/s. Steel specimens were also subjected to thermal ageing in the same test section to facilitate delineation of thermal effects from corrosive effects by sodium.

The chemical composition of stainless steel used as the specimen (in mass %) was Fe—0.054% C, 0.63% Si, 1.85% Mn, 0.045% P, 0.015% S, 16.92% Cr, 2.19% Mo, 11.69% Ni. The initial exposure was interrupted after a period of 16000 h. Discussion below is based on the analysis of the following specimen:

(i) exposed at 823, 723 and 623 K in the hot zone (up-stream region of the loop)
(ii) exposed at 623 and 583 K in the cold zone (down-stream region of the loop)

These specimens were removed from the loop, cleaned free of sodium by using alcohol and samples required for the different investigations were taken out.

Leaching of Metals

The dissolution and leaching of the metals from the sodium-exposed surface of the steel were estimated by Energy Dispersive X-ray Spectrometry (EDS). The sodium-exposed specimens were analyzed for different elements at varying depths from the exposed surface. The elemental concentrations of iron, chromium, nickel and molybdenum were determined in this way and the data given in Figs. 4 (a) to (e) for the different temperatures.

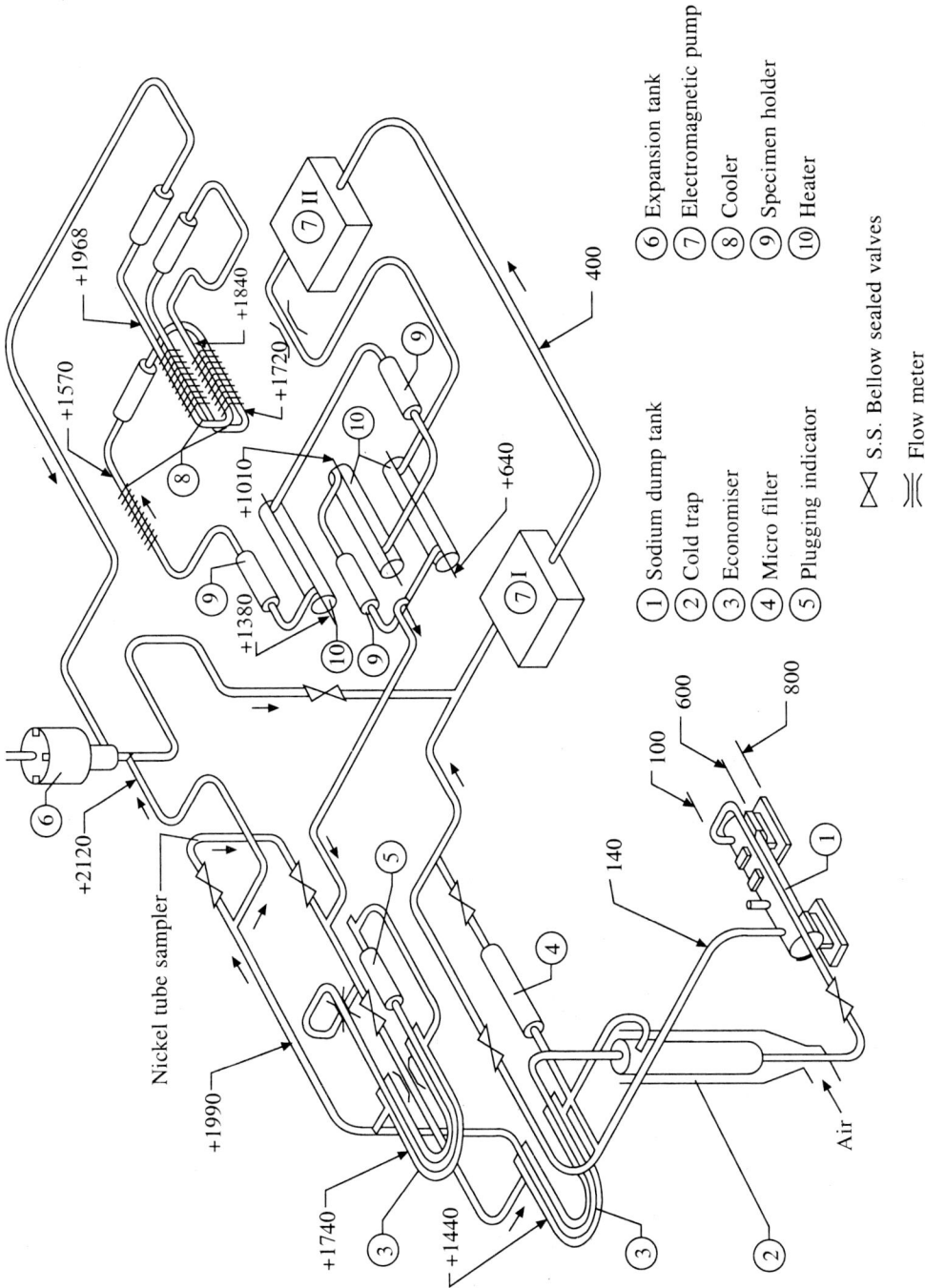

Fig. 2. Flow diagram of the mass transfer loop [50].

① Sodium dump tank
② Cold trap
③ Economiser
④ Micro filter
⑤ Plugging indicator

⑥ Expansion tank
⑦ Electromagnetic pump
⑧ Cooler
⑨ Specimen holder
⑩ Heater

⋈ S.S. Bellow sealed valves
≍ Flow meter

Nickel tube sampler

Air

+2120
+1990
+1968
+1840
+1740
+1720
+1570
+1440
+1380
+1010
+640
100
140
400
600
800

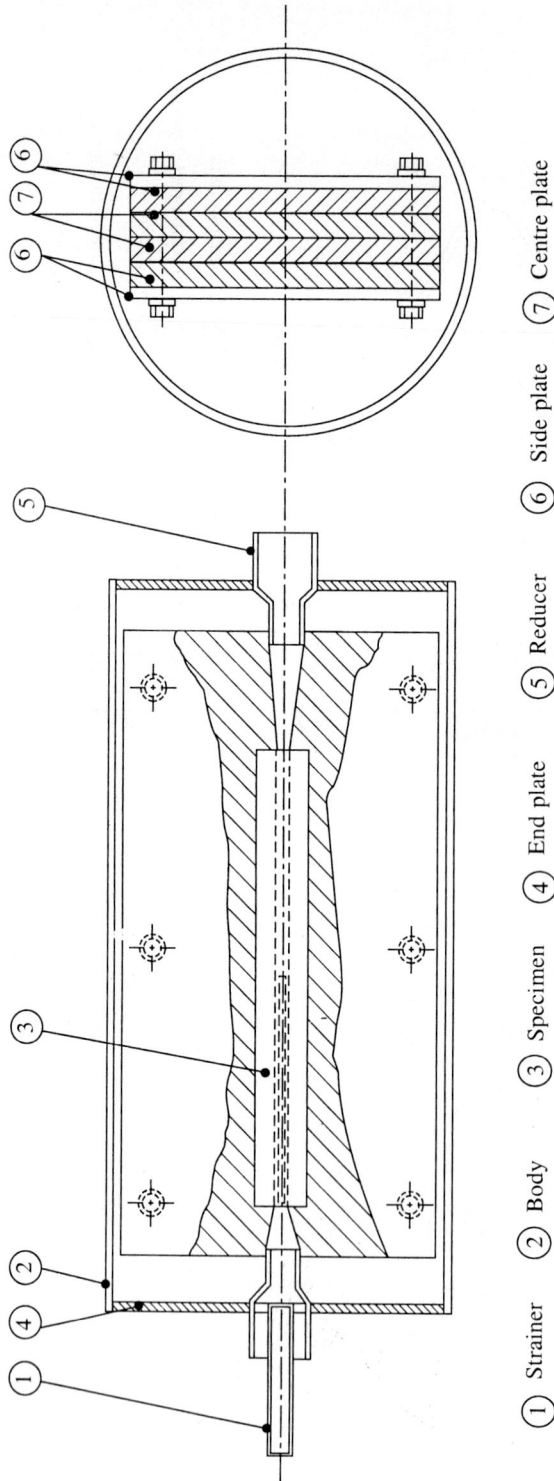

Fig. 3. Schematic of the sample holder in the mass transfer loop [50].

1 Strainer   2 Body   3 Specimen   4 End plate   5 Reducer   6 Side plate   7 Centre plate

The exposure of the specimen at 823 K caused observable change in the concentration of elements such as nickel and chromium and a marginal decrease in the concentration of molybdenum. The concentration of nickel at the surface was nearly 4 mass % when compared to 11.69 mass % originally present in the steel. Up to a depth of 10 μm from the sodium-exposed surface the concentration of nickel was less than 5 mass %. There after, the concentration showed an increase and attained the matrix value at a depth of 15 μm. The concentration of chromium was in the range of 12 to 15 mass % up to a depth of 12 μm from the sodium-exposed surface. Its concentration attained the constant value of the matrix at a distance of 25 μm. A noticeable leaching of molybdenum also occurred at this highest temperature location in the loop. The concentration of molybdenum was less than 1 mass % up to a depth of 7 μm. At approximately 15 μm depth, the concentration attained the constant value of the matrix. On the other hand, at 623 K in the up-stream region there was negligible depletion of elements. The specimen exposed at 723 K in the up-stream region also revealed negligible change in the concentrations of the elements at the surface exposed to sodium. The changes in the concentration of the elements were not high enough to promote the generation of a transformed phase. This behaviour is attributed to the poor kinetics of diffusion of elements in the stainless steel matrix at low temperature coupled with the low solubility at these temperatures. Analysis of the specimen from the down-stream region showed a different behaviour. There was a slight increase in the content of nickel at both the temperatures employed in the present investigation. This behaviour is accounted by the relatively high solubility of nickel in sodium. Nickel dissolved at high temperatures in the up-stream region and was transported to the down-stream position where the sodium temperature was lower. At this lower

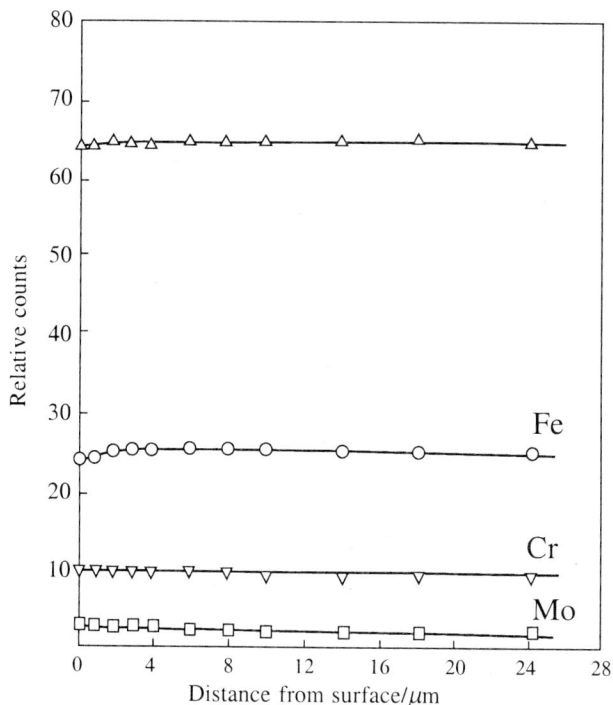

Fig. 4(a).   **EDS analysis of the specimen exposed at 623 K (up-stream) [50].**

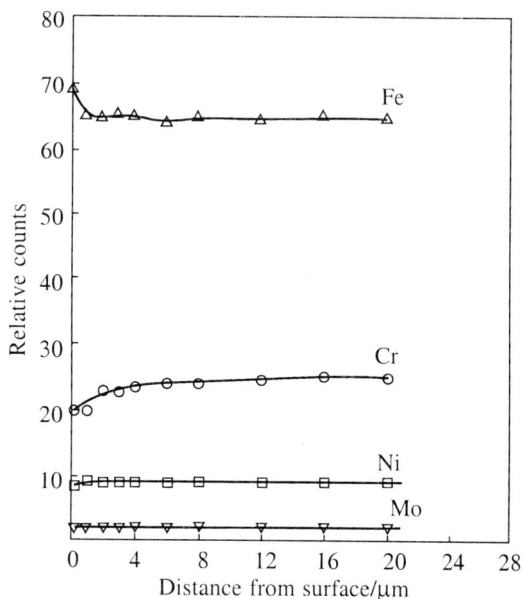

Fig. 4(b).   **EDS analysis of the specimen exposed at 723 K (up-stream) [50].**

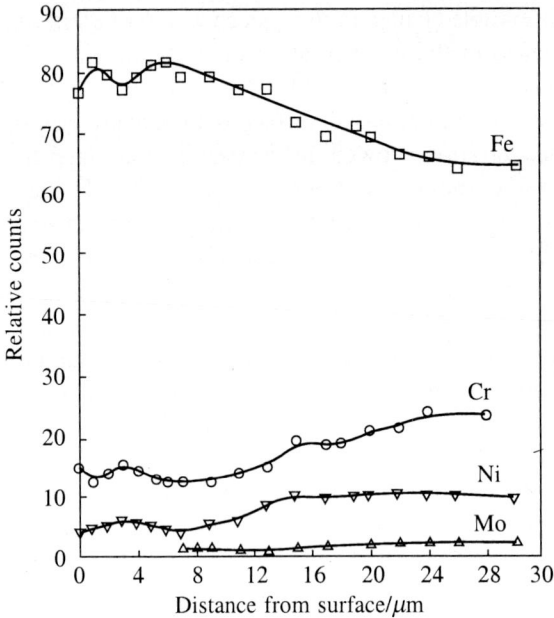

Fig. 4 (c). EDS analysis of the specimen exposed
at 823 K [50].

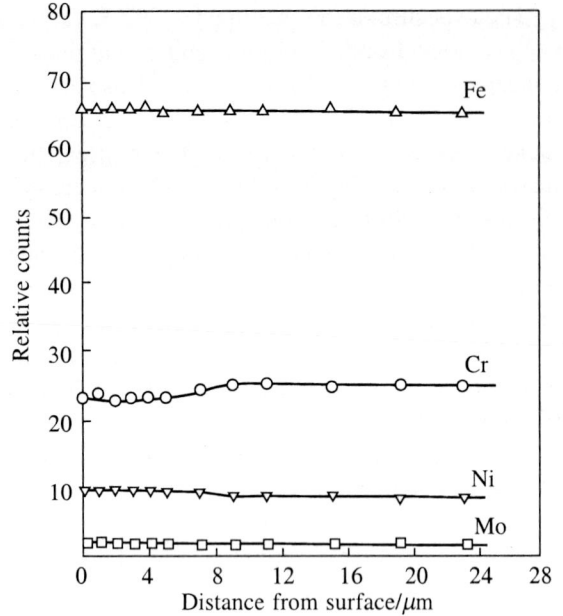

Fig. 4(d).   EDS analysis of the specimen exposed
at 623 K (down-stream) [50].

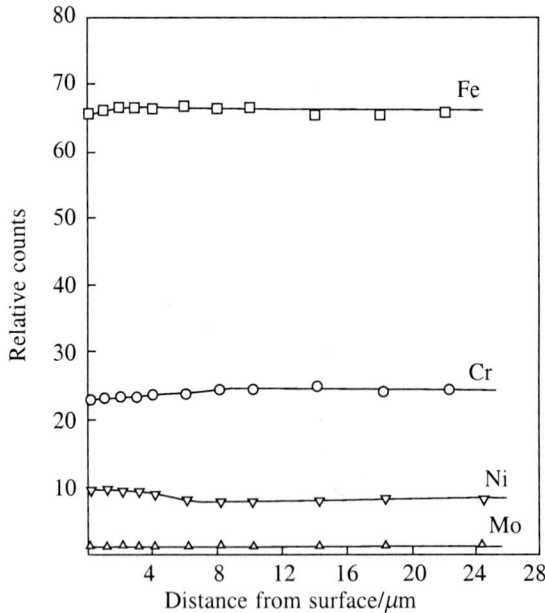

Fig. 4(e). EDS analysis of the specimen exposed at 583 K (down-stream) [50].

temperature, nickel was found to precipitate and deposit on the walls of the specimens. The other
constituents of stainless steel do not show similar leaching and deposition behaviour. Moreover, the
other elements such as chromium, molybdenum and iron form ternary compounds when adequate
quantity of oxygen is available in sodium and they do not exist in the dissolved elemental form.

Leaching of Non-Metals

Carbon is the most important non-metal, the leaching of which causes adverse effect on the mechanical properties of stainless steel. In the present investigation, the change in the concentration of carbon on the sodium-exposed side of the specimen was determined using Secondary Ion Mass Spectrometer (SIMS). The sodium-exposed specimens were polished and analyzed using the SIMS equipment.

The primary ions employed were cesium of energy 10 keV. The intensities of the secondary ions of carbon were determined using a multichannel analyzer. The results of analysis of the carbon profile are given in Fig. 5. Special care was taken to obtain the SIMS spectra without interference from the carbides precipitated at the grain boundaries. The location for the SIMS analysis was identified after imaging the surface and locating regions free from the interference of the carbide particles.

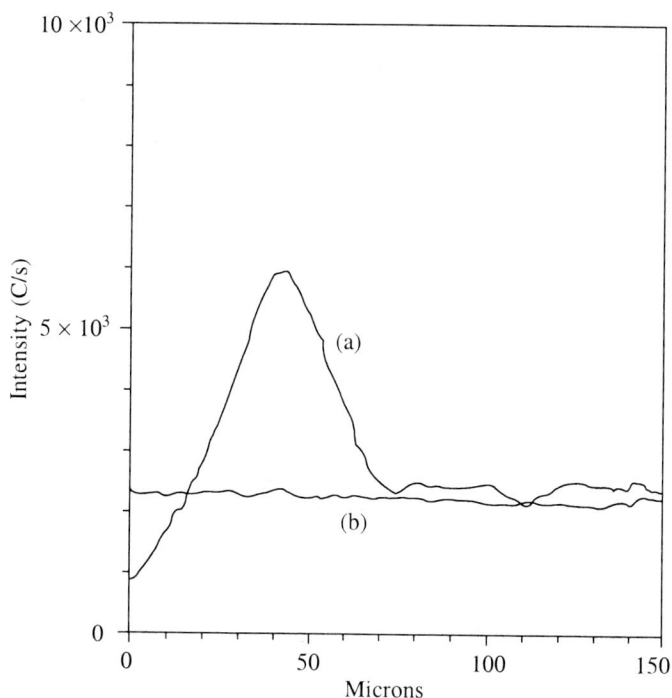

Fig. 5. **Carbon profile of stainless steel by secondary ion mass spectrometry: (a) exposed to sodium at 823 K and (b) exposed at other temperatures [50].**

Even though all the specimens, which were exposed to sodium at the different temperatures, were analyzed for the possible change in the concentration of carbon, a distinct change was observed only in the case of the specimen exposed at 823 K. The profile showed a carbon depleted zone at the surface. The carbon content was less than that of the bulk matrix up to a depth of 20 μm. Thereafter, the concentration of carbon was found to increase and reached a maximum value at a depth of 43 μm from the surface. Subsequently, the concentration of carbon decreased progressively and attained the value of the matrix after a depth of 74 μm. Analysis to higher depths indicated a constant concentration of carbon similar to the as-received material. At all the other temperatures, there were uniform concentrations of carbon from the surface to the interior similar to the case of the as-received material.

One notable observation is that Indian sodium has caused the carburization of stainless steel even at the highest temperature zone. A cross over from carburization to decarburization did not occur (as the temperature of sodium increased) because of the inherently high carbon potential of the sodium. The relatively lower temperature (maximum of 823 K) used in the present investigation also maintained the carbon activity of sodium to a high level causing only carburization to the stainless steel specimens.

Analysis of Corrosion Products Formed on the Surface

The nature of the corrosion products formed on the sodium-exposed surface was analyzed using X-ray Photoelectron Spectroscopy (XPS). The energy of the photoelectron emitted from the surface of the sodium-exposed sample was determined using a multichannel analyzer. From the energy spectrum of the emitted electron, it is possible to gauge the chemical state (oxidation state) of each element on the surface.

The analytical procedure was standardized using an as-received stainless steel specimen. Subsequently, each of the sodium-exposed specimens was mounted inside the vacuum chamber of the equipment with the exposed surface facing the incident X-ray beam. The XPS spectra were recorded in the cases of all the specimens exposed at different temperatures in the loop. The spectra of sodium, chromium, iron, oxygen and carbon were obtained separately to probe the possibility of formation of compounds by reaction between these elements at the different temperatures.

The XPS spectra of carbon for the different samples are compared in Fig. 6. From the analysis of the spectra it is evident that the segregation of carbon on the sodium-exposed specimen has occurred only at 623 K in the up-stream region. On the other hand, the specimen exposed at 723 K in the up-stream region and 583 K in the down-stream region revealed a lower concentration of carbon when compared to the specimens exposed at other temperatures. However, there was an increase in the concentration of carbon at all the temperatures due to carburization by the sodium. There was negligible change in the surface carbon contents in the case of the specimen exposed at 823 K. The increase in

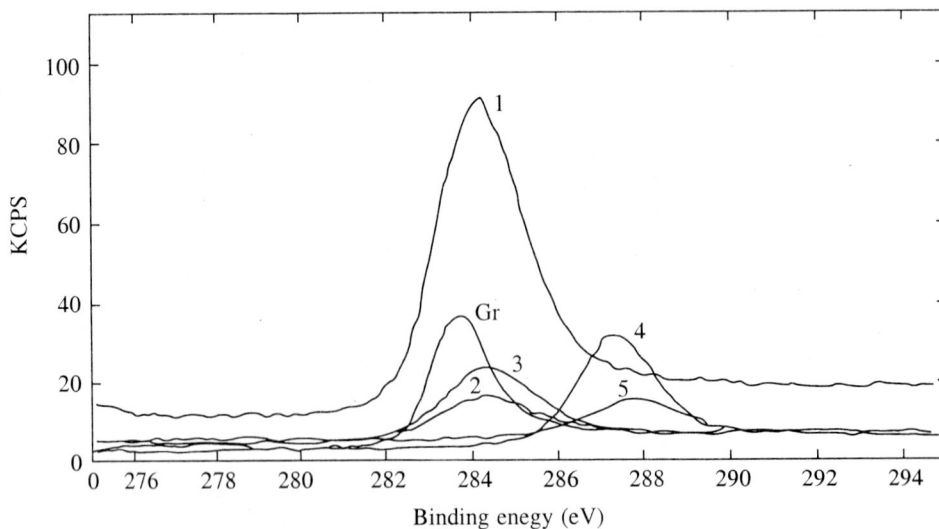

Fig. 6.   The XPS spectra of carbon obtained from the specimens under given conditions. 1—623 K up-stream, 2—723 K up-stream, 3—823 K up-stream, 4—623 K down-stream, 5—583 K down-stream, Gr = graphite dissolved in austenite [50].

the carbon content on the surface is attributed to precipitation of the carbon dissolved in sodium which was carried to this region. Ternary compounds of carbon are also formed on the sodium-exposed surface of the steel. These compounds bear the generic formula $Fe_xCr_{23-x}C_6$ [52–54]. The value of x ranges from 2 to 6 and the carbon potential is controlled by the equilibrium

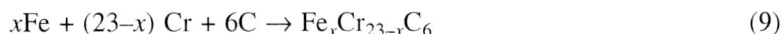

$$xFe + (23-x) Cr + 6C \rightarrow Fe_xCr_{23-x}C_6 \qquad (9)$$

The generation of such compound is the main mode of corrosion by carbon. Carburization increases the mechanical strength of the material. However, excessive carburization makes the material brittle and amenable to brittle failure.

The XPS spectra of sodium did not reveal the formation of a compound containing this element as the corrosion product. It may also be possible that the ternary oxygen compounds of sodium were lost from the surface of the specimen during the process of cleaning. At any rate, under the conditions prevalent in the loop the generation of only $NaCrO_2$ is expected [8]. The formation of ternary compounds of other transition metals requires higher contents of oxygen. For example, the formation of sodium ferrate (Na_4FeO_3) occurs when the oxygen content of sodium is as high as 25 PPM [55].

The spectra of oxygen (Fig. 7) from the sample exposed at the down-stream position revealed a chemical shift in the $1s$ energy level. Similarly, the spectra of iron (Fig. 8) and chromium (Fig. 9) revealed a shift in the $2P_{3/2}$ energy level. These observations indicated the possible formation of compounds containing Fe-O and Cr-O bonds. It was not possible to establish the accurate chemical formulae of these corrosion products from the tables of XPS binding energy data reported in the literature. The corrosion products (sodium-transition metal-oxygen) expected in sodium loops have not been investigated by XPS method. The observed shift in the binding energies did not correspond to any compound whose spectrum has been recorded by XPS study. Formation of compounds containing Cr-O and Fe-O bonds have been reported in the literature [51]. These complexes were probably formed at the high temperature regions of the loop (where favourable kinetics existed) and transported

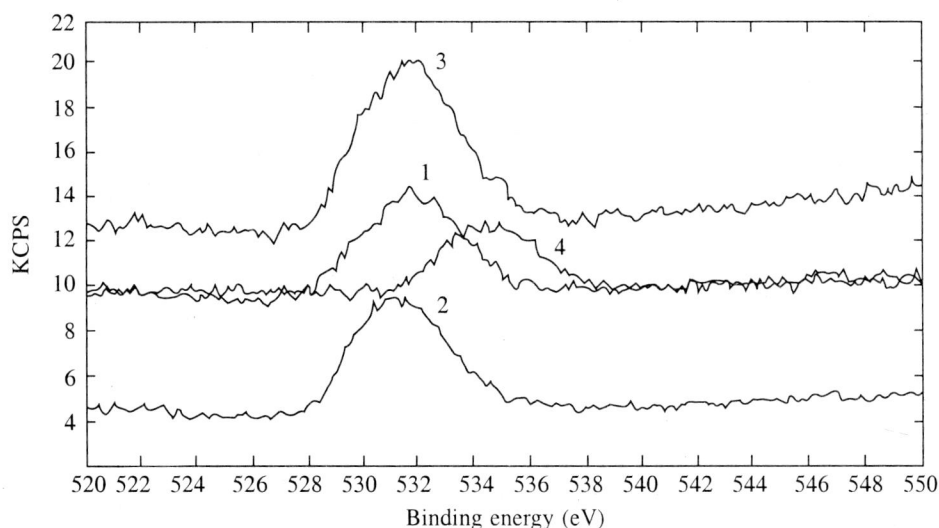

Fig. 7.   XPS spectra of oxygen on the specimens under different conditions. 1—623 K up-stream, 2—723 K up-stream, 3—823 K up-stream, 4—623 K down-stream [50].

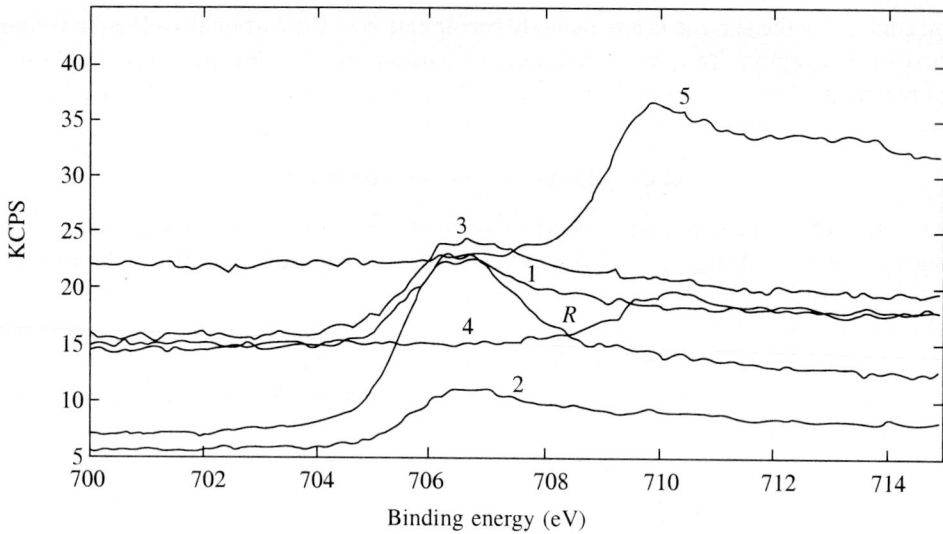

Fig. 8. The XPS spectra of iron obtained on the specimens under different conditions. 1—623 K up-stream, 2—723 K up-stream, 3—823 K up-stream, 4—623 K down-stream, 5—583 K down-stream, *R*: Reference material [50].

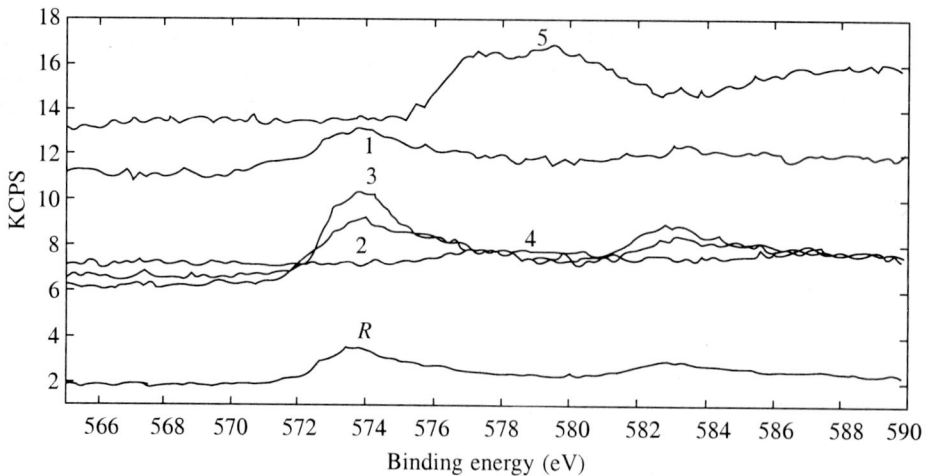

Fig. 9. XPS spectra of chromium obtained from specimens under different conditions. 1—623 K up-stream, 2—723 K up-stream, 3—823 K up-stream, 4—623 K down-stream, 5—583 K: down-stream, *R*: Reference material [50].

to the down-stream regions where these were precipitated and adsorbed on to the walls of the loop piping.

Microstructural Changes

The microstructural changes of the polished and etched specimens were investigated by using an optical microscope.

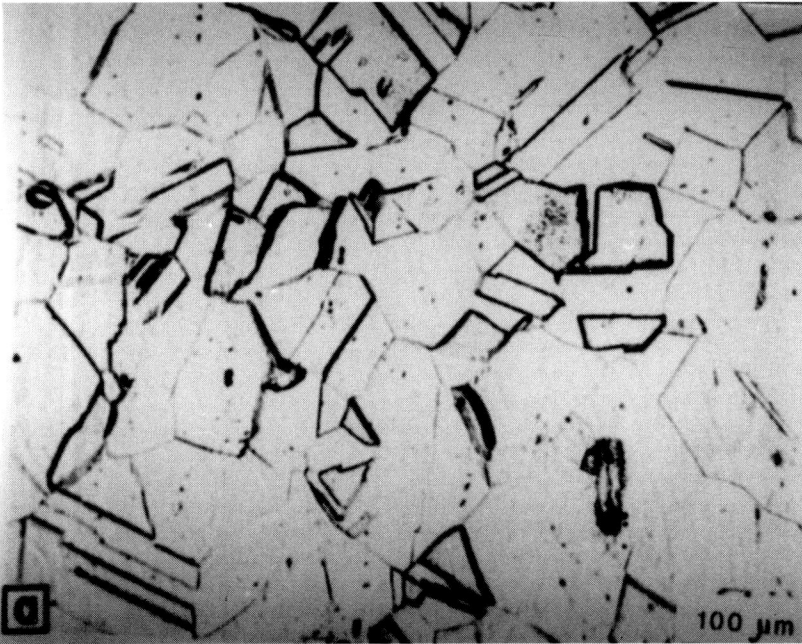

Fig. 10(a). Optical micrograph of the material in the as-received condition [50].

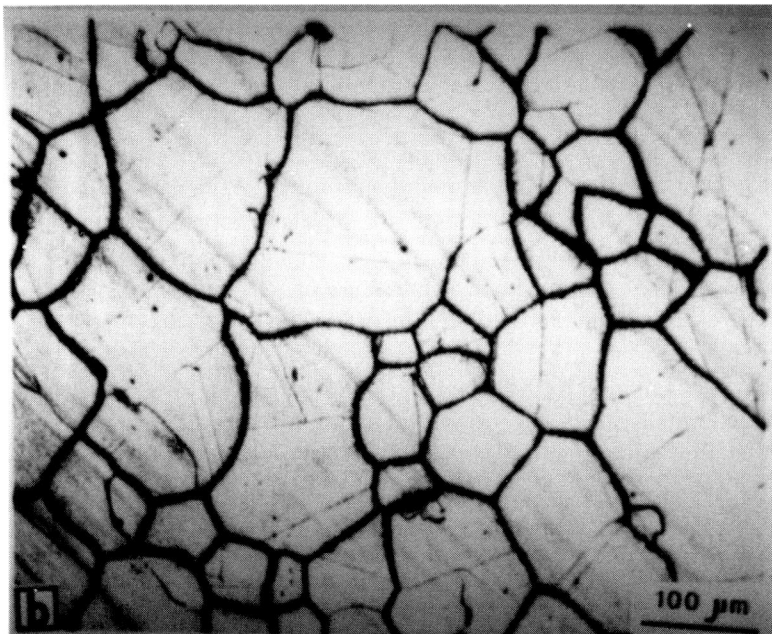

Fig. 10(b). Optical micrograph of the material thermally-aged at 823 K [50].

Fig. 10(c). SEM micrograph of the material exposed to sodium at 823 K [50].

The microstructures of the as-received, thermally-aged and sodium-exposed materials (exposed at the highest temperature of 823 K) are given in Fig. 10. The as-received material was in the mill-annealed condition and thus exhibited a step structure (a structure with clear demarcation of grain boundaries without any carbide precipitate). The specimen subjected to thermal ageing and to sodium exposure at 623 K retained the step structure. The absence of any microstructural variation even on the sodium-exposed surface is attributed to the poor solubility of elements in sodium coupled with poor rate of diffusion of constituents of steel through the matrix at this temperature. The microstructural changes on the specimen exposed at 723 K were also not significant because of kinetic reasons.

The specimens that were experiencing a temperature of 823 K showed a significant change in the microstructure. There was continuous precipitation of carbide at the grain boundaries (ditch structure). In addition there was the generation of a transformed layer on the surface exposed to sodium. The transformed layer was examined by XRD and revealed the presence of ferrite phase (Fig. 11).

The specimens, thermally aged and exposed to sodium in the down-stream region did not reveal any noticeable alteration in the microstructure on account of the kinetic factors already discussed.

DIFFERENT MODES OF CORROSION BY SODIUM

The exposure of stainless steel to sodium brings about different modes of interaction and mass transfer. These modes are expected to be same for all the grades of stainless steels irrespective of minor variation in their composition. The different modes of corrosion and their effects are discussed below in the light of the present investigation and the data reported in the literature.

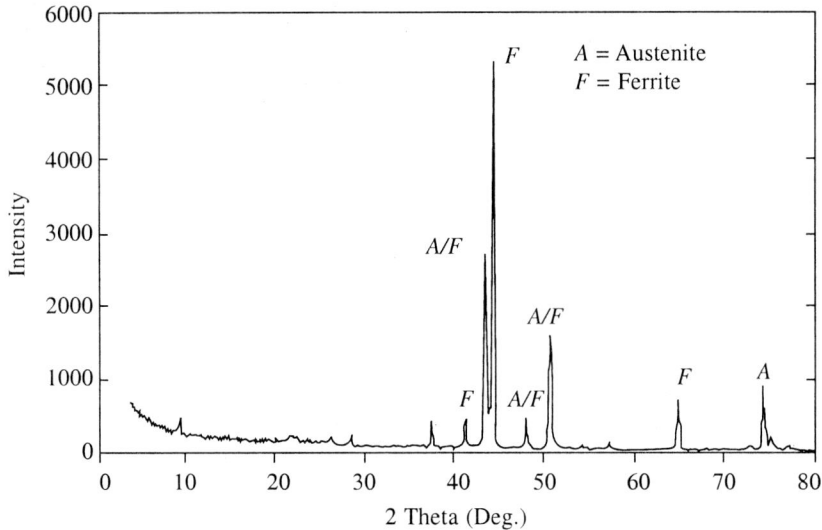

Fig. 11. XRD pattern of ferrite phase formed on the surface of stainless steel exposed to sodium at 823 K [50].

Complete Loss of Material

The dissolution rate of the constituent elements of stainless steel by sodium depends on the amount of oxygen contained in the sodium. Different models are reported in the literature [31, 36, 56, 57] which attempt to relate the rate of corrosion with the amount of oxygen in sodium. Borgstedt [58] has carried out a critical analysis of all these models by examining the extent of fit these models offer to the actual experimental data obtained in a loop experiment. Based on this analysis, the corrosion rate in liquid sodium has been proposed to be best expressed by the model of Thorley and Tyzack [33] for concentration of oxygen in the range of 2 to 9 PPM. The recommended expression for the corrosion loss is

$$\log S \text{ (mL/a)} = 2.44 + 1.5 \log [O] - 7532/2.3 \, RT \tag{10}$$

$$(1 \text{ mL} = 25 \ \mu m)$$

where [O] is the concentration of oxygen in sodium, R the gas constant (J K^{-1} mol^{-1}) and T the temperature in K.

In sodium that contains 5 PPM oxygen, the wall thinning has been estimated to be 1.26 μm/a at 823 K.

The loss in the wall thickness of the sodium-exposed components has been recently estimated by Russian workers [59]. They have reported a loss of 40 μm for components of austenitic stainless steel which have been exposed to sodium at 823 K for 30 years. This data compares exceedingly well with that evaluated using the eq.(10), which gives a loss of 37.8 μm for the same duration.

Formation of Degraded Layer

Our investigations revealed that the exposure of stainless steel specimen at 823 K in the loop for 16000 h resulted in the generation of degraded layer (layer with a composition different from the normal austenitic steel) which has a thickness of 10–15 μm. Yoshida et al. [60] have carried out the analysis of microstructural modification on pipes of AISI 304 stainless steel which has been exposed to sodium for about 100000 h at 823 K. These authors have reported the generation of a surface degraded layer on the walls of the loop piping. They proposed following expression to relate the thickness of the degraded layer Y and duration when stainless steel (grade 304) is exposed to sodium at 823 K

$$Y \, (\mu m) = -97.1 + 25.8 \log_{10}(t) \tag{11}$$

where t is the duration in h.

Above equation has been formulated after analyzing pipes of stainless steels, which have been in contact with sodium that contained 10 PPM of dissolved oxygen. The velocity of sodium in the loop has been maintained as 1 ms^{-1}. This equation is considered to be suitable to evaluate the thickness of the degraded layer formed on 316 stainless steel when exposed to dynamic sodium up to a velocity of 5 ms^{-1} (which is normally encountered in a reactor system). Such an assumption is considered valid because the leaching of elements is predominantly dictated by their rate of diffusion through the austenite matrix rather than the kinetics of dissolution in sodium (which is a fast step). The minor variation in the chemical composition between the different grades of stainless steel is not expected to significantly alter the leaching by sodium. The thicknesses of the degraded layer generated on account of exposure to sodium for different duration are given in Table 3.

Table 3. Estimated thickness of degraded layer of austenitic stainless steel for different duration of exposure to sodium

Duration/h	Thickness of the degraded layer (μm)
2	12.4
5	22.7
10	30.4
20	38.4
30	42.7

The thickness of the degraded layer on exposure to sodium for 16000 h at 823 K has been evaluated by above expression and obtained a value of 11.4 μm. This calculated data are in very good agreement with the thickness of the degraded layer determined (10–15 μm) in the present investigation.

Generation of Carburized Austenite

Analysis of the specimen exposed at 823 K by SIMS revealed existence of a layer of carburized austenite below the surface degraded layer. The concentration of carbon in the degraded layer was very low on account of the generation of ferrite phase which has a poor solubility for carbon.

The rate of diffusion of carbon in stainless steel at the experimental temperature is influenced by factors such as,

(i) nature of the carbide phase that is precipitated
(ii) carbon potential of sodium

Hence, the effective diffusion coefficient of carbon is expected to be different from the data reported in [61]. In the method described in [61] the possibility of precipitation of the carbide was circumvented by employing very high temperature so that the carbon is completely dissolved in the matrix. Moreover, specimens with extremely high grain sizes were employed to circumvent the retarding effect of grain boundary precipitation. Based on the distribution of carbon in the sodium-exposed stainless steel, its effective diffusion coefficient (D_C^{eff}) in the matrix was calculated. This calculation was based on the application of the Fick's law of diffusion by assuming the initial boundary conditions as,

$$C(x, 0) = C_0; \ x > 0 \ t = 0$$

$$C(0, t) = C_s; \ x = 0 \ t > 0$$

$$(C - C_s)/(C_s - C_0) = [1 - \text{erf} \ x \ /\{2(D_C^{eff} \ t)^{1/2}\}] \tag{12}$$

where C is the concentration of carbon at a depth x, C_0 the original uniform concentration of carbon and C_s the surface concentration of carbon after t seconds.

The average of three calculated values (calculated by employing different values of C at the respective depths from the sodium exposed surface) of D_C^{eff} is 3.75×10^{-18} m^{2}s^{-1}. This value is three orders of magnitude lower than that obtained by employing the equation given in [61]. As already discussed the low activity of carbon coupled with the possibility of formation of carbide precipitate have impeded the rate of diffusion significantly. Moreover, the stainless steel on exposure to sodium undergoes phase transformation on the surface (generation of ferrite phase). Because of the poor solubility of carbon in this phase, the flux of carbon into the austenite matrix is reduced. A supporting evidence for the slow diffusivity of carbon when austenitic stainless steel is exposed to sodium has been reported in the literature by Casteels et al. [62]. These authors exposed AISI 316 stainless steel in flowing sodium in a loop at 813 K. The samples were analyzed and the diffusion rate of carbon was estimated to be 1.4×10^{-20} m^{2}s^{-1}. This extremely low value of the diffusivity is attributed to the low activity of carbon in sodium. Obviously, the simultaneous precipitation of the carbon diffusing into the matrix to form carbides also would have impeded the process of diffusion.

Using the estimated value of D_C^{eff} the profile of carbon likely to be attained by 316 stainless steel (at a temperature of 823 K) on exposure to flowing sodium for 2, 5, 10, 20 and 30 years were calculated by employing the diffusion equation. In this calculation, the maximum concentration of carbon in the stainless steel was assumed to have attained a dynamic equilibrium value of 0.135 mass % for all the duration of exposure. This assumption is considered a good approximation as the carbon activity of sodium and the rate of diffusion through the austenite matrix determines the peak concentration of carbon. In the duration employed in the present experiment, a carbon concentration value represented by the dynamic equilibrium value is assumed to have been achieved. Exposure to longer duration caused the generation of carburized austenite of higher thickness.

The overall effects of the three modes of corrosion on AISI 316 stainless steel are shown in Fig. 12. These carburization profiles have been calculated after assuming a constant source of carbon in sodium. However, at longer duration, due to carburization of the components, the contents of the carbon (and its activity) in sodium will alter and the profile will be modified accordingly.

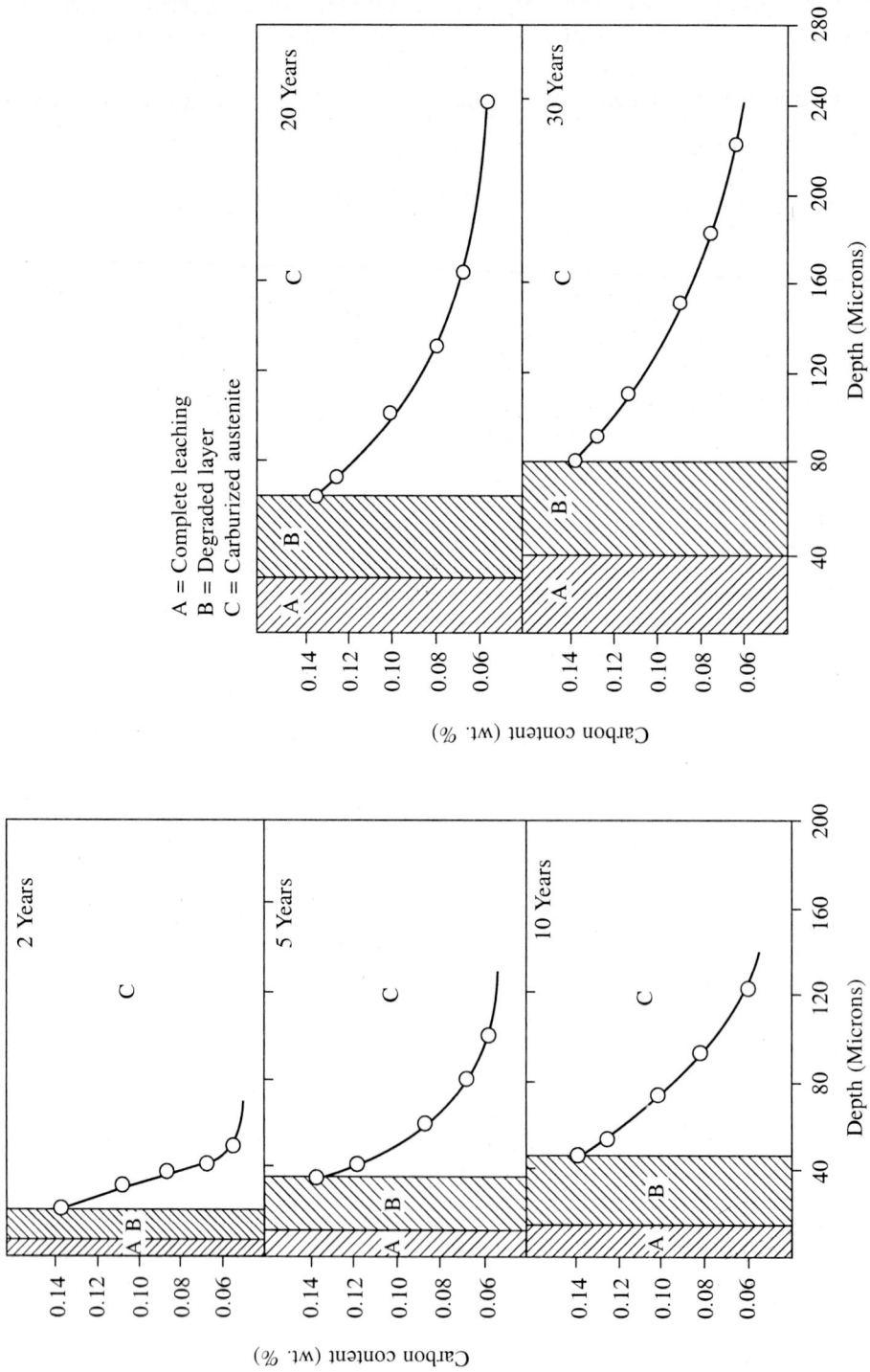

A = Complete leaching
B = Degraded layer
C = Carburized austenite

Fig. 12. Cumulative long-term effect of exposure of stainless steel to sodium at 823 K [45].

REFERENCES

1. Mausteller, J.W., Tepper, F. and Rodgers, S.J., *Alkali Metals Handling and System Operating Techniques*, Gordon Breach Science Publishers, NY (1967).
2. Borgstedt, H.U. and Mathews, C.K., *Applied Chemistry of the Alkali Metals*, Plenum, NY(1986)
3. Claar, T.D., *Reactor Technology*, 13(20) (1970) 124.
4. Awasti, S.P. and Borgstedt, H.U., *Journal of Nuclear Materials*, 116 (1983) 103.
5. Kolster, B.H., Veer, J.V.D. and Bos, L., in: *Materials Behaviour and Physical Chemistry In Liquid Metal Systems*, (Ed. H.U. Borgstedt), Plenum, NY (1982) 37.
6. Barker, M.G. and Wood, D.J., *Journal of Less Common Metals*, 35 (1974) 315.
7. Thorley, A.W., Blundell, A. and Bradsley, J.A., in: *Materials Behaviour and Physical Chemistry in Liquid Metal Systems* (ed. Borgstedt, H.U.), Plenum, NY (1982) 5.
8. Rajendran Pillai, S., Khatak H.S. and Gnanamoorthy, J.B., *Journal of Nuclear Materials* 224 (1995) 17.
9. Gnanasekaran, T. and Mathews, C. K., *Journal of Nuclear Materials*, 140 (1986) 202.
10. Kolster, B.H., Bos, L., in: *Proc. Third Intern. Conf. on Liquid Metal Engineering and Technology*, Oxford, (1984) (BNES, London, 1984) vol. I, p. 235.
11. Khatak, H.S., Hasan Shaik and Gnanamoorthy, J.B., in: *Proc. 4th Intern. Conf. Liquid Metal Engineering and Technology*, Avignon, France (1988) Pub. SFEN, France, Paper 514.
12. Natesan, K., Kassner, T.F. and Che-Yu-Li, *Reactor Technology*, 19 (4) (1972) 244.
13. Suzuki, T., Mutoh, I., Yagiant, T. and Ikenaga, Y., *Journal of Nuclear Materials* 139(1986)97.
14. Lloyd, G.J., *Atomic Energy Review*, 16 (1978) 155.
15. Krankota, J.L., *Journal of Engineering Material and Technology* (Trans. ASME), Jan. (1976) 9.
16. Mishra, M.P., Borgstedt, H. U., Frees, G., Seith, B., Mannan S.L. and Rodriguez, P., *Journal of Nuclear Materials*, 200 (1993) 244.
17. Borgstedt, H.U. and Huthmann, H., *Journal of Nuclear Materials*, 183 (1991) 127.
18. Wada, Y., Asagawa, T. and Komine, R., in: Proc. *Intern Working Group on Fast Reactors*, KFK – 4935 (ed. H.U. Borgstedt, Pub: Forschungszentrum Karlsruhe, Germany) (1991) 149.
19. Singer, R.M. and weeks, J.R., *USAEC Report, ANL-7520*, part 1 (1968) p. 309.
20. Singer, R.M., Fleitman, A.H., Weeks, J.R. and Isaacs, H.S., *Corrosion by Liquid Metals*, eds. Draley J.E. and Weeks, J.R., (Plenum Press, NY, 1970).
21. Stanaway, W.P. and Thompson, R., in: *Proc. Materials Behaviour and Physical Chemistry in Liquid Metal Systems*, (Ed. H.U. Borgstedt) Plenum, NY (1982) p. 421.
22. Eichelberger, R. and Mckisson, R.L., *USAEC Report ANL-7520*, part 1 (1968) p. 278.
23. Jansson, S.A. and Berkey, E., in: *Corrosion by Liquid Metals*, Eds. Draley, J.E. and Weeks, J.R. (Plenum Press, New York, 1970) p. 479.
24. Azad, A.M., Sreedharan, O.M. and Gnanamoorthy, J.B., *Journal of Nuclear Materials*, 144 (1987) 94.
25. Azad, A.M., Sreedharan, O.M. and Gnanamoorthy, J.B., *Journal of Nuclear Materials*, 151 (1988) 293.
26. Sreedharan, O.M. and Gnanamoorthy, J.B., *Journal of Nuclear Materials*, 89 (1980) 113.
27. Noden, J.D., *Journal of British Nuclear Energy Society*, 12 (1973) 32.
28. Thompson, R., *Specialists' Meeting on Carbon in Sodium*, IWGFR/33 (IAEA, Vienna, 1979) p. 6.
29. Vissers, D.R., Holmes, J.T., Bartholme L.G. and Nelson, P.A., *Nuclear Technology* 21 (1974) 235.
30. Weeks, J.R., Isaacs, H.S., in: *Advances in Corrosion Science and Technolog*, vol. 3, Plenum Press, NY (1973).
31. Bagnall, C., Jacobs, D., *Report*, WARD-NA-3045–23 (1975).
32. Roy, P. and Wozadlo, G.P., *Nuclear Technology* 10 (1971) 307.
33. Thorley, A.W., Tyzack, C., *Alkali Metal Coolants* (IAEA, Vienna, 1967) p. 97.
34. Shiels, S.A., Keeton, A.R. and Anantamula, R.P., *Report HEDL-TME-77–71, UC 9b*, Feb (1978).
35. Bagnall, C., Witkowski, R.E., *Report*, WARD-NA-3045–53, Sept (1978).
36. Thorley, A., Longson, B. and Prescott, J., *Report, TRG-1909(C)*, UKAEA, Risley (1969).
37. Borgstedt, H.U. and Grosser, E.D., in *Proc. Conf. Liquid Metal Technology*, (BNES) London (1973) p. 275.
38. Paul, C.S. Wu. and Chang, A.L., in: *Proc. Intern. Conf. Liquid Metals Technology in Energy Production*, Richland, Wa, USA (1980) p. 22.

39. Borgstedt, H.U., Frees, G. and Schneider, H., *Nuclear Technology* 34 (1977) 270.
40. Suzuki, T. and Mutoh, I., *Journal of Nuclear Materials*, 171 (1990) 253.
41. Roy, P., Wozadlo, G.P. and Camprelli, F.A., in: *Proc. Intern. Conf. Sodium Technology*, Argonne, USA, ANL-7520, Part 1 (1968) p. 31.
42. Schrock, S.L., J.N., Miller, R.L. and Lohr, D.E., *Corrosion by Liquid Metals* (Plenum Press, NY, 1970) p. 41.
43. Furukawa, K. and Nehei, I. in: *Proc. Intern. Conf. Sodium Technology*, Argonne (USA), ANL-7520, Part 1 (1968) p. 143.
44. Natesan, K., Chopra, O.K. and Kassner, T.F., *Journal of Nuclear Materials*, 73 (1978) 137.
45. Rajendran Pillai, S., Khatak, H.S. and Gnanamoorthy, J.B., *Materials Transations*, Japan Institute of Metals, 39(3) (1998) 370.
46. Harries, D.R., Journal of British Nuclear Energy Society, 5 (1966) 74.
47. Smith, D.L., Zeeman, G.J., Natesan, K. and Kassner, T.F., in: *Proc. Intern. Conf. Liquid Metal Technol. in Energy Production*, Champion. Penn. USA (1976) p. 365.
48. Dietz, W., Grosser, E.D. and Lorenz, H., in: *Proc. Intern. Conf. on Liquid Metal Technology in Energy Production*, Champion, Penn. (1976) p. 343.
49. Kirschler, L.H. and Andrews, R.C., *Report ANL-7520*, Argonne National Laboratory (Part 1) (1968) p. 41.
50. Rajendran Pillai, S., Khatak, H.S., Gnanamoorthy, J.B., Velmurugan, S., Tyagi, A.K., Kale, R.D., Swaminathan, K., Rajan, M. and Rajan, K.K., *Materials Science and Technology* 13 (1997) 937.
51. Suzuki, T. and Mutoh, I., *Journal of Nuclear Materials*, 140 (1986) 56.
52. Rajendran Pillai, S. and Mathews, C.K., *Journal of Nuclear Materials*, 150 (1987) 31.
53. Rajendran Pillai, S., Ranganathan, R. and Mathews, C.K., *Bulletin of Electrochemistry*, 6(6) (1990) 672.
54. Dai, W., Seetharaman, S. and Staffanson, L.I., *Metallurgical Transactions*, 15B (1984) 319.
55. Thorley, A.W. and Tyzack, T., in: *Proc. Liquid Alkali Metals*, BNES, London, (1973) p. 257.
56. Kolster, B.H. and Bos, L., in: *Proc. Intern. Conf. Liquid Metal Technology in Energy Production*, Champion, USA, ed. Cooper, M.H., Pub: Am. Nucl. Soc. CONF – 760503-P1, Vol 1 (1976) p. 368.
57. Bogers, A.J., Chirrer, E.G. and Borgstedt, H.U., *Fast Reactor Fuel and Fuel Elements, Proc. of an Intern. Meeting*, Ed. Dalle Donne, M., Kummerrer K. and Schroeter, K., Forschungszentrum Karlsruhe (1970) p. 610.
58. Borgstedt, H.U. in: *Proc. 2nd Intern. Conf. Liquid Metal Technology in Energy Production*, Richland, Washington, Ed. Dahlke, J.M., Pub: American Nuclear Society, (1980) p. 7.
59. Trapeznikov, Ju.M., Grishmanovskaja, R.N., Groyvin, I.V., Trojanov, V.M. and Malygin, M.F., in: *Proc. Intern Atomic Energy Specialists' Meeting, KFK-4935, IWGFR/84*, Forschungszentrum Karlsrühe (Pub: IAEA, Vienna) (1991) p. 37.
60. Yoshida, E., Kato, S. and Wada, Y., in: *Proc. LMS-II*, Forschungszentrum Karlsruhe, Ed. Borgstedt, H.U., Plenum, NY (1993) p. 55.
61. Agarwalla, R.P., Naik, M.C., Anand, M.S. and Paul, A.R., *Journal of Nuclear Materials*, 36 (1970) 41.
62. Casteels, F., Tas, H., Dresselaers, J., Cools, A. and Knaeper, L., in: *Proc. Intern. Conf. Liquid Metal Technology in Energy Production*, Champion, CONF-760503–p2, Vol. 2, ed. Cooper, M.H., Pub: American Nuclear Society (1976) p. 577.

11. High Temperature Corrosion of Austenitic Stainless Steels

S. Rajendran Pillai[1]

Abstract One of the important modes of degradation of the materials at high temperature is through reaction with the environment. In this context the important corrosion phenomena are oxidation, carburization, nitridation and sulphidation. This article reviews various studies reported in the literature on corrosion of stainless steel. The thermodynamic and kinetic factors that influence the corrosion process are pointed out. The role of different elements in influencing the corrosion process is discussed. A brief mention is also made of the various methods of surface modification to mitigate or control the high temperature corrosion.

Key Words Oxidation, carburization, nitridation, sulphidation, stainless steel, surface modification

INTRODUCTION

Austenitic stainless steels find increasing application in high temperature systems. The choice of a particular alloy is largely governed by its mechanical strength at high temperature and good compatibility in the desired service environments. Equal emphasis should be placed relating to the interaction between corrosion and mechanical behaviour. Not only the corrosive loss of thickness (by way of oxidation, carburization, sulphidation or nitridation) but the introduction of voids affects the mechanical strength of the material. Depending on the specialised application encountered in each industry, the corrosion involves one or more of these processes such as oxidation, sulphidation, carburization or hot corrosion (corrosion induced by the deposition of molten salts on the surface of the material). In the case of environments with two or more reactants, the corrosion may proceed by more than one mechanism. For example, oxidation and sulphidation may occur simultaneously, the relative rates being determined by the chemical activities of the reacting species.

The stabilities of the oxides, sulphides and carbides are in general higher than the corresponding metals. However, the rate of reaction of the non-metal with the metal is so slow that the corrosion can be ignored at relatively lower temperatures. The rate of reaction increases rapidly with temperature and for alloys in high temperature service, the loss due to oxidation, sulphidation, carburization and nitridation becomes a matter for important attention.

The detrimental effects brought about by high temperature corrosion are broadly classified as follows [1, 2]:

---

[1]Scientific Officer, Corrosion Science and Technology Division, Materials, Characterization Group, Indira Gandhi Centre for Atomic Research, Kalpakkam-603 102, India.

(i) The loss of material through corrosion reduces the load-bearing capability of the construction material.
(ii) Carburization and nitridation can render the material brittle and thus susceptible to brittle failure.
(iii) Damage to material may also result from precipitation induced stress on the components [3].
(iv) Vacancy injection into the alloy matrix through the different corrosion processes. These vacancies coalesce to form voids and even channels and adversely affect the mechanical properties.

In order to facilitate the selection of a particular alloy for service in a specific environment, it is necessary to have a strong background information on the corrosion behaviour of the material.

OXIDATION

With ever increasing need to carry out synthetic and energy conversion processes at higher temperatures, it becomes imperative to understand the corrosion resistance of these alloys at elevated temperatures. Oxidation is the most important mode of high temperature corrosion. Nearly all the metals and alloys undergo oxidation when exposed to air at high temperature. It is nearly impossible to completely obviate oxidative wastage of materials. In majority of the industrial environments the residual oxygen content is high enough to cause oxidation of the metal [4]. In certain cases the initial generation of an oxide scale is considered desirable as it offers protection from further attack by other species present in the corrosive environment. Among the oxide forming elements, chromium, aluminum and silicon are reported to form protective scales. Hence these elements are deliberately added to most of the high temperature alloys. Alternately, the high temperature alloys are modified with a layer of coating primarily consisting of chromium or aluminum.

Thermodynamic Basis of Oxidation

The formation of oxide occurs when the oxygen potential of the medium is higher than the equilibrium oxygen potential of the oxide. This threshold pressure of oxygen can be determined from the free energy data of the metal oxide. Consider the reaction

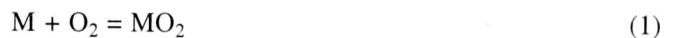

$$M + O_2 = MO_2 \tag{1}$$

The standard free energy of formation of the metal oxide ($\Delta G°$) is given by

$$\Delta G° = - RT \ln a_{MO_2} / a_M P_{O_2} \tag{2}$$

When the metal and the metal oxides coexist in the system, their activities are unity. Therefore

$$\Delta G° = RT \ln P_{O_2} \tag{3}$$

The free energies of formation of some of the technologically important metal oxides are given in the Ellingham diagram (Fig. 1). From the knowledge of the oxygen potential of the environment it is possible to predict if a particular metal oxide will be formed or not.

Apart from molecular oxygen, other gaseous species at equilibrium can control the oxygen potential of a medium. These are H_2/H_2O and CO/CO_2. The reaction involving H_2 and O_2 is written as

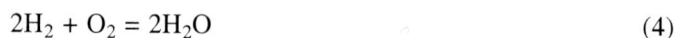

$$2H_2 + O_2 = 2H_2O \tag{4}$$

The standard free energy of formation of the above reaction ($\Delta G°$) may be written as

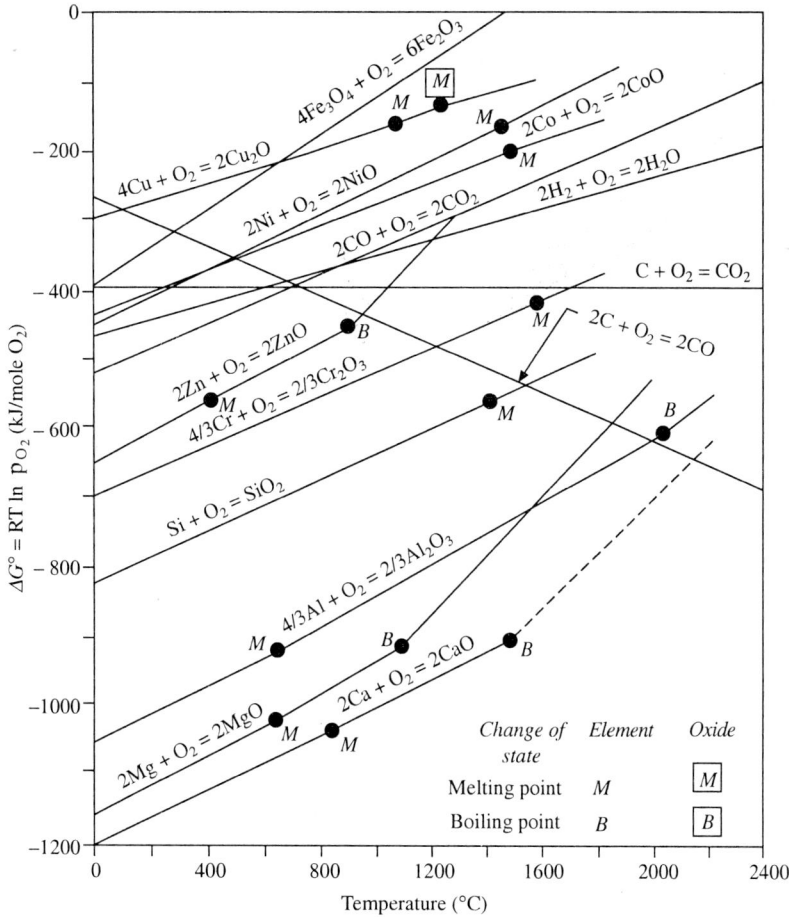

Fig. 1.   Standard free energy of formation of selected oxides at different temperatures.

$$\Delta G^\circ = -RT \ln P_{H_2O}^2 / P_{H_2}^2 P_{O_2} \tag{5}$$

By controlling the ratio of H_2 and H_2O it is possible to impart desired potential of oxygen to the medium. The equilibrium reaction for an environment whose oxygen potential is controlled by the partial pressures of CO and CO_2 is given by

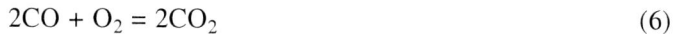

$$2CO + O_2 = 2CO_2 \tag{6}$$

The standard free energy of formation is related to the partial pressures of gases by the expression

$$\Delta G^\circ = -RT \ln P_{CO_2}^2 / P_{CO}^2 P_{O_2} \tag{7}$$

A desired potential of oxygen in the environment may be obtained by controlling the ratio of CO and CO_2 in the gas mixture.

It is evident from the stability diagram of oxides (Fig. 1) that the oxides of iron, nickel and cobalt, which are base metals for majority of the alloys are less stable than the oxides of some of the solute elements (such as chromium, aluminium, silicon etc.) normally present in the alloys. The oxides of these solute elements are preferentially formed. When these elements are present in adequate concentrations, an oxide scale is formed with good adherence, reduced growth rate and hence protects the alloys from further corrosive reaction. For example, the addition of chromium to many of the engineering alloys promotes the generation of an adherent layer of oxide (Cr_2O_3) which imparts resistance against further oxidation. In the case of some of the high temperature alloys, the generation of Al_2O_3 is promoted to inhibit excessive oxidation.

Kinetic Basis of Oxidation

The mechanism of oxidation may be illustrated by considering the reaction of oxygen on the metal surface. This proceeds in several steps as illustrated in Fig. 2. The initial step is the physical adsorption of the oxygen molecule on the surface of the metal followed by its decomposition to oxygen atoms. These atoms are chemisorbed on the metal surface. The adsorbed oxygen atom reacts with the metal. The oxide layer grows laterally which ultimately covers the entire surface. The oxygen atom from the oxide dissolves in the matrix. The rate of dissolution is determined by the solubility and diffusivity of oxygen in the metal. Once a continuous oxide layer is formed the progress of oxidation is governed by the transport of metal or oxygen through the oxide film.

The diffusional transfer of metal or oxide ion takes place by different mechanisms. The diffusion can occur through the lattice, grain boundaries or other short-circuit diffusion paths. Depending on the mode of growth of the oxide scale, vacancies and voids may develop on the scale. These voids may get interconnected to form channels that provided short-circuit diffusion paths for the metal and oxygen. The metal ion diffusing into the oxide layer and participating in the oxidation leaves its vacancy in the metallic phase and thus causes vacancy injection.

The growth of the scale also results in the build up of large stress (owing to the differing metal atom density in the oxide and metal phases). The stress, thus developed may be accommodated through plastic deformation or may cause the development of cracks. Repeated cracking results in the loss of protective property of the scale and further oxidation may be governed by diffusion through a very thin layer. When oxidation involves diffusion through the already generated oxide scale, the reaction follows a parabolic rate equation. On the other hand, when continuous breaking of the oxide scale occurs, the reaction proceeds by a linear rate process.

The ideal model of oxidation, when the oxide scale retains its integrity has been described by Wagner [5]. The transport process is illustrated in Fig. 3. Thermodynamic equilibrium is expected to prevail at the metal/oxide and oxide/gas interfaces. The driving force for the migration is the thermodynamic activity difference induced by the large negative free energy of formation of the oxide. For the above described mechanism of oxidation, the growth of the scale (growth in thickness) is parabolic with time.

$$dx/dt = k_p \cdot 1/x \qquad (8)$$

$$x^2 = 2k_p t + C_o = k_p t + C_0 \qquad (9)$$

where x is a measure of the thickness of the oxide, k_p the parabolic rate constant and C_0 the constant of integration

Wagner derived the expression for the growth of the oxide scale in terms of electrical conductivity σ and the transport number of ions t_i and electrons t_e, respectively, in the oxide which is given by

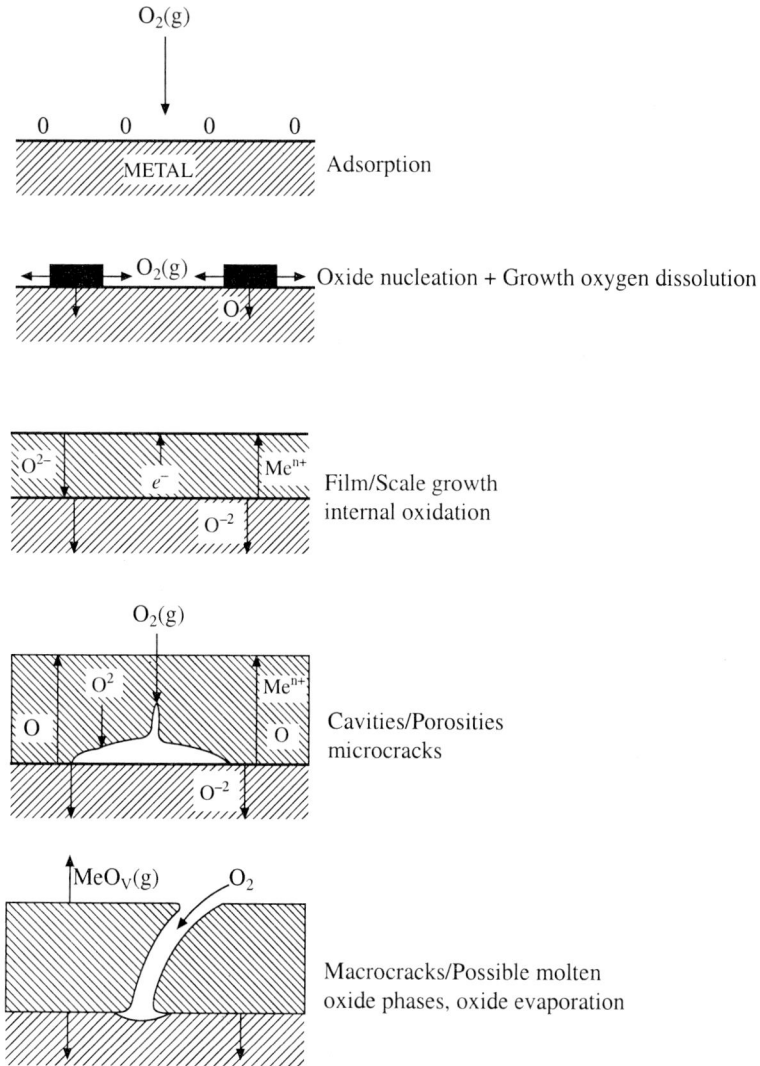

Fig. 2. Schematic of main phenomena in the oxidation of metals.

$$k'_p = (kT/8be^2) \int_{P_{O_2}}^{P^o_{O_2}} \sigma t_i t_e \ln P_{O_2} \tag{10}$$

$P^o_{O_2}$ and P'_{O_2} are the partial pressures of oxygen at the oxide/gas and metal/oxide interfaces, respectively, k is the Boltzman constant, T the absolute temperature, e the electronic charge and b the number of gram atom of oxygen per mol of M_aO_b.

The actual rate of oxidation will be modified due to the generation of voids, stress and deformation of the oxide scale. The layer of oxide scale grows through process governed by diffusion, as described above.

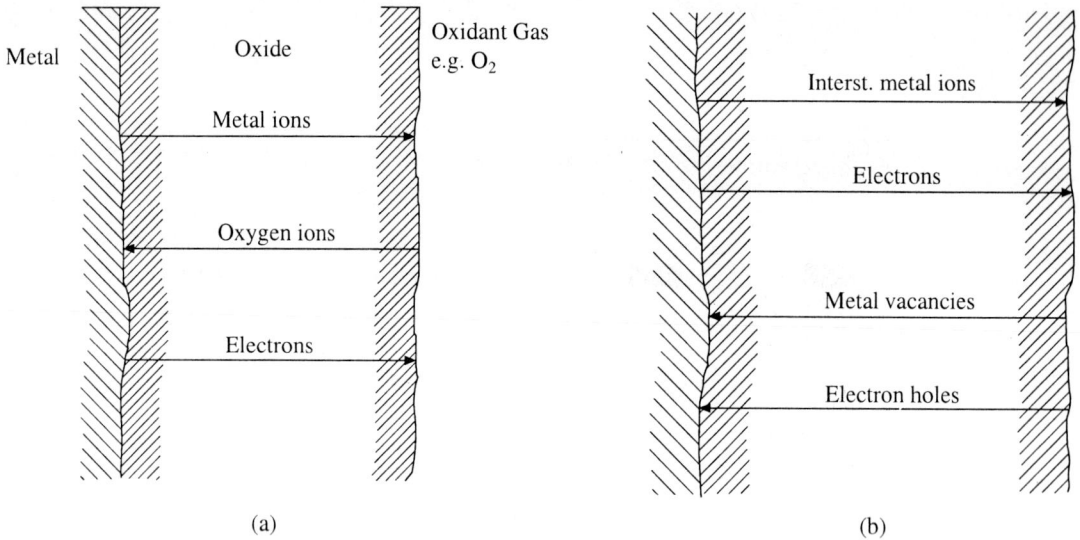

Fig. 3.   Transport process through single phase scales: (a) transport of reacting atoms, ions or electrons are rate determining and (b) transport of vacancies and defects are rate determining [5].

Oxidation of Austenitic Stainless Steel

Austenitic stainless steels contain chromium as an important substitutional element. As already discussed above, chromium readily forms an oxide. If a uniform layer of chromium oxide is formed on the surface, the material is protected from further corrosive effect of the environment. Grodner [6] studied the oxidation resistance of several alloys (composition given in Table 1) in the temperature range of 823 to 1473 K. They have reported that nickel improved the resistance to cyclic oxidation (Fig. 4). Similar results were also reported by Eiselstein and Skinner [7] (Fig. 5). The oxidation resistance of several Fe-Cr-Ni alloys was studied at 1143 to 1473 K by Brasunas et al. [8]. Their study has revealed that the presence of nickel in excess of 10 mass% improves the oxidation resistance of high chromium alloys with chromium content in the range of 11 to 36 mass %. The high nickel alloys, in general, exhibited improved resistance to oxidation than the normal grade of stainless steel.

Table 1.   Concentration (mass %) of major elements in different grades of stainless steels

Grade	Composition (mass %)				
	C	*Cr*	*Ni*	*Fe*	*Other elements*
410	0.15 (max)	11.5–13.5	–	Bal.	–
416	0.15 (max)	12.0–14.0	–	Bal.	S = 0.15 (max)
446	0.20 (max)	23.0–27.0	–	Bal.	N = 0.25 (max)
302	0.15 (max)	17.0–19.0	8.0–10.0	Bal.	–
302B	0.15 (max)	17.0–19.0	8.0–10.0	Bal.	Si = 2.0–3.0
304	0.08 (max)	18.0–20.0	8.0–10.5	Bal.	–
309	0.2 (max)	22.0–24.0	12.0–15.0	Bal.	–
310	0.25 (max)	24.0–26.0	19.0–22.0	Bal.	–
316	0.08 (max)	16.0–18.0	10.0–14.0	Bal.	Mo = 2.0–3.0
347	0.08 (max)	17.0–19.0	9.0–13.0	Bal.	Ti = 5 × C (min)

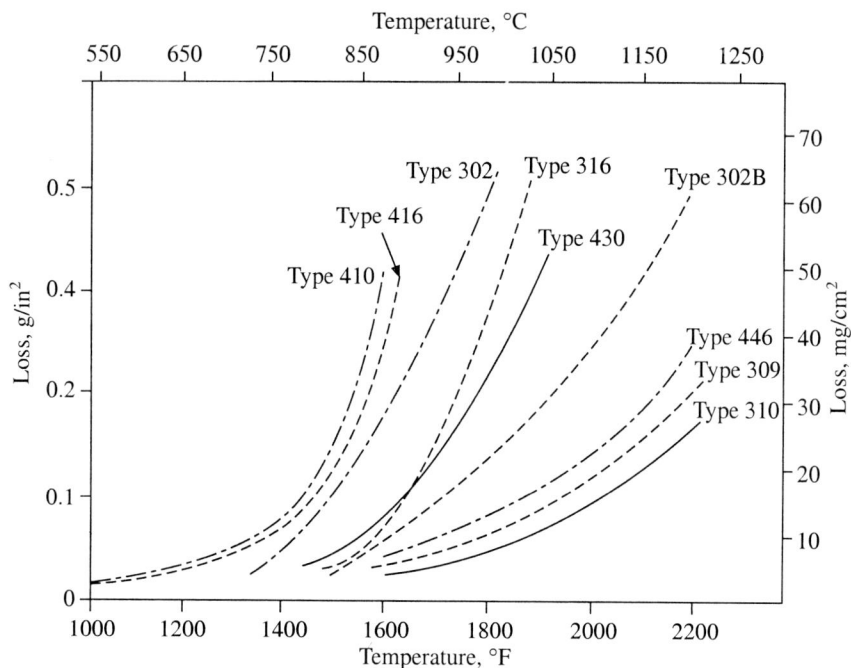

Fig. 4. Oxidation resistance of different grades of stainless steels [6].

The improved resistance to oxidation of alloys containing silicon and aluminium has been demonstrated by the experiments performed by Kado et al. [9]. They have reported that when the alloys were oxidized cyclically 400 times at 1573 K, the DIN 4828 (19Cr-12Ni-2Si-steel) and F1 alloy (Fe-15 Cr-4Al) exhibited superior resistance to oxidation (Fig. 6).

Several studies are reported in the literature to understand the corrosion of stainless steel under the hydrothermal condition experienced in a nuclear reactor [10–13]. A programme has been conducted at NASA-Lewis Research Centre, USA to examine the feasibility of substituting less critical elements for chromium in Type 304 stainless steel, and still retaining adequate corrosion resistance. The study revealed that an optimum concentration of 12% Cr-10% Ni, 1.65% Si, 1% Al and 2% Mo along with other alloying elements normally contained in 304 stainless steel is capable of retaining adequate resistance to corrosion and oxidation [14].

It is well recognized that an atmosphere other than oxygen (or air) also can cause oxidation. In the case of 18/8 steel, a chromium rich sesqui oxide is formed when exposed to an atmosphere of carbon dioxide at elevated temperatures [15–18]. In certain cases a duplex oxide scale is reported with an outer layer of Fe_3O_4 and an inner layer of spinel oxide comprising of the remaining alloying elements as the metallic constituents. The composition of the oxide scale is also influenced by the oxygen potential of the environment. In contact with atmospheric air, the elements present in the bulk at a level of few percentages are incorporated in the oxide scale formed on the surface [19–21]. At low partial pressure of oxygen, minor elements appear to play a significant role in the oxidation characteristics [22–24]. Such very low levels of oxygen contents are likely to be encountered in the interspace between the metal and the oxide phases. Thus, oxides that are normally not encountered when the alloy is oxidized in air are encountered in the interspace. The oxidation behaviour of unstabilized austenitic steel (EN58A) at temperatures between 770 and 1070 K in an oxygen pressure of 10^{-5} Pa was investigated by Wild [25]. This investigation has revealed that the initial incubation period in the oxidation of stainless steel is due to the reaction of the sulphur diffusing from the bulk with oxygen to form sulphur dioxide. The diffusion coefficient of manganese through the oxide scale was found to be two orders of magnitude higher than that of chromium (at 1070 K) and thus the oxide scale contains a large proportion of manganese.

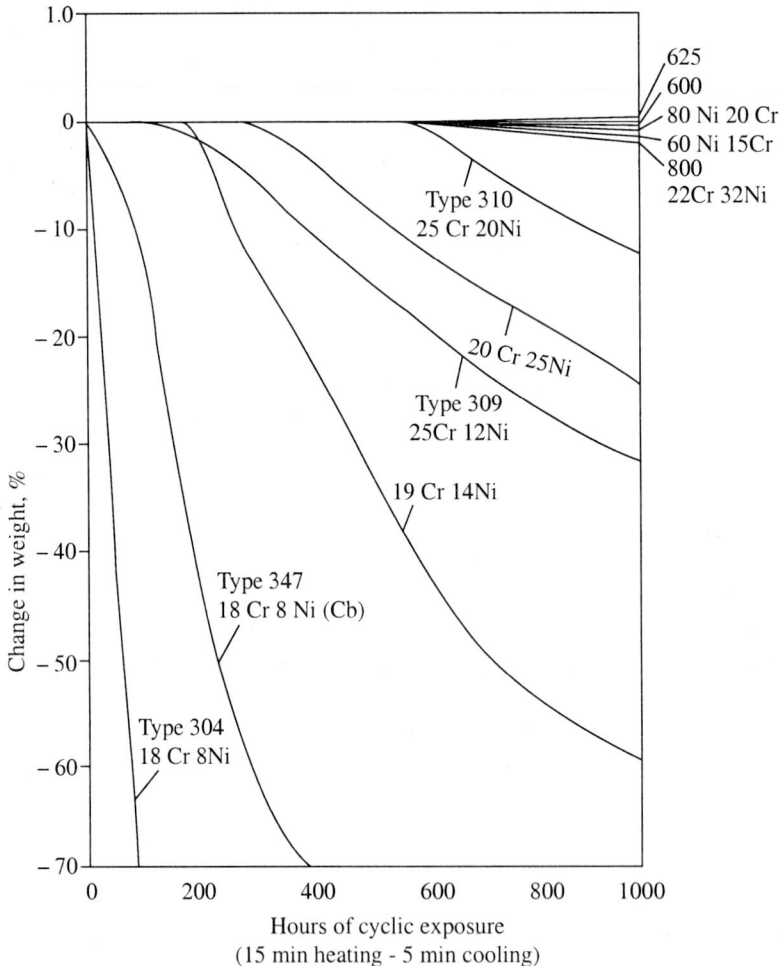

Fig. 5.   Cyclic oxidation resistance of several stainless steels and nickel-base alloys at 1253 K.

There are conflicting reports in the literature [26–32] about the nature of the oxide scale formed on the surface of stainless steel. Even though a protective chromia scale is formed in most cases, some investigations have shown that a certain area of the oxidized surface is characterized by the formation of iron-rich oxides [28, 29]. The enrichment of elements on the surface of the oxidized Type 316 stainless steel was also reported by Tanabe and Imoto [33]. On oxidation of this stainless steel in air at 873 K resulted in the formation of an oxide scale that preserved the composition of the base alloy. Once an oxide layer that is rich in iron is formed, it grows faster than the oxide that is rich in chromium. Nickel plays a minor role in the oxidation reaction even though it is known to impede the kinetics of oxidation. Manganese behaves in a way similar to that of iron and is highly enriched on the surface of the oxide above 1073 K.

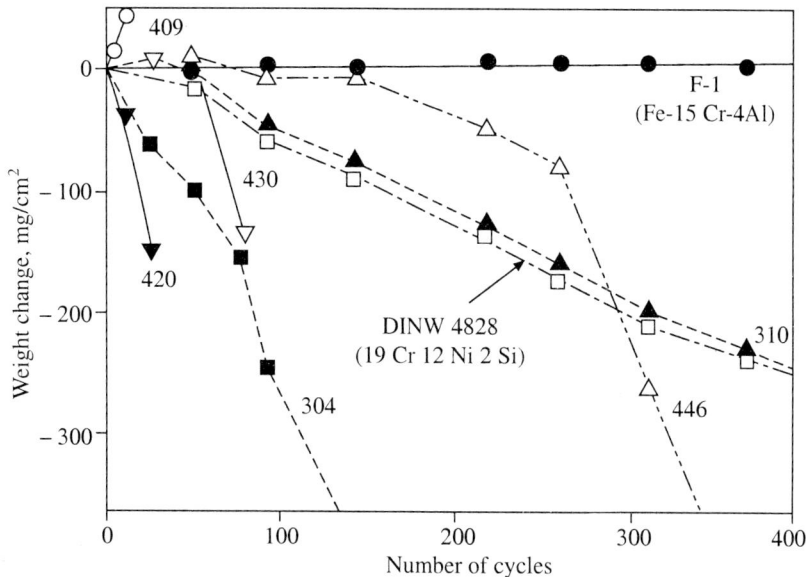

Fig. 6. Cyclic oxidation resistance of several ferritic and austenitic stainless steels at 1473 K (400 cycles, 30 m heating, 30 m cooling) [9].

Investigation on the nature of the oxide scale formed on type 316 stainless steel was carried out by Smith and Hales [34] by employing neutron activation analysis. The annealed specimens (annealed at 1273 K in vacuum of 10^{-5} torr for 24 h) were oxidized in CO_2-2% CO mixture at 600 psi for 750 h at 973 K. They observed an outer layer of Fe_3O_4 and an inner scale of spinel. The concentration profile of chromium (Fig. 7) is indicative of a diffusion-based mechanism of oxidation and also sheds light on the fact that the rate of diffusion of chromium through Fe_3O_4 is very slow. The formation of an outer layer of Fe_3O_4 is attributed to the poor integrity of the Cr-rich sesqui oxide layer that was formed initially. The chromium-depleted layer is exposed to the outside air, which promotes the generation of Fe_3O_4. The slow rate of diffusion of chromium in the duplex oxide layer (Fe_3O_4 and $FeCr_2O_4$) formed on 316 stainless steel was also confirmed by employing tracer diffusion technique. The phases formed in 304 stainless steel were studied by employing Gracing Incidence X-ray Scattering (GIXS) and reported to be comprising of both Cr_2O_3 and $FeCr_2O_4$ on oxidation in air at 1073 K [35].

For commercial application, it is desirable to reduce the vigour of oxidative wastage. Two conditions are required to satisfy an acceptable oxidation behaviour. These are

(i) The alloy must form an oxide, which grows only at a slow rate.
(ii) The oxide layer must remain adherent to the alloy at all conditions.

Of late, different types of surface coatings [36] are also being developed to mitigate the vigour of oxidation. Coating with aluminium or the addition of this element to the matrix of the alloy has been reported to be an effective means of suppressing the vigour of oxidation [37, 38]. Coatings of Al_2O_3, SiO_2, TiN and TiC are also reported to enhance the endurance of austenitic steel to oxidation [39].

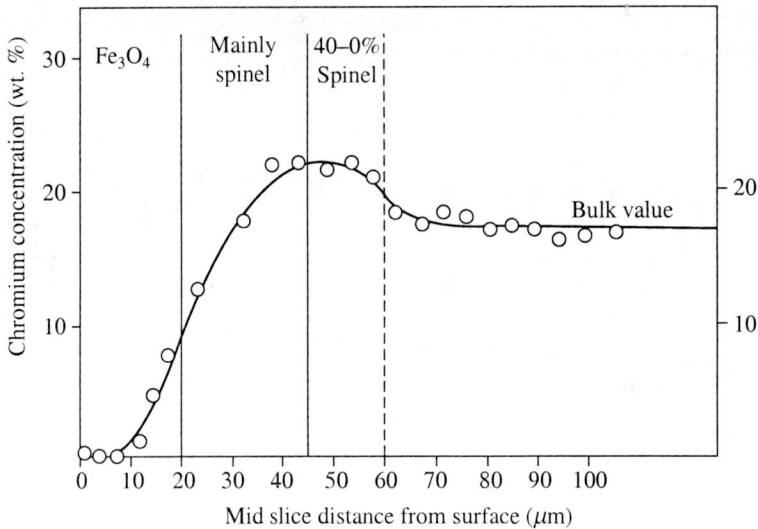

Fig. 7.   Concentration profile of chromium on the oxide scale [34].

Certain alloys with high content of nickel have been developed with better strength for application at high temperatures. As the nickel content of the alloy is increased it becomes progressively resistant to microstructural changes and oxidative loss. The creep strength also reveals a steady increase. These high nickel alloys and super alloys depend on the generation of Cr_2O_3 scale to resist the progress of oxidation. Fe-Cr, Ni-Cr and Co-Cr alloys exhibit the lowest rate of oxidation when the content of chromium is in the range of 16–30 mass % [40]. An important attribute of the Cr_2O_3 scale is its very slow rate of growth (Fig. 8) [41]. The spallation of the Cr_2O_3 scale has been reported to occur only on cooling the specimen [42–44]. At the experimental temperature, the stress on the growing scale has been accommodated through plastic deformation of the scale [45]. Enhanced spallation under compressive stress has been reported by examining the oxidation behaviour of austenitic alloys [46–50].

CARBURIZATION

Most of the metals and alloys are susceptible to carburization when exposed to environments containing carbonaceous species at high temperatures [51]. Carburization leads to the precipitation of carbides, predominantly at the grain boundaries. The nature of the carbide phases formed depends on the carbon potential of the medium [52]. Carburization of austenitic stainless steel causes sensitization because predominantly, chromium reacts with the carbon entering to the matrix through diffusion. The carburization also makes the material brittle and susceptible to failure by this mechanism.

There is yet another consequence of carburization, which results in the wastage of materials through a process similar to pitting. When the environment is highly carburizing (i.e. when the carbon potential is very high) metastable carbides are generated in the matrix. These carbides undergo decomposition in the service condition resulting in the generation of a dust of metal and carbon. This phenomenon is referred to as 'metal dusting'. Metal dusting occurs at temperatures between 703 and 1173 K and has been encountered in heat treating, refining and petrochemical industries [53–57].

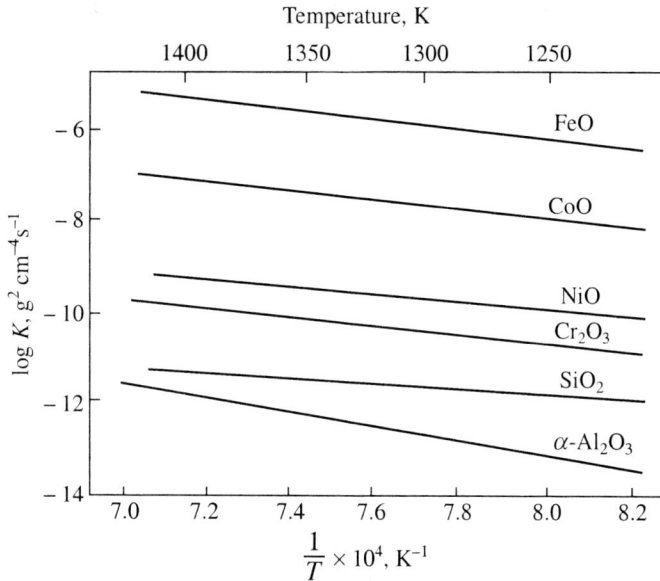

Fig. 8. Parabolic rate constants for the formation of some of the important oxides on stainless steel [41].

Thermodynamic Basis

When an alloy is exposed to a medium, the carburization or decarburization it will undergo, is determined by the carbon activity difference between the environment and the alloy. The transfer of carbon always occurs from a region of higher to lower potential.

The following reactions can fix the carbon potential of an environment

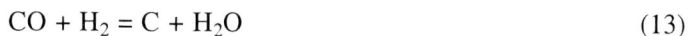

$$2CO = C + CO_2 \tag{11}$$

$$CH_4 = C + 2H_2 \tag{12}$$

$$CO + H_2 = C + H_2O \tag{13}$$

The carbon activities of the different environments may be calculated from the data on standard free energy change of the different reactions. For the above reactions, the standard free energy changes may be written as

$$\Delta G^\circ = -RT \ln (a_C \cdot P_{CO_2} / P_{CO}^2) \tag{14}$$

$$\Delta G^\circ = -RT \ln (a_C \cdot P_{H_2}^2 / P_{CH_4}) \tag{15}$$

$$\Delta G^\circ = -RT \ln (a_C \cdot P_{H_2O} / P_{CO} \cdot P_{H_2}) \tag{16}$$

Austenitic stainless steels contain chromium as an important substitutional element, which is responsible for generating an impervious layer of oxide. The ingress of carbon (from the outside environment) into these alloys promotes the generation of carbides of chromium. Chromium forms three different carbides, $Cr_{23}C_6$, Cr_7C_3 and Cr_3C_2. The nature of the carbides formed depends on the carbon potential of the medium. Relative stabilities of the different carbides and oxides may be best understood from the stability diagram (Fig. 9). From this diagram it is evident that the generation of a particular oxide or carbide is governed by the carbon and oxygen potentials of the medium. As carbon diffuses into the alloy from the external surface it is likely to create a profile of decreasing carbon activity. Such a situation results in the generation of different carbides. The carbides formed in the highest to lowest carbon activity are in the order Cr_3C_2, Cr_7C_3 and $Cr_{23}C_6$.

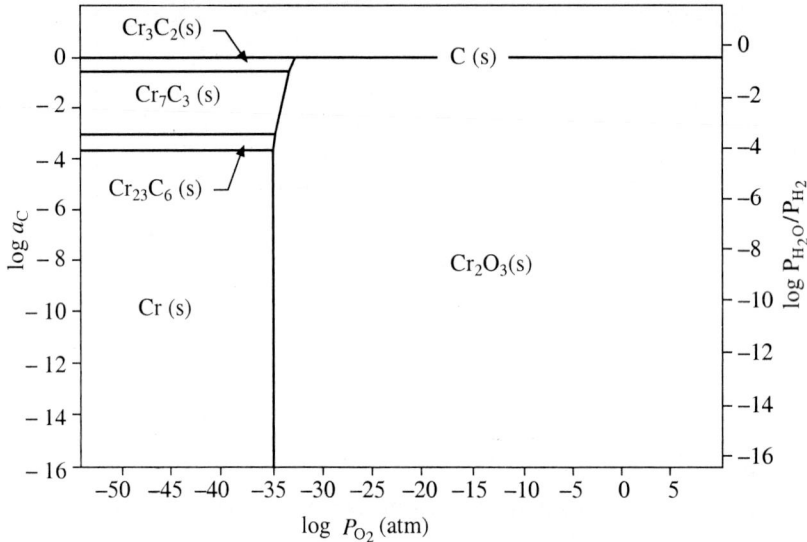

Fig. 9.   Stability diagram of Cr-C-O at 893 K.

For many high temperature alloys, especially super alloys, there are other elements added to them for the purpose of attaining the desired mechanical strength coupled with resistance to corrosion. The commonly added elements are Ti, Ta, Nb, Mo, V, W etc. All these elements are characterized by large negative value for the free energy of formation of their carbides [58]. The stabilities of binary carbides are given in Fig. 10 [59].

Kinetics of Carburization

The kinetics of carbon transport (both carburization and decarburization) is governed by the diffusion process of carbon through the austenite matrix [60, 61]. However, since the solubility of elemental carbon in the matrix is very low [62], (ranging from 21 ppm at 973 K to 700 ppm at 1273 K, in the case of the 18/8 steel) it forms a metastable phase at the normal operating temperature of the alloy, which was initially annealed. Thus, in commercial grades of stainless steel, which contain a maximum carbon concentration of 800 ppm, carbide phases of the type $M_{23}C_6$ (where M is chromium partially replaced by elements such as iron) is most commonly encountered.

For one dimensional diffusion of carbon accompanied by the simultaneous precipitation of carbides of chromium, the Fick's second law of diffusion may be written in the form

$$\partial N_C / \partial t + \partial(v N_{CrCv}) / \partial t = D_C \partial^2_{N_C} / \partial X^2 \qquad (17)$$

$$\partial N_{Cr} / \partial t + \partial(N_{CrCv}) / \partial t = D_{Cr} \partial^2_{N_{Cr}} / \partial X^2 \qquad (18)$$

N_C (N_{Cr}) = atom fractions of carbon (chromium) in the austenite phase.

D_C (D_{Cr}) = the diffusion coefficient of carbon (chromium) in the austenite matrix.

$N_{Cr\,C_v}$ = number of moles of precipitated carbide per mole of the alloy.

v = atom ratio of carbon to chromium in the carbide i.e. 6/23 for $Cr_{23}C_6$.

X = distance from the alloy surface.

t = duration.

In the austenite phase, normally, the precipitation of $M_{23}C_6$ is encountered.

Snyder et al. [63] deduced the following simplified expression from the above diffusion equation to evaluate the diffusion of carbon through austenite matrix

$$d^2_{v_c}/d\eta^2 + 2Bv\eta/N_C\,(dN_C/d\eta) = 0 \qquad (19)$$

where $\eta = X/(4D_Ct)^{1/2}$ and B a constant.

Equation (19) represents the driving force for the migration of carbon in terms of difference in the concentration of carbon and self diffusion coefficient of carbon in the austenite matrix. However, in a precipitating system, the effective diffusion coefficient of carbon has to be experimentally determined to accurately predict the carbon diffusion profiles.

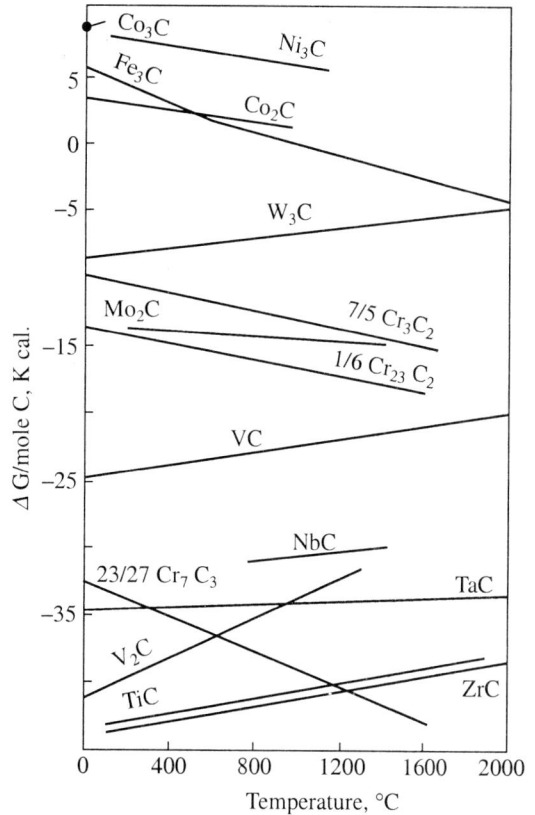

Fig. 10.   Standard free energies of formation of important metal carbides [59].

Carbon Activity of Stainless Steels

In discussing the carbon transport of austenitic stainless steels between different regions, it is necessary to have an accurate understanding of the carbon activity and its relation to composition and temperature.

The most commonly employed method to measure the carbon potentials of metals and alloys is by gas equilibration [64–68]. The gaseous equilibria generally employed are already discussed in the previous section. However, these gas equilibration method suffers from serious difficulties on account of contamination. Hence in majority of the cases the equilibration method is carried out in conjunction with the use of a reference material of known carbon activity-concentration relationship [66–68].

The CH_4/H_2 equilibration method was employed by Tuma et al. [67] for the measurement of carbon activity of stainless steel. These authors co-equilibrated iron foils to find out the equilibrium carbon activity. A similar method was employed by Natesan and Kassner [68] for the measurement of carbon activity of alloys of composition Fe-8 mass % Ni-2-22 mass % Cr. The expression relating carbon activity and composition, proposed by them is given by,

$$\ln a_C = \ln (0.048\% \ C) + (0.525 - 300/T) - 1.845 + 5100/T - (0.021 + 72.4/T) \ \%Ni$$

$$+ (0.248 - 404/T) \ \%Cr - (0.0102 - 9.422/T) \ \%Cr^2 \qquad (20)$$

which is valid in the temperature range of 973–1333 K.

The measurement of carbon activity of stainless steel type AISI 316 was carried out by Handa et al. [66] by employing the CH_4/H_2 equilibration method in conjunction with the use of Si/SiC and Mo/Mo_2C as carbon activity standards. These authors carried out the measurement only at 1273 K. Rajendran Pillai et al. [69] exposed specimens of 18/8 steel in high purity sodium (prepared by vacuum distillation). The equilibrium carbon activity of the steel was determined with the help of an electrochemical carbon meter [70]. The new expression proposed by them based on this investigation (applicable in the range of 860–960 K) is given by

$$\ln a_C = \ln (0.048\%C) + (0.525-300/T) - 1.845 + 5100/T - (0.021 + 72.4/T)\ \%(Ni + Mn)$$

$$+ (0.248 - 404/T)\%Cr - (0.0102 - 9.422/T)\ \%Cr^2 + 0.033\ \%Cr \qquad (21)$$

Consequences of Carburization

Microstructural Changes

Carburization results in the generation of carbides both at the grain boundaries and the matrix. It is very difficult to gauge the extent of carburization of a material from the mass gain data. This uncertainty is on account of the simultaneous oxidation of the material in most of the carburizing environments. This problem is alleviated by determining the mass of carbon (predominantly from a gas phase) absorbed by the carburizing material. The carburized layer is usually characterized by the analysis of carbon concentration profile as a function of distance from the surface of the carburized metal.

Carburization affects the properties of the materials in several ways. Apart from the total amount of carbon absorbed, the maximum concentration attained and the prevailing concentration gradient affect the performance of the alloy [71].

Mechanical Properties

The most pronounced effect of carburization of stainless steel is the reduction in fracture elongation [72]. Usually the yield strength and ultimate tensile strength increase as the concentration of carbon in the material becomes higher. For carbon concentration in the alloy greater than 0.5 mass%, the embrittlement was so severe that fracture occurred with no measurable change of ductility at temperatures between 773 and 1073 K [73].

Several investigations are reported in the literature to examine the creep-rupture behaviour of austenitic stainless steel in liquid sodium environment. If the thickness of the layer is less than 5% of the total thickness of the specimen, then the effects of carburization was not reflected on the mechanical properties [74]. However, if the thickness of the carburized layer is comparable to the thickness of the specimen, significant changes in the properties could be encountered. The long-term effect of carburization of stainless steel type 316 on the changes in ductility has been recently estimated and reported in the literature [75].

In a similar way, decarburization has also its impact on the mechanical properties. Tests carried out in decarburizing environment at 1033 and 1005 K (on both stainless steel types 304 and 316, respectively) have showed a decrease in the stress rupture life [76]. The decrease in the stress rupture property was attributed to changes in the microstructure that resulted from the loss of carbon from the alloy. The initial content of carbon in the alloy was 0.06 mass% which decreased to 0.03 mass % in the case of 304 stainless steel and 0.04 mass% in the case of 316 variety due to the impact of the decarburizing environment. Metallographic examination of the specimen after the test indicated that a significant amount of sigma phase was present in the material. The generation of sigma phase was caused due to decarburization.

Carburization also brings about changes in the fatigue life in the expected line. As carburization embrittles the material, the fatigue life was reduced at all the strain level investigated [77].

NITRIDATION

Most of the metals and alloys are susceptible to nitridation when exposed to environment with high potential of nitrogen. During nitridation, the alloy absorbs nitrogen from the environment. Depending on the potential of nitrogen in the environment and its solubility in the alloy matrix, the absorbed nitrogen may remain in the elemental form or forms nitrides. The formation of nitride precipitate in the matrix and grain boundaries makes the material brittle. The nitridation of metals deserves special attention as atmospheres filled with nitrogen are finding increasing application in processes that requires inert environments. Even though nitrogen is inert at lower temperatures, exposure at high temperature may cause brittleness and premature failure.

Thermodynamic Basis

When a metal or an alloy is exposed to an environment containing nitrogen, nitridation proceeds at elevated temperatures according to the reaction;

$$\frac{1}{2} N_2 (g) = 2N \text{ (dissolved in the metallic matrix)} \qquad (22)$$

The solubility of non-metals in metals may be best understood by employing Sievert's law. The percentage of nitrogen dissolved in the metal may be written as

$$(\%N) = kP_{N_2}^{\frac{1}{2}} \qquad (23)$$

where k is the equilibrium constant for the reaction (22) and P_{N_2} the partial pressure of nitrogen.

As already mentioned, molecular nitrogen is relatively inert to bring about reaction with the metal. However, when the metals and alloys are heated to temperatures as high as 1273 K, nitridation may proceed with appreciable rate.

Nitridation has been reported to occur with high rate when the atmosphere contains nitrogen-bearing compounds rather than molecular nitrogen. Hence ammonia is the preferred choice for most of the cases related to case hardening through nitridation. However, if the ammonia is allowed to decompose and form free nitrogen, the rate of nitridation will be adversely affected. Hence, the decomposition may be allowed to take place on the surface of the steel. The atoms of nitrogen, produced by the decomposition, readily dissolve in the metallic matrix [78, 79]. When the activity of nitrogen is higher than the equilibrium activity of the particular nitride, it starts precipitating in the alloy. The relative stabilities of various nitrides may be compared in terms of their free energies of formation as shown in Fig. 11. Nitridation of the alloy is similar to carburization as both do not cause the wastage of materials. When the temperature of the alloy is low (lower than 773 K), the rate of diffusion of nitrogen is so low that a nitrided layer is formed only at the surface. At temperatures higher than 1273 K, the rate of diffusion of nitrogen is very fast. Nitridation in these cases proceed with the generation of nitrides at the grain boundaries and matrix.

Kinetics of Nitridation

The kinetics of nitridation of various grades of stainless steel was studied by Moran et al. [80]. They carried out experiment in an environment containing ammonia. At a temperature of 773 K, type 304 stainless steel was found to undergo a corrosion rate in the range of 0.02 to 2.5 mm/a as the concentration of ammonia was increased from 5% to 99%. For the same concentration range of ammonia the nitridation rate of type 316 stainless steel was reported to be in the range of 0.012 to 13.21 mm/a. Moran et al. [80] also reported that the rate of nitridation in type 316 stainless steel is higher than that of 304 grade even though the exact reason for this behaviour was not explained by these authors.

Effects of Nitridation

Nitridation, as in the case of carburization, makes the material brittle. Thus the nitridation of the material contributes to the alteration in the mechanical properties, particularly ductility and toughness. Such experimental datas are not adequately reported in the literature. The effects of nitridation on the mechanical properties are expected to be nearly same as that of carburization.

Barnes and Lai [81] reported the nitridation behaviour of several commercial alloys including stainless steel in environment containing ammonia.

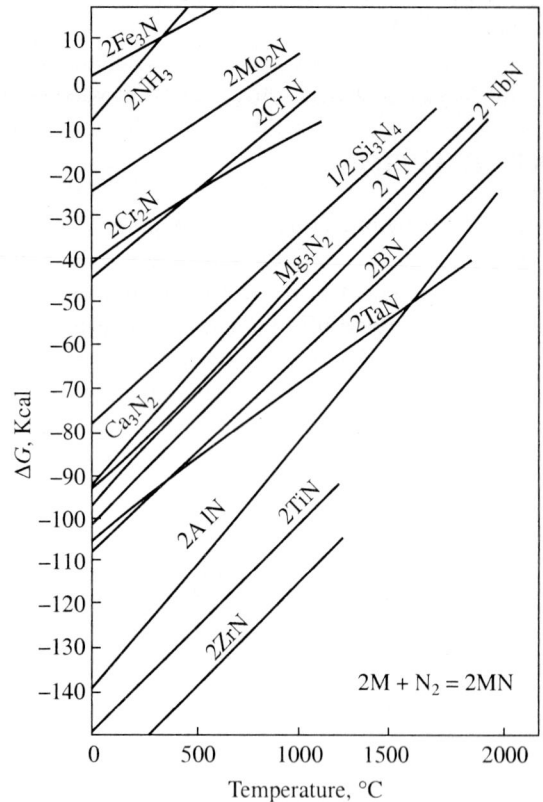

Fig. 11.   **Standard free energies of formation of important metal nitrides.**

In general they observed that an increase of nickel content improved the resistance to nitridation. This behaviour is attributed to the capacity of nickel to reduce the solubility of nitrogen in iron base alloys. The extent of penetration of nitrogen into the stainless steel matrix depends on the temperature. At temperature less than 923 K, a surface nitrided layer, consisting mostly of iron nitride (Fe_2N and Fe_4N) was formed. At higher temperatures internal nitrides such as CrN and Cr_2N were formed. Several factors affect the nitridation resistance of alloys. Elements such as nickel and cobalt is reported to reduce the solubility of nitrogen in iron [82] and thus impart resistance to nitridation.

Nitridation of stainless steel is significantly influenced by the oxygen potential of the environment. Odelstam at al. [83] reported that the penetration of nitrogen is significantly reduced if the oxygen partial pressure of the medium is kept above a threshold value. The oxide scale is expected to reduce the rate of surface absorption of nitrogen molecule. The oxide scale also inhibits the diffusion of nitrogen into the metal.

The nitridation behaviour of stainless steels of different grades were studied by Tjokro et al. [84] at temperatures in the range of 1123 to 1473 K. They reported the generation of CrN with some nitrides of iron. Nitridation of stainless steel is also employed as a possible means for the protection of stainless steel [85, 86] against corrosion by other species present in the medium.

SULPHIDATION

Sulphur and compounds containing sulphur are commonly encountered in high temperature industrial environments. Sulphur is an invariable contaminant in industries that depend on fossil fuels. When combustion occurs in presence of excess oxygen, sulphur forms sulphur dioxide and trioxide. This atmosphere is generally oxidizing and is less corrosive when compared to a reducing environment. Thus, if sulphur is present in the form of H_2S, the sulphidation process will be accelerated to the maximum extent, because of reducing environment.

Thermodynamic Basis

The possibility of formation of sulphides of different metals may be easily gauged from the free energies of formation of the corresponding sulphides. However, most of the sulphidising environments are contaminated with oxygen to exhibit oxygen and sulphur potentials.

Stability diagram can be employed to understand the nature of the various phases formed on the surface of the exposed metals. The metal-sulphur-oxygen stability diagram at 1143 K for iron nickel and chromium are shown in Figs. 12, 13 and 14 [87], respectively. Perkin [88] proposed that in the elemental potentials corresponding to the upper region of the Cr-S-O diagram, both CrS and Cr_2O_3 form initially on the metal surface of chromium and high chromium alloys. The stability of Cr_2O_3 is higher than that of the sulphides of chromium and hence the oxide scale would continue to grow. The constituents of the alloying elements that form stable sulphides have to diffuse through the pre-existing oxide scale to generate a coating of sulphide scale on the oxide scale.

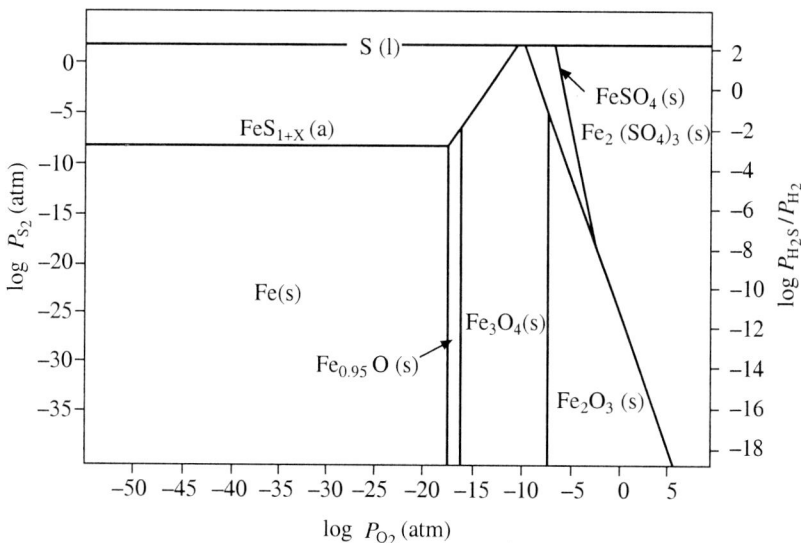

Fig. 12. Stability diagram of Fe-S-O system at 1143 K.

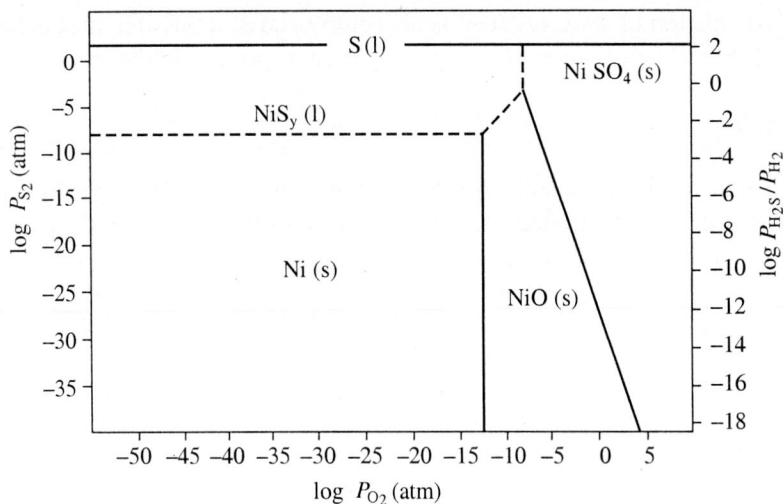

Fig. 13.   Stability diagram of Ni-S-O system at 1143 K.

Another problem associated with the formation of sulphides is the low melting point of some of these compounds. Some of the metal-metal sulphide eutectics have low melting points, which aggravate the corrosion problem [89]. However, the peril against the formation of low melting sulphides can be prevented by promoting the generation of stable chromia layer.

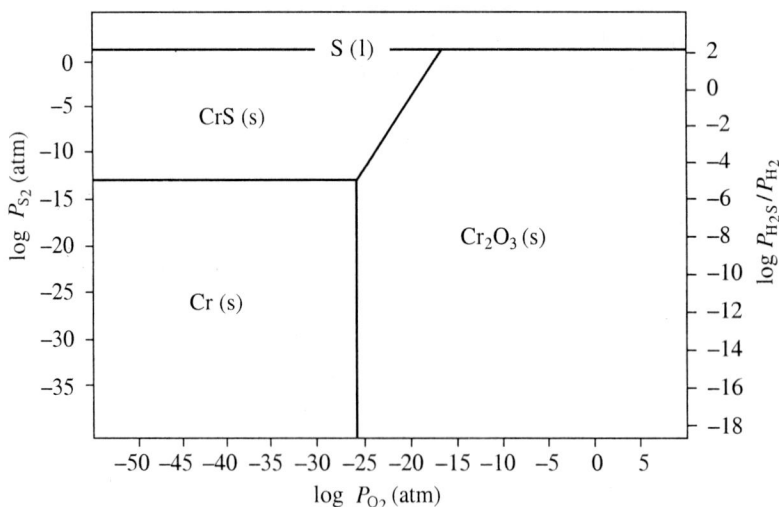

Fig. 14.   Stability diagram of Cr-S-O system at 1143 K.

In environments containing high sulphur potentials, sulphides are formed as stable scale. The environments that favour the formation of sulphide scale are sulphur vapour, H_2/H_2S mixtures with extensively low oxygen activities such that Cr_2O_3 is not expected to be formed.

Kinetics of Sulphidation
Sulphidation normally competes with oxidation and hence the rate of this process follows a complex

kinetics. The reaction rate generally peaks at a particular temperature, then decreases with increase of temperature. The temperature at which the corrosion rate is maximum varies for different metals. For example, in the case of nickel the maximum rate of sulphidation is reported at temperature of 873 K and there after at higher temperatures the rate becomes slower [90]. Chromium, on the other hand, does not form sulphides in environments containing sulphur dioxide at temperatures from 973 to 1273 K, but instead forms stable Cr_2O_3 scale [91]. Thus, chromium as an alloying element is capable of improving the resistance to sulphidation by the formation of an adherent oxide layer.

Sulphidation of Stainless Steel
Sulphidation of stainless steel of different grades have been studied by several investigators [92–94]. These investigations revealed that a slight variation in the oxygen and sulphur potentials can cause a drastic change over from protective scale formation to catastrophic corrosion process. Considerable effort has been centered around developing modified surface that is resistant to sulphidation. An aluminizing process has been reported [95–97] to protect the stainless steel surface against degradation through sulphidation.

METHODS OF MITIGATING HIGH TEMPERATURE CORROSION BY SURFACE MODIFICATION

Service life and performance of many of the engineering alloys can be improved by modifying the composition or microstructure of a very thin layer on the surface. Though surface modification was known since several centuries, the technology was based primarily on artifacts rather than sound scientific understanding. However, in the last 2–3 decades there is an ever increasing realization about the scientific basis of modifying the surface to make the components resistant to corrosion, oxidation, wear and subsequent catastrophic failure. The broad field of surface engineering is an emerging activity, which involves modifying the surface to enhance endurance to the desired service environment. A number of techniques are reported in literature to modify the surface, the prominent among them being

  (i) Electroplating and anodizing
 (ii) Physical and chemical vapour deposition
(iii) Thermal spray and plasma spray
 (iv) Plasma nitriding
  (v) Pack carburization
 (vi) Ion implantation and ion beam mixing
(vii) Laser surface treatment and laser alloying
(viii) Cladding

Coatings are also classified as overlay coating (where the new material is physically adhered to the surface) and diffusion coating (where the coating that was applied initially was allowed to diffuse inside, partially). Some of the methods are employed to modify the surface of stainless steel to render excellent resistance to corrosion. Hannani and Hermiche [98] reported the application of a nitrogen implantation method on stainless steel 304 to improve the corrosion resistance. The wear resistance of Stainless steel AISI 316 was found to be improved when implanted with titanium [99]. The nitrogen implantation technique was also employed by Goel et al. [100] to enhance the microhardness. Sugioka et al. [101] reported a simultaneous method for deposition and diffusion of silicon on the surface of AISI 304 stainless steel by laser melting. They have reported an increase of hardness and better corrosion properties on account of surface modification.

Among the different oxides, alumina scale has been reported to posses the maximum protection against corrosive environment. Thus, the method commonly employed is to alloy the stainless steel with aluminium. However, the formation of alumina causes the depletion of the metal. Thus the alumina layer formed subsequently will possess poor adherence. In order to sustain the corrosion resistant properties an increase in the aluminium content of the alloy is required. However, it is not desirable to add aluminium in excess of 5 mass% due to the decrease of ductility of the alloy. Andoh et al. [38] deposited aluminium from the vapour phase on to the surface of stainless steel and obtained substantial resistance to high temperature oxidation. A change in the microstructural features brought about by the laser melting was reported to have improved the resistance to high temperature oxidation [102]. The oxide layer formed was found to be richer in chromium and thus more protective under oxidizing conditions [103]. The fine grained microstructure that was evolved on laser melting was found to allow a faster diffusion of chromium even though the mechanism is not fully understood.

REFERENCES

1. Rodriguez, P., Transaction of the Indian Institute of Metals, 20 (1967) 213.
2. Kofstad, P., High Temperature Corrosion, NACE-6 (1983) 123.
3. Guttmann, V. and Marriott, J.B., Proc. Conf. Environmental degradation of high temperature materials, Vol. 2 Ser. 3 (1980) Paper No. 13.
4. Kofstad, P., High Temperature Oxidation of Metals, Wiley, NY (1966).
5. Wagner, C. Zeitschrift Physikalische Chemie, Vol. B21 (1933) 25.
6. Grodner, A., Welding Research Council, Bulletin No. 31, 1956.
7. Eiselstein, H.E. and Skinne, E.N., in ASTM STP No. 165, (1954) 162.
8. Brasunas, A.S., Gow, J.T. and Gardner, O.E., Proc. ASTM, 46 (1946) 870.
9. Kado, S., Yamazaki, T., Yamazaki, M., Yoshida, K., Yabe, K. and Kobayashi., H., Transactions of the Iron and Steel Institute of Japan, 18 (No. 7) (1978) 387.
10. Sapiesko, R. and Matijevic, E., Corrosion 37 (1981) 152.
11. Lister, D., Nuclear Science and Engineering, 58 (1975) 239.
12. Lister, D., Nuclear Science and Engineering, 59 (1976) 406.
13. Bart, G., Wasserfallen, K., Haller, M. and Mohos, M., Proc. Intern. Symp. on Water chemistry and corrosion problems of nuclear reactor systems and components, Organized by IAEA, Vienna, Nov. (1982) p. 35.
14. Stephens, J.R., Barrett, C.A. and Chen, W.Y.C., Rev. on coatings and corrosion, Vol. III, No. 4, (1979) p. 211.
15. Garrett, J.C.P., Lister, S.K., Nolan, P.J. and Crook, J.T., BNRES Conf. Corrosion of steel in carbon dioxide, Reading (1974) p. 268.
16. Hales, R., Werkstoff und Korrosion, 29 (1978) 393.
17. Smith, A.F, Werkstoff und Korrosion, 30 (1979) 100.
18. Smith. A.F., Werkstoff und Korrosion, 32(1981) 1.
19. Wood, G.C. and Hobby, M.C., Journal of Iron and Steel Institute, 203 (1965) 54.
20. Francis, J.M., British Corrosion Journal, 3 (1968) 113.
21. Caplan, D. and Cohen, M. J., Metals, (NY) 4 (1952) 1057.
22. Knutsen, A.B., Conde, J.F.G. and Piene K., British Corrosion Journal, 4 (1969) 94.
23. Wild. R.K., Corrosion Science, 13 (1973) 105.
24. Hales, R. and Hill, A.C., Corrosion Science, 14 (1974) 553.
25. Wild, R.K., Corrosion Science, 17 (1977) 87.

26. Wood, C.C., Werkstoff und Korrosion, 6 (1971) 491.
27. Takahashi, N. and Okada, K., Japan Journal of Applied Physics, 11(1972) 1580.
28. Smith, A.F. and Hales, R., Werkstoff und Korrosion, 28 (1977) 405.
29. Storp, S. and Holn, R., Surface Science, 68 (1977) 10.
30. Lumsden, J.B. and Staehle, R.W., Scripta Metallurgica, 6 (1972) 1205.
31. Mathewson, A.G., Vacuum, 24 (1977)505.
32. Schonbert, R., Journal of Vacuum Science and Technology, 12 (1975) 505.
33. Tanabe, T. and Imoto, S., Transactions of the Japan Institute of Metals, 20 (1979) 507.
34. Smith, A.F. and Hales, R., Werkstoff und Korrosion, 29 (1978) 246.
35. Saito, M., Kosaka, T., Matsubara, E. and Waseda, Y., Materials Transactions, Japan Institute of Metals, 36 (1) (1995) 1.
36. Hashimoto, M., Miyamoto, Y., Kubo, Y., Tokumaru, S., Ono, N., Takahashi., T. and Ito, I., Material Science and Engineering, A198 (1995) 75.
37. Benjamin, J.S., Metallurgical Transactions, 1 (1970) 2943.
38. Andoh, A., Taniguchi, S., Shibata, T., Oxidation of Metals, 46 (5-6) (1996) 481.
39. Kohno, M., Ishikawa, S., Ishii, K. and Satoh, S.; in: Proc. Conf. Microscopy of Oxidation–3, Cambridge, UK, (1996) Pub: Institute of Materials, UK (1997) p. 55.
40. Wood, G.C., Wright, I.G., Hodkiess, T. and Whittle, D. P., Werkstoff und Korrosion, 21(1970) 900.
41. Birks, N. and Meier, G.H., " Introduction to high temperature oxidation of metals", John Wiley & Sons, New York, 1966.
42. Asher, J., Sugden, S., Bennett, M.J., Hawes, R.W.M., Savage. D.J. and Price. J.B., Werkstoff und Korrosion, 38(1987) 306.
43. Bennett, M.J., Proc. 10th Intern. Congress on Metallic Corrosion, Pub: Oxford & IBH Publishing Co., New Delhi, 4 (1987) 3761.
44. Tempest, P.A. and Wild, R.K., Oxidation of Metals, 30 (1988) 231.
45. Bennett, M.J., Buttle, D.J., Colledge, P.D., Price, J.B., Scruby, C.B. and Stacey, K.A., Report AERE-R-13464 (1989).
46. Evans, H.E., Materials at High Temperature, 12 (2-3) (1994) 219.
47. Pieraggi, B. and Dabosi, F., Werkstoff und Korrosion, 38 (1987) 584.
48. Goebel, M., Rahmel, A. and Schuetze, M, Oxidation of Metals, 39 (3-4) (1993) 231.
49. Evans, H.E., Nicholls, J.R. and Saunders, S.R.J., Solid State Phenomena, 41 (1995) 137.
50. Schuetze, M., in: Proc. High Temperature Corrosion of Advanced Materials and Protective Coatings, Eds. Y. Saito, B. Oeny and T. Maruyama, Elsevier Science Publisher, (1992) 39.
51. Metals Handbook, 8th edn. Vol. 2, American Society of Metals, Metals Park, Ohio (1964) p. 93.
52. Moller, G.E. and Warren, C.W., Paper No. 237, Corrosion 81, NACE, USA.
53. Schueler, R.C., Hydrocarbon Process, Aug. (1972) p. 73.
54. Lai, G.Y., Rothmann, M.F. and Fluck, D.E., Paper No. 14, Corrosion-85, NACE, USA.
55. Grabke, H.J., Krajak, R. and Mueller-Lorenz, E.M., Werkstoff und Korrosion, 44 (1993) 89.
56. Grabke, H.J., Hemtenmacher, J. and Munker, A., Werkstoff und Korrosion, 35 (1984) 543.
57. Grabke, H.J., Hemtenmacher J. and Munker, A., Werkstoff und Korrosion, 35 (1984) 543.
58. Hultgren, R., Desai, P., Hawkins, D., Gleiser, M. and Kelley, K.K. "Selected values of thermodynamic properties of elements and binary alloys", American Society of Metals, Metals Park, Ohio, USA.
59. Shatyuski, S.R., Oxidation of Metals, 13 (No. 2) (1979) 105.
60. Natesan, K. and Kassner, T.F., Metallurgical Transactions, 37(1970) 223.
61. Snyder, R.B., Natesan, K., and Kassner, T.F., Journal of Nuclear Materials, 50 (1974) 259.
62. Lai, J.K.L., Materials Science and Engineering, 61(1983) 101.
63. Snyder, R.B., Natesan K. and Kassner, T.F., Argonne National Laboratory (USA), Report ANL-8015 (1973).
64. Richardson, F.D., Physical Chemistry of Melts in Metallurgy, Vol. 2 (Academic Press, NY, 1974) p. 342.
65. Smith, R.P., Journal of American Chemical Society, 68 (1946) 1163.
66. Handa, M., Takahashi, I., Tsukada, T. and Iwai, I., Journal of Nuclear Materials, 116 (1983) 178.
67. Tuma, V.H., Groebner, P. and Loebl, K., Archiv Eisenhuttenwesen, 9 (1969) 727.

68. Natesan, K. and Kassner, T.F., Metallurgical Transactions, 4 (1973) 2557.
69. Rajendran Pillai, S. and Mathews, C. K., Journal of Nuclear Materials, 150 (1987) 31.
70. Rajendran Pillai, S. and Mathews, C.K., Journal of Nuclear Materials, 137 (1986) 107.
71. Krikke, R.H., Horing, J. and Smith, K., Materials Performance, Aug. (1976) 9.
72. Krankota, J.L., Journal of Materials Engineering and Technology, Jan. (1976) 9.
73. Thorley, A. and Tyzack, C., in: Proc. Effects of environments on materials properties in nuclear system, July (1971) BNES, London.
74. Andrews, R.C., USAEC Report, MSAR 64–81, MSA Research Corporation, USA (1964).
75. Rajendran Pillai, S., Khatak, H.S. and Gnanamoorthy, J.B., Materials Transactions Japan Institute of Metals, 39 No. 3 (1998) 370.
76. Lee, W.T., USAEC Report, NAA-SR-12353, Atomics International, (1967).
77. Andrews, R.C., Hiltz, R.H., Kirschler, L.H., Rodgers, S.J., Tepper, F.,USACE Report MSAR-65-194, MSA Research Corporation, (1965).
78. Bever, M.B., Floe, C.F., Source book on nitriding, American Society for Metals, Metals Park, Ohio, (1977) p. 125.
79. Lightfoot B.J. and Jack, D.H., Source book of nitriding, American Society for Metals, Metals Park, Ohio (1977), p. 248.
80. Moran, J.J., Mihalisin, J.R. and Skinner, E.N., Corrosion, 17 (No.4) (1961) 191t.
81. Barnes, J.J. and Lai, G.Y., Transactions of the Metallurgical Society, Annual Meeting (1989) Las Vegas, USA.
82. Wriedt, H.A. and Gonzalez, O.D., Transactions of the Metallurgical Society, AIME, Vol. 221 (1961) 532.
83. Odelstam, T., Larsen, B., Martensson, C. and Tynell, M., Corrosion-86, NACE, Houston (1986) paper 367.
84. Tjokro, K. Young, D.J., Johnsson, R., Redmond, J.D., Corrosion-91, NACE, Ohio, USA.
85. Munogaki, M., Ooi, S., Miyazaki, K., Journal of Nuclear Materials, 179–181 (199) 286.
86. Chan, J.J. and Chang, S.C., Journal of Materials Science, 25 (2B), (1990) 1331.
87. Hemmings P.L. and Perkins, R.A., EPRI Report FP-539, (1977) Lockeed Paolo Alto Research Laboratories, Palo Alto, USA.
88. Perkins, R.A., "Environmental Degradation of High Temperature Materials", Series 3, No. 13, vol. 2 (1980) p. 5/1.
89. Hansen, M., and Anderko, K., Constitution of Binary alloys, McGraw Hill, NY (1958).
90. Kofstad, P., High temperature corrosion, Elsevier Applied Science, NY (1988).
91. Asmundis, C.D., Gesmundo, F. and Bottino, C., Oxidation of Metals, Vol. 14, No. 4, (1980) p. 351.
92. Saunders, S.R.J., Gohil, D.D., Osgerby, S. , Materials at High Temperature, 14(3), (1995) 173.
93. Pareek, V.K., Ozekein, A., Mumford, J., Ramanarayanan, T.A., Journal of Materials Science Letters, 14, (3), (1995) 173.
94. Stroosnijder, M.F., Guttmann, V. and de Wit, J.H.W., Corrosion Science, 32 (2), (1991) 151.
95. Green, S.W. and Stott, F.H., Oxidation of Metals, 36 (3–4) (1991) 239.
96. Uihlein, T., Auer, W. and Kaeshe, H., Werkstoff und Korrosion, 41 (10), (1990) 585.
97. Morsinkhof, R.W.J., Frausen, T., Heisenkveld, M.M.D., Gellings, P.J., Materials Science and Engineering, A 120 (1989) 449.
98. Hannani, A. and Kermiche, F., Transactions of the Institute of Metal Finishing, 76(3) (1998) 114.
99. Evans, P.J., Hyvarinen, J. and Samandi, M. in: Proc. Second Australian Intern. Conf. Surf. Engg. Coatings and Surface Treatments in Manufacture, Adelaide, Australia, Pub: Surface Engg. Research Group, Univ. South Australia (1994) C 172.
100. Goel, A.K., Sharma, N.D., Mohindra, R.K., Ghosh P.K. and Bhatnagar, M.C., Indian Journal of Physics, A 64 (6) (1990) 444.
101. Sugioka, K., Tashiro, H. and Toyoda, K., International Journal of Materials Production Technology, 8(2–4) (1993) 316.
102. Ghosh, S., Goswami, G.L., Biswas, A.R., Venkataramani, R. and Garg, S.P., in: Proc. Discussion Meeting on Surface Sci. Engg, Indira Gandhi Centre for Atomic Research, Kalpakkam, (Tamil Nadu, India) (1997) 156.
103. Wade, N., Hoshihama, T. and Hosol, Y., Scripta Metallurgica, 19 (1985) 859.

12. Corrosion Detection and Monitoring in Austenitic Stainless Steels Using Nondestructive Testing and Evaluation Techniques

Baldev Raj[1], P. Kalyanasundaram[1] and S.K. Dewangan[1]

Abstract Corrosion is a term related to material deterioration mechanisms, which may induce flaws affecting the health of a component. If these flaws are not detected at the right time, they may lead to failure of the component resulting in loss in productivity and threat to safety. Detection and monitoring of corrosion is thus very important for any industry. To this effect, nondestructive evaluation (NDE) techniques viz. eddy current, ultrasonic, radiography, acoustic emission, thermography etc. have a significant role to play in the detection and monitoring of corrosion. In this article, the employment of various NDE techniques for the detection and monitoring of corrosion in austenitic stainless steels are discussed.
Key Words Corrosion monitoring, NDT, ECT, UT, acoustic emission, radiography, thermography, laser technique.

INTRODUCTION

Nondestructive testing (NDT) by definition refers to test methods used to examine or inspect a part, material or system without impairing its future usefulness. NDT gives vital information for material characterization including quantitative determination of the size, shape and location of a defect or anomaly thus enabling structural integrity assessment of a component. It has a number of important roles to play in ensuring the quality and reliability of many important products whose integrity is of paramount importance. The traditional NDT in quality control during manufacture, predominantly defect detection, has been complemented in recent years with material characterization, stress management, material degradation assessment and in-service inspections. The correct application of NDT can prevent accidents, save lives, protect the environment and avoid economic loss.

Austenitic stainless steels find extensive applications in numerous industries due to their unique combination of properties such as very good corrosion resistance, high mechanical strength and also excellent weldability. It is a candidate material for various strategic structural components of nuclear power plants and is also widely used in petrochemical, fertilizer plants and pharmaceutical industries. The components operating in aggressive environment like corrosive fluids, high temperature and stresses undergo corrosion and corrosion related degradation. This can take many forms such as

[1]Metallurgy and Materials Group, Indira Gandhi Centre for Atomic Research, Kalpakkam-603 102, India.

localized damage, generalized attack, environment assisted cracking, high temperature oxidation, etc. The mechanisms by which corrosion damage occurs are also varied, but can be classified as electrochemical, chemical or physical. Prompt detection and assessment of corrosion before failure takes place is most important. NDT plays a very important role in providing detection of the early signs of corrosion so that corrective action can be taken before damage becomes severe. Figure 1 summarizes various mechanisms of degradation of materials/components operating in demanding service conditions and the role of NDT for damage assessment and life extension. As the cost of repair or replacement continues to increase, demands on NDT particularly for early detection of corrosion have increased. While every attempt is made in selection of material, proper design and operating environment, corrosion related degradation of components is inevitable. Therefore, it becomes essential to monitor the performance of the components in service to assess the progress of corrosion related degradation to be within the expected and acceptable rate and also to detect any accelerated/unanticipated corrosion related degradation. Such approach is aimed at avoidance of failures, adoption of methodologies for prevention/retardation of degradation, and life assessment and extension of the components. This can be achieved by employing various NDT techniques.

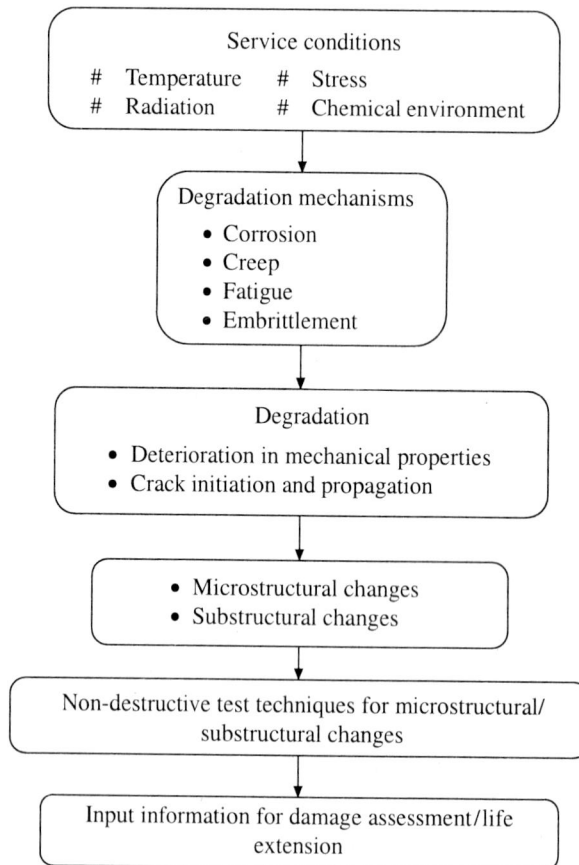

Service conditions
Temperature   # Stress
Radiation   # Chemical environment

Degradation mechanisms
• Corrosion
• Creep
• Fatigue
• Embrittlement

Degradation
• Deterioration in mechanical properties
• Crack initiation and propagation

• Microstructural changes
• Substructural changes

Non-destructive test techniques for microstructural/substructural changes

Input information for damage assessment/life extension

Fig. 1. Material degradation in service and role of non-destructive testing for damage assessment and life extension.

There are a host of corrosion types, which can take place in austenitic stainless steel components, such as uniform corrosion, pitting corrosion, crevice corrosion, stress corrosion cracking (SCC), oxidation, corrosion fatigue, erosion-corrosion etc. The physical manifestation of these various types of corrosion are different and hence their detection and evaluation demand application of different NDT techniques. Thus, a careful choice of an appropriate technique or a combination of techniques, depending on the type of component/structure and the type of corrosion damage to be monitored, are very important. Different NDT techniques like eddy current, ultrasonic, radiography, acoustic emission, thermography etc. are widely used for characterizing corrosion in materials and their damage during service. Table 1 shows the list of NDT techniques applied for monitoring different types of corrosion damage in austenitic stainless steels. While the techniques like eddy current, ultrasonic and radiography are used to detect the static damage, acoustic emission technique offers potential use to evaluate the damage in real-time.

Table 1. NDT techniques applied for monitoring different types of corrosion damage in austenitic stainless steel

Corrosion mechanism	NDT technique
Uniform	UT, ECT, RT, OPTICAL/LASER, IRT
Localized	UT, ECT, RT, OPTICAL/LASER, IRT
Stress Corrosion Cracking	PT, AE, UT, ECT
Corrosion Fatigue	PT, AE, UT, ECT
High Temperature Oxidation	AE, ECT

AE	: Acoustic Emission	PT	: Penetrant Testing
ECT	: Eddy Current Testing	RT	: Radiographic Testing
IRT	: Infra- red Thermography	UT	: Ultrasonic Testing

NDT methods range from the simple to the intricate. Visual inspection is the simplest of all. Surface imperfections invisible to the eye may be revealed by penetrant or magnetic methods. Unless serious surface defects are found, there is often little point in proceeding further to more complicated examination of the interior by other methods like ultrasonics or radiography. Some of the important NDT methods are visual or optical inspection, dye penetrant testing, magnetic particle testing, eddy current testing, radiographic testing and ultrasonic testing [1]. The details of the principles and applications of these techniques can be found in the above reference. Though there are a number of NDT techniques available for corrosion monitoring applications, the use of some of the popular NDT techniques for detection and monitoring of corrosion in austenitic stainless steels are discussed in this chapter. Although, application of some of the techniques like magneto- optic imaging, remote field eddy current testing, random decrement vibratory analysis etc. are not reported for corrosion monitoring in austenitic stainless steel components, they have been included in this chapter as they are recent advancements in the field of NDT, offering promising applications for corrosion detection and monitoring in austenitic stainless steel. A few case studies relating to corrosion monitoring applications in materials other than austenitic stainless steels are also covered in this chapter, as the method used is independent of the material property e.g., wall thickness loss measurement by ultrasonic testing, impedance change in eddy current testing etc. The basic principle of each technique is presented first and then its application for the detection and monitoring of corrosion is given in subsequent sections.

EDDY CURRENT TESTING (ECT)

Basic Principle

Although the evolution of the eddy current theory can be traced back to the initial discovery of both electricity and magnetism, the scientific development of eddy current theory started with the discovery of the law of electromagnetic induction by Faraday in 1832. Faraday's law states that when a magnetic field cuts a conductor, by physical motion of either the magnetic field or the conductor, an electrical current will flow through the conductor if a closed path is provided. From Oersted's discovery, a magnetic flux exists around a coil carrying current proportional to the number of turns in the coil and the current.

ECT normally involves measurement of impedance changes in a test coil. When alternating current is passed through a coil, an alternating primary magnetic field results. Figure 2 shows the generation of eddy current in a conductor placed near varying primary magnetic field. These eddy currents in turn have an associated magnetic field which opposes the primary field according to Lenz's law. Impedance of the coil changes with the secondary field, and its magnitude depends on the eddy current density. It also depends on a number of factors viz. coil dimensions, exciting frequency, conductivity and permeability of the material, presence of discontinuity, and the geometry of the test object. Penetration of eddy currents to large depths is generally impaired by the phenomenon known as "skin depth", thus making it more suitable for inspection of surface and sub-surface defects [2].

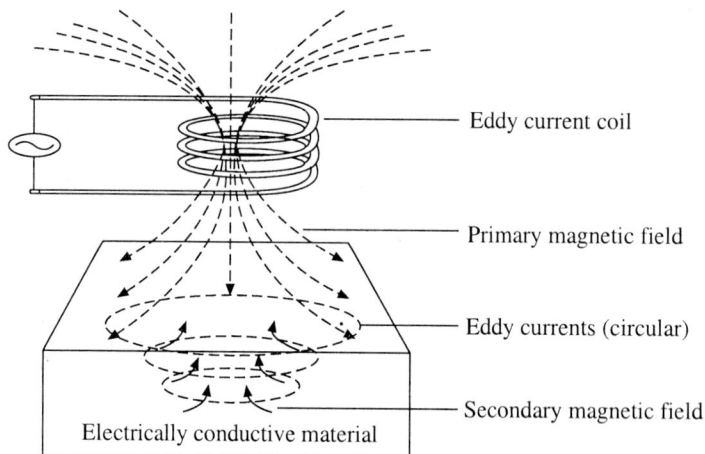

Fig. 2.   Principle of generation of eddy current.

A number of different types of probes are available for use with eddy current testing depending on the purpose, accessibility and desired results. Special probes can also be designed for specific applications to provide optimum performance. All eddy current inspections require a calibration/reference standard containing artificial defects. These standards can be used to optimize the test parameters and also to establish calibration graphs.

Multifrequency ECT uses more than one frequency unit with a common coil, enabling elimination of unwanted parameters from the component under test. The basic approach relies on the skin effect phenomenon of eddy current flowing in the specimen that allows one to obtain independent information at different frequencies. The test signals from different individual frequencies can be combined in real time so as to obtain output signals, which are free from certain parameters but preserve the desired test data. These manipulations are done on actual multifrequency equipment by real time mixing of output channels. There can be a wide variety of ways to mix outputs from various frequencies to cancel a given parameter, but not all of these combinations will provide meaningful or high sensitivity information on desired specimen parameters. Hence care is required in selecting a suitable combination of signals for a given test requirement.

One of the recent advancements in the area of eddy current inspection of tubes for corrosion monitoring in addition to evaluation of other aspects of integrity assessment is Remote Field Eddy Current Testing (RFECT). The primary advantages of this technique are: (i) ability to inspect tubular products with equal sensitivity to both internal and external metal loss or other anomalies, (ii) linear relationship between wall thickness and measured phase lag and (iii) absence of lift-off problems [3]. This technique can be applied to both ferromagnetic and non-ferromagnetic materials for corrosion monitoring applications. In case of non-ferromagnetic materials, the cracks and other discontinuities due to corrosion can be detected because of change in reluctance. For the ferromagnetic materials, this technique is more effective for detection of wall thickness loss due to the large difference in the permeability of air gap/corrosion products and the base metal part. Theoretical and experimental work has been carried out in authors' laboratory and wall thickness loss down to 15% has been detected using an indigenously developed RFECT instrument. The presence of transition and remote field zone and the effect of tube diameter and wall thickness on them have been studied using a 2D-FEM code.

ECT has found popular applications in: (i) measurement or identification of conditions and properties such as electrical conductivity and heat treatment conditions, (ii) detection of seams, laps, cracks, voids and inclusions, (iii) material sorting and (iv) measurement of thickness of non-conductive coating on a conductive material. It is the most sensitive non-destructive testing technique for detection of surface and subsurface discontinuities.

Application of ECT for Corrosion Monitoring

ECT is one of the important techniques for accurate detection and sizing of different types of defects related to corrosion damage. In addition to its well known applications to crack and pit detection, this technique can be used to measure thickness changes caused by corrosion, build up of corrosion products in certain situations, and some changes in material properties such as conductivity degradation caused by intergranular corrosion. ECT applications vary from simple ones with well-established inspection procedures to advanced techniques based on latest developments in eddy current research.

In one of the more advanced developments in eddy current technology, Dodd et al. [4] have used arrays of coils, as many as 16 in one application to record multifrequency complex impedance data as a function of the probe position. These data were used in a least square fit algorithm to determine multiple properties such as thickness, conductivity and flaw size which were related to corrosion damage. The method has been successfully applied to the identification of flaws in various structures such as multilayered plates and tubing in the presence of welds and supporting structure.

Corrosion Crack Detection

While the principle of eddy current crack detection is same, the nature of corrosion related cracks can be quite different from isolated fatigue cracks. Intergranular stress corrosion cracks (IGSCC) for example are often characterized by multiple branched cracks in the region where damage has occurred. The interaction of an eddy current field with such a region is more complex than the interaction with a single crack of simple geometry. An eddy current scan over a region with IGSCC can produce an impedance plane trajectory that closely resembles the signal from a region of low conductivity than the signal from a crack. Intergranular corrosion without stress related cracking can produce a similar signal and may be indistinguishable from IGSCC by the eddy current method.

Corrosion Pit Detection

Both amplitude and phase analysis techniques are used in corrosion pit detection and sizing. With the amplitude method, one assumes that the amplitude of an eddy current signal is proportional to the depth of the pit. The phase sensitive technique assumes that remaining wall thickness can be related to the phase of the signal from a pit. Depth of penetration of eddy current depends on the excitation frequency. If multiple frequencies are mixed then response for different frequency components would correspond to different depths, thus location and size of corrosion pits can be identified even if the specimen is covered with a nonconducting coating. Baron et al. [5] took a different approach to estimate the size of crevice corrosion pits. They started with a theoretical prediction of the signal from a corrosion pit modelled as rectangular slot and used it for relating amplitude data as a function of the probe position to pit dimensions.

Material Loss Assessment

In corrosion monitoring applications in austenitic stainless steel components, measurement of wall thinning due to loss of material is probably the most common use of eddy current testing. If the thickness of a part is of the order of the skin depth, the phase lag of the eddy current probe impedance relative to the phase of the excitation current can be related to thickness, and the linear relationship between phase and wall thickness forms the basis for most measurements of this kind. If the structure to be inspected consists of more than one layer, interpretation of phase shift data becomes more complicated. When corrosion occurs on the outside surface of the inner material, corrosion product build-up can cause an increase in the separation of the layers accompanied by a decrease in the thickness of the inner (second) layer. If on the other hand corrosion occurs on the inner surface of the second layer, its thickness decreases; but there is no change in the air gap between the layers. It is possible to distinguish these cases by comparing phase shifts of low and high-frequency components. Typical application is for detection of corrosion at the interface when austenitic stainless steel is used as cladding material on steam generator nozzle.

Assessment of Corrosion by Measurement of Changes in Material Properties

During the early stages of corrosion damage, changes in near surface properties can occur as a result of intergranular corrosion, formation of corrosion products, or other oxidation and reduction processes. In certain instances, these material property changes can be observed during eddy current testing through an accompanying change in the conductivity or permeability in the surface layer exposed to the environment.

Many studies on corrosion related property changes are concerned with electrical conductivity degradation due to intergranular attack (IGA). As noted earlier, sometimes IGA and IGSCC cannot be distinguished by the eddy current technique because signals from individual cracks cannot be resolved and both IGA and IGSCC are observed as a decrease in the effective conductivity in the damaged region. The theoretical relationship between depth of IGA and eddy current response was investigated by Wait et al. [6]. The damaged region was modelled as a layer of material with conductivity that differs from the values in the undamaged material. They concluded that with proper optimization of the measuring system, the eddy current technique is capable of determination of the depth of IGA. Brown [7] directed his work towards improved detection of volumetric IGA. Through the use of computer models and supporting experiments, he showed that a pancake type coil with its axis in the radial direction is better for IGA detection than a coil with its axis along the tube axis. He also concluded that the multifrequency technique using both amplitude and phase would fare better in IGA depth determination.

Imaging techniques in NDT offer the best promise towards enhanced detection and characterization of corrosion in components. Imaging can be done with many different media, including optical, infrared, X-ray, gamma-ray, ultrasound, eddy current, thermal wave and magnetic resonance. Thus, there is a broad spectrum of techniques from which a selection can be made. The imaging format provides a global perspective of the inspected region and allows a balanced interpretation. In addition, imaging techniques have the potential of automating the measurement process, providing estimates of defect size from the image data, producing accurate characterization of defects and improving the probability of detection.

A computer-based eddy current imaging (ECI) system has been developed at the authors' Lab, to scan the object surface and create impedance images in the form of grey levels or pseudo colours. This consists of a PC controlled X-Y scanner which scans the component point by point in a raster fashion with an eddy current probe, acquires data using a 12 bit analog-to-digital converter, and finally processes and displays data in the form of images. Using ECI, it is possible to obtain images of defects in two dimension enhancing the defect detection capability. Fatigue cracks, corrosion pits, electro-discharge machined (EDM) notches and other types of defects have been imaged and characterized [8]. In the case of fatigue cracks, ECI is found to be capable of revealing the orientation of the crack, which is an important feature for fatigue crack growth and life extension studies. ECI has also been used to detect weld centre line in stainless steel welds with an accuracy of 0.1 mm. Figure 3 shows an eddy current image of a corrosion pit (3 mm dia., 0.5 mm deep) in a stainless steel plate. This image is produced using an absolute surface probe operating at 150 kHz.

15 mm

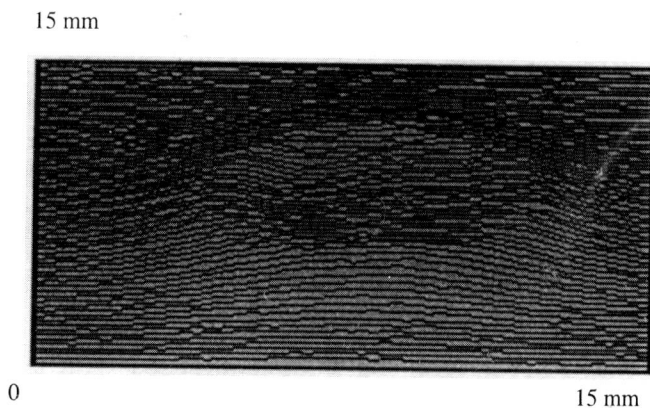

0   15 mm

Fig. 3.   Eddy current image of a corrosion pit (3 mm dia., 0.5 mm deep) in stainless steel plate [8].

ULTRASONIC TESTING (UT)

Basic Principle and General Applications

In ultrasonic testing, high frequency elastic waves are sent into the material being inspected to detect internal flaws (defects) and to study the properties of the material. Most commercial ultrasonic testing is done at frequencies between 1 and 25 MHz. The elastic waves travel into the material with some loss of energy due to attenuation and are reflected at interfaces. The reflected beam is analyzed to detect and locate the defects and for their quantitative evaluation. These waves are almost completely reflected at solid-gas (air) interfaces, partial reflection occurs at solid-liquid or solid-solid interfaces. The reflected energy depends mainly on the ratio of acoustic impedance of the materials at the interface, and the acoustic impedance of a given material is the product of density and velocity.

Ultrasonic testing is used for quality control and material inspection in many industries. Defects like cracks, shrinkage cavities, lack of fusion, pores and bonding faults can be easily detected by this method. Inclusions and other inhomogeneties in the metal can also be detected due to partial reflection or scattering of the ultrasonic waves. This widely used NDT method has a lot of applications like defining bond characteristics, measurement of thickness of the components, estimation of corrosion and determination of physical properties, structure, grain size and elastic constants. In the case of austenitic stainless steel welds, because of the problems associated with high scattering and attenuation, ultrasonic waves at lower frequencies (below 5 MHz) are employed. For successful application of ultrasonic inspection, the testing system must be suitable for the type of inspection being done and the operator must be sufficiently trained and experienced.

Types of Ultrasonic Waves

Ultrasonic waves are classified on the basis of the mode of vibration of the particles of the medium with respect to the direction of propagation of the waves, namely longitudinal, transverse and surface waves.

Longitudinal waves

These are called compression waves. As shown in Fig. 4, in this type of ultrasonic wave, alternate compression and rarefaction zones are produced by the vibration of the particles parallel to the direction of propagation of the waves. Because of its easy generation and detection, this type of ultrasonic waves is most widely used in ultrasonic testing. Almost all of the ultrasonic energy used for the testing of materials originates in this mode and is then converted to other modes for special test applications. This type of waves can propagate in solids, liquids and gases.

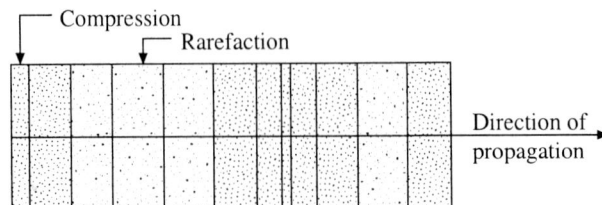

Fig. 4.   Longitudinal wave mode.

Transverse Waves

This type of ultrasonic wave is called transverse or shear wave because the direction of particle displacement is at right angles to the direction of propagation. It is schematically represented in Fig. 5. For all practical purposes, transverse waves can propagate only in solids.

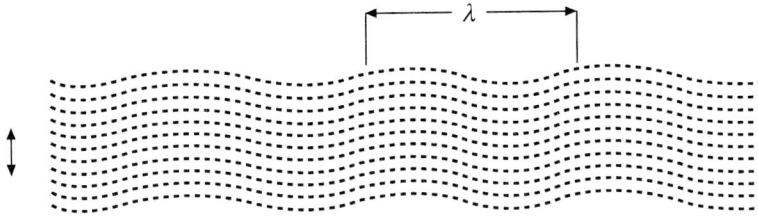

Fig. 5.   Transverse wave mode.

Surface or Rayleigh Waves

Surface waves were first described by Lord Rayleigh and that is why they are called Rayleigh waves. This type of waves can travel only along the surface bounded on one side by strong elastic forces of the solid and on the other side by nearly non-existent elastic forces between gas molecules. The waves have a velocity of approximately 90% that of an equivalent shear wave in the same material and these can propagate only in a region where the thickness is less than about one wavelength below the surface of the material. In surface waves, particle vibrations generally follow an elliptical orbit as shown in Fig. 6. The major axis of the ellipse is perpendicular to the surface along which the waves travel and the minor axis is parallel to the direction of propagation.

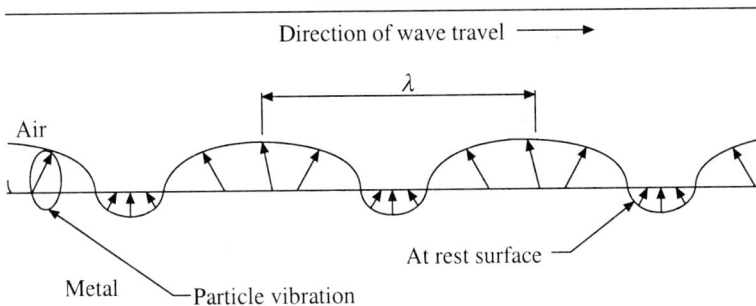

Fig. 6.   Surface or rayleigh waves.

Transducers for Ultrasonic Testing

Conventionally ultrasonic waves are generated by piezoelectric transducers, which convert high frequency electrical signals into mechanical vibrations. These mechanical vibrations form a wavefront, which is coupled to the component being inspected through the use of a suitable medium (couplant). The energy transmitted or reflected is received by receiver probe, which converts mechanical vibrations to electrical signals, these, in turn, are amplified and displayed on cathode ray tube (CRT) screen. Several wave modes can be used for inspection depending upon the orientation and location of the discontinuities. The three most common piezoelectric materials used are quartz, lithium sulphate and polarized ceramics. The most common ceramics are barium titanate, lead metaniobate and lead zirconate titanate. The important characteristics of ultrasonic probes are sensitivity, resolution, dead zone and near field effects.

Basic Methods of Ultrasonic Testing

Transmission and Reflection Techniques: Ultrasonic testing is typically performed in two ways. A beam of ultrasonic energy is directed into the test object and the energy transmitted through or reflected from discontinuities in the object is measured. Such tests are possible because an ultrasonic beam travels with little loss through a homogenous material. Energy loss occurs when the ultrasonic beam is intercepted and reflected by grain boundaries or discontinuities in the elastic continuum. Figure 7 shows these two basic techniques [9]. The discontinuity is detected by the decrease in the energy incident on the receiving transducer in Fig. 7(a) and by the energy reflected to the receiving transducer kept on the same side as the transmitting transducer in Fig. 7(b).

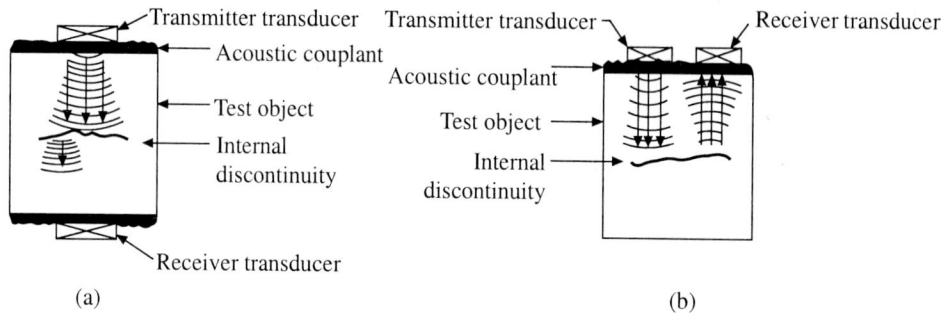

Fig. 7. Basic ultrasonic testing methods [9].

Ultrasonic Test Systems

Basic ultrasonic testing methods shown in Fig. 7 include: (i) a transmitting transducer, (ii) couplant to transfer acoustic energy to the test object, (iii) the test object, (iv) couplant to transfer acoustic energy to the receiver and (v) a receiving transducer. Equipment selection, design and arrangement depend primarily on the specific characteristics of ultrasonic wave propagation being used for detection and measurement of test object properties. The phenomenon involved may include: (i) velocity of wave propagation, (ii) beam geometry (focusing field pattern or dual transducer systems), (iii) energy transfer (reflection, refraction or mode conversion) or (iv) energy losses (scattering and absorption).

The most common technique employed is the pulse echo technique. The basic equipment comprises of an ultrasound pulse generator, a receiver and its signal amplification and display system. Depending on the display of the information, the pulse echo equipment can be sub-divided into three groups, viz. A-scan, B-scan and C-scan. Ultrasonic testing is carried out either in the contact mode or immersion mode. In the contact mode, the probe is placed in direct contact with the test specimen with a thin liquid film used as a couplant for better transmission of ultrasonic waves into it. In the immersion mode, a waterproof probe is used at some distance from the test specimen and the ultrasonic beam is transmitted into the material through a water path or water column. Contact type techniques use ultrasonic waves in normal beam, angle beam and surface wave modes. Immersion testing techniques are used in the laboratory and also in certain production lines for automatic ultrasonic testing.

Current trend in various industries/disciplines indicate that there is an increasing emphasis on defect/flaw characterization and sizing. Quantitative evaluation of components requires the knowledge of crack size and possible growth rate. These information help in estimating the remaining life of a component. Conventional techniques are very good as far as detection of a defect/flaw is concerned.

However, characterizing and sizing the defect requires sophisticated instrumentation and computer based automation. A few techniques/concepts that can meet these requirements quite satisfactorily are discussed below.

In the case of ultrasonic phased array probe system [10], by using a number of elements excited with different time delays, it is possible to steer and focus the beam and to obtain an image on its reception. The amplitude and time of flight locus curves method ALOK [10] incorporates all the features required for automated inspection of pressurized components in any industry viz. high sensitivity and resolution, reliability and reproducibility, quick information on the condition of the component and ability to analyze a defect. These inspection systems have capabilities to detect flaws even in complex geometries of pipe elbows and nozzles. However, this technique is seldom used after the emergence of synthetic aperture focusing technique (SAFT) and time-of-flight-diffraction (TOFD) techniques. The principle of SAFT [11] is to measure the complete sound field scattered/reflected over a certain solid angle, which depends on the beam opening angle viz. the cone of divergence of the beam and the aperture synthesized by this technique. During reconstruction, the sound field is imaged over the scanned region by superposition of the above scattered/reflected waves. The result is a three dimensional amplitude distribution of the sound field. SAFT reconstruction methodology images the defect more accurately with respect to size, shape and orientation as compared to the B- scan image of the defect in view of higher signal-to-noise ratio due to spatial averaging. TOFD [12] technique is used for quantitative assessment of vertical and near vertical defects. When an acoustic wave interacts with a crack like defect, diffracted waves are generated at the crack tips. By processing of these diffracted waves, it is possible to locate the origin of the waves, or in other words, to locate the tips of the defect and hence the defect can be imaged. This technique was originally developed for detection of surface breaking defects. But, as data processing techniques improved, it has been possible to evaluate hidden defects.

Application of UT for Corrosion Monitoring
UT provides a sensitive detection capability for corrosion damage when access is not available to the surface exposed to corrosion. Stress corrosion cracking occurs by production of a new interface within the material, which causes reflections earlier than those from the back surface. Ultrasonic thickness gauges are commonly used for detection of stress corrosion cracking and general material thinning. Pitting and intergranular corrosion cause scattering of the ultrasound and can be detected by the use of shear waves in an angular incidence. In addition, this scattering can result in attenuation of longitudinal waves commonly referred to as loss of back-surface signal. This phenomenon serves as a means of corrosion detection in thick structures.

For detection of IGSCC in austenitic stainless steel welds using ultrasonic flaw detection methods, it is important to identify the form of the echoes from root of the weld and IGSCC echoes. A number of methods are available to measure crack depth by ultrasonic testing. These are (i) defect echo height method, (ii) decibel drop method, (iii) edge echo method, (iv) scattering method and (v) composite aperture method. However, in the case of IGSCC, the cracks tend to branch out and therefore measurements become difficult. Advanced signal analysis methods are useful for the evaluation of defects and damages, which tender small reflectivity.

Ultrasonic Testing of Welds in the By-Pass Lines of Boiling Water Reactor
One of the typical applications of ultrasonic testing is the in-service inspection of the coolant circuit

pipes and the by-pass lines in a boiling water reactor and the pipings are made of AISI type 316 stainless steel. They are intended to serve as lines to keep minimum flow on recirculation pumps outlet during start up of pumps. Detailed survey conducted earlier on various boiling water reactors around the world had indicated that the failures of the by-pass line welds are due to intergranular stress corrosion cracking (IGSCC) at weld heat affected zones (Fig. 8) mainly due to stagnant water legs conditions in bypass piping which used to be kept closed after pump start up for flow control on main recirculation piping. In one of the operating reactors, after 10 years of operation, leaks were observed during hydro test at 3 nos. of joint in one loop and in one joint in the other loop.

In order to estimate the extent of the defects and their locations in the leaky welded joints and also to evaluate the integrity of the other 18 joints in the bypass lines, ultrasonic test was carried

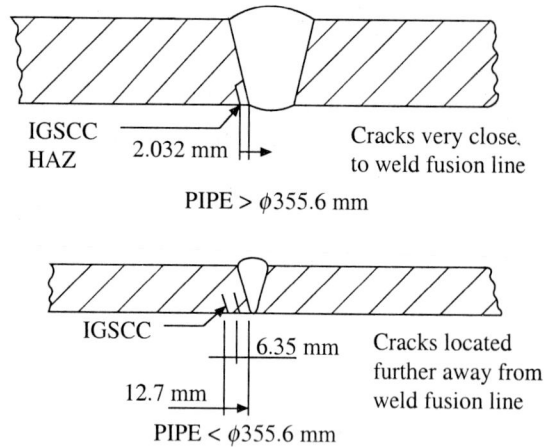

Fig. 8.   Expected location of IGSCC in BWR piping.

out. The ultrasonic test procedure was standardized earlier in the laboratory with mock up welded pipe standards having reference defects. An ultrasonic flaw detector with a 2.25 MHz frequency, 45 degrees angle shear wave probe was used for testing purpose. Clear defect echo patterns were observed from all the artificial defects by choosing suitable test parameters. At the standard sensitivity level, in-situ ultrasonic testing confirmed the hydro test observations and the extent and nature of defects were determined. Based on the ultrasonic observations, it was decided to replace the 4 welded joints (where leaks were observed) with new joints. The consequent metallography of the removed joints confirmed the UT observations.

The analysis of the study indicated that the cracks are of intergranular stress corrosion type. The failures are due to fabrication-induced stresses combined with service-induced corrosion from inside surface of the piping.

INTERNAL ROTARY INSPECTION SYSTEM (IRIS)

Industries like petrochemicals, fertilizers, power, etc. are equipped with various heat exchangers and steam condensers for effective heat transfer and also as part of process requirement. Non-destructive evaluation and condition monitoring of these heat exchanger/steam condenser tubes is generally carried out by single or multi frequency eddy current testing. However, ECT is not sensitive enough for detection of localized pitting (corrosion) type defects, which are isolated, and also defects present under support plates.

IRIS [13] is a relatively new methodology for inspection of pipelines and tubes of heat exchangers and steam generators. Figure 9 shows the plan view of this system. In this system an ultrasonic transducer lies axially along the tube and the pulse emitted impinges upon a mirror angled at 45° to the axis of the tube. The ultrasonic pulse is deflected to penetrate the wall of the tube in a radial direction. An echo of the signal is returned from both the inside and the outside wall surfaces. The signal is displayed on a computer screen. The mirror rotates 360° and successive indications from

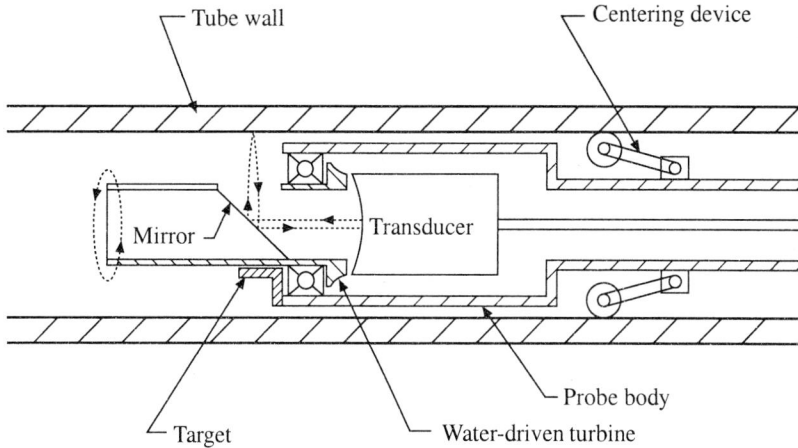

Fig. 9. Plan of internal rotary inspection system.

each pulse are displayed sequentially on the screen. All the data from one circumferential sample is displayed on the screen at a time. As the probe is withdrawn at a controlled rate from the tube, the overlapping footprint of the ultrasonic scan covers every 25 mm square of the tube circumferentially and longitudinally, giving a complete recording of the tube's condition. The technique has the capability of detecting the wall thinning and pitting due to corrosion in tubes of heat exchangers and steam condensers. The advantages of this technique are (i) it can measure the remaining wall thickness up to 500 μ of the tubes and pipes, (ii) it can indicate the reduction in wall thickness that has taken place either from outer surface or inner surface of the tubes/pipes and (iii) it also reveals the circumferential position of the defects such as localized pitting and defects under support plate. Figure 10 ((a) and (b)) shows the CRT pattern obtained by IRIS from a good and a corroded tube. The left boundary of the display represents the condition of the inner surface of the tube and the right extreme trace represents the condition of the outer surface. The height of the display represents the circumference and the wall thickness of the tube is represented by the width of the display as shown in Fig. 10(a).

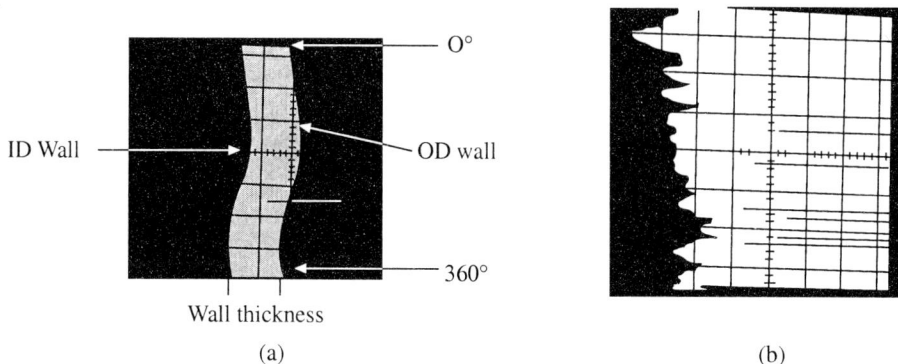

Fig. 10   (a) CRT pattern from a good tube. (b) CRT pattern from a corroded tube.

RANDOM DECREMENT ANALYSIS METHOD (RANDOMDEC)

This is a vibratory signature analysis method for corrosion monitoring based on modal vibration analysis and is capable of providing accurate thickness measurements at specific locations where corrosion is suspected. RANDOMDEC is a vibration monitoring technique for extracting information concerning structural conditions from analysis of the response of structure to random input excitation [14]. When an object is subjected to random excitation, the transducer attached to it detects the vibration response and a signature is obtained which can be used to measure changes related to damping and natural frequency of the structural component. The technique is particularly suited to field measurements on the structural components since excitation is provided naturally by acoustic noise inputs produced by wind, seismic disturbances, traffic loads, etc. It requires the measurement of the dynamic response of the structure subjected to the above random inputs, and after analysis free vibration response or signature of the mechanical structure is obtained.

Typical experimental results of RANDOMDEC application on a six- meter long and 100 mm diameter pipe are given in Fig. 11. Internal corrosion in this pipe section was simulated by incrementally machining a centrally located ID groove in the 6 mm pipe wall. A mechanical shaker attached to the pipe served as a vibration source. Accelerometers mounted in the pipe served as sensors. Analysis involves extraction of spectral density vs. frequency using FFT software and the RANDOMDEC signature. Discernible difference in these signatures between the baseline reference conditions and after simulated corrosion/metal loss in the pipe sample were observed.

As it may be seen in Fig. 11, the sinusoidal signal decays due to random vibration excitation and the position of the first peak shifts distinctly due to change in the simulated corrosion pit depth. This phase shift can be observed distinctly in the frequency domain, or by observing modulation in the cross-power spectrum of the signal with simulated corrosion pit and the reference signal from virgin pipe.

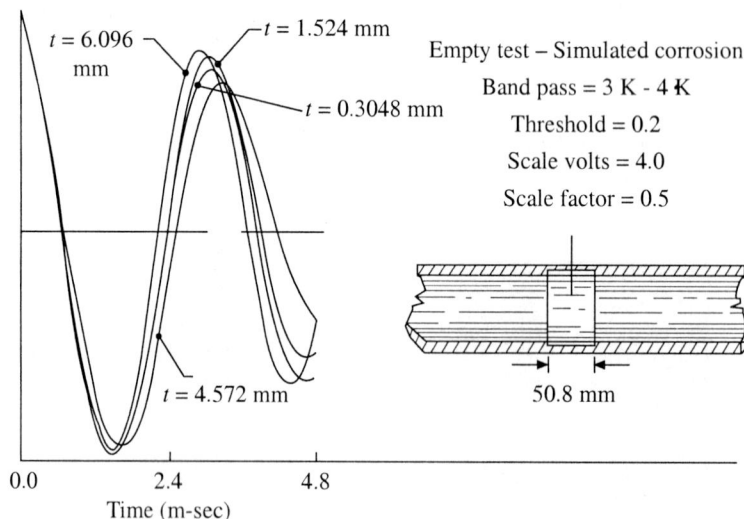

Fig. 11. RANDOMDEC signature: 100 mm diameter test pipe.

ACOUSTIC EMISSION TESTING (AET)

Basic Principle

Acoustic emission (AE) is defined as the class of phenomenon where transient elastic waves are generated by the rapid release of energy from localized sources within a material. It arises from the energy redistribution within a system as a release of a series of short impulsive energy packets. The energy thus released travels as spherical wave front and can be picked up from the surface of a material using highly sensitive transducers, usually electro-mechanical in nature placed on the surface of the material. The wave thus picked up is converted into electrical signal, which on suitable processing and analysis can reveal valuable information about the source causing the energy release. Acoustic emission waveform characteristics that are most commonly used for characterization of the sources are shown in Fig. 12.

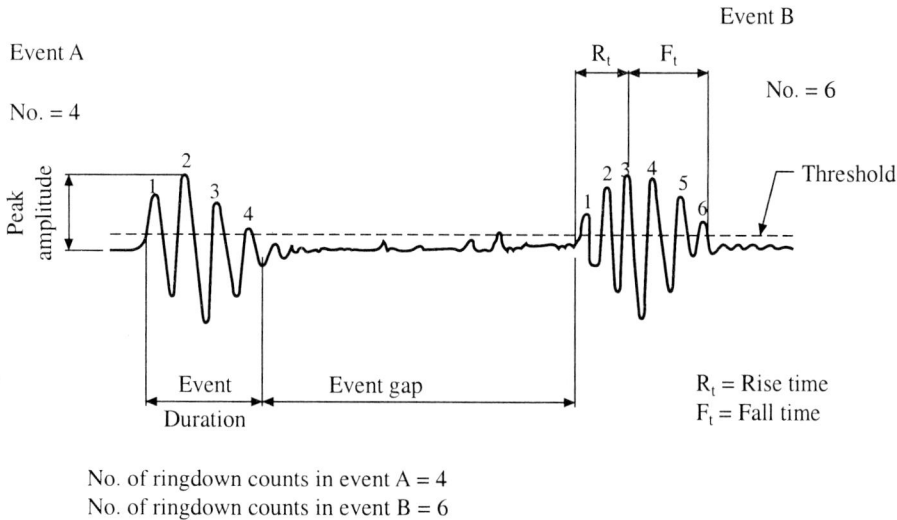

No. of ringdown counts in event A = 4
No. of ringdown counts in event B = 6

Fig. 12.   Basic parameters of acoustic emission signal.

Sources of acoustic emission include many different mechanisms of deformation and fracture. The naturally occurring sources are earthquakes and rockbursts in mines. In metals the different sources are generation and propagation of cracks, movement of dislocations and grain boundaries, formation and growth of twins, fracture and decohesion of brittle inclusions, phase transformations etc. In composites the sources are matrix cracking, debonding and fracture of fibers. There are also secondary or pseudo sources of acoustic emission, which include leaks and cavitation, friction, realignment and growth of magnetic domains (Barkhausen effect), liquefaction and solidification, solid-solid phase transformation. Metallic corrosion involves some chemical reactions, which result into a redistribution of energy. Any corrosion process thus, is a potential source of acoustic emission. Corrosion crack initiation and growth produces a variety of distinct and discernible signals such as electrochemical noise, electrochemical current transients, acoustic emission and load drops. Time domain and frequency domain characteristics of acoustic emission signals generated by different corrosion mechanisms are enlisted in Table 2.

In-situ detection of corrosion crack initiation and growth is an attractive prospect for safety and operation of various plant machinery and equipment especially in nuclear power plants and petrochemical industries where hazardous corrosive media are handled in adverse operating conditions. Early detection of corrosion damage initiation would improve safety while flaw location could reduce effective downtime during maintenance. Of the signals emanating from corrosion sources, only AE can be remotely

Table 2. **Typical characteristics of acoustic emission signals generated by different corrosion mechanisms**

AE type	Source mechanism	AE signal characteristics		
		Time domain		Frequency domain
		Rise time	Amplitude	
I	Pitting	Moderate	Low	Medium frequency (150-300 kHz) with single peak
II	Pitting and Simultaneous Particle Erosion	Slow	High	Wideband frequency (50-650 kHz) with multiple peaks
III	Crater Formation	As above	As above	As above
IV	Stress Corrosion Cracking	Moderate	High	Wideband frequency (50-450 kHz) with multiple peaks
V	Erosion of Secondary Particles to Drive General Corrosion	Fast	Low	High frequency (500-650 kHz) with single peak

detected without prior knowledge of the source. Electrochemical noise and current transients require electrochemical probe placement close to the localized corrosion or crack. Corrosion initiation and cracking can be localized with AE by the use of detector arrays and time of arrival analysis. AE also has the potential to discriminate between signals emanating from different types of sources through signal analysis. These characteristics give AE monitoring of the corrosion behavior a definite edge over the other techniques.

Monitoring Stress Corrosion Cracking Using AET

The cracking of a metal alloy by conjoint action of a tensile stress and a corrosive environment is known as stress corrosion cracking. This is one of the most dangerous corrosion mechanism as per the extent and severity of damage and is a cause of concern for all high-pressure vessels handling corrosive media. AET has been applied for monitoring stress corrosion cracking in a wide variety of materials.

In one such study, AE characteristics during SCC of sensitized AISI type 304 stainless steel were studied [15]. Tests were conducted in high temperature water under three kinds of biaxial stresses. It was found that the AE characteristics during SCC may be divided into three stages (Fig. 13). The first stage corresponds to non-stationary state mainly related to the elevation of the specimen temperature. The AE detected was attributed to thermal expansion of the specimen. The generation of AE signals in the 1st stage was most active during this test. The 2nd and 3rd stages correspond to the stationary state of the test. In the 2nd stage, AE activity was found to be very weak. The total energy (E) and the total emission number (N) in this stage were only several percent of the maximum values in the 1st stage. In the 3rd stage, AE activity was found to be relatively high. The values of E and N in the 3rd stage were several times larger than that in the 2nd stage. The most probable AE sources in the third stage may be the phenomena associated with inter-granular stress corrosion cracking during the propagation period. It is felt that the thermal expansion, that takes place during 1st stage of this test, would lead to a change in stress distribution in the specimen and this will be reflected in a distinct acoustic emission activity. Since the SCC behaviour is strongly dependent on the stress developed, it would be possible to predict the time to failure due to SCC by the AE characteristics in the 1st stage.

It was also found that at the onset of leakage due to SCC, cumulative counts, signal waveforms and peak-amplitude distributions show a great change and the leakage is readily detectable by AET.

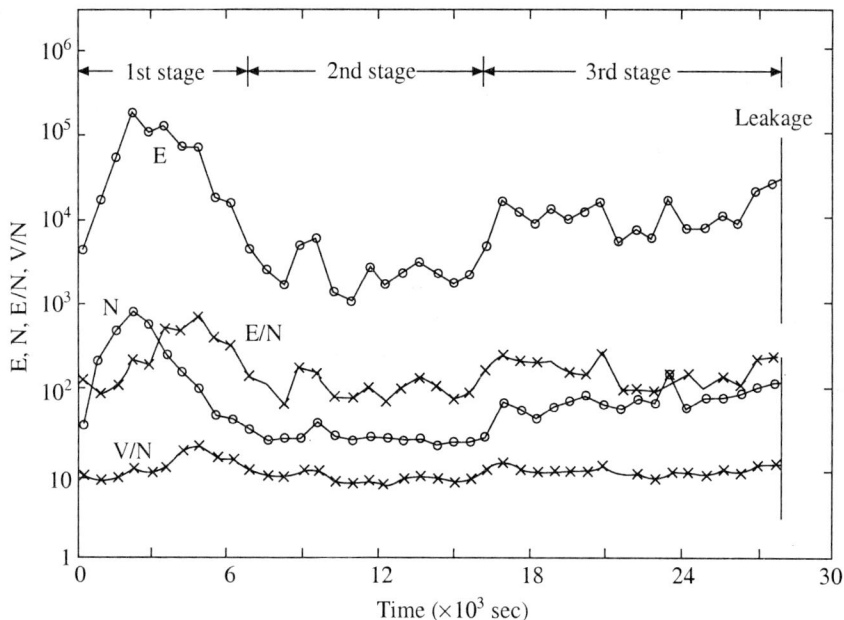

Fig. 13. **AE peak amplitude derivative parameters in the SCC test where** σ_θ/σ_z = 157/182 [15].

Yuyama et al. [16] examined AISI type 304 stainless steel during corrosion, SCC, and corrosion fatigue. In this study, AE was observed to be generated by hydrogen gas evolution arising during passive film breakdown at cathodic potential during crevice corrosion, but no detectable emission accompanied thin passive film rupture and anodic dissolution. Cleavage like regions observed on the fracture surface during SCC were proposed as sources of AE arising during crack initiation.

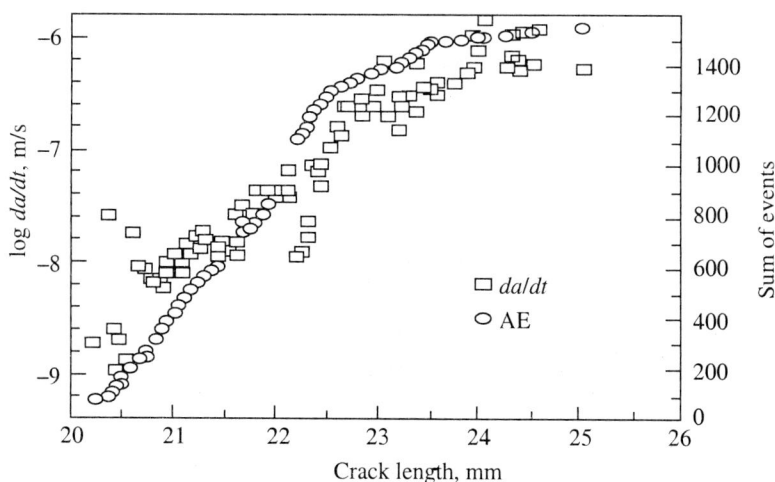

Fig. 14. **Crack velocity and total acoustic emission events vs. crack length for AISI type 304 SS tested in** $Na_2S_2O_3$ **at 90°C [17].**

Jones et al. [17] studied the acoustic emission during intergranular subcritical crack growth in AISI type 304 stainless steel with varying composition and grain sizes in different aqueous environments. Figure 14 shows the variation of total AE events and crack velocity vs. crack length for this specimen tested in $Na_2S_2O_3$ at 90°C. The total AE events due to crack extension was observed to be higher in the initial stage and saturated at higher values of the crack length. The AE result was correlated with the extent of transgranular fracture accompanying the intergranular subcritical crack growth. It was postulated that the transgranular fracture surfaces were the result of ligaments which fracture behind the advancing intergranular crack front. The AE could have been emitted from dislocations, particle fracture, or fracture of the ligaments.

6.3   Acoustic Emission Monitoring During Pitting Corrosion

Pitting is a form of extremely localized attack, the rate of attack being greater at some areas than at others. It is the most destructive form of corrosion and results in sudden failure of equipment due to formation of holes. Under certain conditions of service, stainless steels, which are apparently immune to attack in a given environment, will fail by pitting. Various environmental and metallurgical factors affect the pitting behavior and this includes appreciable concentrations of Cl-, Br- or $S_2O_3^{-2}$ in environments, increase in pH, presence of pitting enhancing elements like Si, Ti, Ce, Nb, C (especially in sensitized condition) etc. Only a few attempts have been made so far to study pitting corrosion by acoustic emission technique.

In one of the early works in this field, the investigators were able to detect pits inside piping systems by AET [18]. Acoustic emission signals were supposed to be generated by secondary effects such as scale cracking or oxide fracture on corrosion defects when a slow steady hydrostatic pressurization was applied to each section of the tested pipes. To study the development of pitting corrosion on AISI type 316L austenitic stainless steel in laboratory experiments, and to establish an acoustic emission system which could detect on-going pitting corrosion and quantitatively monitor the pitting corrosion rates of various vessels or chemical equipments in service, a series of investigations was conducted by Mazille et al. [19]. Tests were conducted at room temperature in 3% NaCl solution acidified to pH 2, at the free corrosion potential or with applied anodic polarization. AE signals were easily detected during pitting corrosion and a good correlation was observed between AE activity and pitting rate. The exact nature of AE sources within pits could not be identified clearly but the results demonstrate that acoustic emission technique can be used to detect and even monitor the occurrence of such phenomena. Figure 15 shows the number of detected AE events vs. the calculated cumulative pit size in terms of total corroded volume of pit cells, or total area of the inner active surface of the cells. In another work AE response of a low-carbon type 304 stainless steel during pitting and transgranular stress corrosion cracking (TGSCC) was measured [20]. Tests were conducted in 0.01 and 1 M NaCl with the pH adjusted to 1 with HCl at potentials of –380 mV and 0 mV (SCE) for no zero stress and with a stress equal to 75% of the yield strength of the material. In this study a correlation with corrosion was suggested by the higher AE rate observed for the 1 M vs. the 0.01 M NaCl solutions and the increase in AE rate with increasing pit frequency and current. Both stress and potential affected the transition from low to high AE rate regimes. From the above case studies, it can be comprehended that there is a good correlation between AE activity and pitting corrosion damage.

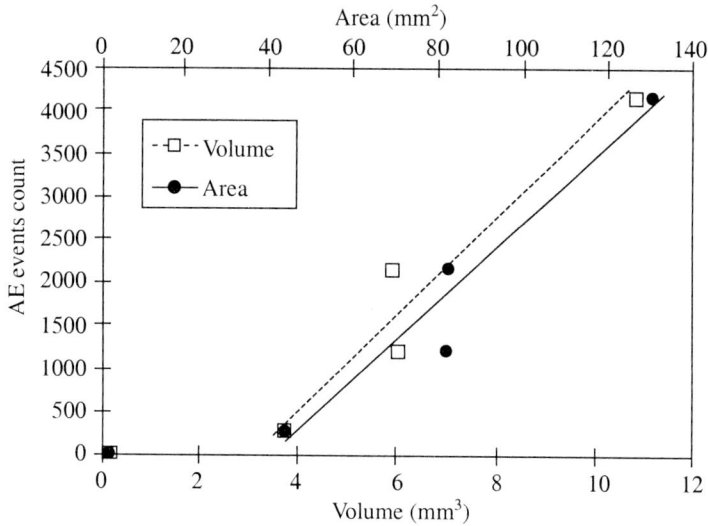

Fig. 15. **Number of detected AE events vs. calculated pit size [18].**

RADIOGRAPHY TESTING (RT)

Basic Principle

Radiography is based on the differential absorption of short wavelength radiations such as X- and gamma-rays on their passage through matter. Because of differences in density and variations in thickness of the component or differences in absorption characteristics caused by variations in composition, different portions of a component absorb different amounts of penetrating radiation. A shadow projection is obtained on a detector which is normally a grey level image with varying grey tones depending on the quantum of radiation received at a given point. The schematic setup for radiographic testing is shown in figure 16. The source of radiation can be X-rays, gamma rays, neutrons, protons or electrons. X-rays and gamma rays are the most commonly used sources of radiation. The detector can be

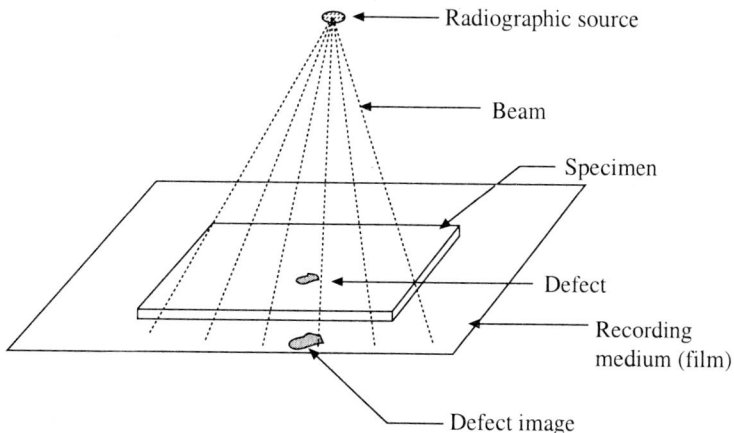

Fig. 16. **Schematic setup for radiographic testing.**

photographic films, image intensifiers or scintillator screens/counters. However, double coated, fine grain, high contrast industrial X-ray films are the most widely used means of detecting the transmitted radiation.

Because a radiograph is a two dimensional representation of a three dimensional object, the radiographic images of most components are some what distorted in size and shape as compared to the actual size and shape of the test piece. The severity of the distortion depends primarily on source size (focal spot size for X-ray sources), source to object and source to film distances and position and orientation of the component with respect to the source and the film. In conventional radiography, the position of a flaw within the volume of a component cannot be determined exactly with a single radiograph, because the depth parallel to the radiation beam is not recorded. However, techniques like stereo radiography, tomography and double exposure parallax methods can be used to locate flaws more exactly within the component volume.

There are several advanced techniques in radiography like high resolution X-radiography, high-energy radiography, neutron radiography and real time radiography. High-resolution radiography has been developed to offer an edge over conventional radiography, in better definition of defects and detectability of small defects like microcracks in components having thin sections and complex geometries. Radiography using X-ray systems of 1 MeV or more is commonly termed as high-energy radiography. The basic principles of this technique are similar to those of conventional radiography. Its advantages are: (i) examination of thicker sections economically due to the greater penetration of the high energy photons, (ii) possibilities of large distance to thickness ratios (D/T) with correspondingly low geometrical distortion and (iii) short exposure times and high production rates. Neutrons, though are not ionizing radiations by themselves, can produce ionizing radiations by their interactions with foil/film/screen, which enable recording of radiography. Neutron radiography is a valuable NDT technique, which compliments conventional X-radiography. Real time radiography or fluoroscopy differs from conventional radiography in that the X-ray image is observed on a fluorescent screen rather than recorded on a film. Fluoroscopy has the advantages of high speed and low cost of inspection.

Application of Radiography for Corrosion Detection

The use of radiographic technique for the detection of corrosion has been reported [21]. The key benefit offered by this technique is the ability to determine the extent of corrosion attack by means of measuring material loss through pipelines with insulation. Conventional X-ray film or digital flexible reusable imaging plates can be used to generate a radiographic image using the standard radiography procedures and exposure techniques. These captured radiographs are then digitized in order to produce a digital electronic radiographic image of the object being examined. This electronic radiograph is a 3-D map of the material loss and spatial resolution of the object. By using image processing density profile techniques on the digital electronic radiographic image, the material loss at specific points can be gauged and graphically plotted. From the density profiles it is analytically possible to display the calibrated results. Further, the technique enables the material loss differentiation to be color mapped after the radiographic profile compensation has been applied. The information gained can be saved into a database that can be used to correlate results and analyze trends.

The use of digital irradiation imaging inspection system instead of conventional industrial X-ray film for carrying out maintenance inspection of pipes [22] utilizes an imaging plate (IP) in which a high sensitivity special luminescent material is applied and as a result sharp images are obtained. In the measurement of these pipes using this technique, the amount of relative penetrating beam from healthy portions and those from damaged portions are measured and compared. The important features of this digital irradiation imaging inspection system are: (i) it is highly sensitive, (ii) has a wide dynamic range, (iii) image processing capabilities due to digital nature of images, (iv) very good linearity between radiated beam and obtained data.

Radiographic wall-thickness measurements of insulated piping in the chemical and petrochemical industry are carried out by exposing film-cassettes with Ir-192 as the X-ray source. The radiographic image of the tube is then evaluated off-line after the film is developed. A new mobile planar array-sensor (Radiation Image Detector) is capable of generating radiographic on-line images even when common Ir 192 sources are used [23]. These images are shown on-line on a PC-screen and can be evaluated directly on-site during the examination. By using the sensors with 16-bit-resolution, both corrosion monitoring and accurate wall-thickness measurements can be carried out. Exposure-time is only a few seconds. This new development is also of great interest for many applications.

Internal Corrosion and Wall Thickness Monitoring Using Radiography

Any chemical plant has characteristic of the internal pipe corrosion, which is a result of materials (metallurgy), procedures (time, temperature and speed of production) and products (chemistry). These characteristics also vary among different plants. Accuracy in establishing the corrosion rate and its location in the critical areas of pipe has been a problem in all refineries and in most of the chemical plants.

Ultrasonic wall thickness measurements for instance, taken periodically in the same areas of pipe provide information on the rate of progress of internal corrosion or erosion. The most significant limitation of this technique, however, is the difficulty of obtaining a proper, reliable reading in the weld heat affected zone where the corrosion damage and its rate are usually highest and the necessity for removal of insulation. This limitation has been overcome by application of the improved radiographic technique described below.

The technique requires accurately calibrated reference block and comprises (i) the film density variation technique that uses comparison between the images of the known stepped wedge and the corroded wall of the pipe and (ii) the popular "Canadian" technique that uses a lead ball of known diameter as the reference standard.

The combination of these techniques modified by application of the calibrated reference blocks is new. It reduces or eliminates "undercutting" and permits to obtain measurements with accuracy between 0.005 and 0.0075 mm. The concept of a calibrated block of the lead based alloy or similar high density metal for measurement of pipe walls by the radiographic method is recent. Accuracy is essential in the fabrication and the calibration of the blocks. The slots should be as narrow as possible.

This requirement becomes more important with the increase in the outside diameter (OD) of examined pipe. In practical field applications this method has been proven most effective on pipes of 200 mm OD and with a wall thickness of 19 mm.

The exposure time calculation should be based on the original wall thickness of the pipe regardless of anticipated corrosion. If the pipe is in use, a correction may be needed for the liquid inside the pipe. To establish the correct attenuation coefficient in such a case may be quite difficult, particularly in oil refineries where high-density additives (lead for instance) are being added to the product. The best results have been obtained in practice when the radiographs were slightly overexposed and developing time was normal or slightly shorter. The method provides the ability to see the results of internal corrosion and accurate measurements of wall thickness in pipe. This has provided useful information for corrosion engineers and inspectors in refineries and chemical plant.

INFRARED THERMOGRAPHY (IRT)

Basic Principle

Thermography is based on the principle of detection and measurement of infrared radiations (IR) arising from the natural or stimulated thermal radiation of an object. It has extended the ability of human vision to image an object in the infrared region. The technique is being extensively used for inspection of cement kilns, condition monitoring of petrochemical and refinery complexes, assessment of insulation efficiencies of refractory and other materials, and detection of hot spots in electric distribution lines and transformers. Scientists today are considering a variety of novel applications such as detection of fatigue failures at inception, on-line monitoring of weld pools and delaminations in bonded materials.

All objects around us emit electromagnetic radiations. At ambient temperatures and above, these are predominantly infrared radiations. The properties of IR are similar to those of other electromagnetic radiations, like visible light, except that their transmission and absorption behaviour are different from those of visible light. Thermography makes use of the infrared spectral band of the electromagnetic spectrum. The boundary of the infrared spectra, at the shorter wavelength end, lies at the lower limit of visual perception i.e. in the deep red, whereas at the long wavelength end, it merges with the radio wavelengths in the millimeter range.

The IR is invisible to human eye. With the aid of a suitable detector, IR emitted from an object can be converted into a visible image called a thermogram/thermal image. The intensity of the emitted spectrum is dependent upon the absolute temperature of the body. Variations in the temperature of the surface of the object can be visualized as grey level variations in the thermal image. This technique exploits this advantage and any deviation from normal temperature of the component can be readily detected.

Techniques in Thermography

Thermoraphy can be classified basically into two categories namely passive and active. In the passive technique, the emitted heat distribution is measured over the surface of a hot structure. In the active technique, heating or cooling is induced or applied to a part or the complete surface of the object and the dynamic redistribution of heat across the test surface is measured. Important examples of passive technique are monitoring of hot spots in electrical installations such as transformers, furnace insulations etc. Example of active thermography can be found in the detection of internal defects in structural materials.

Equipment for Thermography

A non-contact thermography system consists of an infrared scanner, monitor, control unit and a calculator for field calculations. The output can be stored in a video thermal recorder. The image can be analyzed frame by frame later using a personal computer with image processing facilities. The infrared scanner is the heart of the thermal imaging system and essentially consists of an IR lens system, which collimates the incoming infrared radiation onto the detector. The detectors, which are conventionally used, include-mercury cadmium telluride and indium antimonide. The thermography units available today operate in the wavelength range of 0.75-14 μm. Temperature up to 500°C can be measured which can be extended upto 1600°C using filters. A variety of lenses such as 7°, 40° and microscopic lenses are available which can be used depending on the type of application, resolution and coverage required.

Application of Thermography for Understanding Material Degradation and Corrosion

Thermal imaging techniques have been used for the detection of wall thinning in stainless steel pipes [24]. Wall thinning due to corrosion is one of the prevalent problems. Wall thinning of the order of 30% could be detected in stainless steel pipes. This technique provides a qualitative alternative to commonly used techniques such as ultrasonics, x-ray, and inspection via borescope.

Results of studies to assess the effect of several key experimental variables on the detection of blisters and corrosion spots at the interface of organic coatings and steel using infrared thermography are presented [25]. Coating defects, such as blisters and corrosion spots, are detected in thermographic images as a result of differences in the thermal properties of degraded and nondegraded areas of a coated panel when a temperature gradient is induced through the thickness of the coated panel. The resulting thermographic image is then analyzed using image processing to determine size, location, and extent of degradation. Therefore, the method provides a quantitative, nondestructive procedure for determining the extent of deterioration on coated metal panels prepared in the laboratory. The results show that experimental variables, such as the method of heating, the specimen temperature, the flow rate of air in the surroundings, the time between initiation of heating and acquisition of the thermographic image, and the thickness of the nonmetallic coatings, can affect the results.

MAGNETO-OPTIC/EDDY CURRENT IMAGING TECHNIQUE FOR DETECTION OF CORROSION

Magneto-optic/eddy current imaging (MOI) technique has been recently developed for the detection of corrosion [26]. This is based on the eddy current induction and magneto-optic sensing. Faraday's law of induction states that a time varying magnetic field in the vicinity of an electrical conductor will induce a time varying electric field and hence a conduction current in the same conductor. This is the basis of both conventional eddy current and the MOI techniques (Fig. 17). The major difference is that in the eddy current technique current flows in the coils, while in this technique, the current flows in a thin, planar foil placed near, and parallel to, the surface of the test piece. One of the key requirements

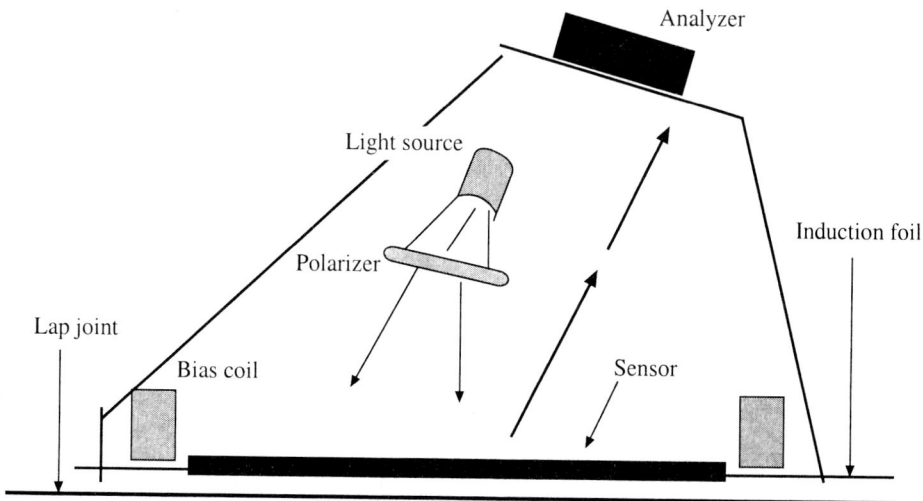

Fig. 17. Schematic setup for magneto-optic imaging [26].

for this technique is the ability to induce uniform currents in a work piece. Since the required induced currents are not circular but planar and linear in nature, these are termed as sheet currents.

When a plane polarized light is transmitted through a glass in a direction parallel to an applied magnetic field, the plane of polarization of linearly polarized light gets rotated. This phenomenon is known as the Faraday magneto optic effect. This technique employs special magnetic garnet materials instead of glass and exploits this effect to make magnetic fields visible. Bismuth doped iron garnets possess very large specific Faraday rotation, up to 30,000 degrees of rotation per centimeter thickness, which makes it possible both transmission and reflection mode magneto optic displays. This capability provides a means to directly image disturbances in the magnetic fields associated with distortions produced by the eddy currents in situations such as the presence of fatigue cracks and corrosion. The advantages of this new visually based technique include increased inspection speed, more intuitive and easily interpreted information, elimination of false calls, elimination of the need for paint and decal removal, and the ability to easily document an examination with videotape. The technique is capable of providing images of a relatively large area compared to that covered by conventional eddy current technique. Because of this, this method is more appropriate for large, flat or convex, relatively unobstructed areas. This technique has proved to be highly reliable and accurate for detection and characterization of fatigue crack and corrosion below the rivets and other subsurface defects.

In the presence of corrosion, the eddy current distribution in the specimen becomes distorted and this disturbs the secondary magnetic field in the vicinity. This distortion in the magnetic field can be readily detected by the MOI detector. Thus, in austenitic stainless steels, this technique can be applied for detection, mapping and characterization of corrosion damage.

LASER TECHNIQUES FOR DETECTION OF CORROSION

Since corrosion is a surface phenomenon, any change in the surface due to corrosion leads to change in the surface morphology. Thus optical methods offer the possibility to detect and monitor corrosion at an early stage. The phenomenon of intergranular corrosion (IGC) in an AISI type 316 stainless steel using laser technique has been studied in the authors' laboratory [27]. The objective of this study was to explore the possibility of using laser scattering parameters to measure corrosion. For this, the laser scattering parameters were calibrated with different known extents of intergranular corrosion. The AISI type 316 stainless steel was used to induce intergranular corrosion by electrolytic etching. Different extents of IGC were obtained with different etching times. Laser scattering parameters were studied as a function of time of etching. The variation of specular reflection as a function of etching time showed that the specular reflection decreases as the time is increased. The full width at half maximum (FWHM) of the scattering pattern was observed to decrease with increasing etching time and vice versa. The monotonous behaviour of the laser scattering parameters and the severity of corrosion in these test specimens indicate the feasibility of this technique for corrosion monitoring.

In industries, there has been a continuous demand for the inspection of inside surface of tubes for pitting and other forms of corrosion. Precise dimensional measurements is quite important in the industry. The recent technological advances in the area of optics and electronics have enabled the development of a system based on principles of optical triangulation called "Laser Optic Tubing Inspection System (LOTIS)" [28]. This system essentially consists of three articulated modules combined to form a probe viz. a laser source, optics and photo detector; rotational drive system; and associated electronics. In this system, a 40 μ diameter laser beam is projected at near normal incidence onto the tube inner surface using an accurate rotational drive system, and the receiving optical system images this spot of light onto a single axis lateral effect photo detector. Using this system, it is possible to obtain the condition of the tube inner surface. The optical images formed by this system can be stored and retrieved for comparison and evaluation in a meaningful way, thus, aiding a reliable damage assessment of components. Since the technique works on the optical principle, which is independent of the type of the material, corrosion at the inner side of the austenitic stainless steel tubes can be detected.

INTELLIGENT PIGGING FOR PIPING INSPECTION

Intelligent pigs based on magnetic flux leakage (MFL) and/or ultrasonics are frequently used for inspecting pipelines for corrosion defects [28]. This is a very useful device to carry NDT sensors inside the pipelines for inspection/structural integrity assessments. Ultrasonic pigs utilize ultrasonic transducers that have a stand- off distance to the pipe wall. Any increase in the stand- off distance in combination with decrease in wall thickness indicates internal metal loss. For a given stand- off distance external metal loss, laminations or inclusions will be detected as decrease in wall thickness. The outer wall echo cannot be distinguished from the inner wall echo for too small a wall thickness (below 3 to 4 mm). Ultrasonic pigs have the advantage that they provide a good quantification of the size of the defect. The interpretation of the signals is optimised by good presentation software (interactive C-scans and B- scans). Depth sizing accuracy of the remaining wall thickness is in the order of ± 1 mm for pits and ± 0.5 mm for general corrosion at a confidence level of about 90%. Though, pigs are generally used for large diameter pipes, this concept can be extended for inspection of austenitic stainless steel pipes.

A new development for the detection and sizing of internal corrosion in small diameter, heavy wall pipelines is the high frequency eddy current pigs. In view of its accuracy and reproducibility, this technique can be used in successive inspections to determine corrosion rates. Eddy current proximity sensors are used for two different types of measurement, global and local. The combination of the measurements from global and local sensors provide the internal profile of the pipeline by which both general and pitting corrosion can be determined. The measurement performance obtained within the laboratory is that all pits exceeding the surface diameter of 10 mm and with a depth exceeding 1 mm are detected and that the sizing accuracy of pit depth and internal diameter is within ± 1 mm. Advantage of this technology is that it can be used for gas and liquid lines up to propulsion speeds of about 5 m/s. Significant improvements have been made by the pigging industry in evolving intelligent pigs, making them mechanically more reliable and providing better inspection results.

CONCLUSIONS

The significant developments, which have taken place in the area of nondestructive testing techniques for the detection and monitoring of corrosion, have been presented in this chapter. Non-destructive testing covers a wide spectrum of capabilities and sensitivity limits for corrosion detection and monitoring. The employment of a technique or a combination of complementary techniques for any given task requires careful analysis and maturity. Multidisciplinary efforts are essential for exploitation of tremendous benefits offered by NDT in ensuring cost effective fitness for purpose of components and plants. Increase in the use of non-destructive testing techniques would lead to better availability of plants and help in extending life of the plants.

ACKNOWLEDGEMENTS

Authors are thankful to C.K. Mukhopadhyay, B. Venkataraman, C.V. Subramanian, B.P.C. Rao, C. Babu Rao and T. Jayakumar, Division for PIE and NDT Development (DPEND), Indira Gandhi Centre for Atomic Research (IGCAR) for help in preparation of the manuscript. Authors also thank many colleagues of DPEND for many useful discussions.

REFERENCES

1. Raj, Baldev, Jayakumar, T. and Thavasimuthu M., Practical Non- Destructive Testing (2/e), Narosa Publishing House, New Delhi, India, 1997.
2. Shyamsunder, M.T. and Raj, B., Eddy Current Testing—Principles and Applications, Quality Evaluation, Vol. 8, No. 2, June 1988, pp. 3–6.
3. Schmidtz, T.R., The Remote Field Eddy Current Inspection Techniques, Materials Evaluation, 42, 1984, pp. 225–230.
4. Dodd, C.V, Deeds, W.E, Smith, J.H. and McLung, R.W, Eddy Current Inspection for Steam Generator Tubing Program: Annual Progress Report, ORNL Report, ORNL/TM-9339, June 1985.
5. Baron, J.A, Leemans, D.V. and Dolbey, M.P., Corrosion Monitoring in Industrial Plants using Nondestructive Testing and Electrochemical Methods, ASTM STP 908, 1986, p. 124.
6. Wait, J.R. and Gardner, R.L., Electromagnetic NDT of Cylindrically Layered Conductors, IEEE Trans. Instru. and Meas. IM-28, 1979, page 159.
7. Brown, S.D., Eddy Current NDE on Intergranular Attack, Final Report, EPRI RPS 201–1, EPRI, 1983.
8. B.P.C. Rao, C.B. Rao, T. Jayakumar and Baldev Raj, Delineation of Features in Eddy Current Images, Acta. Stereologica, Vol. 16, No. 1, March, 1997, pp. 69–74.
9. Nondestructive Testing Handbook, Vol. 7, Ultrasonic Testing, ASNT, pp. 1–20.
10. Bohn, H. et. al., Proving the Capabilities of the Phased Array Probe/ALOK Inspection Technique, Nuclear Engineering and Design, Vol. 102, 1987, pp. 361–372.
11. Muller, W., Schmidt, V. and Schafer, G., Reconstruction by the Synthetic Aperture Focusing Technique (SAFT), Nuclear Engineering and Design, Vol. 94, 1986, pp. 393–404.
12. Silk, M.G., The Use of Diffraction Based Time- of- Flight Measurements to Locate and Size Defects, British Journal of NDT, May 1984, Vol. 26, pp. 208–213.
13. Raj, Baldev, Subramanian, C.V. and Jayakumar, T., Nondestructive Testing of Welds, Narosa Publishing House, New Delhi, India, 2000, pp. 195–198.
14. C.S. Yuang and C.H. Yeh, Some Properties of Randomdec Signatures, Mechanical Systems and Signal Processing, May 1999, Vol. 13, No. 3, pp. 491–508.
15. Hideo Kusanagi, Hiroyasu Nakasa, Tadao Ishihara and Shigeo Ohashi, Proc. of the 4th Acoustic Emission Symp., Tokyo, 18–20 Sept., 1978.
16. Yuyama, S., Nisamatsu, Y., Kishi, T. and Nakasa, H., Proc. Fifth Int. AE Symp., Tokyo, 1980, pp. 115–124.
17. Jones, R.H., Friesel, M.A. and Gerberich, W.W., Met. Transactions, Vol. 20A, 1989, pp. 637–648.
18. Parry, D.L., 14th Conf. On NDE, USA, p. 388 (1983).
19. Mazille, H., Rothea, R. and Tronel, C., Corrosion Science, Vol. 37, No. 9, 1995, pp. 1365–1375.
20. Jones, R.H. and Friesel, M.A., Corrosion, Vol. 48, 1992, p. 751.
21. W.S. Burkle, Application of the Tangential Radiographic Technique for Evaluating Pipe System Erosion/ Corrosion, Materials Evaluation. Vol. 47, 1184, Oct. 1989.
22. P. Willems, B. Vaessen, W. Hueck, U. Ewert, "Applicability of computer radiography for corrosion and wall thickness measurements", INSIGHT, Vol. 41 No. 10, October 1999, pp. 635–637.
23. Hecht, A., Bauer, R. and Lindemeier, F., Jr. of Nondestructive Testing and Ultrasonics, Vol. 3, 1998, p. 1435.
24. Maldague, X., Pipe Inspection by Infrared Thermography, Mat. Eval., Vol. 57, Sept. 1999, pp. 899–902.
25. McKnight, M.E. and Martin, J.W., Detection and Quantitative Characterization of Blistering and Corrosion of Coatings on Steel using Infrared Thermography, J. Coatings Technology, Vol. 61, 1989, pp. 57–62.

26. D.J. Hagemaier, A.H. Wendelbo, Jr., and Y. Bar-Cohen, Aircraft Corrosion and Detection Methods, Materials Evaluation, Vol. 43, 1985, pp. 426–430.

27. Lakshmi, A.V.A, Djamouna, Kamachi Mudali, Babu Rao, C. and Baldev Raj, Study of Corrosion Pits using Laser Scattering, in Optics and Optoelectronics- Theory, Devices and Applications, Vol. 1, pp. 321–324, Eds. Nijhawan, O.P., Gupta, A.K., Musla, A.K. and Kehar Singh, Narosa Publishing House, New Delhi-1999.

28. Raj, B. and Jayakumar, T., Recent Developments in the use of Non Destructive Testing Techniques for Monitoring Industrial Corrosion, Proceedings of International Conference on Corrosion CONCORN'97, pp. 117–127, Eds. Khanna, A.S., Totlani, M.K., Singh, S.K., Elsevier Publications, Mumbai-1997.

13. Corrosion Related Failures of Austenitic Stainless Steel Components

K.V. Kasiviswanathan[1], N.G. Muralidharan[1], N. Raghu[1],
R.K. Dayal[1] and Hasan Shaikh[1]

Abstract Corrosion related failures form a substantial portion of the total failures of engineering components using austenitic stainless steels. Detailed failure analysis is important to understand the root cause of the failures and to take remedial measures to avoid such failures. Most of the corrosion related failures could be avoided by judicious choice of materials and closer control of process variables. In this article, failure analysis of austenitic stainless steel components carried out at our center is discussed and the probable causes of failures and the recommendation to avoid such failures are also highlighted to serve as a feedback to designers, fabricators and the plant engineers.

Key Words Corrosion, stainless steel, welding, failure, bellow, dished end, tank, heating elements, door, sensitization, weld defect, intergranular corrosion, stress corrosion, pitting.

Austenitic stainless steel is considered to be superior for many of the engineering applications due to its capacity to perform well in a very hostile and demanding environment. Austenitic stainless steels retain a substantial portion of their strength and ductility even at elevated temperatures. They also exhibit very good corrosion resistance due to the presence of protective chromium oxide coating. The failures of austenitic stainless steel engineering components are seen to be by and large due to stress corrosion cracking (SCC), intergranular corrosion (IGC), pitting corrosion and crevice corrosion.

Most of the corrosion related failures could be avoided by judicious choice of material and closer control of process variables. However, many a time, these failures result from unexpected quarters like improper design (giving rise to stress concentration regions or crevice), improper storage conditions, faulty fabrication procedure (producing surface residual tensile stresses, microstructural degradation) and abuse in service. These factors indirectly make the material susceptible to corrosion and cracking. Analysis of failures in austenitic stainless steel components has clearly demonstrated such shortcomings in weld design, fabrication and quality assurance aspects. The information on systematic analysis of failures and the lessons learnt from the analysis has to be channeled back to the designers and operators for corrective actions. This feedback approach goes a long way to contribute to the reliability and safety, by way of reduced incidences of failures.

---

[1]Metallurgy and Materials Group, Indira Gandhi Centre for Atomic Research, Kalpakkam-603102, India.

For bringing down the incidence of premature failure, developments are taking place at a fast pace in the science and technology of non-destructive evaluation (NDE). New methods, ingenious techniques for usage of known NDE techniques in remote, hostile environments are being developed for in service inspection of plant and machinery for ensuring integrity in service and detecting incipient failures before they turnout to be catastrophic. In many cases, these have been initiated after failures had occurred and thus necessitated strong need to prevent failures in similar equipment/structures in use in a number of existing operating plants.

STAGES INVOLVED IN FAILURE INVESTIGATION

During the failure analysis, inspection of the failed components has to be carried out at the site where the failure has taken place. The evidences in and around the failed location have to be preserved till the failure investigation team arrives. Detailed discussion with the concerned operator will give the background and circumstances leading to failure. The fracture surface should be preserved by applying grease/lacquer and the corrosion products, if any, must be collected before disturbing the failed component. The following stages are generally involved during the failure investigation. However, judicious choice of combination of various investigation techniques, including simulated studies, mechanical testing etc. is done to arrive at an unambiguous conclusion. The preparation of a final report pin-pointing the reasons for failure is a must to derive maximum benefit from the investigations. The feedback is beneficial to the design, material selection, fabrication and operating agencies, resulting in avoiding similar failures:

- Collection of background data
- Preliminary examination of the failed part
- Preservation of the fracture surface
- Non-destructive testing/examination
- Mechanical testing
- Macroscopic examination
- Microscopic examination
- Fractographic examination
- Chemical analysis
- Determination of failure mechanism
- Analysis of fracture mechanics
- Testing under simulated conditions
- Analysis and synthesis of all the evidences and formulation of conclusion
- Preparation of detailed failure analysis report.

Case studies of corrosion related failures of austenitic stainless steel components, carried out at the authors' laboratory are discussed briefly.

FAILURE OF STAINLESS STEEL BELLOWS

Component A number of stainless steel bellows intended for use in the control rod drive mechanism of a fast breeder reactor exhibited leaks during helium leak testing before being placed in service. These bellows are of welded construction using 150 μm thick multiple pre-formed diaphragms. The individual diaphragms are welded to each other on the outer diameter and the inner diameter alternatively using autogenous gas tungsten arc welding (GTAW) process, to form the bellows. These bellows are terminated with suitable end fittings facilitating easy installation. The size of the bellow was 210 mm in length and 40 mm in diameter.

Material AISI type 347 stainless steel.

History A number of such stainless steel bellows were stored in PVC bags for some years after procurement in a seacoast environment. Before installing these bellows they were subjected to helium leak testing. Very minute leaks of the order of 10^{-6} std. cc/sec at some locations were noticed. Even though the size of the leaks observed was very small they were still unacceptable for service, as per the design specifications. Figure 1 shows the photograph of one of such defective bellow with red marks showing the defect regions detected by helium leak testing.

Fig. 1.   The defective bellow.

Test and results Specimens were cut from the regions where defects were detected for metallographic investigations. The mounted samples were sequentially polished very carefully removing only of the order of 10 μm of surface each time to locate the minute leaks of the order of 10^{-6} std. cc/sec. The microstructure of the bellow material was typically austenitic and with a high degree of cold work (Fig. 2). Several cracks across the wall thickness of the convolute were observed at different locations (Fig. 3). Figure 4 shows origination of a crack from the bottom of a pit. Some of the cracks were seen to be branched (Fig. 5). A number of pits were observed on the surface of the bellow convolutes. A very fine transgranular crack originating at the bottom of a pit was also observed as shown in Fig. 5. The diaphragm material did not reveal any significant amount of inclusion and only isolated inclusions of submicron size were observed. Microhardness measurements on the bellow cross section indicated the hardness to be around 385–395 VHN (25g load); whereas the microhardness value of AISI 347 stainless steel in the annealed condition should be about 170–200 VHN only [1].

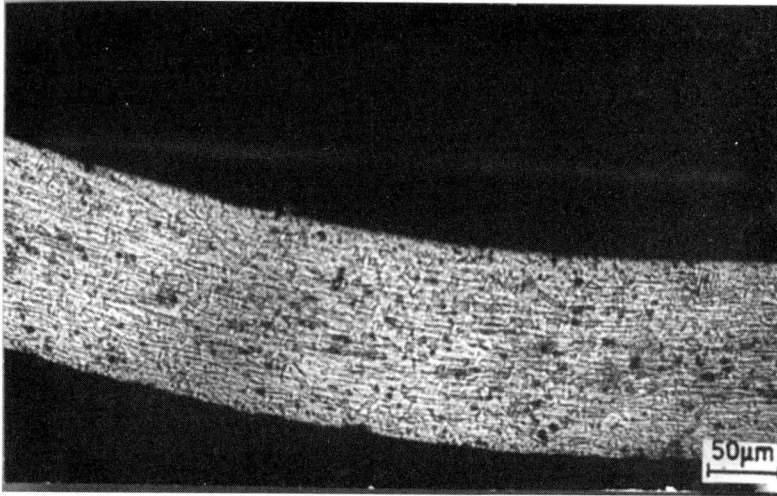

Fig. 2   Microstructure showing heavy coldworked structure.

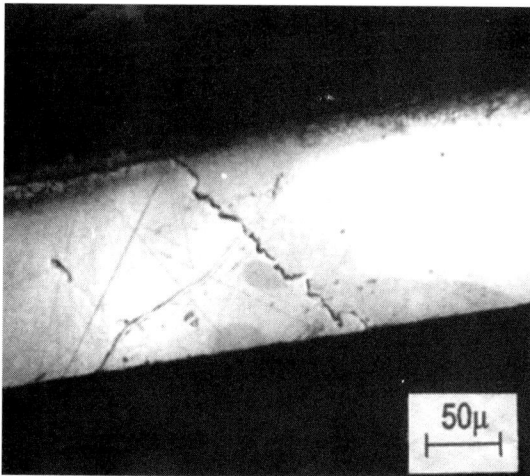

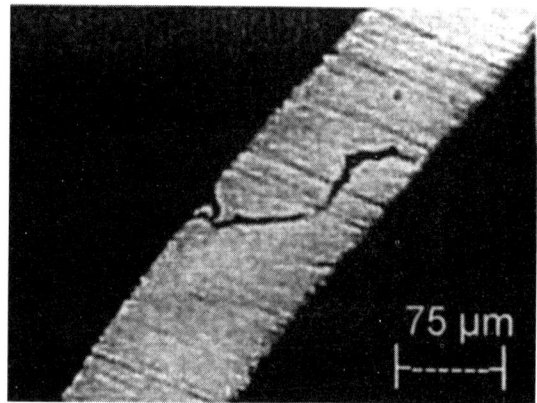

Fig. 3   Through and through cracks across the thickness of the bellow.

Discussion Optical metallographic studies revealed several pits and fine cracks originating from the bottom of these pits and extending across the thickness of the diaphragm. This observation suggests that the pits have acted as sites for crack nucleation. The cracks were seen to be transgranular in nature. The high level of hardness of the bellow material indicated that the bellows had not been stress relieved leaving behind residual stresses generated due to metal forming process. The bellows were also not stored in proper condition. Chloride ions are known to be potent corrosion enhancers and localized adsorption of chloride ions can act as prenuclei for pitting. Pits can also nucleate at carbides, grain boundaries and other material inhomogenities on the metal surface. The presence of moisture in the environment can facilitate the electrolytic path for the chloride ions. These pits act as sites for the nucleation of stress corrosion cracks, extending through the thickness of the convolute under the combined influence of residual stresses and environment [1, 2].

Fig. 4   Crack starting from a bottom of a pit.

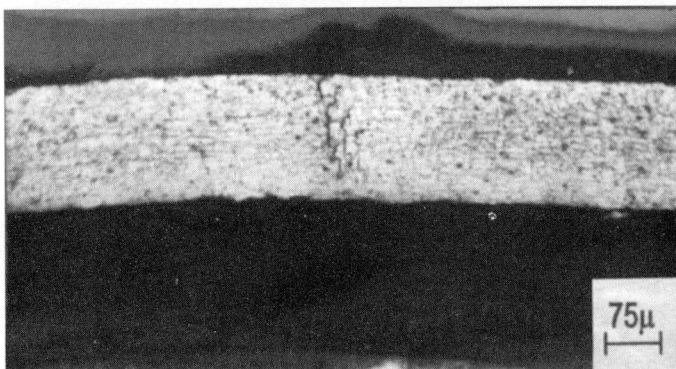

Fig. 5   Branched crack along the bellow convolute.

Conclusion The stainless steel bellows failed by stress-corrosion cracking caused by the presence of residual stresses in the material and improper storage conditions that led to contact with chloride ions in the ambient atmosphere.

Remedial action Such failures can be avoided by proper stress relieving of the components after fabrication. More pitting corrosion resistant stainless steel like type AISI 316 L can also be selected for fabricating these bellows provided other conditions are satisfied. Storage conditions should be improved to avoid direct contact of chloride ions and moisture to the finished components. The bellows have to be stored inside sealed double PVC cover with a dehumidifying agent kept inside.

FAILURES OF STAINLESS STEEL DISHED ENDS DURING STORAGE

Component Stainless steel dished ends made by cold spinning process are used to fabricate cylindrical vessels. The straight portion of the dished ends is to be subsequently welded to the cylindrical shells to fabricate vessels used as storage tank to store process liquids.

Material AISI 304L stainless steel.

History Several dished ends made out of type AISI 304 L stainless steel developed extensive cracking during storage. All the dished ends had been procured from a single vendor and belonged to the same batch of material. Visual inspection before welding revealed extensive cracks in the inside surface on many of the dished ends. However, no cracks were seen on the outside surface. Rust marks were also visible at several locations on the inside surface, may be due to use of carbon steel implements during material handling operations. The dished ends were stored in a coastal environment with an average temperature of 32°C, humidity ranging from 70 to 80% and an atmospheric NaCl content ranging from 8 to 45 mg/m²/day.

Testing Liquid penetrant testing was used to determine the extent and nature of the crack on the surface. *In-situ* Metallographic inspection (Fig. 6) on the inside surface indicated that the cracks were transgranular in nature and were also heavily branched with macro and micro branches. The morphology of the crack is indicative of transgranular stress corrosion cracking (TGSCC), as revealed in Fig. 7.

Fig. 6 *In-situ* **metallography at the inner surface of the failed dished end.**

The presence and qualitative nature of the residual stress in the dished ends were determined by making two saw cuts in the straight portion of the dished end. Considerable movement towards the axis of rotation and the splitting away of the adjoining surface was observed, when the side of the dished end was given a straight hacksaw cut (Fig. 8). This reshuffling indicates that the nature of the residual stress on the inside surface was highly tensile in nature.

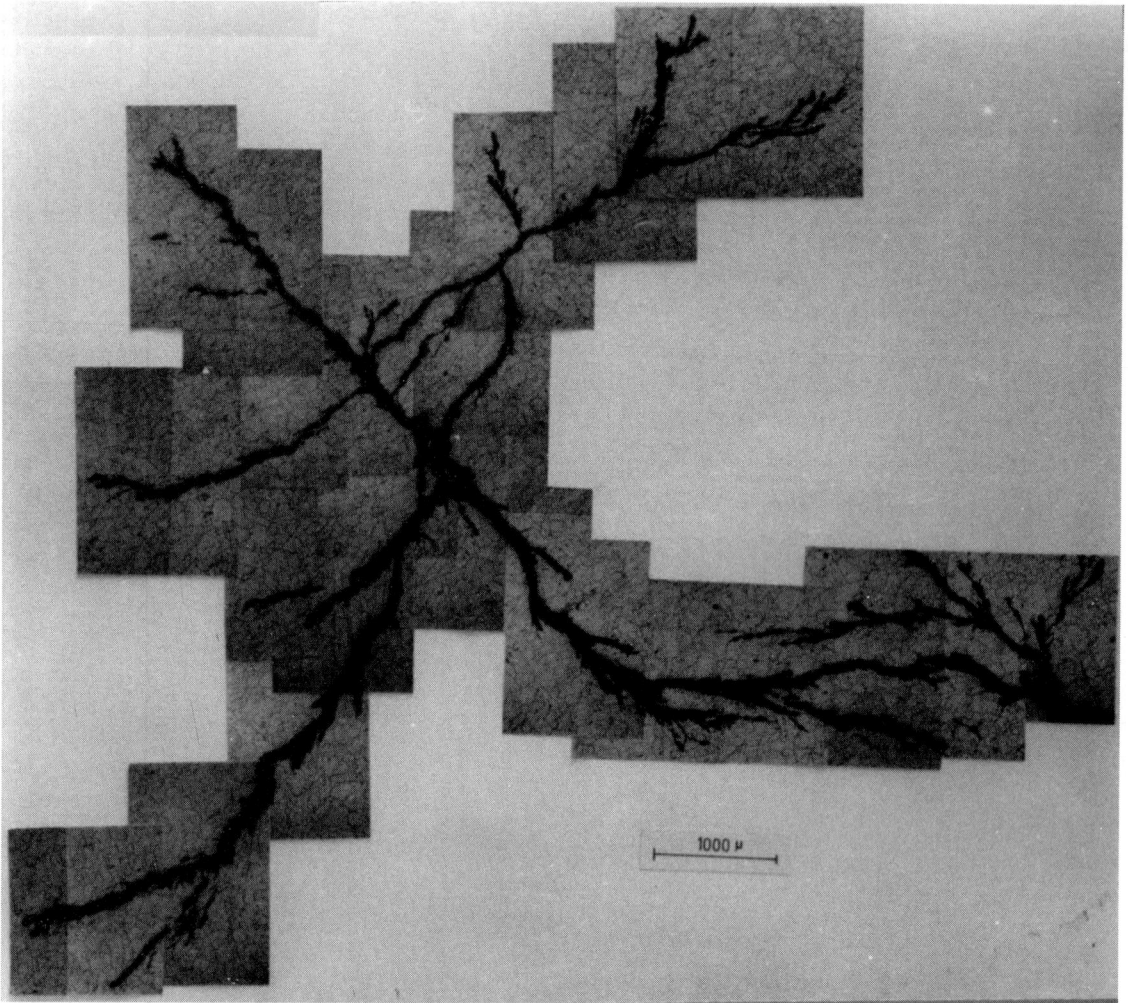

Fig. 7 TGSCC in the inner surface of the dished end.

Discussion The TGSCC occurred due to significant tensile stresses on the inside of the dished ends, presence of surface contamination seen as rust, and improper storage in a hot and humid coastal environment. In the hot and humid coastal atmosphere, long exposures of stainless steel components containing iron surface contaminants could result in the rusting of iron debris. Such rust particles absorb moisture and chlorides from the environment, which leads to SCC of the components [3].

Conclusion Conjoint action of residual stresses and chloride ions in the environment contributed to the cracking of the dished ends.

Recommendation Extreme care is needed to avoid pickup of iron contaminants by the stainless steel components during fabrication and subsequent handling. Pickling and passivation of the stainless steel components prior to final dispatch and storage is also required to remove surface iron contaminants

Fig. 8   Reshuffling of locked up tensile stresses after making two hacksaw cuts in the straight portion of dished end.

and provide adequately thick passive chromium oxide coating. Appropriate stress relieving treatments to relieve residual stresses is also recommended to avoid SCC.

FAILURE OF BELLOW IN A BELLOW SEALED VALVE

Component A bellow sealed valve employed as on-off valve used to control the flow of liquid sodium, used as a coolant in the secondary circuit in a fast breeder test reactor, was found to have failed in service. The valve exhibited high torque to be applied for closing and opening. This particular valve is generally not very frequently operated and the torque required to operate the valve was seen to be quite small during previous operations. This valve has a stainless steel bellow made with two plies to provide adequate leak tightness across the valve stem and at the same time allow movement of the spindle to control flow of liquid sodium. The outside surface of the bellow during operation would be in contact with molten sodium at 573 K while the inside surface of bellow is in contact with argon cover gas.

Material The material of construction of the bellow in the bellow sealed valve as specified was AISI 316L type stainless steel. The bellow is 56 mm long with inside and outside diameter being 24 and 35 mm respectively. The bellow is made of two plies, each 0.150 mm thick, and is welded to the end flanges. The defective bellow sealed valve in dismantled condition is shown in Fig. 9.

History The bellow sealed valve was stored in the hot and humid coastal climate for about three years before putting them into operation. After about eight years of smooth operations, the concerned valve was found to be quite difficult to operate. As valve cleaning and lubrication of the valve stem did not show any improvement it was decided to inspect the valve by removing it from the line. The component was then dismantled to check for the internal damages, by cutting open the bellow from end flange.

Substantial amount of sodium was found trapped inside the stainless steel bellow of the bellow sealed valve indicating that the bellow has failed in service.

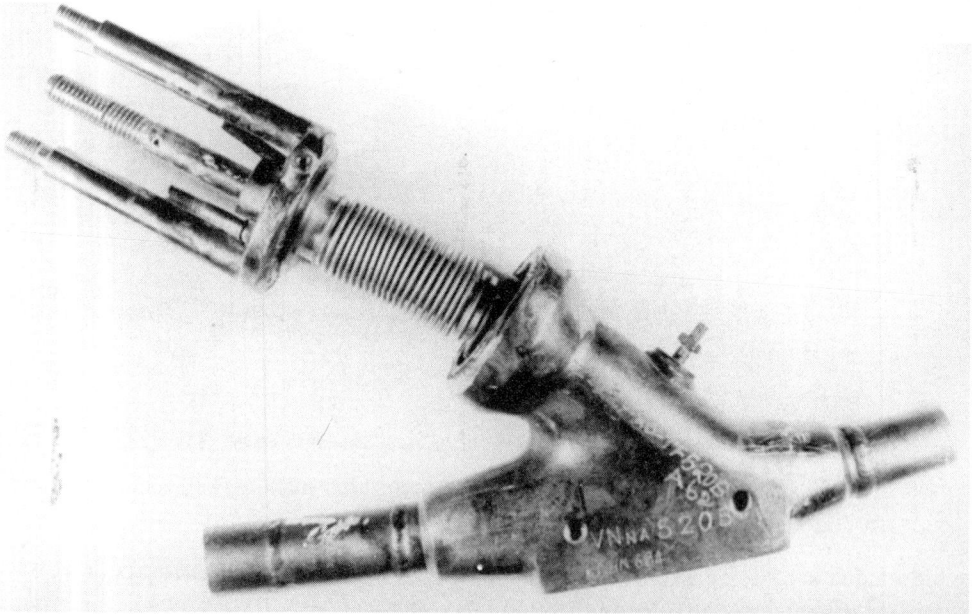

Fig. 9   Defective bellow in the bellow sealed valve.

Testing Helium leak testing carried out on the dismantled bellow after it was cut open from the end flange did not reveal any leak. Microstructural examination of the bellow material from the central portion did not reveal abnormality. A few hairline cracks were noticed near the upper end of the bellow close to the weld region (Fig. 10). Metallographic specimens were selected from these regions for

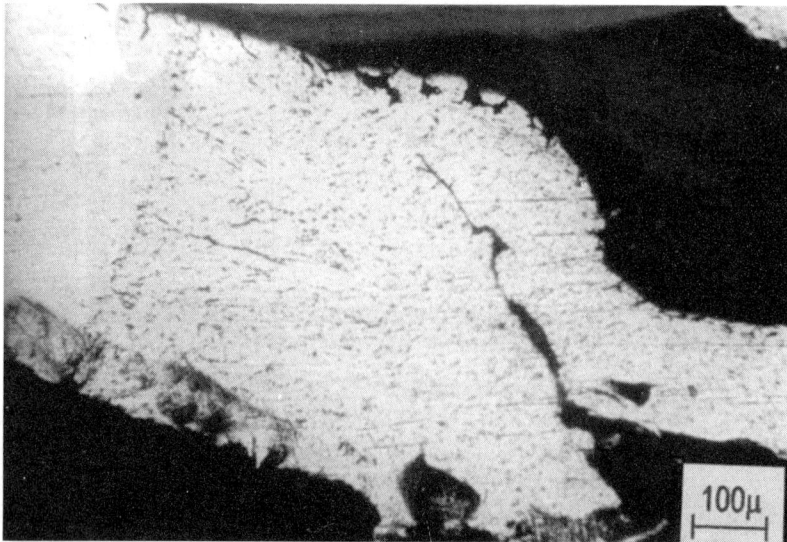

Fig. 10   Crack in the weld zone.

detailed investigations. Several intergranular cracks and grain boundary carbide precipitation was noticed in a very localized area near the weld (Fig. 11). Microhardness measurements revealed a hardness of about 290-320 VHN in these locations Since the microstructure at the heat affected zone (HAZ) of the bellow to end flange weld joint exhibited a sensitized structure, it was decided to carry out chemical analysis of the bellow material. The chemical analysis revealed that the carbon content of the bellow material was high and not conforming to AISI 316L specifications.

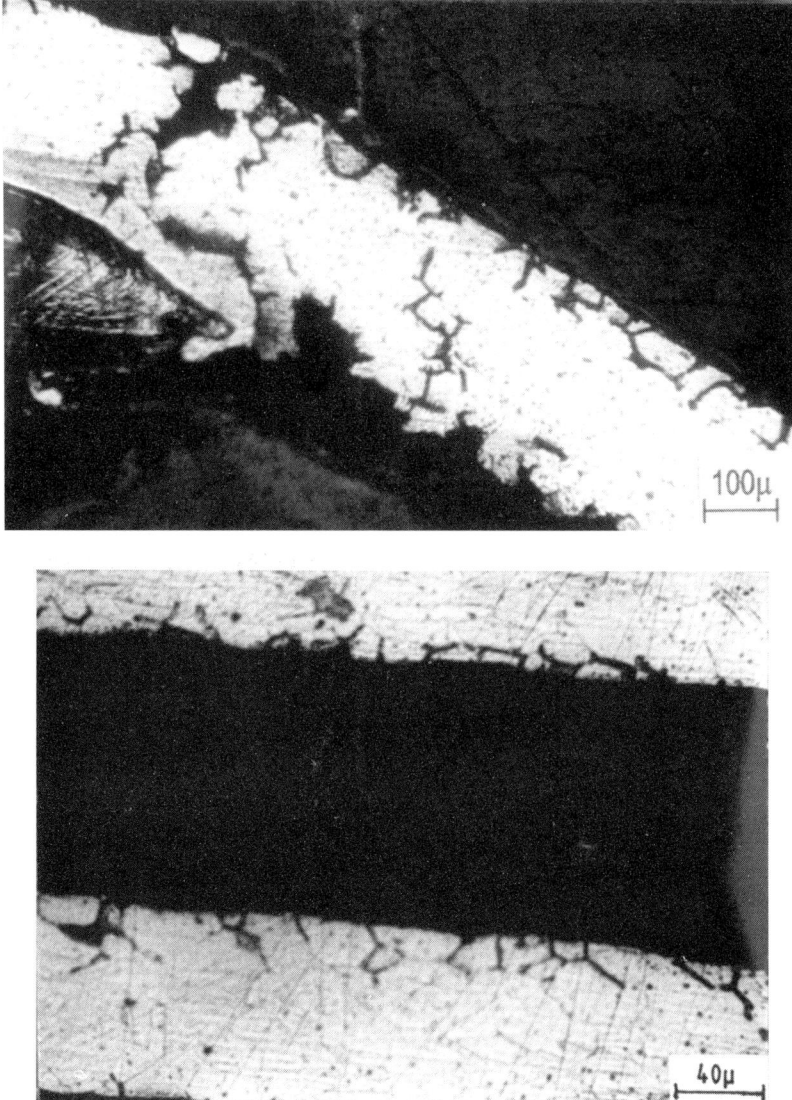

Fig. 11.   Photomicrostructure showing intergranular cracks and carbide precipitation.

Discussion The observation of normal microstructure in the central portion of the bellow and intergranular crack near the weld zone, associated with intergranular precipitation of carbides, suggests that the bellow material had undergone sensitization due to welding. The higher carbon content is responsible for the sensitization of the bellow material during welding. This sensitized region of the bellow is quite likely to undergo intergranular corrosion during storage in humid coastal climate. This would have gone unnoticed until both the plies of the bellow cracked leading to leakage of sodium into the bellow [4].

Conclusion The cause of the failure of the bellow in the bellow sealed valve is attributed to the sensitization of the material during welding and subsequent corrosion of the material during storage and operation. High carbon content of the material assisted in the sensitization of material during welding.

Recommendation To avoid such failures it was recommended that the bellow material chosen should have been a low carbon stainless steel material like AISI 316L as specified. Welding parameters should also be optimized to avoid sensitization during welding. During storage, the valves should be sealed in double PVC cover with dehumidifying agent to avoid direct contact with the coastal environment.

FAILURE OF DELAYED NEUTRON MONITORING TUBE

Component A number of AISI 304L stainless steel seamless tubes, which form a part of a delayed neutron monitoring system in the heavy water circuit of a pressurized heavy water reactor, were found to be leaking during routine inspection. The failed tubes were having an outside diameter of 9.6 mm and wall thickness of 1.4 mm.

History The stainless steel delayed neutron monitoring (DNM) tubes were insulated with mineral wool. The heavy water flowing inside the tube are continuously analyzed to monitor the presence of certain elements in the heavy water circuit of the reactor system. The tubes were found to be leaking at several locations in a single length of the tube. The mineral wool insulation cover was found to be broken at some locations. The mineral wool insulation was also found to be soaked with heavy water to a considerable length on the leaking tubes. The leaking tubes were identified by the plant personnels and removed for detailed investigations.

Testing Visual examination of the DNM tubes (approximately 1050 mm in length) revealed pit marks several locations. A circumferential weld was found in the middle of the tube. One tube was selected for detailed investigations. Dye penetrant testing carried out on the tube, revealed several micro cracks and a crack for about 20 mm in length, which was around 210 mm away from the weld.

Metallography In-situ metallography was carried out on the failed tube near the crack. The examination was carried out inside a fume hood, due to the presence of gamma activity on the tubes. Figure 12 shows the failed location in the tube. Cracks were emanating from the pits, which were found to be branching as TGSCC as shown in Fig. 13. The metallographic examination of a specimen containing a cut cross section revealed TGSCC as shown in Fig. 14(a). Presence of a long seam weld with welding defects is shown in Fig. 14(b), even though the tubes were supposed to be seamless as per design drawings. The microstructure was typical of austenitic stainless steel with a grain size corresponding to ASTM grain size number 10.

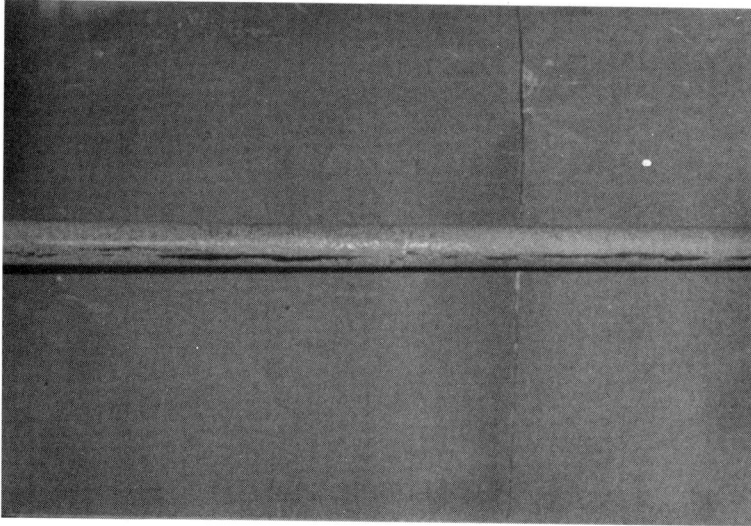

Fig. 12. Failed location of the DNM tube.

Residual stress measurement Residual stress measurement by x-ray diffraction technique revealed presence of tensile stresses on the circumference of the tube, probably due to the presence of the long seam weld. The extent of residual stresses were not uniform and was found be varying from location to location and also along the circumference at a single location. Hence, quantitative estimation of the stresses was difficult.

Chemical analysis on mineral wool Chemical analysis of the insulating material was carried out to determine the amount of leachable chlorine content. This revealed higher amount of chlorine content in the material [5].

Conclusion The failure of the AISI type 316L stainless steel tubes are due to TGSCC, mainly starting from surface pits. The potent corrosive medium may be chloride ion leached from the insulating material due to the leakage water in contact with it. The leached chloride ions in combination with residual stress in the material could have resulted in stress corrosion cracking.

Recommendation Proper selection of the insulating material and proper storing of the insulating material is quite important for avoiding such failures. The procedure for insulation should be such that ingress of moisture to the insulating material should be avoided. The chloride content should be less than specified in appropriate standards. Quality control and inspection procedure during fabrication to be strictly monitored to avoid mix up of seamless tubes and seam welded tubes.

FAILURE OF STAINLESS STEEL BELLOWS

Component Bellow sealed valves used to control the flow of sodium in a component testing facility was seen to be failing prematurely. The bellows in the valves effectively isolate the liquid sodium from the spindle to provide leak tightness and at the same time allows the movement of the spindle to control the flow of the liquid sodium.

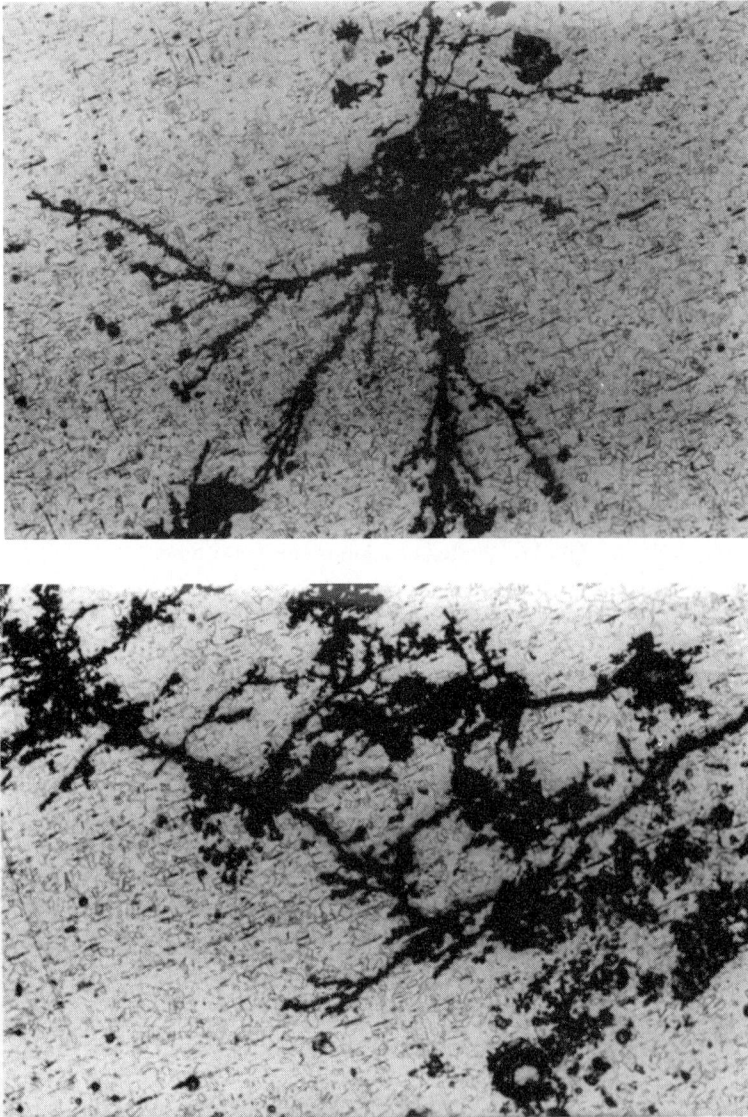

Fig. 13 Branched TGSCC initiation on corrosion pits observed on the outer surface of the SS tube.

History The failed bellows had seen approximately 3 years of service in the testing facility. During the service the bellows were exposed to liquid sodium in the temperature range of 723 to 823 K. The bellows were in compressed condition during the valve in open position and in normal state otherwise. Frequency of operation was once in every two days.

Material The bellows were made of AISI type 316 stainless steel and fabricated by cold forming from a cylindrical sheet. The bellows were of double ply construction.

Testing The failed bellows were subjected to visual inspection, metallographic examination, hardness measurements and fractographic examination with scanning electron microscope (SEM).

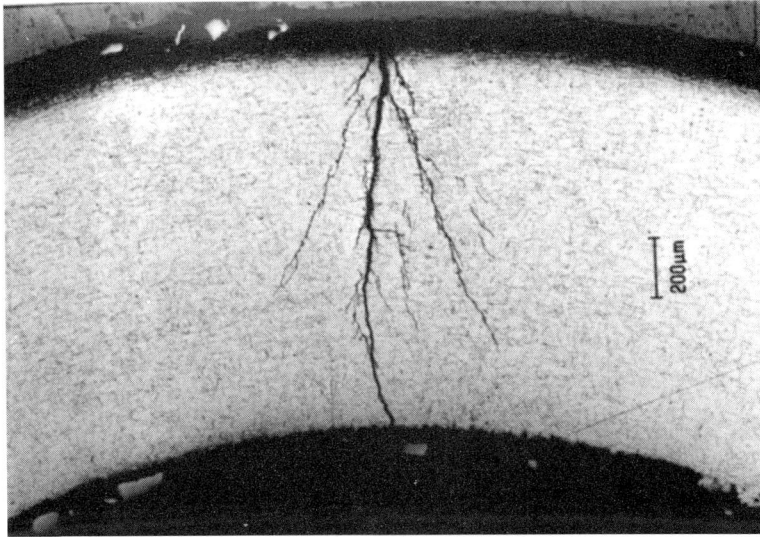

(a)

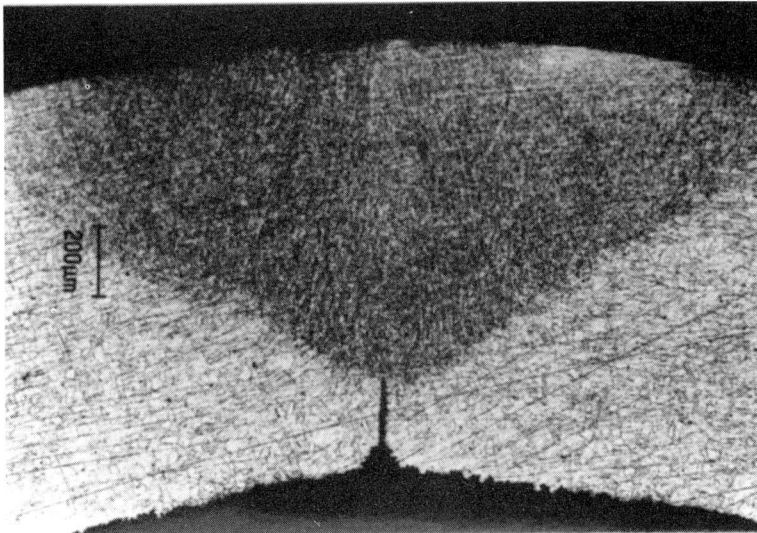

(b)

**Fig. 14. Microstructure at the tube cross section showing propagation of
TGSCC and a long seam weld with weld defect.**

Results Visual examination revealed the failure at the end of the bellow in a circumferential direction
towards the outer side of the convolute (Fig. 15) and the edges of the bellows were covered with
corrosion products. Scanning microscopic studies on the fractured surface indicated that the failure
had occurred in intergranular mode (Fig. 16). Optical microscopic examination indicated a ditched
structure (Fig. 17) and the gap between the two convolutes. The microhardness of the bellows varied
between 250 VHN and 310 VHN, whereas the microhardness of annealed material is about 180 VHN.

Discussion The presence of corrosion product on the fracture surface and the edges of the bellows suggested failure was due to environmentally induced cracking. The intergranular fracture seen on the fracture surface indicates that the failure is due to IGSCC. The ditched structure at the failed location proves that the weld convolutes has undergone sensitization. The higher microhardness observed is indicative of presence of high residual stresses due to cold forming operations adopted during fabrication of the bellows. Environmental species such as chloride and caustic are known to cause IGSCC in austenitic stainless steel components. As the component was in contact with sodium, the possibility of caustic being the cause for IGSCC of the bellows assumes

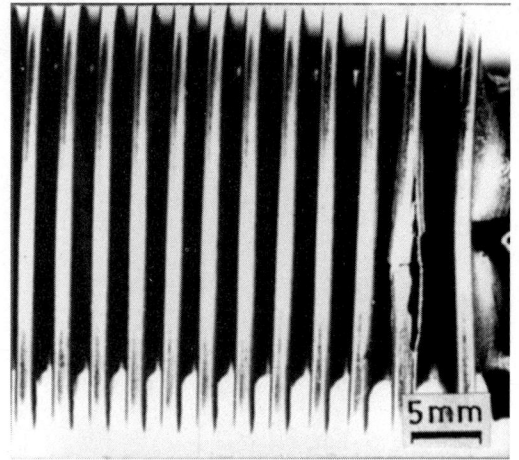

Fig. 15   Failed region in the bellow.

significance. The formation of NaOH would have occurred due to reaction of Na with moist argon during shutdown period in the gap between the walls of the bellows.

Fig. 16   SEM micrograph showing the IGSCC.

Conclusion The conjoint action of residual stress and NaOH, formed by reaction between the sodium and moisture in the cover gas, has resulted in SCC of the bellows.

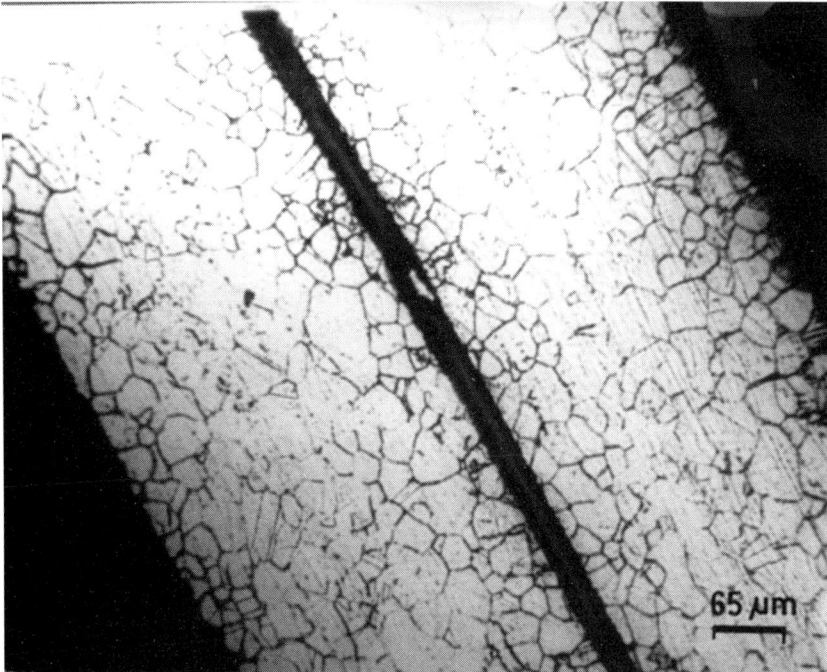

Fig. 17 Optical photomicrograph showing sensitized microstructure.

Recommendation Solution annealing of the bellows after cold forming is recommended to eliminate the residual stresses. Use of low carbon variety of stainless steels such as AISI type 316L material should have avoided sensitization of the HAZ of the weld.

FAILURE OF CONTAINMENT BUILDING DOOR BELLOWS

Component The containment building in a nuclear reactor is always maintained at a negative pressure with respect to ambient atmosphere. This helps to contain the radioactive dust within the boundary of the building, in case of an accident. In order to maintain leak tightness, and at the same time allow for thermal expansion all wall penetrations in the containment building are provided with metallic bellows. One of such bellows failed prematurely in a nuclear establishment in a coastal environment.

Material The bellows are made of AISI type 304 stainless steel fabricated by welding the pre formed 1.6 mm thick convolutes. The convolute pitch of the bellow is about 250 mm. A 3 mm thick and 25 mm wide backup strips made of AISI type 304 stainless steel was used for proper fit up. (Fig. 18). The backup strip were tack welded to the two pre formed convolutes to hold them in place.

Testing Visual examination of the failed bellows indicated several branched, through-thickness cracks, both in the weld region and the base metal. Some cracks initiated in severely deformed regions of the bellows at the weld metal and in the fusion line. The surface of the through-thickness crack showed black corrosion deposit with patches of yellowish brown colour near the crack edge. A cavity of about 15 mm dia was observed in the weld metal at the deformed region of the bellow. This cavity was formed by localized corrosion of the weld metal. Figure 19 shows details of the cavity and its adjoining areas. Major cracks emanating from the cavity can also be clearly observed. The failed bellows were subjected to detailed metallographic examination and microhardness testing. The chemical analysis of the bellows and the backup strip were found to be nearly the same and corresponding to the composition of the AISI type 304 stainless steel. However, microhardness varied from 160 VHN in the undeformed region to a maximum of 220 VHN in the deformed areas of the convolute.

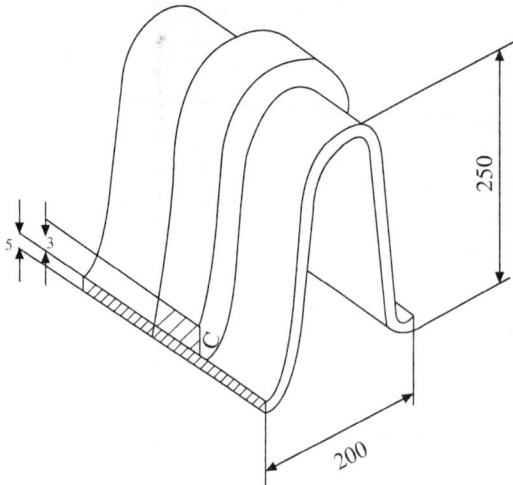

Fig. 18 Schematic diagram of the bellow.

Fig. 19 Section of the failed bellow showing weld cavity and cracks.

Optical microscopic examination of the weld metal indicated the presence of duplex microstructure containing delta ferrite in the austenitic matrix. Microstructure also revealed that the cracks were highly branched and transgranular in nature. Figure 20 shows the transgranular cracks emanating from a crevice. Figure 21 shows transgranular cracking in the weld fusion line while Figure 22 shows the

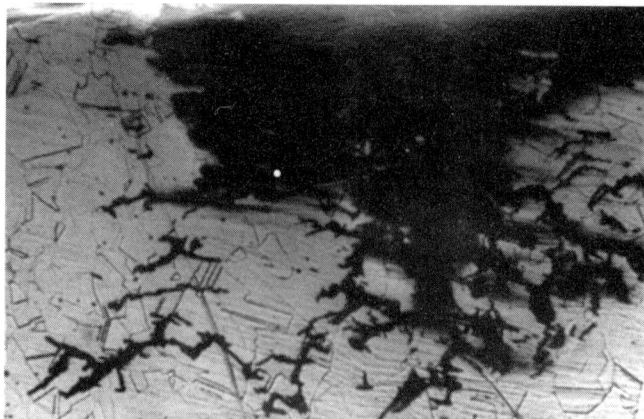

Fig. 20 Transgranular cracks emanating from a crevice.

microstructure of the crack which propagates along austenitic-δ ferrite interface. SEM examination revealed transgranular fan shaped fracture surface.

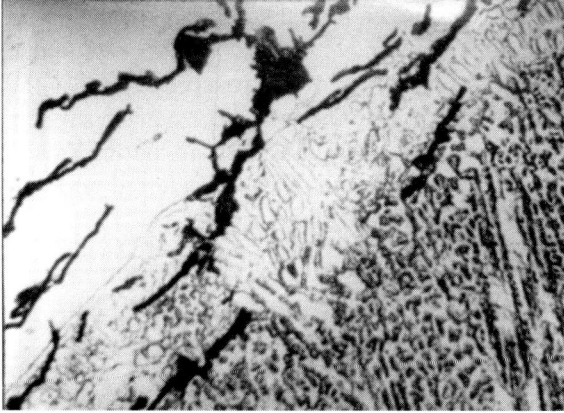

Fig. 21   TGSCC cracking in the weld fusion line.

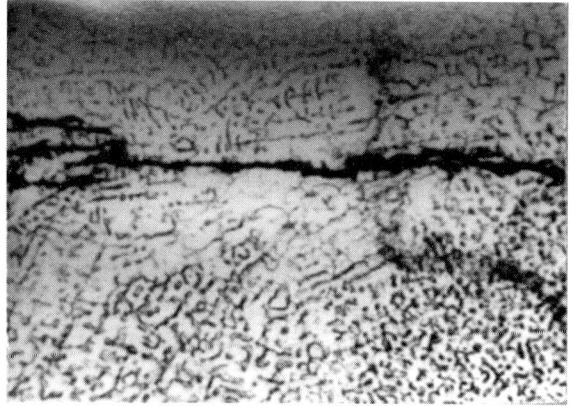

Fig. 22   Photomicrostructure revealing crack along δ/γ-interface.

Discussion The presence of a large corroded cavity along with corrosion product at the fracture surface points to an environment induced failure. The yellowish coloured corrosion products are generally observed in iron-based alloys, which are exposed to stagnant aqueous environment for a long time. Since the backup strip was tag welded, a crevice was created between the backup strip and the bellows. Exposure of this face of the bellows to atmosphere caused crevice corrosion. The crevice corrosion is due to two reasons. The root of the crevice was in the weld metal leading to aggloramation of the corroding species in the weld metal. The other reason, the weld metal of the austenitic stainless steel is more susceptible to pitting and crevice corrosion than the parent metal due to depletion of chromium and molybdenum that occurs in the austenite matrix due to enrichment of these elements in the delta ferrite at the weld metal. Higher microhardness values indicate that proper solution annealing was not given to the formed bellows [6].

Conclusion The failure of the containment building stainless steel bellows is mainly due to simultaneous occurrence of crevice corrosion and stress corrosion crack. The gap between the backup strip and bellow acted as a crevice. Accumulation of chloride ion in the crevices in conjunction with the residual stresses in the bellows led to stress corrosion failure. The use of backup strip for welding caused the premature failure of the bellows.

CORROSION FAILURE OF STAINLESS STEEL REDUCER DURING SURFACE PRETREATMENT

Component Two stainless steel reducer section fabricated from AISI type 316 stainless steel were found to be severely corroded—one reducer soon after pickling and other after passivation treatments.

Material AISI type 316 stainless steel.

History A conical reducer has to be made using two half sections so as to fit it inside a critical pipe line by welding. To meet stringent weld fit-up tolererances, each half section was to be made from a full conical reducer by slitting it into two unequal parts and using the larger half after machining to the required size. The conical reducer were made from a 135 mm dia forged bar (component A) with a trepanned round from a 110 mm thick plate (component B). To relieve residual stresses caused due to machining, the components were annealed. The parts were pickled and passivated as per the procedure. The components B developed severe pitting attack on both the inside and outside surface, but no pitting occurred on the cross-section area. The component A did not suffer any pitting attack, but the surface developed a dull gray finish after this. The components A and B referred above is shown in Fig. 23.

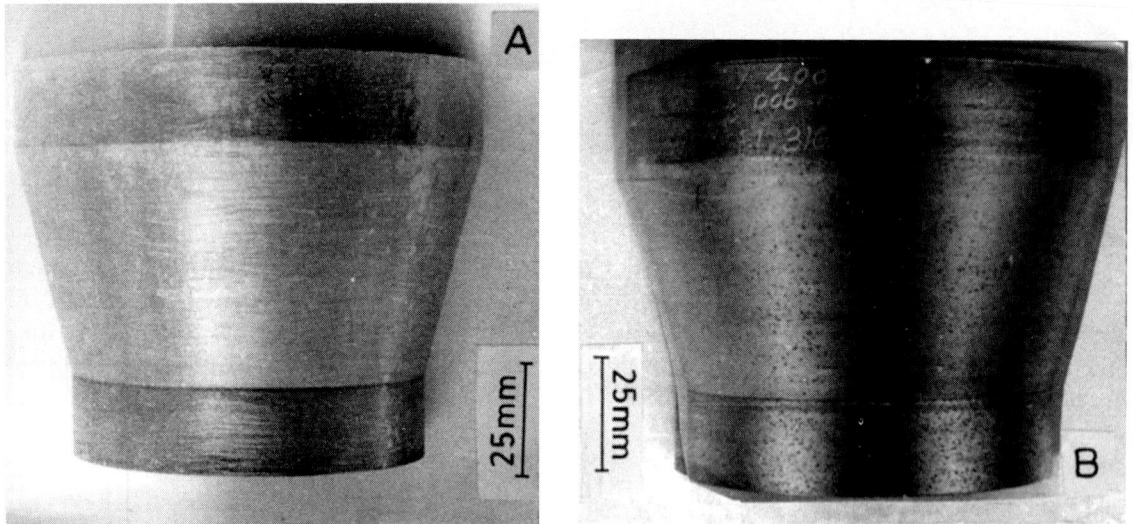

Fig. 23   Components A and B.

Testing Optical microscopic examination of etched cross sections of the two components revealed that component A was in a sensitized condition and that intergranular corrosion attack up to about 20 μm in depth had occurred on both components, from inside and outside surfaces. However, component B did not show any sign of sensitization. Figure 24 shows the photomicrostructure at the thickness crossection in components A and B. The grain size of component B was significantly larger than that of component A. The inclusion content of component B, carried out as per ASTM E-45 standard, was found to be low.

Discussion Both the components were attacked by corrosion after the surface treatment, though in different forms. Component A was in a sensitized condition, underwent intergranular attack, as the cooling rate of 120 C/h after annealing was not adequate to prevent sensitization. Component B was air cooled after annealing treatment, did not undergo sensitization. This component, machined from the cross section of a thick plate, experienced severe pitting, primarily because of the exposure of smaller end grains [7].

Conclusion and recommendation Component A was severely sensitized and the pickling and passivation resulted in severe intergranular corrosion. Component B, fabricated from the transverse section of the thick plate resulted in severe pitting due to end grain effect.

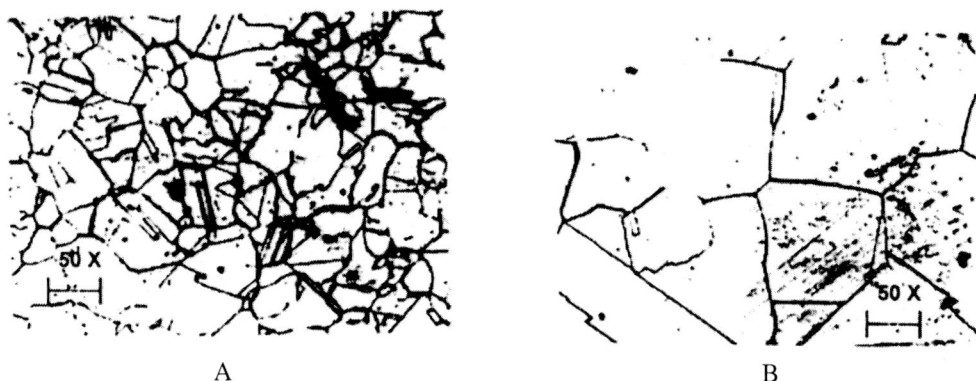

A   B

Fig. 24   Photo microstructure at the thickness cross section in components A and B.

The component A was reused after removing the intergranular attacked layer and re-solution annealing. The component B was reused after polishing to remove the pits and the machining marks. Subsequently pickling and passivation treatment were carried out by careful selection of composition, temperature and time as per the specification. The heat treatment parameters should be strictly controlled avoid sensitization, and pickling modified composition and time duration was recommended for pickling and passivation treatments.

CORROSION FAILURE OF STAINLESS STEEL TANKS

Component Four fabricated stainless steel tanks were found to be leaking before their commissioning, though they had passed a similar test at the manufacturer's site 12 months before.

Material 3 mm thick AISI type 304L stainless steel.

History A few AISI type 304L stainless steel tanks which had passed helium leak tests (under pressure) and hydrostatic testing at the fabrication site failed during hydrostatic testing before its commissioning at site after a period of 12 months. The 700 mm dia and 1000 mm long tank was also found to be corroded on the outside surface due to atmospheric corrosion during the pre-commissioning stage; crevice and pitting corrosion due to the presence of crevices formed due to weld spatter and uneven weld deposit during fabrication, and excessive pitting due to non-draining of stagnant service water for a period of 12 months after hydro tests. The sketch of the tank is shown in Fig. 25.

Testing and Results The outside surfaces of all the tanks showed evidence of improper passivation. At a number of places on the outside surface, atmospheric corrosion had occurred due to the coastal atmosphere. Excess penetration of weld deposit on the inside surface was noticed where the dished end had been welded. Pitting and crevice corrosion on the weld deposit and on the base metal were clearly seen as shown in Fig. 26. Most of the attack had taken place on the bottom side of the tank where service water was stagnant for about one year. At a number of places crevices had formed due

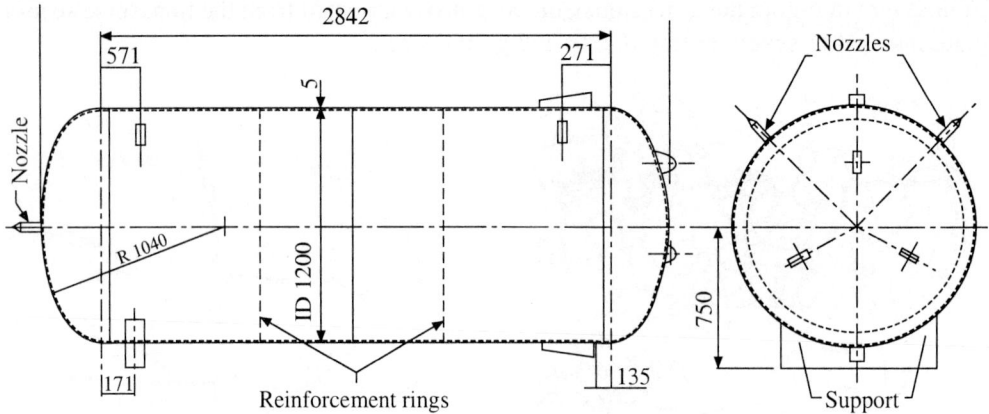

Fig. 25   Schematic sketch of the failed tank.

to weld spatter and excess penetration of weld deposit and gap between reinforcement ring and shell surface. Corrosion was also observed on the reinforcement ring itself.

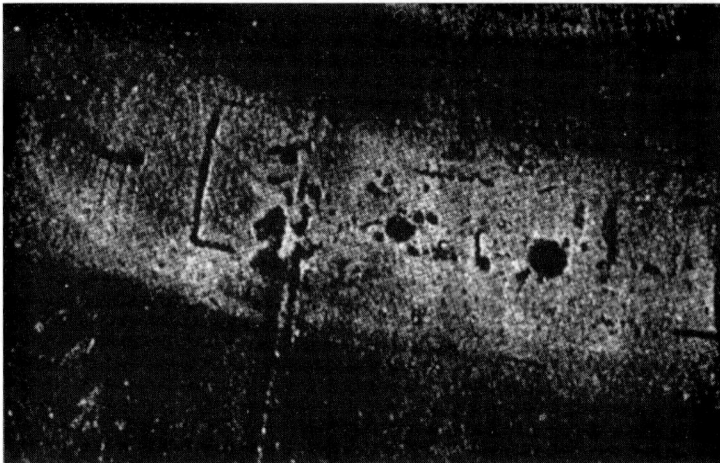

Fig. 26   The inside surface of the tank containing pitting attack, crevice attack on the weld deposit and base metal.

When the other three tanks were examined, two of them were found to contain water, which was supposed to have been drained out after the hydrostatic test. In these cases, water was stagnant in the tank. Samples of water were analysed for Cl and F content, which indicated Cl = 65 ppm, F < 0.5 ppm and pH = 4.5. The water was almost saturated with rust particles.

Discussion The chemical composition and metallography of the tank material did not reveal any adverse aspect, which could give rise to any abnormal corrosion. The condition of the outside surface of the tanks indicated that the passivation treatment had not been properly done. This led to atmospheric corrosion during storage period at the site due to the presence of chloride in the coastal environment. Besides this, service water with high chloride content had been used and had not been drained out from the tanks after hydrostatic test. This water had been stagnant for more than one year inside the tanks. Corrosion attack was found only at the bottom of the inside surface of the tanks where water was stagnant. In the presence of water containing chlorides, corrosion had taken place at crevices [8].

Conclusion and recommendation The failure could have been prevented if: (1) the crevices formed due to weld splatter and uneven weld deposit on the tanks could have been removed by grinding and smoothening the surfaces, (2) stainless steel surfaces could have been properly passivated for resisting atmospheric corrosion and (3) the tank had been immediately rinsed with demineralised water after passivation and after hydrostatic testing. The water could have been drained completely and tanks dried thoroughly before despatch.

FAILURE OF STAINLESS STEEL TANK DISHED ENDS USED IN HEAVY WATER/HELIUM STORAGE

Component The dished ends of a heavy water/helium storage tanks manufactured from 8 mm thick AISI type 304 stainless plate was used to fabricate this storage tank which failed during hydro testing.

History The heavy water/helium storage tanks was designed for a pressure of 0.1 MPa and for a temperature of 67°C to contain helium gas at a maximum pressure of 0.035 MPa and a temperature of 40°C as well as to provide emergency storage of heavy water at atmospheric pressure and at a maximum temperature of 67°C. The tank had a capacity of 1,32,000 liters with an outside diameter of 4.42 m. The wall thickness of the shell was 6 mm and the thickness of the dished end was 8 mm. The sketch of the stainless steel tank is shown in Fig. 27.

During hydro test, leaks were observed in the bottom dished end. Dye-penetrant inspection revealed numerous cracks confined to the HAZ. These were ground and weld repaired. Subsequent pneumatic soap bubble test at 0.10 MPa revealed very minute leaks. Repeated attempts at repair welding of the dished ends did not alleviate the problem. Examination of samples from the dished end revealed that cracking was confined at the HAZ surrounding the circumferential welds and to a lesser extent, radial welds that were part of the original constructions. Most of the cracks initiated and propagated from the inside surface of the dished ends.

Testing and results A sample of size $120 \times 15 \times 8$ mm was cut from one dished end to study the mode of the failure and its possible causes. This sample included about 40 mm long weld deposit at one end. During metallographic examination, an island of basemetal in between the weld passes was observed. The microstructure of the base metal in this island, as well as for a portion of 10 mm from the weld metal was found to be sensitized. The portion away from this did not indicate sensitization. The above sample was subjected to intergranular corrosion testing as per ASTM Standard A 262 Practice E, followed by a U-bend test. During the bend test, the sample fractured very easily in the HAZ.

Discussion Severe sensitization had occurred in the HAZ, perhaps the result of use of an improper fabrication technique. With the joint design, pit-up problems could have caused higher metal deposition, as well as possible weld repair. The excessive heat input to the low wall thickness plate had resulted in sensitization. The dished end which had been in storage a long time before assembly was left unattended during the storage. Thus the HAZ had experienced severe sensitization and residual stress.

The intergranular susceptibility has accelerated the corrosion cracking, resulting in premature failures [9].

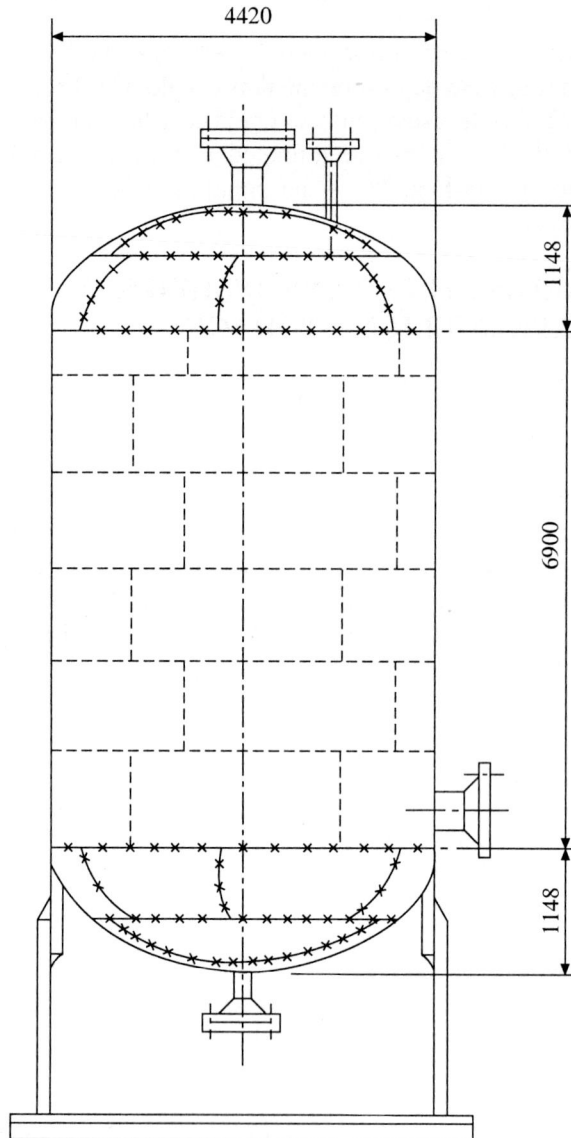

Fig. 27   Schematic sketch showing the construction of the tank.

Conclusion and Recommendation The large heat input occurred during the repeated repair weld passes, has led to severe sensitization of the HAZ. The presence of fit-up and welding stresses combined with the exposure to the coastal atmosphere provided favorable condition for stress corrosion cracking to occur. Repairs were successfully carried out using a new procedure that eliminated all cracks in the weld repair zone as given in Fig. 28.

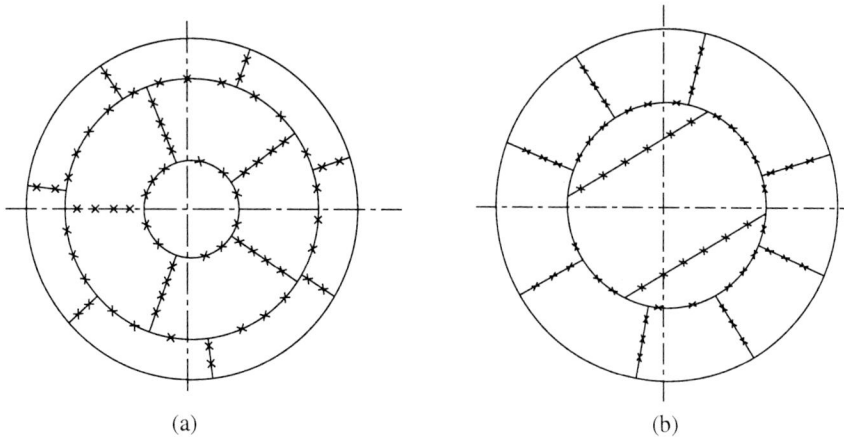

(a)   (b)

Fig. 28   Weld details of leaking dished ends and the modified dished end.

The fabrication procedure for such critical circumferential joints were eliminated and modified design was incorporated for fabricating these tanks.

STRESS-CORROSION CRACKING IN STAINLESS STEEL HEATER SHEATHING

Component 304L type stainless steel sheaths on nichrome wire to be used as a heater was supplied in the form of coils in polyethylene packing material and was transported by sea. The cracks were discovered when the heater coils were removed from storage in their original polyethylene packing materials and straightened for use. The cracks originated from rusted areas on the cladding under iron particles left on the surface during manufacture.

Material AISI type 304L stainless sheaths on nichrome wire heater that are wound over stainless pipes for heating. The heater use nichrome wire as the heating elements, magnesia powder as an insulating material and the AISI type 304L stainless steel as sheathing material. The heater was in coils in varying lengths from 5 to 30 m and the sheath thickness varied from 0.2 to 0.3 mm. The heater was specified to be annealed, pickled and passivated at the final stage of manufacturing.

Testing and results SEM Fractograph of the cracking is shown in Fig. 29. The nature of cracking was typical of TGSCC with fan shaped patterns. The polished and etched sample taken near the failed location revealed step structure indicating that the material was not sensitized. The crack revealed under deposit attack, which initiated from the corroded surface and penetrated across the wall thickness as shown in Fig. 30.

EDAX analysis on the corroded brown spots on the heater surfaces was found to composed of iron and chlorine. This indicated that the brown spots were formed after the iron particles embedded on the surface during manufacture rusted.

Microhardness measurements were found to be in the range of 190–232 HV and it was high on the outer surface. This measured hardness was higher than the reported value for solution annealed type 304L stainless steel, which is 150 HV.

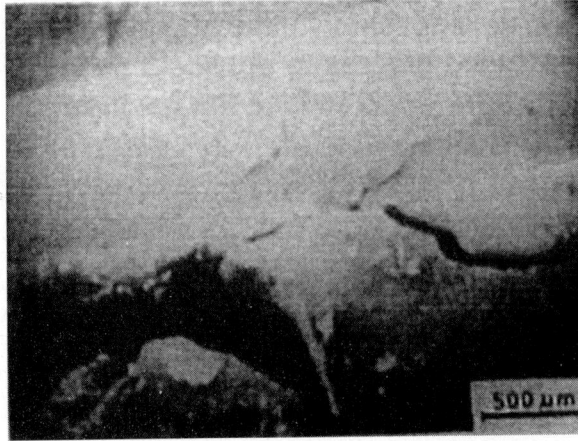

Fig. 29   SEM fractograph showing severe cracking on the
failed heater.

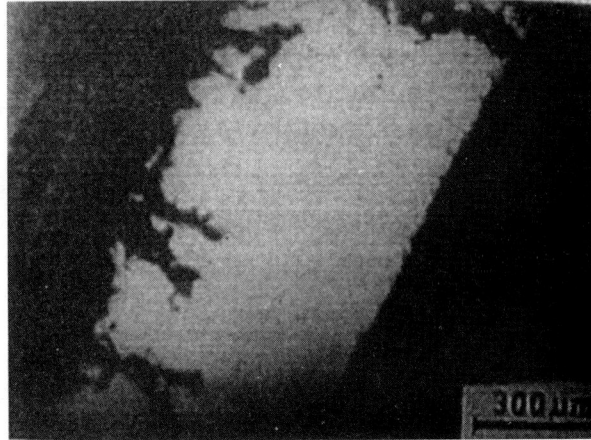

Fig. 30   Optical microstructure showing crack initiated
in a region of underdeposit attack.

Discussion Fractography established that the fracture occurred by SCC originating at regions where the cladding had rusted. Underdeposit attack had also occurred. The rusting resulted from incomplete removal of iron particles embedded on the heater surface during manufacture. The material was not sensitized and the higher hardness observed indicate that the sheathing was cold worked—whereas solution annealing was specified. Pickling and passivation were not carried out properly which is evident from the iron impurities on the outer surface. These iron particles got rusted during its exposure to humid atmosphere. Rust generally contains ferric ions and also collects moisture and chloride ions from the atmosphere. The presence of ferric ions raised the electrochemical potential of the stainless steel, creating conditions for the underdeposit attack and SCC. Stress present in the material because of improper annealing may have been augmented by stress arising from the wedging action of the corrosion products during the underdeposit attack [10].

Conclusion and recommendation The primary cause of the fracture was SCC originating at rusted regions on the surface of the heaters. The presence of ferric ions in the rust and absorption of chloride ions and moisture from a coastal atmosphere created the necessary environmental conditions for underdeposit attack and SCC. Residual stress in the material in conjunction with stresses resulting from the wedging action of corrosion products facilitated SCC.

It is recommended that proper pickling and passivation of the heaters be conducted as the final stage in the manufacturing process to provide a surface free from surface impurities. Annealing of the heaters would reduce the hardness of the material and relieve the residual stress.

REFERENCES

1. N.G. Muralidharan, Rakesh Kaul, K.V. Kasiviswanathan and Baldev Raj, Practical Metallography, Vol. 28, 12/91, pp. 662–668.
2. N.G. Muralidharan, Rakesh Kaul, K.V. Kasiviswanathan and Baldev Raj, Handbook of Case Histories in Failure Analysis, Edited by Khlefa A. Esaklul, ASM International, Vol. 2, 1998 pp. 259–261.
3. D.K. Bhattacharya, J.B. Gnanamoorty and Baldev Raj, Handbook of Case Histories in Failure Analysis, Edited by Khlefa A. Esaklul, ASM International, Vol. 2, 1998 pp. 135–137.
4. Rakesh Kaul, N.G. Muralidharan, N. Raghu, K.V. Kasiviswanathan and Baldev Raj, Practical Metallography, 33, (1996), pp. 315–321.
5. U.S. Atomic Energy Commission, Regulatory Guide 1.36 pp. 1.36.1–1.36.3.
6. T.V. Vinoy, H. Shaikh, H.S. Khatak, J.B. Gnanamoorthy and Baldev Raj, Practical Metallography, 34, 1997, pp. 527–534.
7. R.K. Dayal, J.B.Gnanamoorthy and G. Srinivasan, Handbook of Case Histories in Failure Analysis, ASM International, edited by Khlefa A. Esaklul, Vol. II, 1993, pp. 506–508.
8. R.K. Dayal, J.B. Gnanamoorthy, Handbook of Case Histories in Failure Analysis, ASM International, edited by Khlefa A. Esaklul, Vol. I, 1992, pp. 194–197.
9. R.K. Dayal, J.B. Gnanamoorthy, Handbook of Case Histories in Failure Analysis, ASM International, edited by Khlefa A. Esaklul, Vol. II, 1993, pp. 253–255.
10. P. Muraleedharan, H.S. Khatak and J.B. Gnanamoorthy, Handbook of Case Histories in Failure Analysis, ASM International, edited by Khlefa A. Esaklul, Vol. II, 1993, pp. 427–429.

14. Surface Modification for Corrosion Protection of Austenitic Stainless Steels

P. Shankar[1] and U. Kamachi Mudali[1]

Abstract Corrosion of engineering components can be suitably controlled by applying protecting coatings or by surface modification. Austenitic stainless steels can be applied with several types of coatings through thermochemical methods like ion nitriding, Kolsterisation, sol-gel process, thermal spraying and physical vapour deposition for corrosion protection purposes. In addition, surface modification of components made of austenitic stainless steels through laser surface melting and alloying, and ion implantation can yield corrosion resistant surfaces. This article highlights the above methods and techniques and their usefulness in providing corrosion protection for austenitic stainless steels.

Key Words Coatings, surface modification, laser, ion implantation, corrosion protection.

1. GENERAL INTRODUCTION

Selection of materials for engineering applications is mostly based on requirements of bulk and macroscopic properties like strength, ease of fabrication, availability, cost etc. However, most engineering failures due to fatigue, corrosion etc. are known to be initiated at the surface. Initiation of cracks/ defects at the surface is the life-limiting factor. This is because the surfaces of the components are exposed to severe thermal, chemical and/or stress conditions. Methods to selectively enhance the surface properties of the materials can therefore result in significant benefits. However, for service environment leading to corrosive wear and erosion, apart from the selection of the material, suitable surface modification procedures that can withstand various types of stresses are required.

Mechanical, microstructural and/or microchemical alterations of the surface, without affecting the bulk, to tailor the material properties at the surface for specific end applications is called surface engineering. This involves the application of traditional and innovative surface treatments to engineering components and materials in order to produce a composite surface with properties matching with the substrate or the surface layer. Surface modifications are frequently employed to counter problems like friction, wear, erosion, corrosion, high temperature oxidation and fatigue to ensure optimum component service. Surface engineering may therefore be considered as designing of a composite system (coating plus substrate) having properties, not otherwise achievable by either one alone [1, 2]. For example, in

[1]Scientific Offficer, Materials Characterisation Group, Indira Gandhi Center for Atomic Research, Kalpakkam-603 102, India.

the modern aero engines, base alloys have been designed primarily for high temperature strength. Surface coatings have been applied to impart additional corrosion resistance to them.

Before deciding a specific surface modification treatment, alternative options like (i) changing the entire component with a better material, (ii) changing the design of the component, (iii) changing the operating conditions etc., have to be considered. A detailed cost/benefit analysis is required to be done based on the available options and the risks involved. This should also include considerations of the entire consequential costs of failure like (i) cost of material replacement, (ii) labour cost of replacement (iii) production losses, (iv) cost of consequential damage to other equipments, (v) damage to human life in case of critical components etc. The choice of surface treatment should be made only if it is cost-effective and technically, the most effective and viable option.

The exact surface treatment, chemistry and surface property desired would, however, depend on the end application. The range of surface treatments available to modify surface properties are numerous and listed in Fig. 1. The selection criteria for choosing one technique over another should include the following consideration: (i) the surface treatment should not impair the bulk properties, (ii) deposition process must be capable of coating the components with complexity in size and shape. Thicknesses of the surface layers play a major role as they have to withstand the contact stresses throughout the operating lifetime of the component. Figure 2 shows the variations in the thickness of surface layers produced by various surface modification methods.

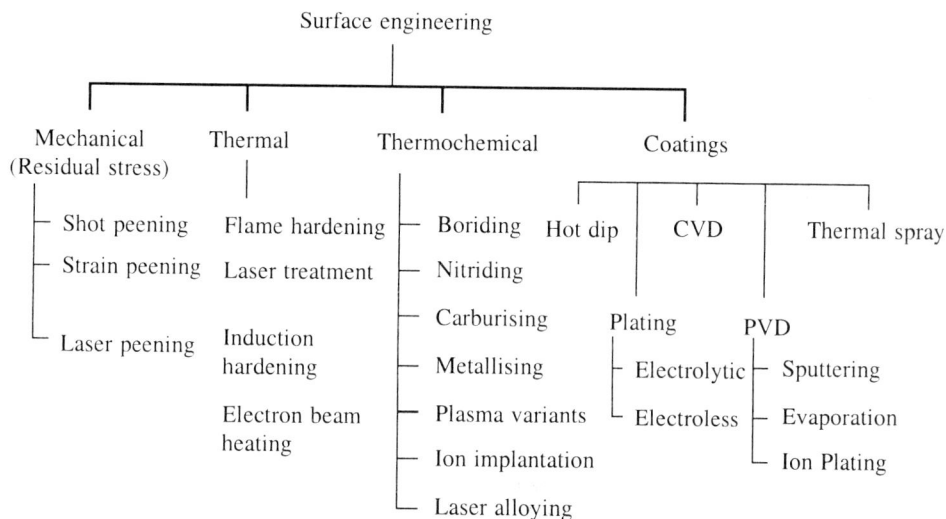

Fig. 1. Various treatments available for surface modification of austenitic stainless steels.

Generally austenitic stainless steels find wide applications in engineering industries, by virtue of their good mechanical and corrosion resistance properties. It may be difficult to apprehend as to why austenitic stainless steels need surface modification. This is because, austenitic stainless steels suffer from the following considerations: (i) poor wear resistance and risk of galling, (ii) susceptibility to sensitisation, (iii) inferior cavitation erosion resistance, (iv) sensitivity towards localised corrosion attack etc. Cavitation erosion is a form of erosive wear occurring on the metal surface, which encounters high fluid flow velocity. Typical examples of cavitation erosion can be found on high-speed impellers, pump casings, ultrasonic mixers in the food and pharmaceutical industries, etc. Austenitic stainless

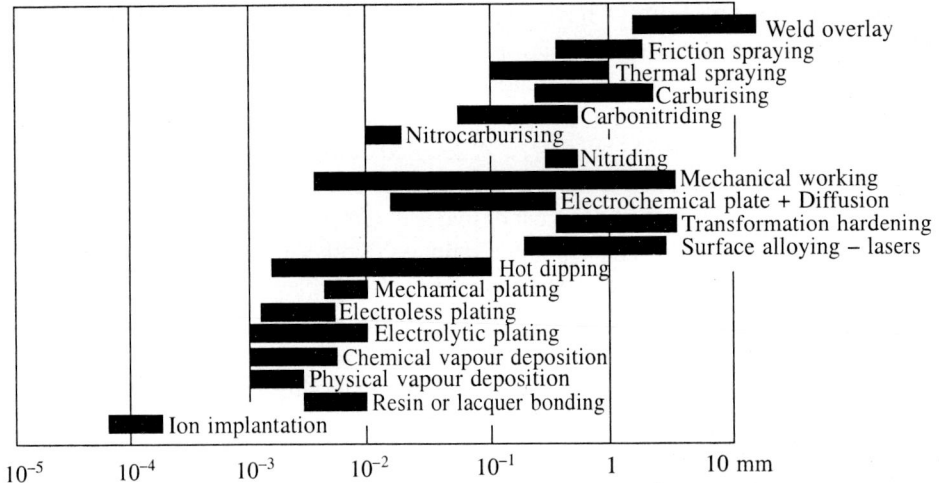

Fig. 2. Thickness changes during various surface modification procedures [2].

steels are widely used as pumping components for marine and urban water supply systems, because of their good pitting corrosion resistance in marine condition. However, type 316L stainless steel (SS) suffers from poor cavitation erosion resistance. Hence, in applications requiring improved wear resistance, cavitation erosion resistance, resistance to sensitisation etc., suitable surface modification techniques are necessary for improved service performance. One of the most commonly used techniques for improved wear resistance is thermochemical treatment like carburising, nitriding and boriding. More recent developments for modifying austenitic stainless steel surfaces using lasers, ion implantation and coatings are also discussed in this chapter.

2.   THERMOCHEMICAL METHODS

Thermochemical methods involve a change in both the surface chemistry as well as the microstructure by thermal treatments. The most common thermo-chemical surface hardening methods involve diffusion of interstitials like carbon, nitrogen or boron on the surface. Hard case results from formation of carbides, nitrides or borides respectively. Many methods including solid-state pack diffusion, molten bath, gaseous diffusion and electrolytic routes have been established for carburising, nitriding or boriding austenitic stainless steels. Sufficient literature is available on these techniques [2] and is not elaborated in this section.

2.1   Ion Nitriding
Glow-discharge plasma is increasingly utilised to introduce interstitial elements into surface of metals. A typical ion nitriding set up is shown in Fig. 3. Because of the large number of process parameters that can be controlled, ion nitriding is more versatile than gas or salt bath nitriding methods. Power density and absolute pressure, control the rate of sputtering. Sputtering can be used to limit the compound layer thickness or to clean the surface before nitriding. The energetic positive ions can effectively remove the oxide film from the surface of austenitic stainless steels and thereby accelerate nitrogen mass transfer [3]. Glow discharge nitrogen plasma contains a high concentration of nitrogen ions and thus a high concentration of nascent nitrogen exists at the surface of the workpiece. As a

result, the ion nitrided surface has a compound layer (white layer) that is a single phase, unlike the compound layer formed in other nitriding methods. Below this layer are the precipitated alloy nitrides of the diffused case. The important advantages of ion nitriding in comparison to gas and salt bath nitriding are well known.

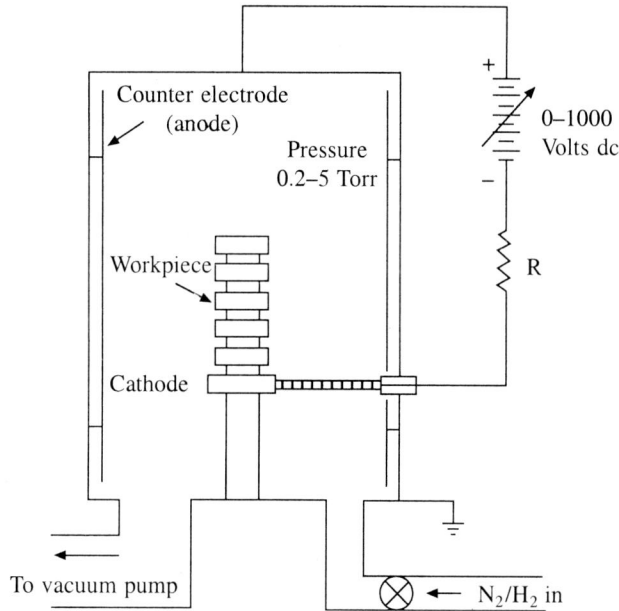

Fig. 3   Ion nitriding set up for surface modification purposes.

Ion nitriding uses simple mixtures of nitrogen and hydrogen, while in gas nitriding, ammonia or nitrogen diluted ammonia is used. In salt bath nitriding also, the disposal of the spent salts has become a considerable environmental concern. Broader temperature ranges can be used in ion nitriding that cannot be obained with other methods. Nitriding is possible even at relatively lower temperatures. High surface concentration of nitrogen ions also results in enhanced kinetics. Cycle times as much as 50% shorter have been documented. Plasma processes provide an easier method for preventing or masking certain regions of the work piece from getting nitrided. Since the plasma covers the entire cathode (wrap around effect), the surface reaction and hence the composition alteration will occur uniformly all over the surface, resulting in uniform case hardening at all points.

By gaseous nitriding it is often difficult to achieve a good case depth over a deep depression on the surface or inside a blind hole. However, by ion nitriding, the plasma follows the contour of the hole, and hence case hardening will occur wherever the plasma sheath contacts the surface. However, there is a lower limit, to the size of blind hole that can be treated. In general holes having a diameter of less than twice the thickness of the plasma sheath, cannot be penetrated because the plasma may be shorted out as it enter the hole.

Sun and Bell [4] have reported an increase in wear resistance of plasma nitrided type 316 SS by more than two orders of magnitude when sliding against bearing steel. Also the wear behavior of plasma nitrided type 316 stainless steel has been found to depend sensitively on the nitriding temperature, counterface material and testing conditions (Fig. 4). However, a major limitation associated with nitriding of austenitic stainless steels is that the precipitation of chromium nitrides in the case results in the depletion of chromium from the adjacent austenitic matrix, thus leading to a significant reduction in the intergranular corrosion resistance of the nitrided layer [5]. Hence, an increase in surface hardness and wear resistance of austenitic stainless steels by nitriding is usually accompanied by a loss in corrosion resistance.

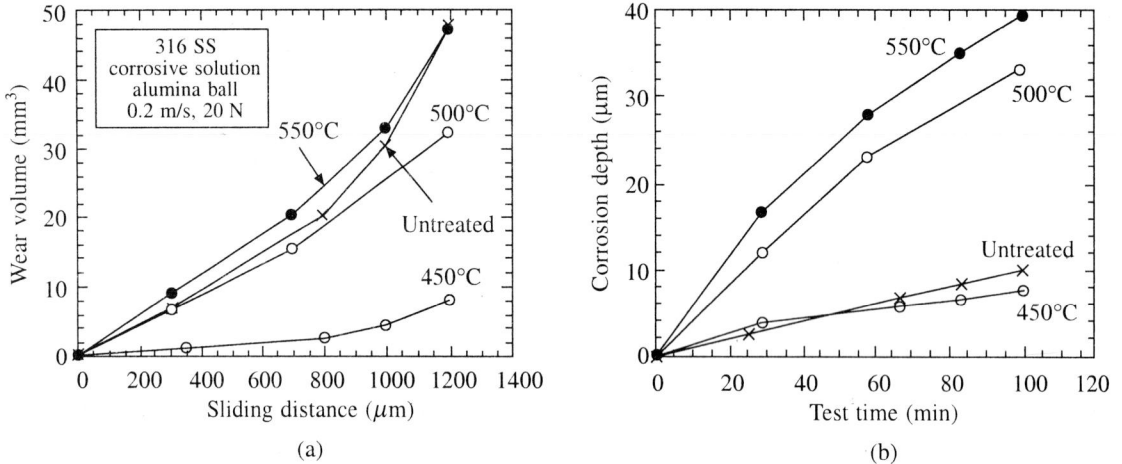

Fig. 4   **Corrosion wear volume in the wear track as a function of sliding distance (a) and corrosion depth outside the wear track as a function of testing time (b) for various 316 SS specimens plasma nitrided at different temperatures [4].**

To improve the corrosion resistance of nitrided austenitic stainless steels, a low temperature (below 723 K) plasma nitriding technique has been suggested [6]. X-ray diffraction analysis has shown the presence of a supersaturated (up to 10 wt. % nitrogen) austenite matrix at the surface of a low temperature plasma nitrided austenitic stainless steel [7]. This phase is often designated as "S" phase and is highly stressed and distorted. Nitriding at 723 K produced a thicker layer, which is predominantly chromium nitride phase. The "S" phase has been found to possess excellent wear resistance as well as corrosive wear resistance as can be seen in Fig. 4. Hence combined corrosion and wear resistance of type 316 stainless steels has been achieved by plasma nitriding at 723 K [4]. It can also be seen from Fig. 4 that when sliding against an alumina slider in the tested corrosive solution, the high temperature (723 to 823 K) nitrided layer does not provide any improvement in corrosive wear resistance, due to its deteriorated corrosion resistance. In contrast, plasma nitriding at low temperatures (723 K) can maintain or even improve the corrosion resistance and hence the corrosion wear behaviour of type 316 SS.

Recentaly, a new technique of electron cyclotron resonance microwave plasma source ion nitriding has been introduced for the surface modification of austenitic stainless steel for combined wear and corrosion resistance enhancement [8]. This nitriding effect is similar to that by using plasma immersion ion implantation. The new method with special features of low temperature and low working pressure provides great experimental flexibility, low unit cost and technologically simple apparatus design, thus leading to industrial application.

2.2   Kolsterization

In order to improve the corrosion resistance of carburised types 304 and 316 austenitic stainless steels, a Kolsterizing process has been developed [9]. This process results in a supersaturated austenite matrix containing carbon at the surface without any precipitation of chromium carbides. Hence the Kolsterized components do not suffer from loss of corrosion resistance. The case depths of the treated austenitic stainless steels are about 22-33 μm. The Kolsterizing process induces surface compressive stresses and thereby enhances the fatigue strength, stress corrosion cracking resistance and pitting resistance of the steel. Further, Kolsterization does not change the dimensions, shape or color of the products. The process can therefore be applied to completely machined and finished components without the need for any further treatment. Even fits with tolerances of a few microns remain unaltered and polished surfaces retain practically the same finish. The process offers special benefits to the food, chemical and pharmaceutical industries. Other examples of Kolsterized parts are rotors and casings of positive displacement pumps, gears and casings of gear pumps, pistons and casing of piston pumps and the lobes and casing of lobe pumps. It has been demonstrated that the life of austenitic stainless steels in moving and fast rotating machine components, depending on the loading, can be increased by a factor of more than 100 by Kolsterization.

3.   ION IMPLANTATION

Ion implantation is a technique of changing the surface chemistry by bombardment with elemental ions of sufficient energy (50 to 200 KeV), such that they can penetrate the surface and get embedded. The penetration depth is usually in the range of 0.01 to 1 μm. Since the atoms are injected into the surface mechanically, there is no need for application of high temperature to produce a thermal diffusion of the incident atoms into the surface. Implanted atom concentration and depth profiles are determined by ballistic collision physics. An ion implantation system consists of an ion source, an extractor to remove the ions from the source, a device to separate the ions to recover only those desired for implantation, an accelerator to impart necessary energy for implantation into lattice, a target chamber in which component to be implantated can be adequately scanned and maneuvered to achieve proper surface coverage. The entire set-up is maintained at high vacuum (10^{-6} torr) to maintain beam purity and to prevent beam defocusing. Atoms to be implanted are fed into the ion source assembly where they are ionised in an electrical discharge. If the alloying element is in a gaseous state, like nitrogen, the gas is introduced directly into the ion source and in the presence of electrons accelerated from a hot filament, the nitrogen atoms are ionised and plasma is formed. The positive ions are extracted from the ion source through a narrow slit and accelerated in an electric field before being directed to the surface of the workpiece.

The advantage of ion implantation is that specimen is usually at room temperature and bulk properties are not changed. Also the dimensional changes are negligible and have little effect on surface finish. Several metastable structures not possible by other methods can be produced by ion implantation. Also, higher flexibility in tailoring surface alloy characteristics can be provided by ion beam mixing or ion beam enhanced deposition. Ion beam mixing is a two step process that proceeds by forming a thin film of material on the surface of the component to be treated and subsequently mixing that film into the surface of the component with a high energy beam. Films should be less than 100 nm in thickness, due to difficulty in mixing thicker films as a result of sputtering effects. Beam mixing of precious metals such as palladium and platinum have been employed to improve aqueous corrosion resistance of steels. Ion beam enhanced deposition is a compound process involving ion-implantation simultaneously when a thin film is being deposited on the surface. It results in a film that is initially mixed into the surface of the component and then modified, as it is further grown on the surface. By this technique adherent coatings of aluminium on type 316L stainless steel have been prepared and found to have substantially improved fatigue life due to a delay in crack initiation [10].

Ion implantation has been successfully applied for manufacturing small and complex form of precision tools and orthopedic or aerospace components. Low energy ion implantation has been shown to improve wear, corrosion resistance and fatigue life of austenitic stainless steels [11]. Dual implantation of boron and nitrogen in austenitic stainless steels has been reported to result in increased surface hardness and fatigue life time up to 250% [12]. Nitrogen ion implantation in type 316 austenitic stainless steels [11, 13] has been reported to result in improved wear and corrosion resistance. Nitrogen enrichment in the metal film surface and interface is known to play an inhibiting role on the kinetics of anodic dissolution. Ion implantation has been shown to be an effective technique to introduce a large amount of nitrogen into the materials in a controlled fashion to precisely modify the surface properties. Mudali et al. [14] reported an improvement in pitting resistance of type 304 stainless steels on nitrogen ion implantation. In comparison with the solution annealed specimens, ion implanted (1×10^{17} ions/cm^2) specimens showed lower critical current density and passivation current density and higher pitting potential as shown in Fig. 5. This implies that nitrogen ion implantation improves the pitting corrosion resistance which could be attributed to

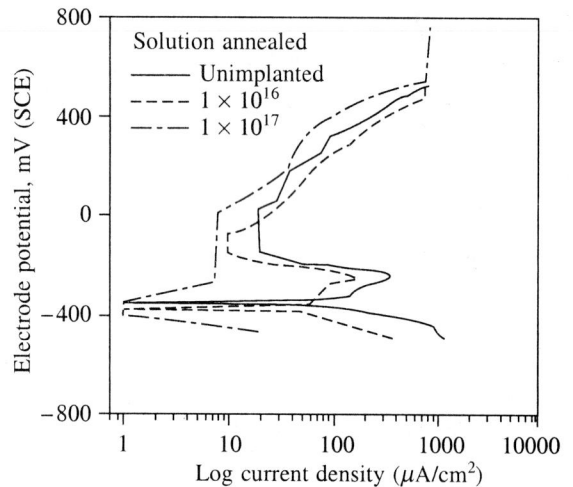

Fig. 5.   Potentiodynamic anodic polarisation curves for nitrogen ion implanted type 304 SS [14].

the beneficial effects of nitrogen. However, contradictory reports have been published on the role of nitrogen ion implantation on the pitting corrosion resistance. Sabot et al. [15] reported that nitrogen ion implantation at 2.5×10^{16} ions/cm^2 significantly lowered the pitting resistance of type 304 SS in chloride medium and they have attributed this to the implantation induced damage generated at the surface and various precipitates formed during ion implantation. Nair et al. [16] reported that implantation of 304 stainless steel with nitrogen ions at 1×10^{17} and 5×10^{17} ions/cm^2 led to deterioration in localized corrosion resistance in chloride medium and this was significant with increase in the dose of nitrogen. The deterioration in the pitting resistance of type 304 SS on nitrogen ion implantation has been attributed to (i) the formation of ferrite at grain boundaries due to the migration of Cr during ion implantation [16] and (ii) the formation of various precipitates and the damages caused due to implantation [15]. In the case of type 316 SS, the presence of molybdenum significantly improved the pitting resistance [17]. The combined effect of molybdenum and nitrogen has been reported to be very significant in improving the pitting corrosion resistance [15]. During implantation it has been reported that an oxide layer, Ni and Cr enriched zone followed by nitrogen enriched zone was present across the implanted zone. During the active dissolution the oxide layer, Cr and Ni enriched layers could have dissolved more in the acidic chloride medium with pH < 1 compared to neutral chloride medium, which spontaneously passivated the surface. Thus during passivation the presence of nitrogen enriched zone promoted a stable passive film.

Plasma immersion ion implantation (PIII) has emerged as a preferred method to implant nitrogen in niche applications. This technique circumvents the line of sight and retained dose limitation inherent to conventional beam-line ion implantation and is thus particularly suitable for large components possessing non-planar and complex geometries. PIII of nitrogen in type 304 SS has been found to show improved pitting resistance in NaCl solution [13]. Nitrogen ion implantation was found to result in a large shift of the equilibrium potential from 110 mV to + 150 mV and is expected to go still higher with increase in implantation depth. Also, a decrease in corrosion current and a broad anodic zone with low dissolution current was observed in implanted specimens as shown in Fig. 6. The improvement originated from the chemical state change induced by ion implantation on the surface. Elemental depth profile analysis using Auger Electron Spectroscopy showed a remarkable redistribution of the chemical constituents (Fig. 7). The concentration of chromium was found to increase at the surface to about 1.8 times that of the bulk. The oxygen profile also showed a peak at the surface corresponding to the Cr-enriched layer (5 nm). Compared to the untreated sample, oxygen was reported to have penetrated to a depth more than 15 nm indicating that low-temperature nitrogen PIII process increased the thickness of the oxide layer. Oxygen in the modified layer stems mainly from the original oxide film on the sample surface recoiled into the substrate by energetic ion bombardment. The increased thickness of the chromium oxide layer and enrichment of the top surface with nitrogen played a synergistic role in enhancing the corrosion resistance of austenitic stainless steel surface by nitrogen PIII process.

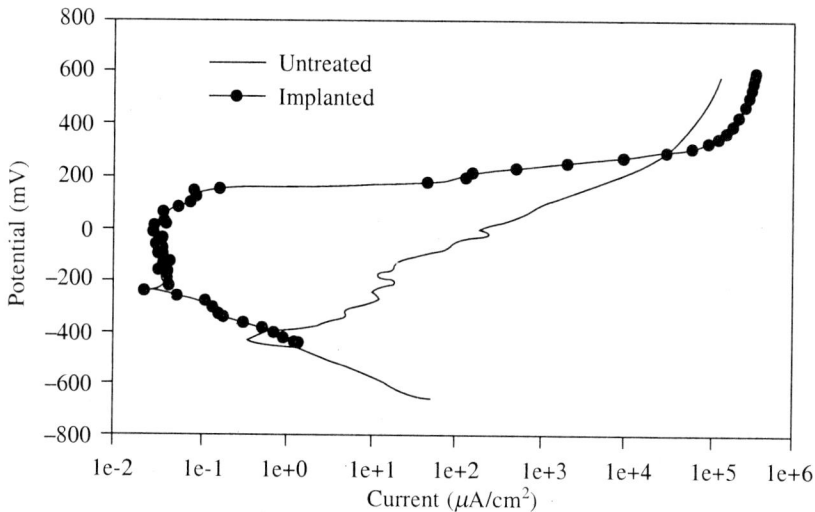

Fig. 6. **Polarisation curves of type 304 SS specimen with and without the PIII treatment in 3% NaCl solution [13].**

4. LASER SURFACE MODIFICATION

The advantage of using lasers for surface treatments is that the entire energy is absorbed within the first atomic layers of opaque materials, such as metals. Further, it can be focused precisely to the specific surface needed only. This makes lasers an ideal tool for surface engineering. It is chemically clean, remote and non-contact process with feasibility of automation. Also, the thermal profile and hence the distortion and shape and location of the heat affected region can be controlled very effectively. Figure 8 shows the various surface modification processes that can be obtained by controlling the

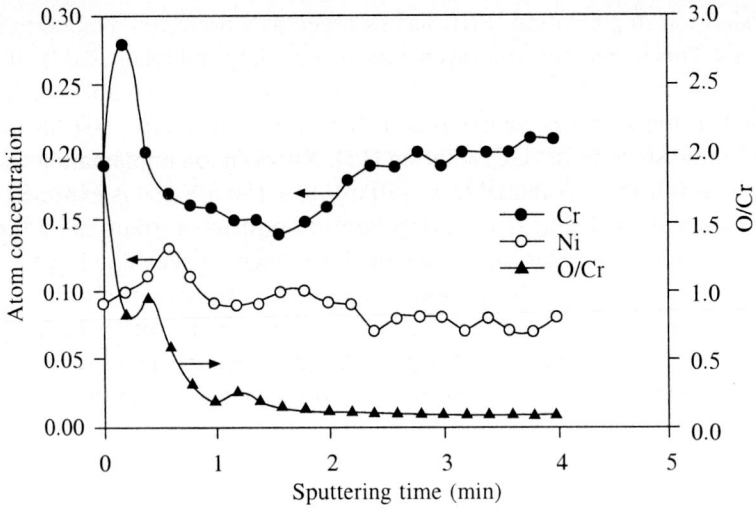

Fig. 7.   AES depth profile of various elements in the PIII treated specimens [13].

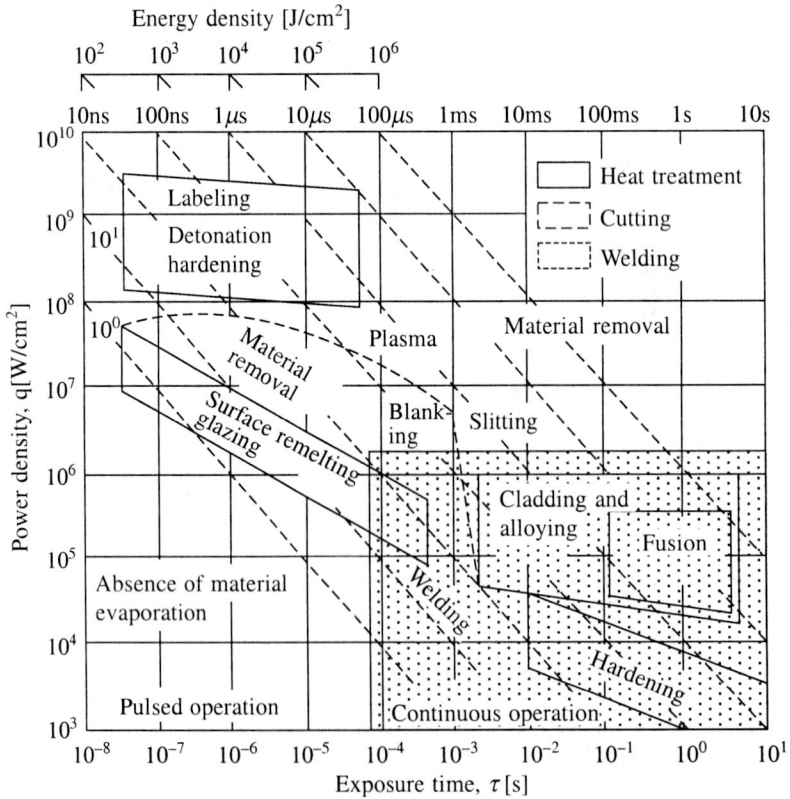

Fig. 8.   Various laser surface treatments using exposure time and power density of laser [2].

interaction time and energy density of the laser. Laser surface melting and laser surface alloying techniques have been employed for surface modification of austenitic stainless steels.

In laser surface melting, the surface to be melted is shrouded in an inert gas atmosphere. The main characteristics of this process are: (i) moderate to rapid solidification rates producing fine and near homogenous structure (ii) little thermal penetration, resulting in less distortion (iii) surface finishes of around 25 μm can easily be obtained. Laser surface alloying is similar to surface melting except that a second material is injected into the molten pool. The alloying elements may be introduced into the melt by a number of methods including powder feed, wire feed, reactive gas shroud and pre-placed powder coating. Laser surface alloying is capable of producing a wide range of surface alloys and in many cases metastable alloys, not possible by other conventional methods. Laser cladding is also similar to laser surface alloying in that if cladding is performed with thinner material for the specific laser power then surface alloying would result. The aim of most cladding operations is to overlay one metal over another to form a sound interfacial bond or weld without diluting the cladding metal with substrate material.

4.1   Laser Surface Melting

When austenitic stainless steels are either heat treated or slowly cooled through the temperature regime of 723–1073 K during welding, heat treatment or fabrication, chromium-rich carbides are formed along the grain boundaries. This leads to a significant depletion of chromium adjacent to these carbides and thus selective IGC or IGSCC is likely to initiate at these locations during service. The formation of such sensitized microstructure should be eliminated in order to improve IGC or IGSCC resistance. Also, cold work which is commonly used for the fabrication of components, can result in the deterioration of the IGC resistance. Laser surface melting has been found to be a useful technique to obtain improved corrosion resistance of the components without affecting the bulk properties.

Laser surface melting has been proved to be an effective technique in improving the pitting corrosion resistance of austenitic stainless steels [18–21]. Laser surface melting of type 316 SS with a pulsed ruby laser at 6 J/pulse energy for two successive passes, was found to result in improved pitting corrosion resistance in acidic chloride medium [19]. The E_{pp} value was 725 mV (SCE) for the laser melted specimen compared to 600 mV (SCE) for the as-received specimen. The critical current density values for the laser melted specimen and as-received specimen have been reported to be 0.70 A/m^2 and 1.95 A/m^2 respectively (Fig. 9). It was reported that [20] the laser melting of stainless steels improved the pitting corrosion resistance by eliminating probable pitting sites like grain boundary precipitates, second phase precipitates, inclusions and segregated interfaces, thus providing a homogeneous surface. The dissolution of such probable pitting sites during melting and subsequent uniform distribution of alloying elements can facilitate the development of a stable passive film with improved pitting corrosion resistance. Also the distribution of carbides and other precipitates in smaller sizes has been reported to improve the pitting corrosion resistance [21]. Apart from improving the pitting corrosion resistance, laser surface melting is also beneficial in enhancing the erosion corrosion resistance of type 316L SS. As mentioned earlier, type 316L stainless steels although have sufficient pitting corrosion resistance in marine conditions, suffer from poor erosion corrosion resistance. Laser surface melting of 316L stainless steel with a 2 kW CW Nd-YAG laser that produced a rapidly quenched surface layer, resulted in a surface with improved cavitation erosion resistance by about 22% at 300 K [22]. The pitting potential of the melted surface also increased from 359 to 452 mV at 300 K in 3.5% NaCl solution.

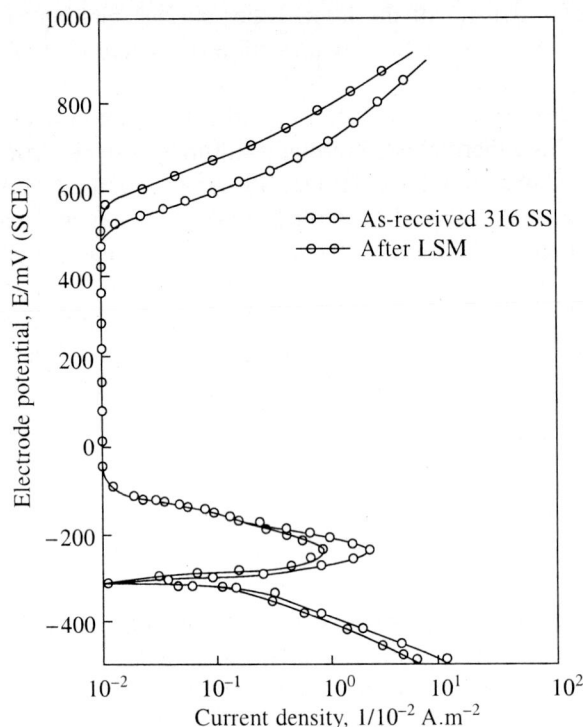

Fig. 9.   Potentiodynamic polarisation curves of type 316 SS (before and after laser melting) [19].

Laser surface melting has also been shown to be an effective technique to improve the IGC resistance of sensitized 304 and 316 stainless steels [23–25]. Laser surface melting of sensitized 304 and 316 type stainless steels using a 300 W Nd-YAG laser [23], produced a dendritic-cellular structure with a heat affected zone free from sensitization. The depth of the desensitized surface including both, the melted as well as the heat affected zone, varied from 13 to 185 μm for type 304 specimens and from 19 to 210 μm for type 316 specimens. The reactivation peak current density (15.8 mA/cm^2) and the charge values (1.03 C/cm^2) from the Electrochemical Potentiokinetic Reactivation (EPR) tests were quite high for the as-sensitized samples [26]. The EPR test results (Fig. 10) showed smaller reactivation peak for the laser surface melted sample compared to the as-sensitized sample. This indicated the formation of a microstructure without any significant chromium depletion after laser surface melting. This also confirmed the total redistribution of the carbides, with no detectable chromium depleted regions. Rapid melting and subsequent quenching of the laser melted region has been reported to dissolve [17, 23, 26] and redistribute the second phase precipitates leading to the elimination of the sensitized microstructure. This modified microstructure at the surface would not initiate any IGC or IGSCC, although the bulk still remained sensitized. The results of the microhardness and impact tests indicated that there was marginal increase in the strength and toughness of the sensitized austenitic steel specimens after laser melting [23]. The in-situ laser treatment of components sensitized during service would help in extending the life of the components by preventing failures due to IGC or IGSCC. Similarly [27], laser surface melting of sensitized nitrogen-bearing type 316 L austenitic stainless steel has also been shown to result in improved pitting resistance.

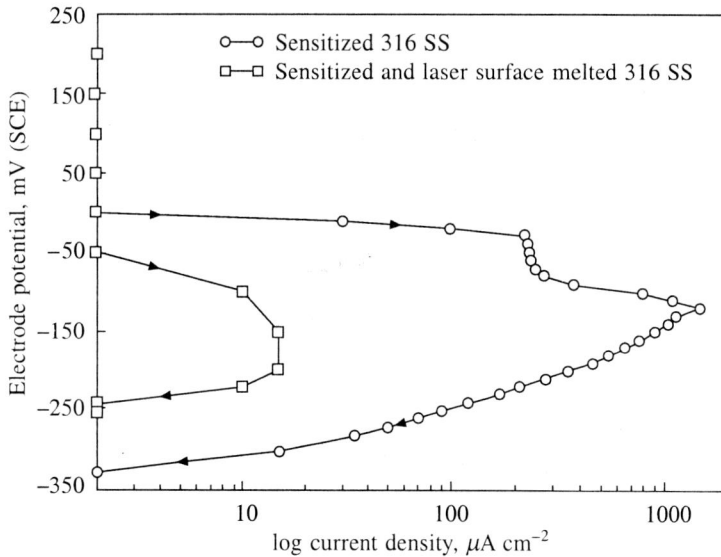

Fig. 10 EPR curves for sensitised and laser melted type 316 SS specimens [26].

4.2 Laser Peening

Laser peening (LP) has emerged as a novel industrial treatment to improve the cracking resistance of turbine blades [28] or the stress corrosion cracking of austenitic stainless steels in power plants [29]. In contrast to laser melting LP does not involve any melting but introduces residual stress at the surface without modifying the microstructure and therefore results in a purely mechanical modification. Recently, it has been shown by Peyre et al. [30] that LP can improve the pitting corrosion behavior of a type 316 L SS in a 0.5 M NaCl solution. LP resulted in increase in rest potentials, reduction in passive current densities and anodic shifts of the pitting potentials evidenced by a stochastic approach of pitting. It is well known that residual compressive stresses on the surface of components can reduce the SCC susceptibility. When material is exposed to intense laser pulse, the material surface absorbs the laser energy and a plasma forms by ablative interaction of the laser pulse with the material through a multi-photon process. The effect of water confinement is to intensify plasma pressure significantly by about ten to hundred times. Consequently, the plasma pressure reaches 1-10 GPa. The impulse of such high-pressure plasma has a similar effect on the material as shot peening, i.e., the impulse generates a plastic wave at the material surface, which loses energy as it propagates to create dynamic strain in the material, as shown in Fig. 11. Schematic of the apparatus for laser material processing in water is shown in Fig. 12. Mudali et al. [17] found that reactive quenching of the cold worked type 316L SS led to the formation of artificial passive films over the surface. LP of type 304 stainless steels was found to result in a residual compressive stress exceeding 400 MPa over 200 μm in depth by the laser irradiation in water, as shown in Fig. 13 [31]. This method has been demonstrated to be successful in enhancing the SCC resistance of type 304 austenitic stainless steel [32].

4.3 Laser Surface Alloying

Laser surface alloying (LSA) is another common and widely used method to modify the surfaces of austenitic stainless steels. LSA of type 316 stainless steel with various elements like Co, Ni, Mn, C, Cr, Mo, Si and alloys/compounds like AlSiFe, Si_3N_4 and NiCrSiB has been attempted to enhance

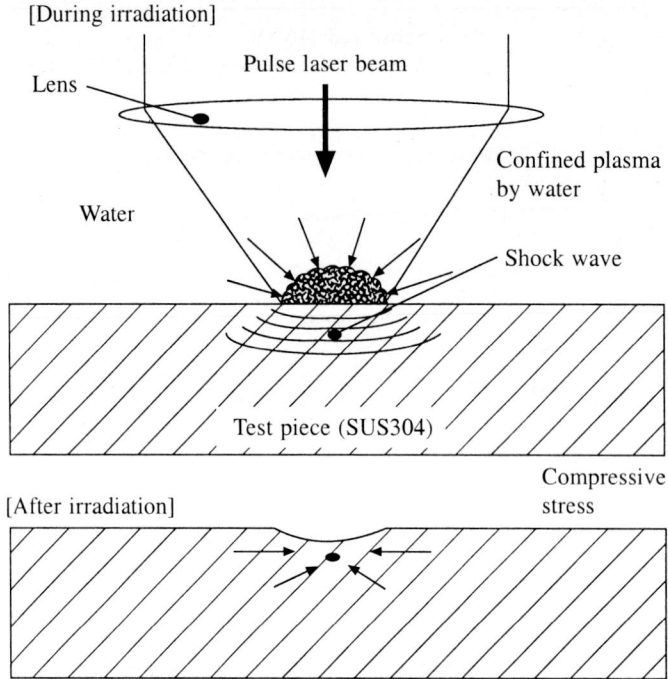

Fig. 11. Principle of stress development by underwater laser irradiation [31].

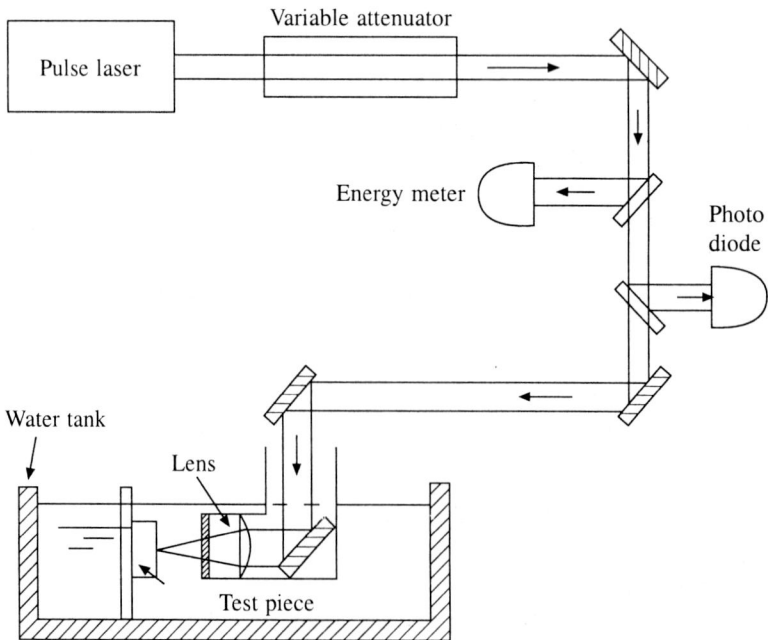

Fig. 12. Experimental setup for underwater laser materials processing [31].

its cavitation erosion resistance [33]. The specimens alloyed with Co, Ni, Mn, C or NiCrSiB were found to contain austenite as the main phase, with carbides and other precipitates as minor phases. For specimens alloyed with Cr or Mo, the major phase was ferrite. In the case of Si or Si_3N_4 the major phase was an intermetallic compound Fe_3Si. The largest improvement in corrosion resistance in 3.5% NaCl solution was achieved with Si and Si_3N_4, leading to a noble shift in the pitting potential of 170 and 211 mV, respectively and a corresponding noble shift in the protection potential of 130 and 221 mV.

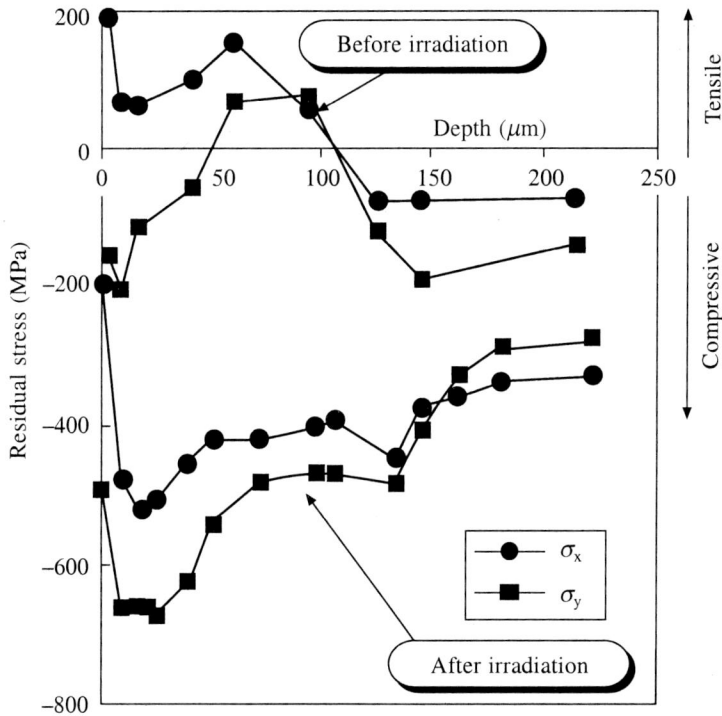

Fig. 13   Change in residual stresses developed during laser treatment under water [31].

Similarly laser surface melting of type 304 SS in a nitrogen and argon atmosphere using a continuous wave CO_2 laser was found to result in improved pitting corrosion resistance [34]. The improvement in localized corrosion behavior was attributed to the increase in nitrogen content, which was incorporated into the surface layer during laser surface melting. Similarly laser implant deposition (LID) of silicon in type 305 stainless steels by irradiating with a KrF excimer laser in a SiH_4 gas ambient has been reported [35] to have improved hardness and anti-corrosion properties. LID has been found to result in a more uniform and adhesive distribution of Si on steel surface, compared with other conventional deposition processes.

4.4   Laser Cladding

A laser cladding method which produced a highly corrosion resistant material coating layer on thinner surface of the type 304 austenitic stainless steel pipe has been reported to prevent SCC occurrence [36]. Since SCC is a serious problem with austenitic stainless steels, laser cladding can be effective in minimizing the damage due to SCC. Laser cladding method involves application of mixed metallic powder paste and subsequent irradiation with a Nd : YAG laser beam on the dried paste. A schematic of the laser cladding process, shown in Fig. 14, has already been employed at several BWR type nuclear power plants to enhance the SCC resistance of austenitic stainless steels [36].

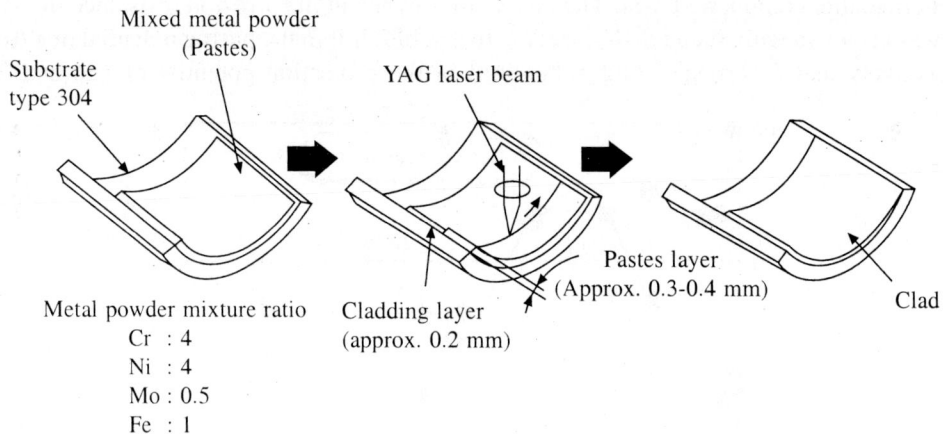

Fig. 14   Schematic of the sequence involved in laser cladding operations [36].

5.   SURFACE COATINGS FOR AUSTENITIC STAINLESS STEELS

All the aforementioned surface modification techniques including thermochemical methods, ion implantation and laser surface modification, result in chemical or structural modification of the base material only. However, for a few applications it may be necessary to develop overlay of a completely different phase like ceramics, intermetallics etc., on the base material. Coating techniques like thermal spraying, chemical vapour deposition, physical vapour deposition etc., are useful in this regard. In the following section, a few techniques are discussed.

5.1   Thermal Spray Coatings

Thermally sprayed coatings are widely used to protect the surface from wear. Particularly austenitic stainless steels are vulnerable to erosion corrosion as discussed earlier. Ceramic-base materials represent attractive possibilities for service in high-wear conditions. Cermet coatings of nickel-chromium-silicon-boron-carbon have been applied to type 316 austenitic stainless steel by high velocity oxy-fuel thermal spraying process for their improved erosion-corrosion resistance applications [37]. The ceramic coatings have high hardness and contribute to the increased wear resistance of the coated steel. The weight loss by erosion of the stainless steel was found to be twice that of the coating. However, the corrosion resistance of the coating was found to be inferior to that of the stainless steel. In contrast, the coating was found to be more resistant than that the stainless steel to erosion-corrosion, as shown in Fig. 15.

5.2   Sol-Gel Coatings

Another method of surface modification of stainless steels is by sol gel coating. Sol-gel processes have been used to apply a variety of thin ceramic coatings to metal. This method involves simple application (dip, spin or spray coating) of metalorganic precursor gels to solid substrates followed by densification at temperatures between 873 and 1273 K. Sol-gel coatings have strong adhesion and involve minimal dimensional and weight change of protected parts. Sol-gel oxide films in general have high resistance to heat, oxidation, friction and wear. Sol-gel SiO_2 coatings on stainless steels have been used as barriers against oxidation and gaseous NH_3 corrosion [38, 39]. Sol-gel coatings also find an attractive way for improving the performance of metallic prosthesis for clinical applications. It is well known that metallic implant materials such as 316 L austenitic stainless steels are widely used as load-bearing prosthesis in orthopedics and dentistry due to their excellent mechanical properties. These implants, however, suffer from limited corrosion resistance in the human body and lack of bioactivity, i.e., they are not able to bond to living tissues without cementation to external fixation devices. Galliano et al. [40] have developed SiO_2 and SiO_2-CaO-P_2O_5 sol-gel coatings on 316 L for such clinical applications.

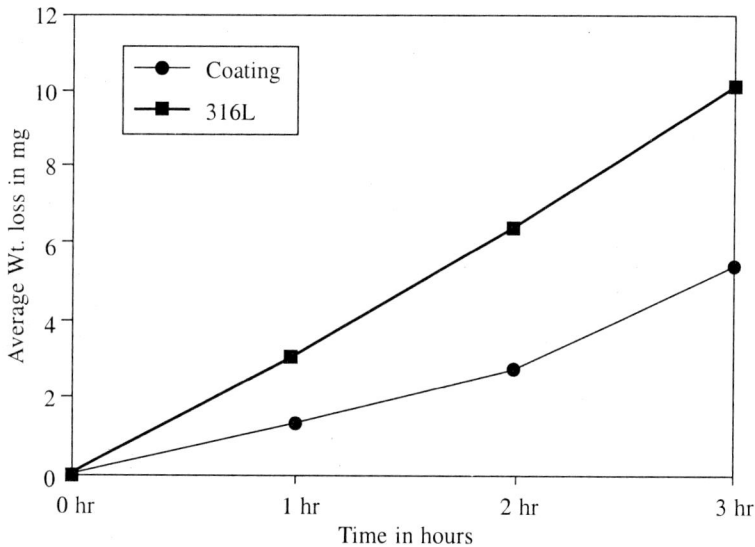

Fig. 15   Average weight loss during erosion-corrosion testing of cermet coated type 316 SS [37].

Zirconia coatings [41, 42, 43, 44] are well known for their thermal barrier and corrosion resistant properties. Zirconia also has high hardness which makes it useful for wear resistance, especially in environments involving high temperature and corrosive solution. Zirconia coatings on steels find applications against environmental attack. Several techniques have been used to produce zirconia coatings, including plasma or thermal spraying, physical vapour deposition and sol gel deposition. The advantage of using sol-gel method is that the applied coatings are uniform and homogeneous at low temperature, without affecting the bulk properties and structure. Zirconia coatings are also found to be very effective in enhancing the high temperature oxidation resistance of the steel [42]. The oxidation rate of coated steel was found to be dependent on the coating thickness as shown in Fig. 16. The weight increase of stainless steel sheets coated with zirconia film of about 130 nm was about one-half of that of the uncoated stainless steels. Electrochemical behaviour of the coated type 316L SS in deaerated 15% sulphuric acid [44] showed that the corrosion rate of the coated stainless steel was about 8.4 times lower than that of the uncoated steel. Detailed analysis of the EPR curves indicated that the film acts as a physical barrier thereby enhancing the corrosion resistance of the steel.

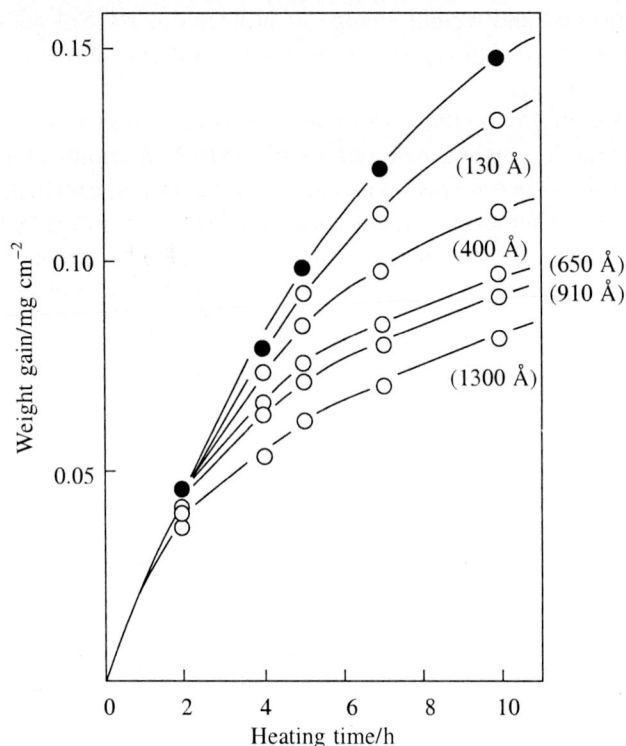

Fig. 16   Effect of ZrO$_2$ thickness on the oxidation behaviour of stainless steel specimens [42].

In another study [45], effect of ceria and ceria-titania sol-gel coatings on the localized corrosion behaviour of type 304 stainless steel in deaerated NaCl was investigated. The ceria coated specimen showed a lower E_{corr} of approximately 0.5 V, an increase in i_p and an increase in E_{pp} in comparison to the uncoated specimen. The decrease in E_{corr} suggested an inhibition of the cathodic reaction, an increase in the anodic reaction rate, or both, under open-circuit conditions. The increase in E_{pp} indicated that the ceria coating improved the pitting resistance of type 304 stainless steel. Application of titania over the ceria coating was found to result in a further increase in E_{pp} and decrease in the passive current. This suggested that the coating provided a more protective barrier for the metallic substrate. This was substantiated by the fact that compared to the uncoated and ceria coated specimens, no crevice corrosion was observed on the duplex coated steel. It was thus inferred that duplex coating of ceria-titania rendered better crevice and pitting corrosion resistance to the type 304 stainless steel.

5.3   Physical Vapour Deposition

Physical vapour deposition is a widely used technique to deposit thin films of a metal or alloy over the substrate for specific end applications. Nb-Cr alloy coatings have been deposited on type 316 L SS [46] using a dual-crucible electron beam evaporation source with simultaneous bombardment by 250 eV Ar ions to obtain a dense and column-free coating. The coated specimens were found to exhibit better corrosion resistance than the uncoated specimens. Investigations over a wide range of compositions showed that Nb-30Cr coatings resulted in the best corrosion resistance. In another study [47], both unbalanced magnetron sputter deposition and high-energy short-pulsed plasma discharge were used to produce a nanocrystalline surface on AISI 310 S stainless steel specimens. The average grain size after surface modification was estimated to be about 100 nm by atomic force microscopy. The cyclic oxidation resistance was found to have improved in the surface modified alloy. The presence of nanocrystalline phases at the surface resulted in beneficial chemical repartitioning and a more adherent oxide layer.

Methods to form aluminide composite coatings on austenitic stainless steels for enhanced erosion-corrosion resistance and for superior high temperature oxidation resistance have been developed [48-50]. Aluminide coatings were formed by diffusion annealing a nitrogen-containing type 316 stainless steel pre-deposited with a thin film of aluminium coated by resistance beam vacuum-evaporation process. Surface alloying resulted in the formation of AlN, $Al_{13}Fe_4$ and $FeAl_3$ phases. To study the effect of matrix nitrogen content on the surface properties, four different types 316 SS specimen with nitrogen contents of 0.015, 0.1, 0.2 and 0.56 wt.% were studied. Formation of the intermetallic phases at the surface resulted in a significant enhancement of the hardness in comparison to the base alloys as shown in Fig. 17. With increasing nitrogen content of the base alloy, the surface hardness of the modified steels was also found to increase. Fig. 18 ((a) and (b)) shows the concentration profiles of the elemental constituents across the diffusion case in a 0.2 and 0.56 wt.% N alloys, respectively, using SIMS. The high concentration of oxygen at the surface indicated the formation of a thin and adherent oxide layer. This oxide layer can be seen to be rich in aluminium and chromium, but low in iron and nitrogen. The sharp decrease in the oxygen concentration with depth signified that the oxides existed only on the surface and were not associated with the deposition process. In the region adjacent to the oxides, an enhancement of nitrogen concentration was observed. No significant chemical partitioning of the substitutional solute elements was seen on diffusion annealing of aluminium into steel. The chromium concentration in the diffusion case was as high as in the matrix. This clearly

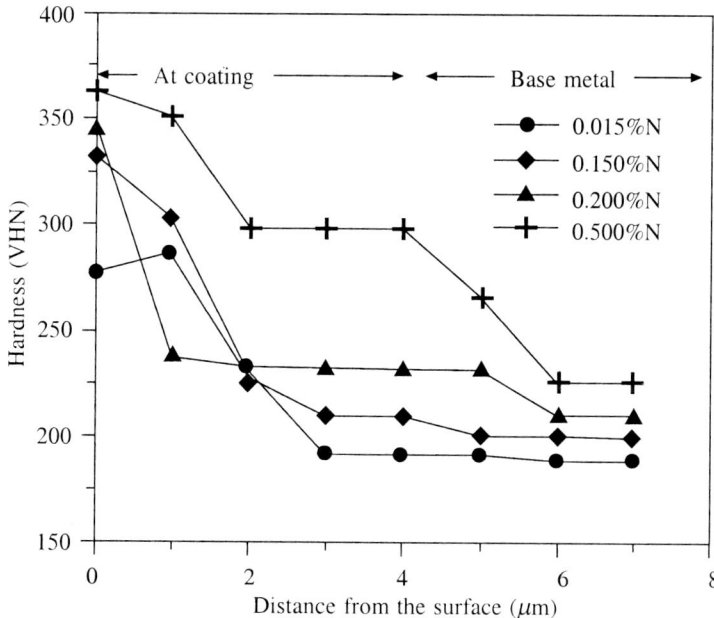

Fig. 17   Variations in the hardness of the surface after aluminide formation [49].

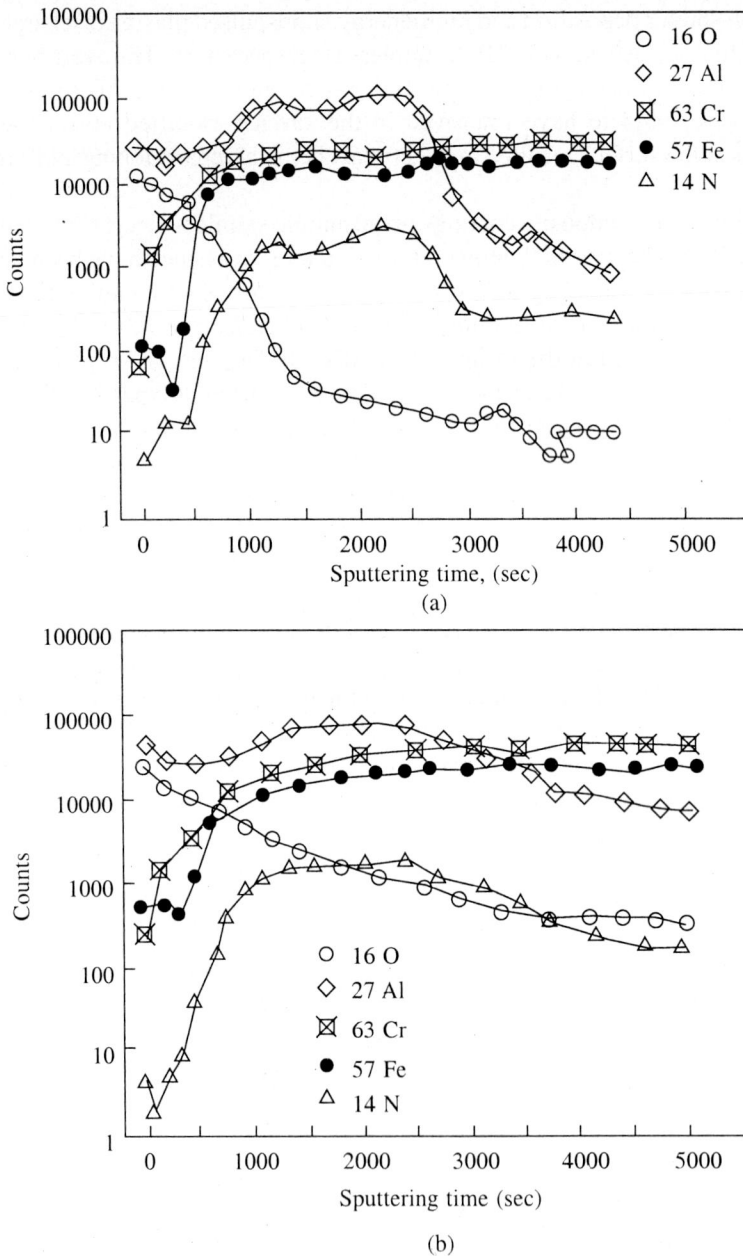

Fig. 18   SIMS depth profile of: (a) alloy with 0.56% N and (b) alloy with 0.2% N [50].

indicated the presence of a significant solid solution of chromium in the surface aluminide phases. It is well known that the presence of chromium in aluminides can significantly enhance their corrosion and oxidation resistance and hence may turn out to be useful in improving the corrosion properties of the surface modified steel. Open Circuit Potential (OCP)-time (Fig. 19a) and potentiodynamic polarization studies (Fig. 19b) in sulphuric acid medium showed that the 0.1wt.% N surface modified alloy exhibited better passive behaviour. Although the differences in corrosion potentials among the various

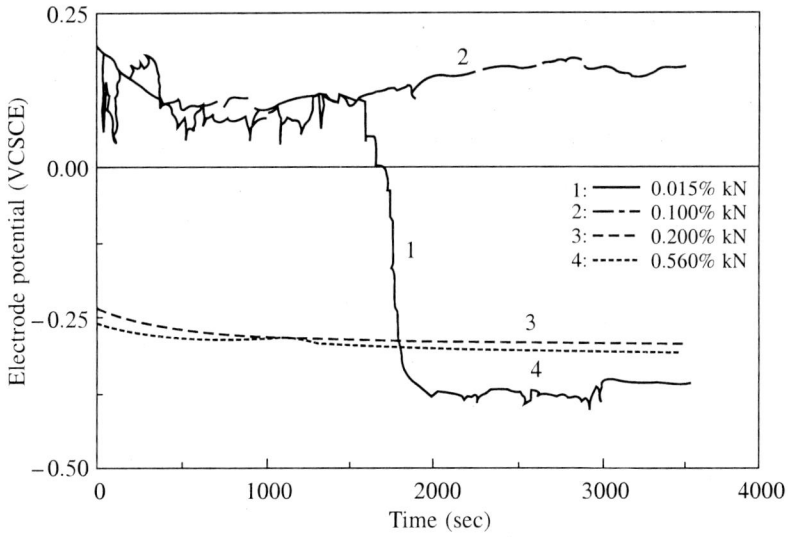

(a)

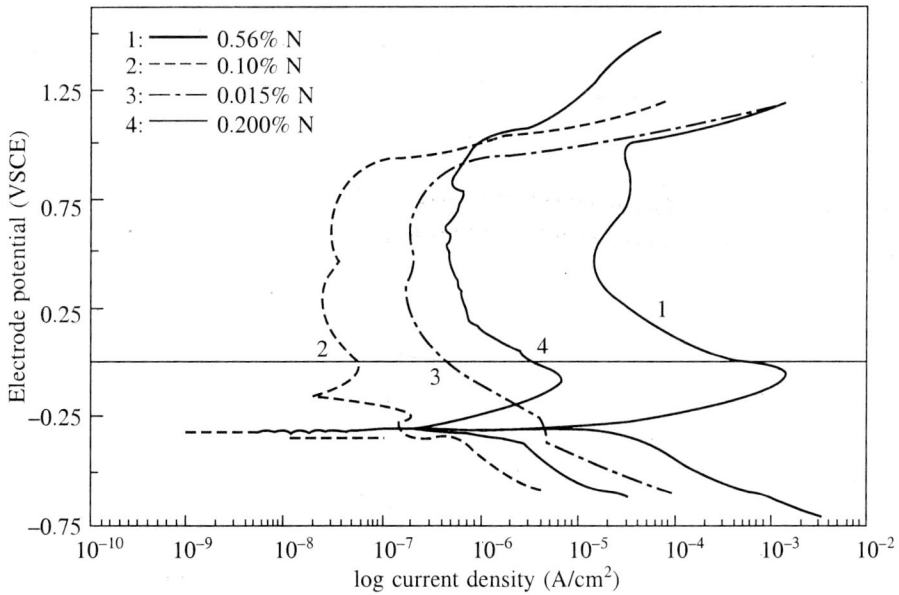

(b)

Fig. 19 **OCP-time measurements (a) and potentiodynamic polarisation curves (b) for type 316 L with aluminide coatings [50].**

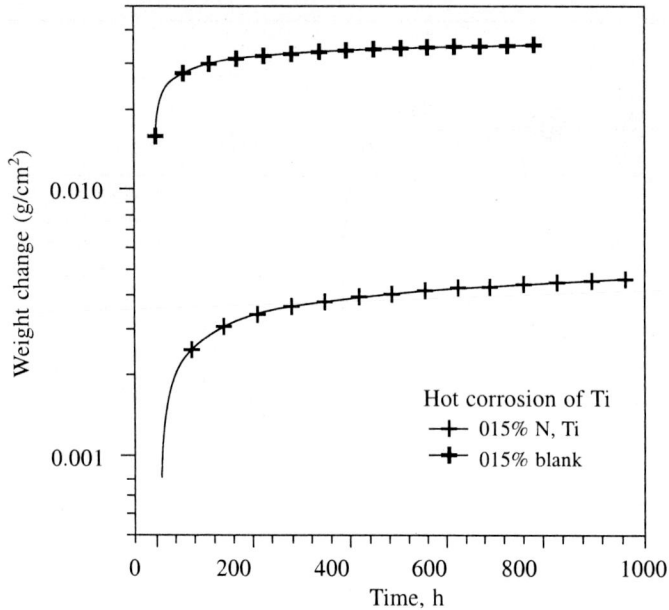

Fig. 20   Hot salt corrosion behaviour of titanium coated type 316 L SS [50].

nitrogen-containing alloys were very small, their passive current densities were found to vary significantly. The surface modified alloy with 0.1wt.% N showed the minimum passive current density. Electrochemical impedance spectroscopic (EIS) studies again showed that the polarization resistance (R_p) was higher and interfacial capacitance was lower for the surface modified alloy with 0.1wt.% nitrogen. The formation of the aluminide composite layer at the surface was also found to be beneficial in enhancing the high temperature oxidation resistance in chloride atmosphere as shown in Fig. 20.

SUMMARY

Austenitic stainless steels find wide applications in industries by virtue of their excellent mechanical properties. However, to tide over their poor resistance to wear, erosion-corrosion, high temperature oxidation and susceptibility to sensitization, specific surface modification techniques are required for certain applications. Thermochemical methods including carburising, nitriding and boriding are the most common methods of surface hardening austenitic stainless steels. Methods including low temperature plasma nitriding and Kolsterisation to avoid the loss in corrosion resistance during surface modification have also been discussed. More recent methods to modify surfaces of austenitic stainless steels include techniques like ion implantation, laser surface modification and other coating techniques.

REFERENCES

1. Advanced Surface Coating, eds. D.S. Rickerby and A. Matthews, Chapman and Hall, NY, USA.
2. Surface Engineering of Metals, eds. T. Burakowaski and T. Wierzchon, CRC Press, Florida, USA, 1998.
3. B. Edenhofer, Heat. Treat. Met., 1 (1) (1974) 23.

4. Y. Sun and T. Bell, Wear, 218 (1998) 34–42.

5. J. Flis, J. Mankowski and E. Rolinski, Surf. Eng. 5 (1989) 151.

6. M. Samandi, B. A. Shedded, T. Bell, G.A. Collins, R. Hutchings, J. Tendys, J. Vas. Sci. Technol., B 12 (1994) 935.

7. E. Menthe, K.T. Rie, J. W. Schultze and S. Simson, Surf. Coat. Tech., 74–75 (1995) 412.

8. Lei. M. Corr. Sci. and Protection Technique, 8 (1) (1996) 64.

9. Marcus J. Bos and Hardiff B. V, World Pumps (1998).

10. P. Villechaise, J. Mendez and J. Delafond, Conf. On Surface Modification Technologies IV, Paris, France, Nov 1990, Publ: The Minerals, Metals and Materials Society, 420 Commonwealth Dr., Warrendale, Pennsylvania 15086, USA, 1991, P. 335–345.

11. H. Pelletier, P. Mille, A. Cornet, J. J. Grob, J.P. Stoquert and D. Muller, Nucl. Inst. And Methods in Phys. Research B 148 (1999) 824.

12. A. Faussemagne, Surface and Coatings Technology, 83 (1996) 70.

13. Xiubo Tian and Paul K. Chu, Scr. Mater., 43 (2000) 417.

14. U. Kamachi Mudali, T. Sundararajan, K.G.M. Nair and R.K. Dayal, Mat. Sci. Forum, 318–320 (1999) 531–538.

15. R. Sabot, R. Devaux, A. M. De Becdelievre and C. Duret-Thunal, Corr. Sci., 33 (1992) 1121.

16. M.R. Nair, S. Venkatraman, D.C. Kothari and K.B. Lal, Nucl Inst. And Meth. In Phys. Res., Vol. B34, (1988) p. 53.

17. U. Kamachi Mudali, R.K. Dayal, J.B. Gnanamoorthy and P. Rodriguez, Materials Transactions JIM, 37 (1996).

18. A. Conde, R. Colaco, R. Vilar and J. de Damborenea, Materials, and Design 21 (2000) 441.

19. U. Kamachi Mudali, R. K. Dayal, J. B. Gnanamoorthy, S. M. Kanetkar and S. B. Ogale; Mater. Trans JIM, 33, No. 9 (1991) 845.

20. E. McCafferty, P.G. Moore, J.D. Ayers and G.K. Hubler; Corrosion of Metals Proceesed by Directded Energy Beams, Ed., by Clive R. Clayton and Carolyn M. Preece, The Metallurgical society of AIME, New York, USA (1982) 1.

21. T. Tsuru and R.M. Latanision, Corrosion and corrosion protection, Ed. R. P. Frankenthal and F. Mansfeld, The electrochemical society, Pennington, USA (1981) 238.

22. C.T. Kwok, F.T. Cheng and H.C. Man; Proc. ICALEO' 97: Laser Materials Processing, Vol. 83 II, San Diego, USA, Nov' 97, Publ. Laser Institute fo America, 12424 Research Parkway, Orlando, FL 3226, USA.

23. U. Kamachi Mudali, R.K. Dayal and G.L. Goswami, Surf Engg., 11 (1995) 331.

24. J. De Damborena, A.J. Vazquiz, J.A. Gonzalez and D.R.F West, Surf. Engg., 5 (1989) 235.

25. J. Stewart, D.B. Wells, P.M. Scott and A.S. Bransden, Corrosion, 46 (1990) 618.

26. U. Kamachi Mudali and R.K. Dayal, J. Mater. Engg. and Performance, 1 (1992) 341.

27. U. Kamachi Mudali, M.G. Pujar and R.K. Dayal, J. Mater. Engg. And Performance, 7 (2) (1998) 214.

28. S. R. Mannava, US Patents US5591009A, US5584662A, US5584586.

29. Y. Sano, N. Mudai, K. Okasaki and M. Ohata, Nucl. Inst. Met. Phys. Res. (Japan) B 121 (1997) 432.

30, P. Peyre, X. Scherpereel, L. Berthe, C. Carboni, R. Fabbro, G. Beranger and C. Lemaitre, Mater. Sci. Engg., A 280 (2000) 294.

31. N. Mukai, N. Aoki, M. Obata, A. Ito, Y. Sano and C. Konagai, paper no. S404–3, p. 1489.

32. W. Kono, S. Kimura, H. Sakamoto, S, Kawano and Y. Tongu, 351.

33. C.T. Kwok, F.T. Cheng and H.C. Man, Mater, Sci. Engg., A 290 (2000) 55.

34. N. Parvathavarthini, R. K. Dayal, R. Sivakumar, U. Kamachi Mudali and A. Bharati, Mater. Sci. Tech., 8 (1992) 1070.

35. K. Sugioka, H. Tashiro and K. Toyoda, Int. J. Mater. Prod. Technol., 8 (2–4) (1994) 316.

36. H. Fujimagari, M. Hagiwara, T. Kojima, Nucl. Engg. Design, 195 (2000) 289.

37. T. Hodgkiess, A. Neville and S. Shrestha, Wear 233–235 (1999) 623.

38. O. de Sanctis, L. Gomez, N. Pellegri, C. Parodi, A. Marajofsky and A. Duran, J. Non-Cryst. Solids, 121 (1990) 338.

39. J. de Damborenea, N. Pellergri, O. de Sanctis and A. Durain, J. Sol-Gel Sci. Tech., 4 (1995) 239.

40. Pablo Galliano, Juan Jose De Damborenea, M. Jesus Pascual and Alicia Duran, J. Sol-Gel Sci. Technol., 13 (1998) 723.

41. Melissa. J. Paterson and Besim Ben-Nissan; Surf. Coat. Tech., 86–87 (1996) 153.
42. K. Izumi, M. Murakami, T. Deguchi, A. Morita, N. Tohge and T. Minami, J. Ame. Ceram. Soc., 72 (8) (1989) 1465.
43. M. Atik, S.H. Messaddeq, M.A. Aegerter and J. Zarzycki, J. Mater. Sci. Lett., 15 (1996) 1868.
44. M. Atik, C.R. Kha, P. De Lima Neto, L.A. Avaca, M.A. Aegerter and J. Zarzycki, J. Mater. Sci. Lett., 14 (1995) 178.
45. P.P. Trzaskoma-Paulette and A. Nazeri, J. Electrochem. Soc., 144 (4) (1997) 1307.
46. J.H. Hsieh, W.Wu, R.A. Erck, G. R. Fenske, Y.Y. Su and M. Marek, Surf, Coat. Tech., 51 (1–3) 212.
47. Z. Liu, Y. He and W. Gao, J. Mater. Eng. Perform., 7 (1) (1998) 88.
48. U. Kamachi Mudali, N. Bhuvaneswaran, P. Shankar and H.S. Khatak, Paper presented at the Gordon Research Conference on Aqueous Corrosion, New London, USA, July 2000.
49. N. Bhuvaneswaran, U. Kamachi Mudali, P. Shankar and S. Rajeswari, Proceedings of the Third International Conference on Advances in Composites ADCOMP 2000, eds. E.S. Dwarakadasa and C.G. Krishnadas Nair, Aug. 2000, Bangalore (India), pp. 257–264.
50. N. Bhuvaneswaran, U. Kamachi Mudali, P. Shankar, S. Rajeswari and H. S. Khatak, Tenth national Congress on Corrosion Control, NCCI (India), Madurai, September 2000.

15. General Guidelines for Corrosion Control

H.S. Khatak[1] and V.R. Raju[1]

Abstract Corrosion can be avoided or minimized by proper selection of materials, proper design of the component or equipment, control of operating conditions within design limits, and employment of suitable corrosion control measures.

Key Words Corrosion control, design guidelines, weld design, fabrication, operation, expert systems, materials selection.

1. INTRODUCTION

A few guidelines that could be useful for protecting components and systems from corrosion during design, fabrication, transport, storage, erection and operation are discussed in this chapter. The success of any corrosion prevention program depends on the constant and continuous familiarization of the personnel involved in the different stages mentioned above on the various aspects of corrosion prevention methods. Engineers should consider the extent to which corrosion protection is to be built in and the extent to which corrosion prevention during maintenance can be depended upon by conditioning the environments or by temporary protection.

Corrosion and its control is an important, but often neglected, element in the practice of engineering. The proper solution to any corrosion problem is the most economical one, provided safety is assured. Taking care right from the material selection to successful operation, adhering to the guidelines is the only possible way to minimize losses due to corrosion. For controlling corrosion, the process should start at design stage itself. Corrosion control at the design stage does not just happen by itself, it should be planned. Due to lack of corrosion awareness, designer is likely to ignore some factors, which leads to failure of the components. These problems can be avoided if designer gives equal importance to corrosion resistance as to strength of materials. Similarly, precautions have to be taken at various stages of component life, e.g. fabrication, storage, commissioning and operation. Based on the experience in the use of austenitic stainless steels, analyzing many failures and the failures documented in the literature, these guidelines have been prepared which will help the engineers to realize the importance of corrosion.

[1]Corrosion Science and Technology Division, Indira Gandhi Centre for Atomic Research, Kalpakkam-603 102, India

2. DESIGN ASPECTS

The embodiment of corrosion control into the design of a product can be achieved most efficiently by captivating this control within the products geometry, i.e., in its three-dimensional form, its layout and its relative and spatial positions. There is no other design effort which can assist so much in prevention of corrosion for such a comparatively small outlay.

Whereas basically the pattern of a utility depends on its functional, material and fabrication requirements, it is within the scope of a good designer to select from the available possibilities only such geometric shapes or combinations of forms that help to reduce corrosion attack in the most efficient and economic manner.

2.1 Selection of materials

For the best equipment or structural design, the materials of construction must be carefully selected from a corrosion resistance standpoint. Design details should preserve the corrosion resistance of the materials. Concise and clearly written specifications should be provided to the supplier to ensure that the material needed is accurately ordered. The equipment should be fabricated properly and adequately inspected to prove compliance with specifications. The equipment must be operated properly. Lastly, the equipment must be maintained properly. All these factors must be considered by the designer to ensure the long life of the equipment he designs.

Selection of the optimum material of construction for a specific application is critical to the designer from the performance, safety and economic standpoints. Conducting tests in a laboratory or pilot plant, relying on published corrosion rates from technical sources or producers literature or basing selection on previous experience are ways in which optimum materials are selected.

Bimetallic corrosion (galvanic corrosion) arises due to the action of a bimetallic cell, i.e. a galvanic cell, where electrodes consists of different materials. It is however essential that an electrolyte is present for galvanic corrosion to take place. To avoid bimetallic corrosion the following guidelines can be followed:

1. Do not connect metals which are well separated in the electrochemical series or in a more representative galvanic series. This requirement must be fulfilled in marine atmosphere and where the metal surfaces are expected to be permanently exposed to moisture.
2. Dissimilar metals should be insulated from each other wherever possible, so that metallic contact does not occur, for example with age-resistant plastics (Fig. 1).
3. Design the construction so that moisture cannot remain at the point of contact.
4. Dielectric materials are good electric insulators. Choice of materials subject to their dielectric strength should be graduated in relation to the environment and function, i.e. the more aggressive the conditions, wider emf potential and more critical functional conditions, the higher is the required ohmic resistance.
5. Various shapes and sizes of faying surfaces demand diverse materials: whilst an extended linear surface may require sealant in the form of sealant tape, multiform or small connection may benefit from an elastomer sealant or caulking compound, applied by a gun or spatula, dielectric gasket or washer.
6. Separation materials should not be porous to such a degree that the absorbed water or other electrolytes will cause uninterrupted conductivity between the metals of the bimetallic couple. For example, in heavily wetted areas dielectric asbestos or other suitable gaskets with dielectric effect and low water absorption should be used, instead of porous asbestos gaskets.

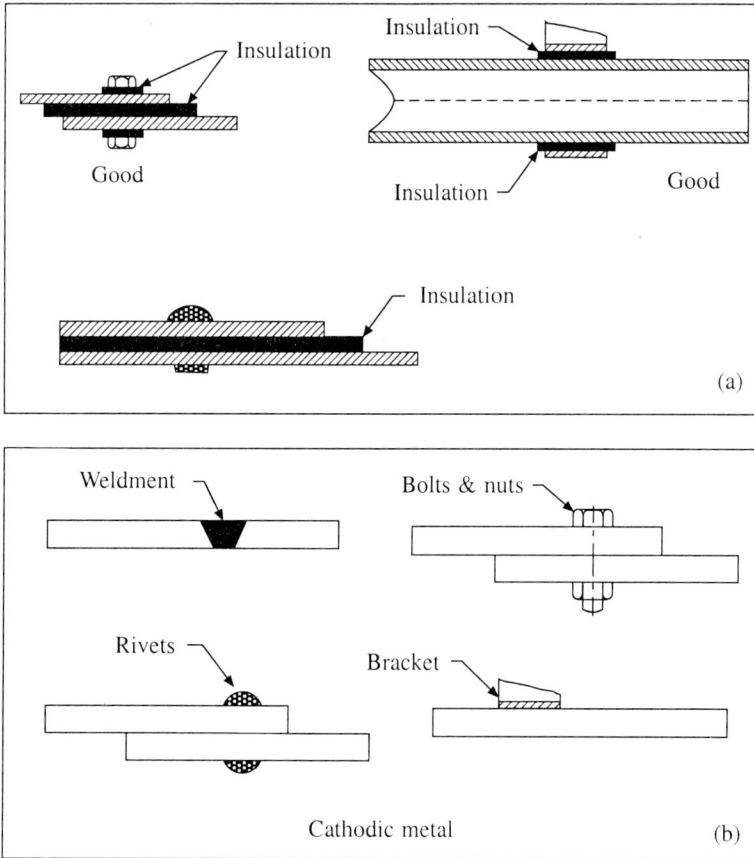

Fig. 1   For: (a) dissimilar metals, insulate one from the other; (b) dissimilar metals that cannot be insulated, make the more noble metal (the cathode) the smaller area.

Following are the guidelines for selection of fasteners:

1. Fasteners should maintain their function of connecting safely two pieces of metal in a way easy to disassemble.
2. Fasteners should not adversely affect the materials of the basic components and should not be affected reciprocally by them.
3. In aggressive conditions fasteners should not be made of metal anodic to both metals of the joint.
4. Fasteners should preferably be made of a material compatible to both metals in the connection, i.e. slightly cathodic.

2.2   Design

In order to avoid the crevice corrosion in any component/part, it should be ensured that the formation of crevices at the welded joints are avoided at the design/fabrication stage itself. The most efficient way of eliminating the crevice corrosion problem in heat exchanger tubes is to weld the tube to the tube-sheet after expansion by rolling. The baffles in the heat exchanger afford crevices in the areas where the tubes pass through. To avoid this, the designer should leave a generous clearance between the tube wall and the hole in the baffle. Just a small amount of bypassing of the solution through the baffle hole will keep the area clean. Also, over-rolling of the tube into the tube sheet can lead to formation of crevice sites (Fig. 2). The best way of avoiding the crevices at the tube to tubesheet joints is by modifying the tubesheet design as shown in Fig. 3.

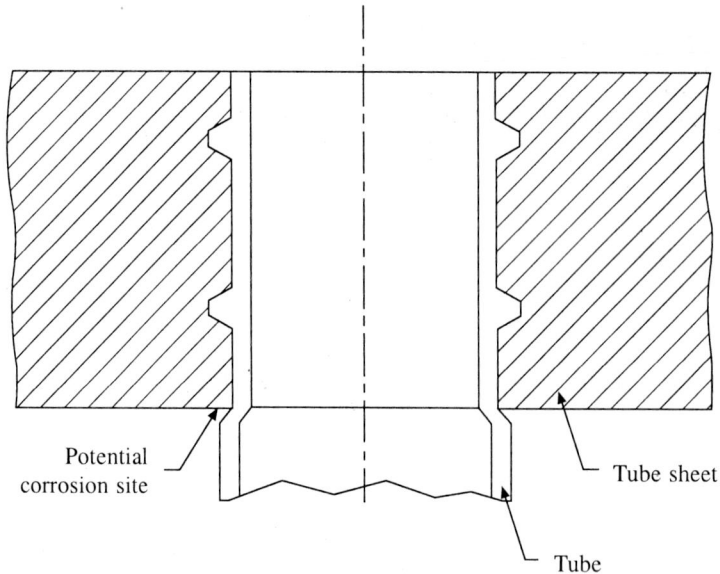

Fig. 2   Over-rolled tube into tube-sheet can foster local corrosion.

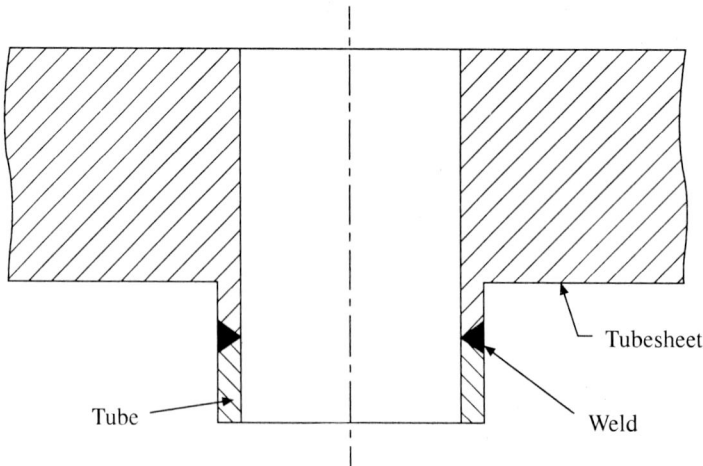

Fig. 3   Modified tube to tubesheet design.

Intermittent or ship welds as shown in Figure 4 should be avoided as these lead to crevice corrosion. Also the supports for the various tanks and pipes shall be so designed to avoid any kind of crevices as shown in Figures 5 and 6.

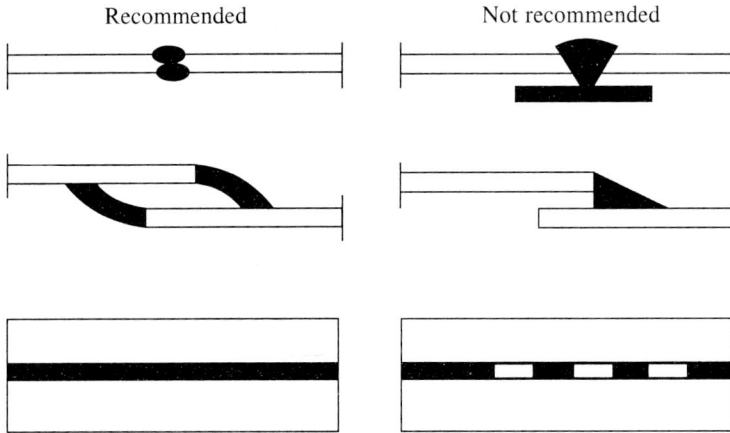

Fig. 4 **Crevices should be avoided at welds, where moisture can accumulate and give rise to crevice corrosion.**

The residual stresses shall be kept to a minimum by employing manufacturing processes that leave minimum residual stress. In the case of austenitic stainless steel components, the presence of residual stress can cause Stress Corrosion Cracking (SCC), only if material is in sensitized condition and the environment is marine. For 316 LN and 304 LN austenitic stainless steels, the sensitization possibility is least during welding and fabrication. However, residual stresses introduced by cold working of materials can result in increased tendency to pitting attack.

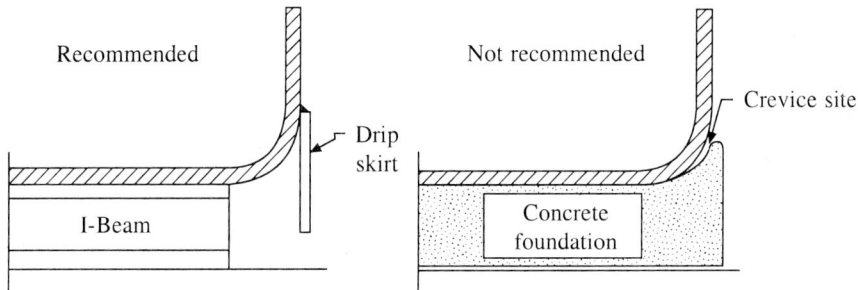

Fig. 5 **Foundations for tanks should be designed so that the risk of crevice corrosion is avoided.**

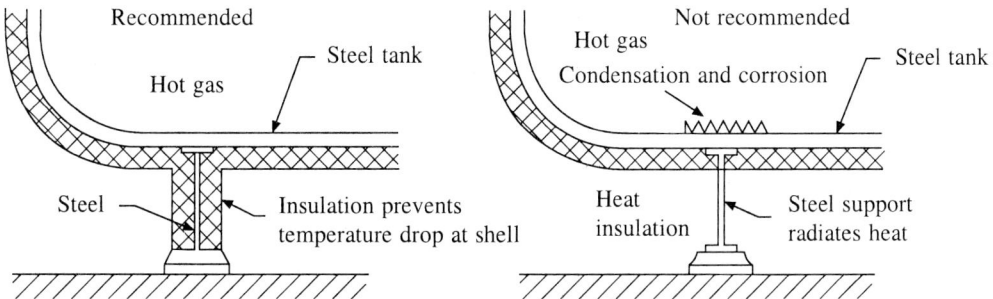

Fig. 6 **Heat-conducting supports can cause condensation and corrosion of tanks containing hot gases.**

It is not possible to define the stress levels to avoid SCC, however based on experiments in authors laboratory, the limits on critical stress levels for austenitic stainless steels can be taken as follows:

- For sensitised materials, the total tensile stress at room temperature should be restricted to a K_I value of 20 MPa.m$^{1/2}$.
- For solution annealed material which will see temperatures of around 100°C (maximum susceptibility region), the total tensile stress level should not exceed K_I value of 20 MPa.m$^{1/2}$ and for sensitized one, 10 MPa.m$^{1/2}$.

This may be applicable for components under insulation. The limits will be higher if proper precautions are taken during manufacture, storage and operation and also during shut downs if the temperature of insulated components is maintained above 200°C.

Design geometries which ensure free drainage of the liquid without stagnation at the corners or pockets should be selected for keeping the vessels dry during storage, transport or long shut downs as shown in Figure 7.

Tight-fits, ill-fitting or over-tightened bolt assembly designs, which cause high stresses are to be avoided.

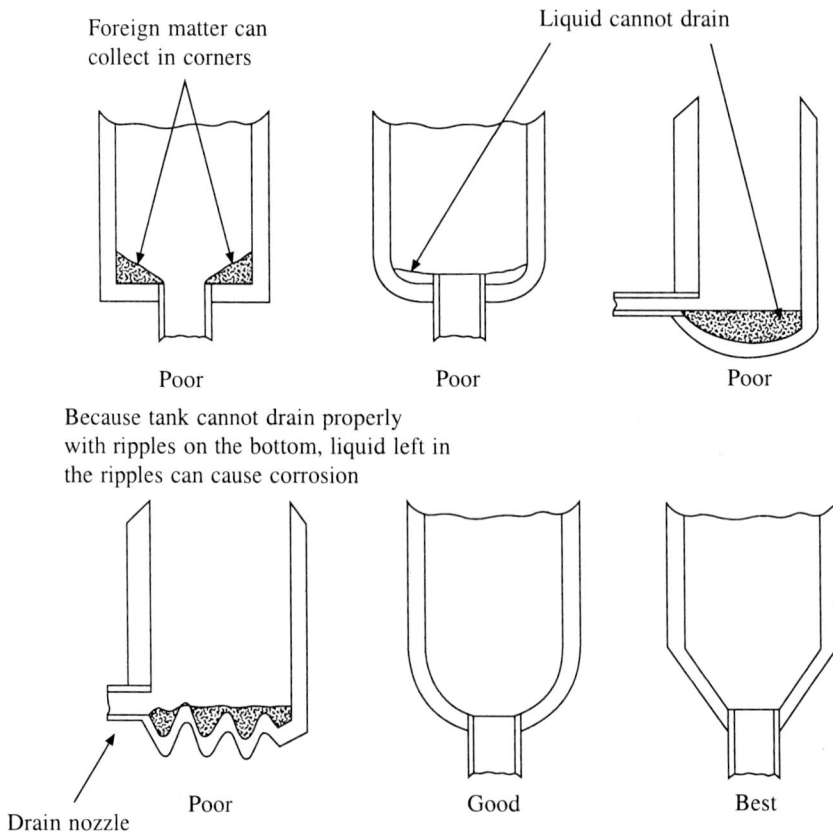

Fig. 7 Poor, good and best tank drainage designs.

In a severely corrosive environment, austenitic stainless steels are susceptible to a form of selective corrosion termed the end grain attack. Figure 8 presents the schematic of a SS plate showing long lines

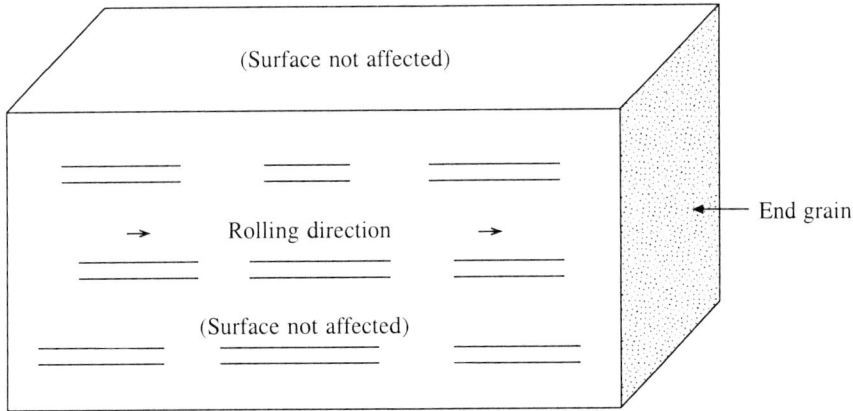

Fig. 8 Schematic diagram of stainless steel plate showing long lines of inclusions and stringers.

of inclusions and stringers. The designer should be alert to the possible exposed end grains in equipment's for use in an aggressive corrosion environment. The simplest solution to end grain problem is to "butter" the exposed end grain with weld metal as illustrated in Fig. 9.

Austenitic stainless steels are also susceptible to fretting corrosion. Hence it is very important that the parts are designed to exhibit less or no relative motion, as also high finishes should be avoided on them. In cases where it is not possible to avoid them, molybdenum sulfide can be used as a lubricant.

2.3 Corrosion Under Insulation

Thermal insulation materials do not in themselves cause Stress Corrosion Cracking (SCC) of stainless steels, but where they are applied to hot equipment above about 80°C, the risk of failure is much greater.

Water entering the outside of insulation on hot equipment may diffuse inwards through the insulation until it reaches a region where the temperature is such that evaporation takes place. In this region of "dry out", soluble salts, contained in the water, or picked up by the water during its passage through the insulation, can concentrate. Immediately outside the region of "dry out" will be a zone where the pores of the lagging are filled with a saturated solution of water-borne salts,

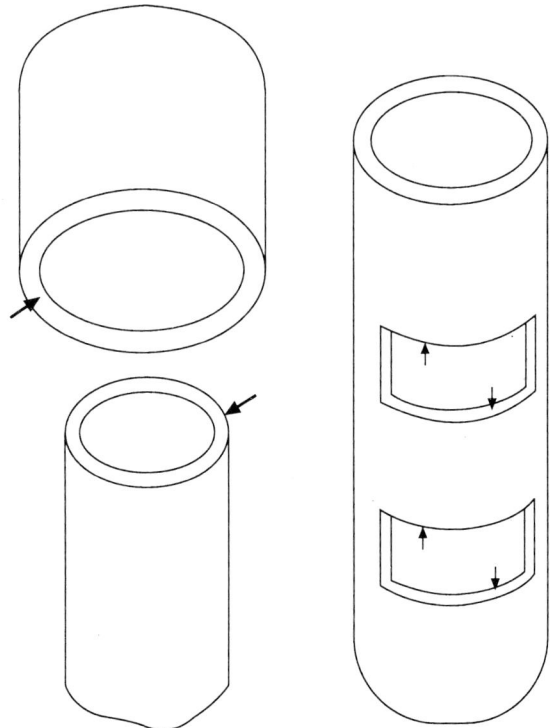

Fig. 9 Example of how end grain susceptibility can be eliminated by "buttering" with weld metal at the arrow locations.

including chloride ions. Besides it can leach out soluble chloride from the insulation also. There is no criteria to define a minimum chloride content in the insulation.

When by reason of shutdown, or other process change, the metal wall temperature falls, the zone of salts saturated solution will move inwards until it contacts the metal surface. On re-heating, the metal will be temporarily in contact with a saturated solution of chloride ions and stress corrosion may be initiated.

An insulation having a chloride content of about 15 ppm and corresponding silicate content as per US regulatory guide—1.36 and passing the test as per ASTM C-692 could be specified as the acceptance criteria for insulation to reduce the risk of SCC. A suitable jacketing such as aluminum foil sheet for minimising the contamination of the insulation from atmosphere should be provided.

3.   FABRICATION AND TESTING

Fabrication processes such as drawing, stretching, rolling, bending, shearing, etc. will introduce residual stresses which should be kept to a minimum as discussed in the previous section. Force fits during welding shall be avoided. Weld spatter and fumes from coated electrodes can leave chlorides or fluorides on the surface. These should be thoroughly cleaned and passivated before the component is put into operation. Even human sweat contains chlorides. Hence contact of finished metal parts with human naked hands should be avoided.

During pickling, it should be ensured that all the surface impurities such as iron particles, mill scales etc. are removed. Iron contamination can lead to SCC of non-sensitized austenitic stainless steels at room temperature. Thorough rinsing with demineralised water after pickling and passivation is to be carried out. Finally the components should be dried off completely. Stamping of identification marks on to a finished metal part should not be encouraged. This will introduce local residual stresses. Scribing may be recommended for such purposes.

For process plant applications, pressure testing with water is the normal method to prove the structural integrity of vessels, equipment and piping after fabrication and installation. With stainless steel this useful practice has the inherent draw-back that any water left behind after the test may easily cause severe pitting corrosion in the presence of chloride ions in water. In the stagnant pools of water, pitting corrosion is often rapid and may lead to perforation before the plant is actually started up.

Low heat input welding procedure should be adopted to avoid sensitization. Sensitized microstructure is prone to SCC in the presence of water and chloride during storage.
Recommended practice for pressure testing SS vessels, equipment and piping:

1. The stainless steel parts must be cleaned prior to final assembly for the hydraulic test.
2. The water to be used, apart from further requirements on chloride content, should be free from sediment, i.e., dissolved solids, whatever the nature thereof.
3. If the equipment or piping will never have a metal temperature higher than 50°C, water with upto 200 ppm chlorides may be used for pressure testing.
4. If the metal temperature of the equipment or piping may be higher than 50°C during commissioning or operation and proper draining is not possible the recommended practice is to utilise water with less than 1ppm chlorides (condensate or DM water).
5. If the temperature may be above 50°C as under (4) above and the equipment or piping is flushed with condensate or DM water immediately after testing, water with upto 200 ppm

chlorides may be used for the pressure test. It is essential to ensure when flushing with DM water that all surfaces previously wetted during the pressure test are flushed.

6. If the temperature may be above 50°C as under (4) above but the water can be fully drained and completely removed, e.g. by mopping-up, water with up to 200 ppm chlorides is acceptable.

7. Steam or electrically traced systems should be tested with condensate or DM water or be properly flushed with condensate or DM water prior to testing the tracing.

8. Shell and tube heat exchangers to be operated at temperatures above 50°C and of such a construction that crevices are present should be tested with condensate or DM water or be properly flushed to remove any remnants of chloride-containing water.

9. SS bellows, as present in many carbon steel and SS systems, should preferably be isolated from the system when pressure testing. If this is not possible, the pressure test must be carried out with DM water or condensate.

10. No attempts should be made to remove remnants of water by blowing with hot air or other gas, unless the system has been tested with or properly flushed with condensate or DM water.

11. The above rules also apply to medium or high nickel alloys and ferritic chromium stainless steels where pitting corrosion is a risk.

12. It should be noted that risk of pitting corrosion is considerably diminished if the time between pressure testing and startup is as short as possible. If the time is anticipated to be long flushing with condensate or DM water is recommended.

13. The components should be dried and filled with inert gas wherever possible.

4. STORAGE OF COMPONENTS

Since most corrosion processes require water or moisture, components should be stored in a dry atmosphere to be free from corrosion. It is found that when the relative humidity is less than 35%, no condensation takes place on metallic surfaces and hence the corrosion is avoided. A separate clean and humidity controlled room for storing critical components can be planned. The relative humidity of the air can be kept less than 35% by using commercially available "air driers". Alternatively, the components can be wrapped and sealed in polythene sheets after keeping desiccants inside. This is applicable during transportation of the components also. Pipe ends or openings of components should be closed with plugs impervious to moisture. Polythene caps could be use for this purpose.

Direct contact of dissimilar metals during storage should be avoided. The need for correct labelling of all materials in stores and a firm ban on material substitution without authorization are to be envisaged. Components stored in humidity controlled room should be inspected at least once in a year. Components stored in polythene sheets should be checked for breakage of sheets, seals, every six months and the component's surface should be checked for any rust spot after every two years. To assess the extent of damage found during inspection and for preventive action, corrosion engineer shall be consulted.

5. ON-SITE ERECTION

Using of temporary supports on components by welding should be avoided, unless it is contemplated in design. Even then, welding dissimilar materials should be avoided. Cutting, bending, drilling, welding etc. not included in design should not be undertaken. After erection, it should be ensured that there is no surface contamination such as dust, weld splatter, iron particles etc. All stainless steel surfaces should be passivated and kept dust free. During erection, care must be taken to avoid deep scratching of stainless steel components. These scratch marks may provide initiation points for localised corrosion attack. Any rust mark noticed should immediately be cleaned and repassivated. During the period between completion of erection and start of operation, the components should not be left unprotected.

6. OPERATION

The chemistry of the process medium should be formulated first and strictly adhered to throughout the operation. No untreated water should be used during the commissioning stage. During shut-down, there should be periodic circulation of water or trickle flow to the extent of at least one change in two weeks for all the water systems. Any deviation from the recommended procedure of chemical control should be brought to the attention of the corrosion engineers. Removal of the corrosion products during planned intervals is a must to avoid hot spots due to poor heat transfer (particularly in heat exchangers and steam generator). A cleaning frequency of once in 5 years is recommended for the equipment systems like steam generators.

7. GENERAL

The National Association of Corrosion Engineers (NACE) and the National Institute of Standards and Technology (NIST) have jointly developed databases covering the extensive corrosion data available within NACE. COR-SUR based on Corrosion Data Survey, Metals section, NACE gives corrosion rate data for the most common metals and alloys exposed to approximately 1000 corrosive environments at different concentrations and temperatures. COR-SUR 2, based on Corrosion Data Survey, Non-Metals Section, NACE gives corrosion rate data for the non metallic materials in 850 corrosive environments. MIC.AB is an economical reference library of 280 citations on microbiologically influenced corrosion (MIC) and covers MIC of numerous alloys, e.g., aluminum, copper, iron, nickel, stainless steel, in cooling water systems, municipal water systems, oil fields, pulp and paper plants, and seawater environments.

One of the major sources of corrosion data in any manufacturing and user organisation is actual monitoring of corrosion on existing products and equipment. To obtain continuous, dependable data on corrosion, efforts have been made to computerise this monitoring. Several such monitoring systems and software programs are available, and most of the monitoring set-ups can be interfaced with computers, thus making available monitored data for the computerised material selection process.

Once the critical property requirements of a particular application are worked out and fed into a computer, with the help of available programs they can be compared with property and corrosion data from different databases and software packages. The properties for some chosen candidate materials after the initial screening, along with processing and cost data, can be fed into the design analysis and optimisation software, and an optimum material-design combination can be selected. Through the use of expert systems, this computer selection can be made quite systematic and convenient, eliminating any possibility of missing essential steps.

In the material selection domain, the expert systems presently available are mostly for restricted situations, but expert systems based on broad comprehensive knowledge are under development. It is a new and developing field, and some of the expert systems available for corrosion control applications in materials selection are discussed as follows.

ESCORT (Expert Software for Corrosion Technology) is an expert system for materials selection in the chemical process industries, where corrosion control is very important. It has the following knowledge bases: (1) materials, (2) equipment type, e.g., heat exchanger, packed column, or pump, (3) environmental characteristics, (4) types of corrosion, and (5) preventive measures. The program asks for certain inputs as regards industry, application environment, process or operation, preferred classes of materials, based on availability, codes, cost, customer, etc. Having completed the information gathering process, the system starts the main reasoning process, performing stepwise "if-then" decisions. The system output is a list of materials that meet the input specifications. The user can question the basis of selection of a particular material, and is given the rules applied and met in making the choice. Additional information, such as on forming, welding, etc. of the recommended materials can also be obtained.

MICPro (Predictive Software for Microbiologically Influenced Corrosion) is a predictive expert system that permits evaluation of conditions under which MIC will occur, which components or systems are the most susceptible, how operating conditions and system variables affect vulnerability, and the effectiveness of control measures. Output is presented on a 1 to 10 scale. Corrosion under abiotic conditions, which is included, enables comparison, and can guide the selection of the control measures.

SOCRATES (Selection of Corrosion Resistant Alloys Through Environment Specifications) is an expert system for selection of martensitic and duplex stainless steel for oil and gas production service, involving general corrosion, pitting, sulfide stress cracking, and anodic SCC, because of individual and synergistic effects of pH, H_2S, and chlorides in a sour gas-oil mixture.

Computer strategies outlined for very specific as well as general applications are available such as the so called expert systems. The use of these as a teaching tool can be tried. For example, the AUSCOR system relates to the corrosion of 28 alloys, including the common ferritic and austenitic stainless steels. Accommodating a wide range of liquid corrodants, it can be used for predicting the corrosion behaviour.

8. CONCLUSION

By taking the precautions listed above in the design, selection of materials, fabrication, storage of the components and the operation it is possible to protect the equipments from the corrosion problems and extend the life of the components.

Appendix
Corrosion Testing Standards

The list include American Society for Testing and Materials (ASTM), Association Franchaise de Normalization (AFNOR), British Standards Institute (BSI), Deutsches Institut für Normung (DIN), International Standards Organization (ISO), Unified Systems of Corrosion and Ageing Protection (USCA) and National Association of Corrosion Engineers (NACE).

Title	ASTM	AFNOR	BSI	DIN	USCAP	ISO	NACE
Standard Practices for Detecting Susceptibility to Intergranular Attack in Austenitic Stainless Steels.	A262-98	A05-160 A05-159	5903	50914	6032	3651-1 & 3651-2: 1998(E)	
Standard Practice for Cleaning, Descaling, and Passivation of Stainless Steel Parts, Equipment, and Systems.	A380-99el						
Standard Test Method for Accelerated Life of Nickel-Chromium and Nickel-Chromium-Iron Alloys for Electrical heating.	B76-90 (1995)						
Standard Test Method for Monitoring Atmospheric Corrosion Tests by Electrical Resistance Probes.	B826-97						
Test Method for Evaluating the Influence of Thermal Insulations on External Stress Corrosion Cracking Tendency of Austenitic Stainless Steel.	C692-00						
Practice for Preparing, Cleaning, and Evaluating Corrosion Test Specimens.	G1-90 (Reapproved 1999)		7545		9.907	8407: 1991(E)	
Practice for Conventions Applicable to Electrochemical Measurements in Corrosion Testing.	G3-89 (Reapproved 1999)						
Standard Guide for Conducting Corrosion Coupon Tests in Field Applications.	G4-95						
Standard Reference Test Method for Making Potentiostatic and Potentiodynamic Anodic Polarization Measurements.	G5-94 (Reapproved 1999)						

Title	ASTM	AFNOR	BSI	DIN	USCAP	ISO	NACE
Standard Terminology Relating to Corrosion and Corrosion Testing.	G15-99b	A05-001				8044:1999 (E/F/R)	
Standard Guide for Applying Statistics to Analysis of Corrosion Data.	G16-95 (Reapproved 1999)						
Test Methods of Detecting Susceptibility to Intergranular Corrosion in Wrought, Nickel-Rich, Chromium-Bearing Alloys.	G28-97					9400:1990 (E)	
Practice for Making and Using U-Bend Stress-Corrosion Test Specimens.	G30-97	AC5-501-3	6980(3)		9.901.3	7539: 1989(E)	
Standard Practice for Laboratory Immersion Corrosion Testing of Metals.	G31-72 (Reapproved 1999)	A05-300				11845:/ 1995(E)	
Practice for Determining the Susceptibility of Stainless Steels and Related Nickel-Chromium-Iron Alloys to Stress Corrosion Cracking in Polythionic Acids.	G35-98						
Practice for Evaluating Stress-Corrosion-Cracking Resistance of Metals and Alloys in a Boiling Magnesium Chloride Solution.	G36-94 (Reapproved 2000)						
Practice for Making and Using C-Ring Stress Corrosion Test Specimens.	G38-73 (1995)	A05-501-5	6980(5)		9.901.5	7539-5: 1989(E)	
Practice for Preparation and Use of Bent-Beam Stress Corrosion Test Specimens.	G39-99	A05-501-2	6980(2)		9.901.2	7359-2:/ 1989(E)	
Practice for Determining Cracking Susceptibility of Metals Exposed Under Stress to a Hot Salt Environment.	G41-90 (Reapproved 2000)						
Standard Practice for Exposure of Metals and Alloys by Alternate Immersion in Neutral 3.5% Sodium Chloride Solution	G44-99					11130: 1999(E)	
Standard Guide for Examination and Evaluation of Pitting Corrosion.	G46-94 (Reapproved 1999)					11463: 1995(E)	
Test Methods for Pitting and Crevice Corrosion Resistance of Stainless Steels and Related Alloys by use of Ferric Chloride Solution.	G48-00						

(Contd)

Title	ASTM	AFNOR	BSI	DIN	USCAP	ISO	NACE
Practice for Preparation and Use of Direct Tension Stress-Corrosion Test Specimens.	G49-85 (Reapproved 2000)	A05-501-4	6980(4)		9.901.4	7359-4 1989(E)	
Practice for Conducting Atmospheric Corrosion Tests on Metals.	G50-76 (1997)					8565: 1992(E)	
Practice for Simple Static Oxidation Testing.	G54-84 (1996)						
Practice for Preparation of Stress-Corrosion Test Specimens for Weldments.	G58-85 (Reapproved 1999)		6980(1)	50915	9.901.1	7539-8: 2000(E)	
Standard Test Method for Conducting Potentiodynamic Polarization Resistance measurements.	G59-97						
Test method for Conducting Cyclic Potentiodynmic Polarization Measurements for Localized Corrosion Susceptibility of Iron-, Nickel-, or Cobalt-Based Alloys.	G61-86 (Reapproved 1998)						
Guide for Conducting and Evaluating Galvanic Corrosion Tests in Electrolytes.	G71-81 (Reapproved 1998)el						
Guide for Crevice Corrosion Testing of Iron-Base and Nickel-Base Stainless Alloys in Seawater and Other Chloride-Containing Aqueous Environments.	G78-95					11306: 1998(E)	
Practice for Evaluation of Metals Exposed to Carburization Environments.	G79-83 (1996)						
Guide for Development and Use of a Galvanic Series for Predicting Galvanic Corrosion Performance.	G82-98						
Guide for On-Line Monitoring of Corrosion in Plant Equipment (Electrical and Electrochemical Methods).	G96-90 (1996)						
Method for Conducting Cyclic Galvanostaircase Polarization.	G100-89 (Reapproved 1999)						
Practice for Calculation of Corrosion Rates and Related Information from Electrochemical Measurements.	G102-89 (Reapproved 1999)						

Title	ASTM	AFNOR	BSI	DIN	USCAP	ISO	NACE
Test method for Electrochemical Reactivation (EPR) for Detecting Sensitization of AISI Type 304 and 304L Stainless Steels.	G108-94 (Reapproved 1999)						
Guide for Corrosion Tests in High Temperature or High-Pressure Environment, or Both.	G111-97						
Test Method for Evaluating Stress-Corrosion Cracking of Stainless Alloys with Different Nickel Content in Boiling Acidified Sodium Chloride Solution.	G123-96						
Practice for Slow Strain Rate Testing to Evaluate the Susceptibility of Metallic Materials to Environmentally Assisted Cracking.	G129-95						
Guide for Computerized Exchange of Corrosion Data for Metals.	G135-95						
Corrosion of Metals and Alloys-Stress Corrosion Testing-Part-1: General Guidance on Testing Procedures.		A05-501-1	6980(1)	50915	9.901.1	7539-1: 1987(E)	
Corrosion of Metals and Alloys-Stress Corrosion Testing-Part-6: Preparation and Use of Precracked Specimens.		A05-501-06	6980(6)		9.901.6	7539-6: 1989(E)	
Corrosion of Metals and Alloys-Stress Corrosion Testing-Part 7: Slow Strain Rate Testing.		A05-501-7	6980(7)		9.901.7	7539-7: 1989(E)	
Corrosion of Metals and Alloys-Stress Corrosion Testing-Part 8: Preparation and Use of Specimens to Evaluate Weldments.						7539-8: 2000(E)	
Corrosion of metals and alloys-Corrosion fatigue testing-Part 1: Cycles to failure testing.						11782-1: 1998(E)	
Corrosion of metals and alloys-Corrosion fatigue testing-Part 2: Crack propagation testing using precracked specimens.						11782-2: 1998(E)	
Corrosion of metals and alloys-Guidelines for applying statistics to analysis of corrosion data.							
Corrosion of metals and alloys-Evaluation of stress corrosion cracking by the drop evaluation test.							

(Contd)

Title	ASTM	AFNOR	BSI	DIN	USCAP	ISO	NACE
Protective Coatings for Carbon Steel and Austenitic Stainless Steel Surfaces Under Thermal Insulation and Cementitious Fireproofing (corrosion under insulation).							NACE EG2
Protection of Austenitic Stainless Steel in Refineries Against Stress Corrosion Cracking by the Use of Neutralizing Solutions During Shut Down.							RP0170
Collection and identification of Corrosion Products.							RP0173
Standard Format for Computerized Electrochemical Polarization Curve Data Files.							RP0197
Laboratory Corrosion Testing of Metals for the Process Industries (similar to ASTM G31)							TM0169
Autoclave Corrosion Testing of Metals in High-Temperature Water.							TM0171
Laboratory Methods for the Evaluation of Protective Coatings and Lining Materials in Immersion Service.							TM0174
Method of Conducting Controlled Velocity Laboratory Corrosion Tests.							TM0270
Dynamic Corrosion Testing of Metals in High-Temperature Water.							TM0274

Subject Index